Unmanned Air Vehicles: Design, Development and Applications

Unmanned Air Vehicles: Design, Development and Applications

Edited by
Rowan Reyes

Willford Press
www.willfordpress.com

Published by Willford Press,
118-35 Queens Blvd., Suite 400,
Forest Hills, NY 11375, USA

ISBN: 978-1-68285-779-3

Cataloging-in-Publication Data

Unmanned air vehicles : design, development and applications / edited by Rowan Reyes.
p. cm.
Includes bibliographical references and index.
ISBN 978-1-68285-779-3
1. Drone aircraft. 2. Drone aircraft--Design and construction. 3. Vehicles, Remotely piloted.
4. Flying-machines. I. Reyes, Rowan.

TL685.35 .U56 2020

629.132 6--dc23

For information on all Willford Press publications
visit our website at www.willfordpress.com

Willford Press

Printed in China.

Contents

Preface

Aircrafts that work without any human pilot on board are known as unmanned air vehicles. These are the component of an unmanned aircraft system. They can be classified into six broad categories that are target and decoy, reconnaissance, combat, logistics, research and development, and civil and commercial. The flights of such vehicles can either be controlled by a remote control that is operated by a human, or they work autonomously with the help of the computers present on board. Unmanned air vehicles are often used for the missions that are too dangerous for humans. Initially, they were mostly used in dangerous military operations, now they are used to fulfill commercial, agricultural and scientific needs such as policing, peacekeeping, drone racing, aerial photography or product deliveries. This book explores all the important aspects of unmanned aerial vehicles in the present day scenario. The various advancements in the field are glanced at and their applications, as well as ramifications, are looked at in detail. This book attempts to assist those with a goal of delving into this field.

The information contained in this book is the result of intensive hard work done by researchers in this field. All due efforts have been made to make this book serve as a complete guiding source for students and researchers. The topics in this book have been comprehensively explained to help readers understand the growing trends in the field.

I would like to thank the entire group of writers who made sincere efforts in this book and my family who supported me in my efforts of working on this book. I take this opportunity to thank all those who have been a guiding force throughout my life.

Editor

Design of a Control System for an Organic Flight Array based on a Neural Network Controller

Bokyoung Oh,[1] Junho Jeong,[2] Jinyoung Suk,[2] and Seungkeun Kim[2]

[1]*Agency for Defense Development, Daejeon, Republic of Korea*
[2]*Department of Aerospace Engineering, Chungnam National University, Daejeon, Republic of Korea*

Correspondence should be addressed to Seungkeun Kim; skim78@cnu.ac.kr

Academic Editor: Kenneth M. Sobel

This paper presents a flight control system for an organic flight array (OFA) with a new configuration consisting of multimodularized ducted-fan unmanned aerial vehicles. The OFA has a distinguished advantage of assembling or separating with respect to its missions or operational conditions because of its reconfigurable structure. Therefore, designing a controller that can be flexibly applied in each situation is necessary. First, a dynamic modeling of the OFA based on a single ducted-fan vehicle is performed. Second, the inner loop for attitude control is designed through dynamic model inversion and a PD controller. However, an adaptive control component is needed to flexibly cope with the uncertainty because the operating environment of the OFA is varied, and uncertainty exists depending on the number of modules to be assembled and disturbances. In addition, the performance of the neural network adaptive controller is verified through a numerical simulation according to two scenarios.

1. Introduction

Unmanned aerial vehicles (UAVs) with many types, sizes, and ways of flight have been used in various fields during the past decades. Interest in vertical take-off and landing (VTOL) vehicles, in particular, greatly increased because having a space for a runway is not necessary, and hovering is easier than in fixed-wing UAVs. These features have recently been used for military purposes, such as surveillance of urban warfare and near-field reconnaissance. In addition, performing missions (e.g., communication relay, wide-area reconnaissance, and radar jamming) is possible because the field of application is diversified and because of the growth of technology. The VTOL UAVs are also utilized in the civil field for image photographing and in the aviation transport industry because of their advantages. Moreover, the VTOL UAVs have also been actively utilized for hobby and leisure and are classified into multirotor [1], coaxial [2], Coanda [3], and ducted-fan type [4].

Although various VTOL UAVs are capable of hovering, their operation is restricted in the environment, where gusts or strong winds are constantly blown. Moreover, a disadvantage of having the thrust-to-weight ratio lower than that of a fixed-wing also exists. Therefore, many studies were conducted to overcome the disadvantages of the VTOL UAVs. Accordingly, studies have also been recently conducted to overcome the instability of the vehicle by using a controller that can more robustly cope with gusts and disturbances [5, 6]. Several studies were also performed to complement the disadvantages and overcome the payload limitation through the cooperation of several VTOL UAVs [7]. However, there is a limit in overcoming the disadvantages through the existing flight module in the situation when the mission becomes more diverse. Hence, a study is proposed to overcome the disadvantages by constructing a vehicle through physical assembling. In this way, the vehicle can have a higher robustness than when it is operated as a single vehicle. The vehicle can also have an advantage of increasing the payload according to the number of assembled vehicles. Moreover, organically coping with the mission becomes possible by changing the assembled shape in an environment, where reconnaissance is difficult in the field, such as indoor or in

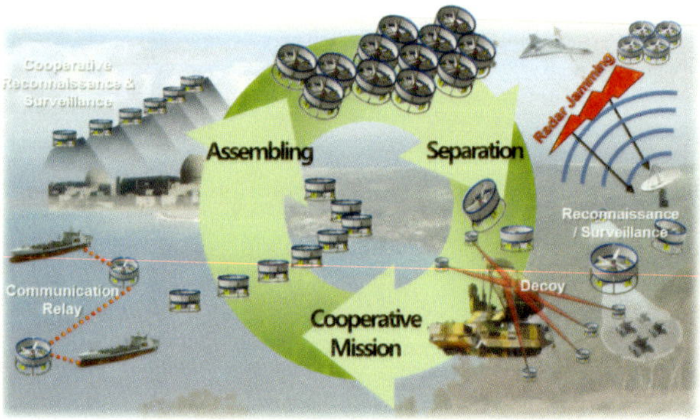

FIGURE 1: Operational concept of the OFA.

a narrow road. One of these studies is on the distributed flight array (DFA) conducted by the ETH Zurich of Switzerland. The vehicle used consisted of autonomous hexagonal single-rotor modules that are able to drive, dock with its peers, and fly in a coordinated fashion. The DFA can generate the thrust but cannot fly with a single module because it cannot remove the antitorque. Hence, the single-rotor modules must be combined to fly with other modules [8, 9]. The International Islamic University of Malaysia also conducted a study on autonomous flight vehicles with distributed flight array concepts. This module had triangular frames that can be combined with each module. In flight mode, the control gains are changed based on the in-flight data [10]. In this study, a similar concept with the DFA was used to overcome various disadvantages by assembling a ducted-fan vehicle. The operational concept that can be assembled and separated freely during flight was defined as shown in Figure 1 to adapt to the diversity of the mission [11]. However, the control performance greatly deteriorated because of the model uncertainty and the nonlinear disturbance when the assembling and separation modes were performed in flight. So, the model uncertainty that can occur in accordance with the operational concept of the organic flight array composed of ducted-fan vehicles is defined as the model inversion error of the process of DMI, and it is significant to verify whether such uncertainty is mitigated through the radial basis function neural network controller which enables control. This study proposed an adaptive control method to cope with the disadvantages and limitations of other studies on flight array by using the radial basis function neural network that can consider the nonlinear characteristics and model uncertainties. The dynamic model of the OFA in this process was defined based on the single ducted-fan vehicle [4]. A dynamic model of the OFA was then transformed into the affine system form to apply the dynamic model inversion technique [12]. Subsequently, the attitude control system was designed through a PD controller based on the affine system [13]. However, the model error was present between the model simulated by the model inversion process and the actual model. In addition, the adaptive control system that can compensate for the model error by applying the neural network control technique was also designed and added to the existing controller [14]. A radial basis function neural network with a simple structure that can guarantee the performance and minimize the amount of computation was applied because generating the adaptive control input in real time was necessary. The activation function of the hidden layer in this process was constructed by using a Gaussian function [15]. This paper is structured as follows: Section 2 describes the detailed modeling of a ducted-fan vehicle adopted in the OFA. The OFA dynamic equation of the affine system was then derived. Section 3 describes the design of the controller through a dynamic model inversion and a PD controller, which is the basic controller for attitude control. The adaptive controller of the neural network to cope with the uncertainty was also derived. Section 4 presents the numerical simulation results for each scenario. Section 5 summarizes the research conclusions.

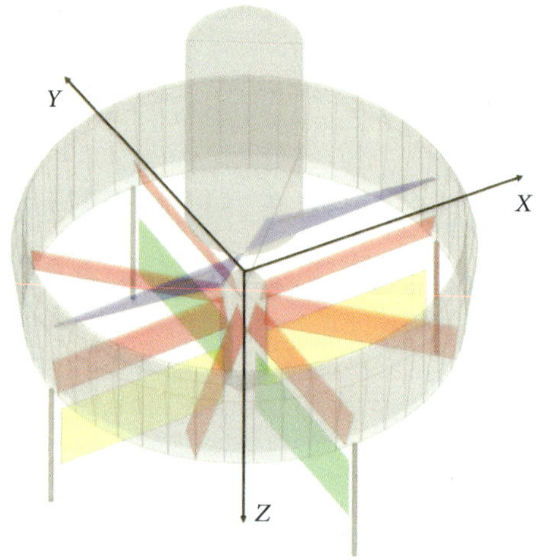

FIGURE 2: Simulation model.

2. Dynamic Model

2.1. Ducted-Fan Model Dynamics. Figure 2 shows the simulation model defined based on the ducted-fan vehicle. The x-axis is defined as the direction similar to that of the aileron control surface, whereas the y-axis is defined as the direction similar to that of the elevator control surface. The z-axis is defined as the downward direction in which the yaw angle of the clockwise direction is set to a plus sign. The simulation model is basically composed of the duct, fuselage, rotor, stator, and control surfaces. The rotor that generates the thrust and the stator that counteracts the antitorque are placed inside the duct. Thus, the yaw axis stability is secured using the stator. Each control surface located directly under the duct structure is designed in a cross shape around the z-axis. This design creates a moment for the control of the roll, pitch, and yaw Euler angles. Defining the translational motion direction in advance is necessary because the ducted-fan vehicle has a symmetrical structure with respect to the z-axis. The coordinate axes are defined herein in accordance with the direction of the control surface. Table 1 shows the control surface direction and effect.

The acceleration and angular acceleration can generally be defined based on the coordinate system of Figure 2 as follows. In particular, the coupling components J_{xy}, J_{yx}, J_{xz}, J_{zx},

TABLE 1: Control surface sign conventions.

Definition	Sense	Flaps	Effect
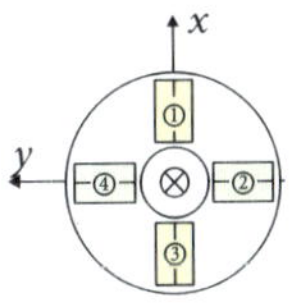	$+\delta_{\text{ail}}$	1, 3	$+M_x$
	$+\delta_{\text{ele}}$	2, 4	$+M_y$
	$+\delta_{\text{rud}}$	1, 2, 3, 4	$+M_z$

TABLE 2: Coefficients of the rotor component by DFDC.

Component	Parameter	Value
Thrust	$b_{\text{thr},1}$	$-1.5601e-3$
	$b_{\text{thr},2}$	$4.1367e-6$
Torque	$b_{\text{tor},1}$	$9.0261e-5$
	$b_{\text{tor},2}$	$-2.6851e-8$

J_{yz}, and J_{zy} of the moment of inertia are not considered in this research because of the symmetry characteristic [16].

$$\begin{aligned}
\dot{u} &= vr - wq + \frac{F_x}{m},\\
\dot{v} &= wp - ur + \frac{F_y}{m},\\
\dot{w} &= uq - vp + \frac{F_z}{m},\\
\dot{p} &= \frac{\{qr(J_{yy} - J_{zz}) + \mathbf{M}_y\}}{J_{xx}},\\
\dot{q} &= \frac{\{pr(J_{zz} - J_{xx}) + \mathbf{M}_y\}}{J_{yy}},\\
\dot{r} &= \frac{\{pq(J_{xx} - J_{yy}) + \mathbf{M}_z\}}{J_{zz}},
\end{aligned} \tag{1}$$

where u, v, and w are the linear velocities and p, q, and r are the angular velocities in the body frame. In (1), F_x, F_y, and F_z are the total forces in the x, y, and z directions, respectively. The moment vectors $\mathbf{M}_x$, $\mathbf{M}_y$, and $\mathbf{M}_z$ are a combination of the moments produced by each of the control surfaces. These forces and moments are generated by the duct, fuselage, rotor, stator, and control surfaces in the respective directions as follows:

$$F = F_{\text{fuse}} + F_{\text{rotor}} + F_{\text{duct}} + F_{\text{cs}} + F_{\text{grav}}, \tag{2}$$

$$M = M_{\text{fuse}} + M_{\text{rotor}} + M_{\text{duct}} + M_{\text{cs}} + M_{\text{gyro}}. \tag{3}$$

The moment variables of (2) must be considered by the entire center of gravity if two or more ducted-fan vehicles are assembled into the OFA. In this study, modeling was performed considering the moment arm that changes as the OFA shape changes. Each component of (2) is defined accordingly [4].

(1) Rotor:

$$\begin{aligned}
F_{\text{rotor}} &= \begin{bmatrix} 0 \\ 0 \\ -T \end{bmatrix} = \begin{bmatrix} 0 \\ 0 \\ -b_{\text{thr},1}\text{RPM} - b_{\text{thr},2}\text{RPM}^2 \end{bmatrix},\\
M_{\text{rotor}} &= \begin{bmatrix} 0 \\ 0 \\ \tau \end{bmatrix} = \begin{bmatrix} 0 \\ 0 \\ b_{\text{tor},1}\text{RPM} + b_{\text{tor},2}\text{RPM}^2 \end{bmatrix},
\end{aligned} \tag{4}$$

In (4), $b_{\text{thr},1}$, $b_{\text{thr},2}$, $b_{\text{tor},1}$ and $b_{\text{tor},2}$ are the output values via the DFDC (Table 2).

(2) Fuselage: the fuselage generates a drag force proportional to the area along each axis. The force and the moment by the fuselage are defined as follows:

$$\begin{aligned}
F_{\text{fuse}} &= -\frac{1}{2}\rho_\infty \begin{bmatrix} C_{D,x^u}|u|S_{\text{top}} \\ C_{D,y^v}|v|S_{\text{top}} \\ C_{D,z^w}|w|S_{\text{side}} \end{bmatrix},\\
M_{\text{fuse}} &= -\frac{1}{2}\rho_\infty \begin{bmatrix} -F_{\text{fuse},y}l_{\text{fuse}} \\ F_{\text{fuse},x}l_{\text{fuse}} \\ 0 \end{bmatrix}.
\end{aligned} \tag{5}$$

Here, S_{top} and S_{side} are the cross-sectional areas along the direction of the fuselage, and ρ_∞ is the density. The drag coefficients along each axis are defined as C_D. The distance between the aerodynamic center and the center of gravity is defined as l_{fuse}.

(3) Duct: the duct has a circular symmetrical shape around the z-axis. Therefore, duct modeling is performed by integrating the lift and the drag forces generated from the unit duct from 0 to 360°. The flow velocity around the duct surface is defined as the function of θ_{duct} as follows:

$$\begin{aligned}
V_r(\theta_{\text{duct}}) &= -u\cos(\theta_{\text{duct}}) - v\sin(\theta_{\text{duct}}),\\
V_z(\theta_{\text{duct}}) &= v_{\text{ind}} - w,
\end{aligned} \tag{6}$$

where V_r is the flow velocity of the radial direction and V_z is the flow velocity in the direction of the rotation axis. The rotation angle about the z-axis, which rotates along the duct plane, is defined as θ_{duct}. The dynamic pressure and angle of attack can be written as follows using the flow velocity defined in (6):

$$\begin{aligned}
q_{\text{duct}}(\theta_{\text{duct}}) &= \frac{1}{2}\rho_\infty\left(V_r^2 + V_z^2\right),\\
\alpha_{\text{duct}}(\theta_{\text{duct}}) &= \tan^{-1}\left(\frac{V_r}{V_z}\right).
\end{aligned} \tag{7}$$

The force and the drag per unit angle generated by the airfoil using (7) are as follows:

$$l_{x,y,z}(\theta_{\text{duct}}) = C_{L,\text{duct}}(\alpha_{\text{duct}}) q_{\text{duct}} c_{\text{duct}},$$
$$d_{x,y,z}(\theta_{\text{duct}}) = C_{D,\text{duct}}(\alpha_{\text{duct}}) q_{\text{duct}} c_{\text{duct}}, \tag{8}$$

where $C_{L,\text{duct}}$ and $C_{D,\text{duct}}$ represent the lift and drag coefficients for the duct airfoil, respectively. These parameters are derived from the empirical formulas. The total lift and drag can be defined by integrating the lift and the drag per unit angle from 0 to 2π as follows:

$$L_{x,y,z,} = r_{\text{duct}} \int_0^{2\pi} l_x(\theta_{\text{duct}}) d\theta_{\text{duct}},$$
$$D_{x,y,z,} = r_{\text{duct}} \int_0^{2\pi} d_x(\theta_{\text{duct}}) d\theta_{\text{duct}}. \tag{9}$$

The lip moment induced by the lift imbalance phenomenon and the momentum drag generated by the reaction of the duct during the translational motion are included in the duct component.

Each component can be expressed as follows:

$$F_{\text{mdrag}} = -\dot{m} \begin{bmatrix} v_x \\ v_y \\ 0 \end{bmatrix} = -v_{\text{ind}} \rho_\infty \pi r^2{}_{\text{duct}} \begin{bmatrix} u \\ v \\ 0 \end{bmatrix},$$
$$M_{\text{lip}} = C_{m,\text{duct}} \rho_\infty r_{\text{duct}} \begin{bmatrix} v|v| \\ -u|u| \\ 0 \end{bmatrix}, \tag{10}$$

where $\dot{m}$ is the mass flow through the rotor and $C_{m,\text{duct}}$ is the duct moment coefficient. The moments and forces generated by the duct considering (9) and (10) are as follows:

$$F_{\text{duct}} = \begin{bmatrix} L_x + D_x \\ L_y + D_y \\ L_z + D_z \end{bmatrix} + F_{m,\text{drag}},$$
$$M_{\text{duct}} = \begin{bmatrix} -F_{\text{duct},y} l_{\text{duct}} \\ -F_{\text{duct},x} l_{\text{duct}} \\ 0 \end{bmatrix} + M_{\text{lip}}. \tag{11}$$

The l_{duct} of (11) is the distance between the aerodynamic center of the duct having an airfoil shape and the center of the gravity point.

(4) Control surfaces: the control surfaces, which change the attitude of the vehicle by adjusting the direction of the wake generated by the rotor, are located directly below the duct. They perform the roles of an aileron, elevator, and rudder. The forces and moments generated by the control surfaces are defined as follows:

$$F_{\text{cs}} = \begin{bmatrix} \text{sgn}\,(v_{\text{ind}} - v_z) q_e S_e C_{Le} \delta_e \\ -\text{sgn}\,(v_{\text{ind}} - v_z) q_a S_a C_L L_a \delta_a \\ 0 \end{bmatrix},$$
$$M_{\text{cs}} \begin{bmatrix} -F_{\text{cs},y} l_a \\ F_{\text{cs},x} l_e \\ \text{sgn}\,(v_{\text{ind}} - v_z) qr C_{Lr} \delta_r S_r l_r \end{bmatrix}, \tag{12}$$

where δ_r, S_r, l_r, C_L, and Q_e represent the deflection angle, area, distance between the aerodynamic center and the center of the gravity point, lift coefficient of the control surface, and dynamic pressure, respectively.

(5) Gravity: the gravity generates a force based on the body-fixed coordinate system whenever the attitude of the vehicle changes. The forces generated by the roll and pitch motion of the vehicle are defined as follows:

$$F_{\text{grav}} = mg \begin{bmatrix} -\sin\theta \\ \cos\theta \sin\phi \\ \cos\theta \cos\phi \end{bmatrix}, \tag{13}$$

(6) Gyroscopic moment: the gyroscopic moment is defined as a moment that is incidentally induced to the vehicle by the rotating rotor. Each component can be expressed as follows:

$$M_{\text{gyro}} = N_{\text{b}} J_{\text{b}} \omega_{\text{r}} \begin{bmatrix} -q \\ p \\ 0 \end{bmatrix},$$
$$\omega_{\text{r}} = \text{RPM} \frac{2\pi}{60}. \tag{14}$$

The gyroscopic moment is a value proportional to the angular velocity and the RPM of the rotor. In (14), N_{b}, J_{b}, and ω_{r} denote the number of blades, moment of inertia, and angular velocity of the rotor.

2.2. Organic Flight Array Model Dynamics. The OFA generally means a combined vehicle of two or more single ducted-fan modules. In this research, each ducted-fan vehicle was assembled to have the same control surface direction for the simplification of the dynamic modeling. In addition, all modules were located on the same plane, so as not to consider the z-axis moment arm in the assembling mode. Figure 3 shows the assembling concept of the OFA. The dynamic model in the assembling mode must simultaneously consider the forces and moments generated by each module. The force and moment defined in (1) are replaced by the sum of each component as follows:

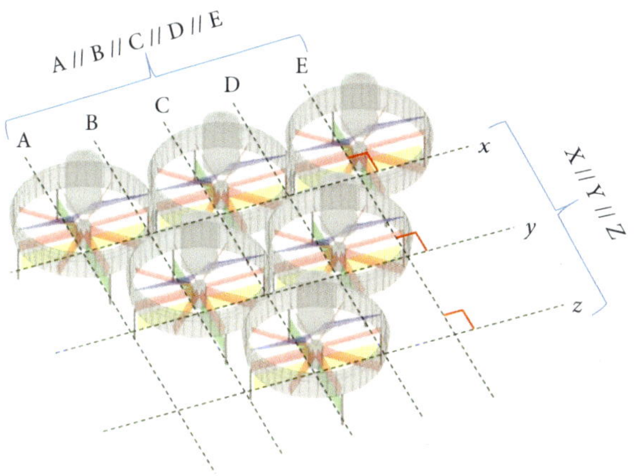

FIGURE 3: Arrangement of the control surfaces in the assembling mode.

$$
\begin{aligned}
F_x &= \sum_{i=1}^{N} F_{x_i},\\
F_y &= \sum_{i=1}^{N} F_{y_i},\\
F_z &= \sum_{i=1}^{N} F_{z_i},\\
M_x &= \sum_{i=1}^{N}\left(M_{x_i} - l_{z_i}F_{y_i} + l_{y_i}F_{z_i}\right),\\
M_y &= \sum_{i=1}^{N}\left(M_{y_i} - l_{x_i}F_{z_i} + l_{z_i}F_{x_i}\right),\\
M_z &= \sum_{i=1}^{N}\left(M_{z_i} - l_{y_i}F_{x_i} + l_{x_i}F_{y_i}\right),
\end{aligned}
\tag{15}
$$

where l denotes the moment arm from each module to the center of the gravity point and i is the module number of each vehicle. l_{z_i} is defined as zero because we assumed that the force generated by each module acts at the same height. Equation (1) shows that designing a controller by directly using the nonlinear dynamic equation is not generally easy. Moreover, the OFA is characterized by a high nonlinearity depending on the operation mode. Hence, the nonlinear dynamic equation is transformed into an affine system to apply the dynamic model inversion method that eliminates the nonlinearity. The affine system is a form that makes it easy to control the state variables according to the control input. The system consists of the state matrix, control matrix, and input and is presented as follows:

$$
\dot{\underline{x}} = \mathbf{F}(\underline{x}) + \mathbf{G}(\underline{x})(\underline{u}), \tag{16}
$$

where the variable x means the translational velocities u, v, and w and the rotational angular velocities p, q, and r. Here, if two or more modules are assembled into the OFA, the variable x is defined as the translational velocity and angular velocity of the combined whole modules. The state matrix **F** is a component generated by the fuselage, duct, gyroscopic moment, and gravity. The state matrix **F** is defined as follows:

$$
\mathbf{F}(\underline{x}) = \begin{bmatrix}
vr - wq + \dfrac{X_a}{m}\\[2mm]
wp - ur + \dfrac{Y_a}{m}\\[2mm]
uq - vp + \dfrac{Z_a}{m}\\[2mm]
\dfrac{\{qr(J_{yy} - J_{zz}) + L_a\}}{J_{xx}}\\[2mm]
\dfrac{\{pr(J_{zz} - J_{xx}) + M_a\}}{J_{yy}}\\[2mm]
\dfrac{\{pq(J_{xx} - J_{yy}) + N_a\}}{J_{zz}}
\end{bmatrix}, \tag{17}
$$

where the matrix **F** has a matrix size of $\mathbf{F}(\underline{x}) \in R^{6\times 1}$ regardless of the number of modules that make up the OFA. The control matrix induced by the control surfaces and the RPM of the rotor is expressed as follows:

$$
G(\underline{x}) = \begin{bmatrix}
0 & 0 & G(1,3) & 0 & \cdots\\
0 & G(2,2) & 0 & 0 & \cdots\\
G(3,1) & 0 & 0 & 0 & \cdots\\
G(4,1) & G(4,2) & 0 & 0 & \cdots\\
G(5,1) & 0 & G(5,3) & 0 & \cdots\\
G(6,1) & G(6,2) & G(6,3) & G(6,4) & \cdots
\end{bmatrix},
$$

$$
\begin{aligned}
G(1, 4(i-1)+3) &= \frac{\operatorname{sgn}(v_{\text{ind}} - \omega) q_{\text{ele}} C_{L_{\delta_{\text{ele}}}} S_{\text{ele}}}{m},\\
G(2, 4(i-1)+2) &= \frac{-\operatorname{sgn}(v_{\text{ind}} - \omega) q_{\text{ele}} C_{L_{\delta_{\text{ail}}}} S_{\text{ail}}}{m},\\
G(3, 4(i-1)+1) &= \frac{c_{\text{thr}}}{m},\\
G(4, 4(i-1)+1) &= \frac{l_{y_i} c_{\text{thr}}}{J_{xx}},\\
G(4, 4(i-1)+2) &= \frac{\operatorname{sgn}(v_{\text{ind}} - \omega) q_{\text{ail}} C_{L_{\delta_{\text{ail}}}} S_{\text{ail}} l_{\text{ail}}}{J_{xx}},\\
G(5, 4(i-1)+1) &= \frac{-l_{x_i} c_{\text{thr}}}{J_{yy}},\\
G(5, 4(i-1)+3) &= \frac{\operatorname{sgn}(v_{\text{ind}} - \omega) q_{\text{ele}} C_{L_{\delta_{\text{ele}}}} S_{\text{ele}} l_{\text{ele}}}{J_{yy}},\\
G(6, 4(i-1)+1) &= \frac{c_{\text{tor}}}{J_{zz}},\\
G(6, 4(i-1)+2) &= \frac{-l_{x_i} \operatorname{sgn}(v_{\text{ind}} - \omega) q_{\text{ail}} C_{L_{\delta_{ail}}} S_{\text{ail}}}{J_{zz}},\\
G(6, 4(i-1)+3) &= \frac{-l_{y_i} \operatorname{sgn}(v_{\text{ind}} - \omega) q_{\text{ele}} C_{L_{\delta_{\text{ele}}}} S_{\text{ele}}/J_{zz}}{m},
\end{aligned}
$$

$$G(6, 4(i-1)+4) = \frac{\text{sgn}\,(\nu_{\text{ind}} - \omega) q_{\text{rud}} C_{L_{\delta_{\text{rud}}}} S_{\text{rud}} l_{\text{rud}}}{J_{zz}}. \tag{18}$$

The control matrix **G** is defined by the area of the control surface of each module, induced velocity, and vertical velocity of the vehicle. The size of the matrix **G** varies depending on the number of combined vehicles, unlike the **F** matrix. Hence, the matrix size of **G** can be defined as $\mathbf{G}(\underline{x}) \in R^{6\times 4N}$.

3. Design of the Control System

3.1. Dynamic Model Inversion. The model dynamic inversion technique can linearly convert nonlinear dynamic equations to cancel the nonlinearities. Applying a linear controller to such nonlinear engineering models is not easy because most engineering models have nonlinearities. The dynamic equation can be expressed as follows if the state variables of the dynamic model are measurable [12]:

$$\begin{aligned} \ddot{\underline{x}} &= \underline{\nu}, \\ \underline{\nu} &= f(\underline{x}, \dot{\underline{x}}, \underline{u}), \end{aligned} \tag{19}$$

where $\underline{u}$ denotes the control input of the dynamic system and $\underline{\nu}$ is the pseudo control input. If $f(x, \dot{x}, u)$ is invertible and all states are measurable, the inverse conversion of the dynamic system in accordance with the control input is presented as follows:

$$\underline{u} = f^{-1}(\underline{x}, \dot{\underline{x}}, \underline{\nu}). \tag{20}$$

The nonlinearity of the system can be canceled, and the system can be controlled in a stable state if the pseudo control input can be appropriately designed. Furthermore, the system can be controlled using a linear controller if the nonlinearity is removed by applying the model inversion technique to a nonlinear system. However, perfectly implementing the same model as that of the actual aircraft is impossible because the aircraft model can be changed due to the mission environment or arbitrary influence even in actual flight situations. Therefore, a complete model cannot be implemented, and an error occurs when applying the DMI technique. This process is called the model inversion error. The dynamic model considering the model error can be expressed as follows:

$$\begin{aligned} \ddot{\underline{x}} &= \underline{\nu} + \Delta'(\underline{x}, \dot{\underline{x}}, \underline{\nu}), \\ \Delta'(\underline{x}, \dot{\underline{x}}, \underline{\nu}) &= f(\underline{x}, \dot{\underline{x}}, \hat{\underline{u}}) - f(\underline{x}, \dot{\underline{x}}, \underline{u}), \end{aligned} \tag{21}$$

where $\Delta'(x, \dot{x}, \nu)$ is the model error. Therefore, considering an adaptive control input that can compensate for the model error is necessary when designing a pseudo input. The pseudo input including the PD input, command, and adaptive input is defined herein as follows:

$$\nu_i = \nu_{\text{pd},i}(t) + \ddot{\mathbf{x}}_{\text{c},i}(t) - \hat{\nu}_{\text{ad},i}(t), \tag{22}$$

where the vector **x** is composed of the roll, pitch, and yaw Euler angles. Each axis of the Euler angles is considered separately for ease of designing the adaptive controller. Each $\nu_{\text{pd},i}(t)$ is determined by using the PD control theory as follows:

$$\begin{aligned} \nu_{\text{ail}} &= \ddot{\phi}_{\text{cmd}} + k_{\varphi,d}\left(\dot{\phi}_{\text{cmd}} - \dot{\phi}\right) + k_{\phi,p}(\phi_{\text{cmd}} - \phi), \\ \nu_{\text{ele}} &= \ddot{\theta}_{\text{cmd}} + k_{\theta,d}\left(\dot{\theta}_{\text{cmd}} - \dot{\theta}\right) + k_{\theta,p}(\theta_{\text{cmd}} - \theta), \\ \nu_{\text{rud}} &= \ddot{\psi}_{\text{cmd}} + k_{\psi,d}(\dot{\psi}_{\text{cmd}} - \dot{\psi}) + k_{\psi,p}(\psi_{\text{cmd}} - \psi), \end{aligned} \tag{23}$$

As a result, the error equations can now be defined by (21) and (23).

$$\ddot{\tilde{x}}_i + k_{i,d}\dot{\tilde{x}}_i + k_{i,p}\tilde{x}_i = \hat{\nu}_{ad,i} - \Delta'\left(\underline{x}, \dot{\underline{x}}, \underline{\nu}\right). \tag{24}$$

The objective of the adaptive control input is to design $\nu_{\text{ad},i}$ that can make the right side zero (i.e., $\nu_{\text{ad},i}$ and $\Delta'(x, \dot{x}, \nu)$) to be the same value. Equation (24) can be expressed in the state space form as follows:

$$\begin{aligned} &\dot{\underline{e}} = \mathbf{A}\underline{e} + B\left(\hat{\nu}_{\text{ad},i}\Delta'(\underline{x}, \dot{\underline{x}}, \underline{\nu})\right), \\ &\mathbf{A} = \begin{bmatrix} 0 & 1 \\ -k_{i,d} & -k_{i,p} \end{bmatrix}, B = \begin{bmatrix} 0 \\ 1 \end{bmatrix}, \underline{e} = \begin{bmatrix} e_1 \\ e_2 \end{bmatrix} = \begin{bmatrix} \tilde{x} \\ \dot{\tilde{x}} \end{bmatrix}, \end{aligned} \tag{25}$$

3.2. Adaptive Control System of the Neural Network. The adaptive control input was designed to compensate for the model error, which changes in real time. A model error can generally be implemented by the neural network with a finite number of basis functions as follows:

$$\hat{\Delta}'(\underline{x}, \dot{\underline{x}}, \underline{\nu}) = \sum_{j=1}^{N} \hat{\omega}_{ij}\beta_i'(x, \dot{x}, \underline{\nu}) = \hat{\omega}_i^T \beta_i'(x, \dot{x}, \underline{\nu}). \tag{26}$$

In this case, a value can be defined as follows if it is obtained by the weight that can generate an optimal adaptive control input on $d' = (x, \dot{x}, \nu)$ domain as ω_i^*:

$$\omega_i^* = \arg\left\{\min_{\hat{\omega}} \max_{d' \in D'} \left|\Delta'\left(d'\right) - \hat{\Delta}'\left(d'\right)\right|\right\}. \tag{27}$$

The absolute difference between the estimated and actual values of the model error is given by (28) if $\epsilon_i > 0$ is defined as the boundary of the abovementioned optimal problem.

$$\left|\Delta'\left(d'\right) - \hat{\Delta}^*\left(d'\right)\right|, \quad \forall d' \in D'. \tag{28}$$

On the compact domain, it is possible to assume that $\hat{\nu}_{\text{ad}}$ that always satisfies Assumption 1 is present.

$$\nu_{\text{ad}} - \hat{\Delta}\left(\omega, \nu_{-1}, \nu_{\text{pd}} + \ddot{x}_{\text{c}} - \nu_{\text{ad}}\right) = 0. \tag{29}$$

Assumption 1. There is a fixed point ν_{ad} of $\hat{\Delta}$ at all $d' \in D'$.

As a result, the model error can be canceled through the adaptive input and can be defined as $\hat{\Delta}'(\underline{x}, \dot{\underline{x}}, \underline{\nu}) = \hat{\nu}_{\text{ad},i}$ using the components of (26) and (29).

$\hat{\omega}_i(t) \in R^N$ denotes the weight updated in real time by connecting the hidden and output layers and is defined as follows [14, 17]:

$$\dot{\hat{\omega}}_i = \begin{cases} -\gamma_i e_i^T P_{0i} b \beta' i(x, \dot{x}, \nu), & \text{when } e_{i2} > e_0, \\ 0, & \text{when } e_{i2} < e_0, \end{cases} \quad (30)$$

The weights designed in (30) are defined through the Lyapunov stability analysis involving the adaptive control input applied only to the dead zone, where the norm of the state error vector is defined as e_0. In (30), γ_i is the learning rate or adaptation gain, and β_i is the basis function of each node comprising the hidden layer. $\mathbf{P}_0$ is a symmetric, positive, and definite matrix satisfying the Lyapunov equations as follows:

$$\mathbf{P}_0 A_i + A_i^T \mathbf{P}_0 = -I, \quad i = 1, 2, \ldots, n. \quad (31)$$

The $\mathbf{P}_0$ that satisfies (31) can be defined in the Lyapunov analysis process and include the control gain of the PD controller as follows:

$$\mathbf{P}_0 = \begin{bmatrix} \frac{K_{i,d}}{2} + \frac{K_{i,p}}{2K_{i,p}}\left(1 + \frac{1}{K_{i,p}}\right) & \frac{1}{2K_{i,d}} \\ \frac{1}{K_{i,p}} & \frac{1}{2K_{i,d}}\left(1 + \frac{1}{K_{i,p}}\right) \end{bmatrix}. \quad (32)$$

3.3. Basis Function. Figure 4 shows the control system designed using a radial basis function neural network with a single hidden layer. In the process of designing a neural network, the basis function $\beta_i'(\underline{x}, \underline{\dot{x}}, \nu)$ must be defined. The Gaussian function and the general combination of the Gaussian function were applied as the basis function in the same way as Figure 5 in this study. In the neural network, the basis function played a role of activating the variable value coming into the hidden layer. The Gaussian function has the advantage of a variable entering through the centers. The deviation can also be efficiently activated. The Gaussian functions used as basis functions are generally expressed as follows:

$$\beta_{ij}'(\underline{x}, \underline{\dot{x}}, \nu) = \Phi(\|\underline{x} - c_i\|_2), \quad \Phi(r) = \exp\left(\frac{-r^2}{2\sigma^2}\right), \quad (33)$$

where c is the center of the Gaussian function and σ is the deviation. These values are variables that determined the degree of activation of the values input to the hidden layer. Tuning these values can change the performance of the adaptive controller. The radial basis function neural network has the feature of an input value directly entering the hidden layer without considering the weight. The input variables to the neural network in the attitude control system of the OFA were divided into roll, pitch, and yaw channels in the same manner as the PD controller. Table 3 basically shows the parameters input to the input layer. The variables input to each channel were angular velocity, Euler angle, Euler angular error, Euler angular error differential value, and each control surface deflection angle. The number of the variables input to the neural network according to each channel was $4 + N$. The number of the input variables increased as the number of the assembled modules increased.

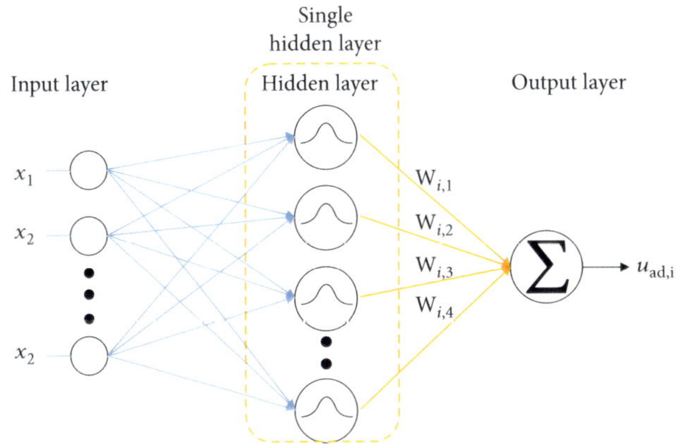

FIGURE 4: Schematic of the radial basis function neural network.

3.4. Control System Structure. A PD controller was used, and the DMI method was applied herein as an inner loop for attitude control. A controller structure with an adaptive control input based on the neural network was additionally considered in a situation where a model error occurs. Figure 6 shows the control structure for the attitude control of the OFA.

4. Numerical Simulation

A numerical simulation was performed herein by applying the dynamic model inversion technique and the PD controller. The simulation result was analyzed according to the presence or absence of the NN by defining the scenario in which the uncertainty or the disturbance occurred. Each of the scenarios simulated the attitude control assuming a situation, in which an arbitrary model uncertainty is given, and simulated the path control in the situation where the strong disturbance continuously influences. Figure 7 shows the OFA model used in each scenario.

4.1. Scenario 1. The first scenario assumed that the model uncertainty increases as the number of ducted-fan vehicles increases. It is assumed that 0 to 30% model uncertainty is generated in real time when two modules are added to the existing module. A simulation was performed to verify the control performance according to the application of the neural network. Figure 8 shows a relatively large overshoot and a slow follow-up when the neural network was not applied because the PD controller alone cannot compensate for the model error. On the contrary, each attitude response followed the command relatively quickly when the neural network was applied. However, as shown in Figures 9 and 10, the control input was excessively generated when the neural network was considered. The control allocation problem was not considered separately in this study. Hence, the

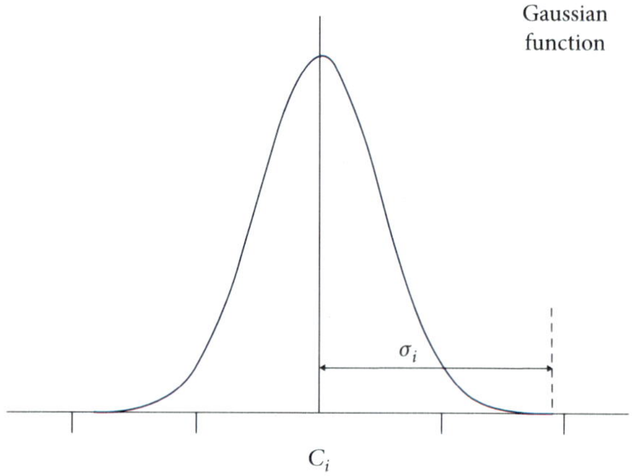

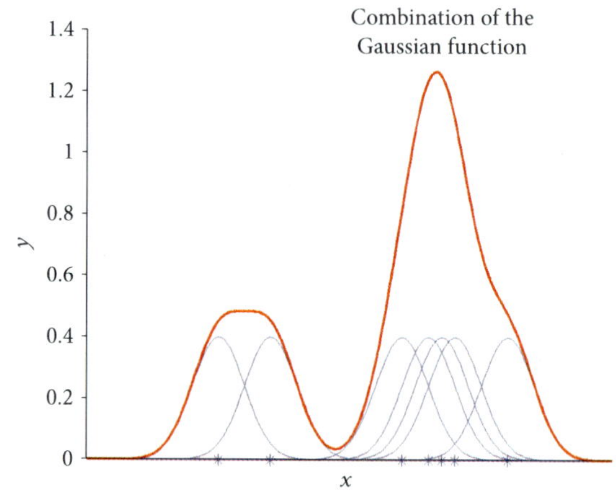

FIGURE 5: Gaussian function used as basis.

TABLE 3: Input variables of the neural network.

Channel	Input variables
Phi (ϕ)	$p, \phi, \tilde{\phi}, \dot{\tilde{\phi}}, \delta_{\text{ail},1-N}$
Theta (θ)	$q, \theta, \tilde{\theta}, \dot{\tilde{\theta}}, \delta_{\text{ele},1-N}$
Psi (ψ)	$r, \psi, \tilde{\psi}, \dot{\tilde{\psi}}, \delta_{\text{rud},1-N}$

control surfaces of the three modules moved in the same angle and direction. Each control surface deflection response was illustrated in a common graph. Meanwhile, the RPM control inputs showed different values to make the thrust difference by the center of the gravity. Figure 11 shows the weight of each channel when the neural network was applied. The weight change amount becomes large at the time of the command change, such that a deficient control input is generated.

4.2. Scenario 2. The second scenario presented a situation where the OFA consisting of three modules performed path control to combine with the other arrays in a strong disturbance environment. The attitude controller of the inner loop was generally classified according to whether the neural network was applied.

The outer loop for the path control adopted the PID controller. The disturbance was also assumed to be a uniform wind of 1.5 m/s along the x-axis of the body-fixed coordinate. As a result, the path control was not possible when the NN was not considered, as shown in Figure 12.

However, when the neural network controller is applied, it is observed that the position command is slightly slower but follows normally. Figure 13 shows the changes in the attitude of the UAV when the path control is normally performed in Figure 12. Through this result, it is confirmed that the attitude control system applying the neural network guarantees stability of the UAV in a strong disturbance environment. Figures 14 and 15 represent the control inputs when the neural network controller is applied. The oscillation of the elevator control surface in Figure 15 is caused to control the pitch angle in order to secure stability of the UAV and to follow the x-axis position command under the strong disturbance. This oscillation is caused by the changes in the cross-sectional force component, induced velocity, and z-axis translational velocity depending on the RPM, as shown in (18). This implies that the simulation results presented in this paper are under marginally stable conditions in which the ducted-fan UAV can operate.

Since harsh disturbance is injected from the beginning of the simulation, such oscillation may be observed until the weights between the neural network layers are adjusted with sufficient samples of the system inputs/outputs. Furthermore, we considered the HS-5070MH servo motor as the control surface actuator that can deflect 60 deg for 0.12 sec at no load condition. Even if the full actuator performance cannot be guaranteed due to the aerodynamic load to the servo during the flight, there is still a sufficient margin to accommodate the maximum angular velocity (200 deg/s) observed in the simulation results. Figure 16 presents the weight update value of the neural network. The weight update value was confirmed to converge to a constant value after the position control is normally performed.

5. Conclusion

This study presented a concept of the organic flight array that can simultaneously perform various missions with a single ducted-fan vehicle through the assembling, separation, and cooperation modes. First, this study proposed a dynamic model for a single ducted-fan vehicle and defined the OFA dynamic model of the affine form, which is specified by the number of assembling modules. A control system was designed based on this through a dynamic model inversion and a PD controller. The adaptive controller based on the radial basis function neural network was also designed for the control system that can avoid model uncertainty or disturbance. Subsequently, the integrated simulation environment was constructed to verify the controller performance. The simulation results confirmed that the control performance was improved when the adaptive controller using the radial basis function neural network was considered in the case of model uncertainty or disturbance. Consequently,

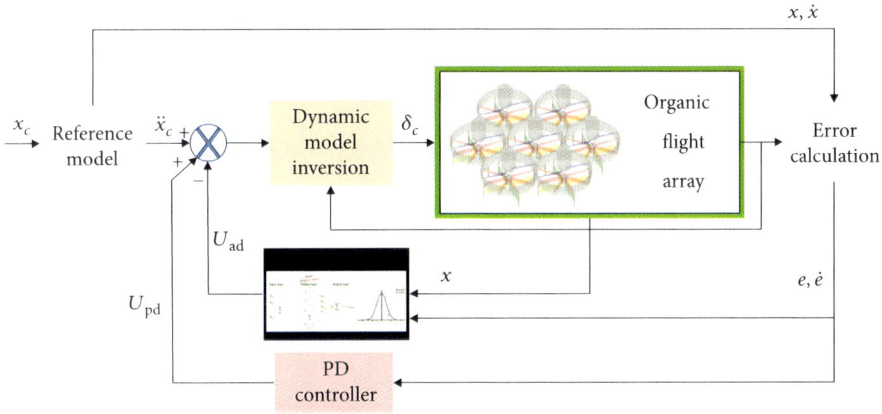

FIGURE 6: Controller structure of the adaptive neural network.

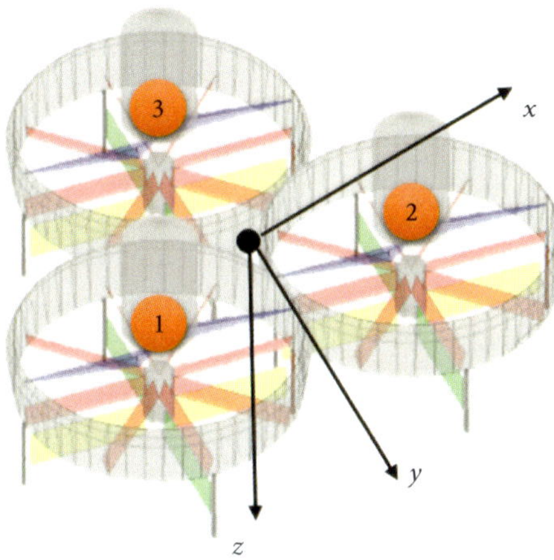

FIGURE 7: Simulation model consisting of three single modules.

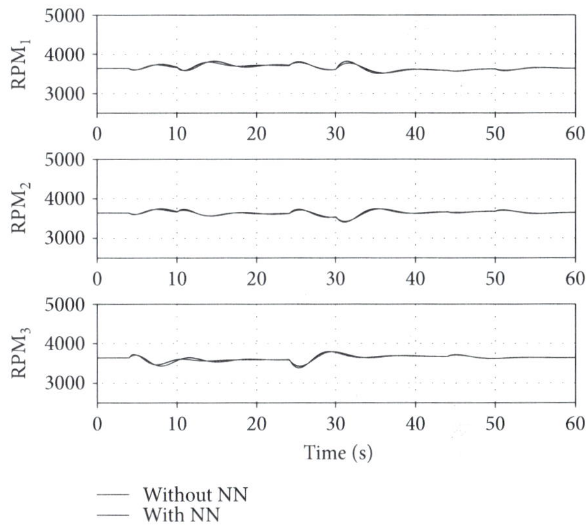

FIGURE 9: RPM control input in scenario 1.

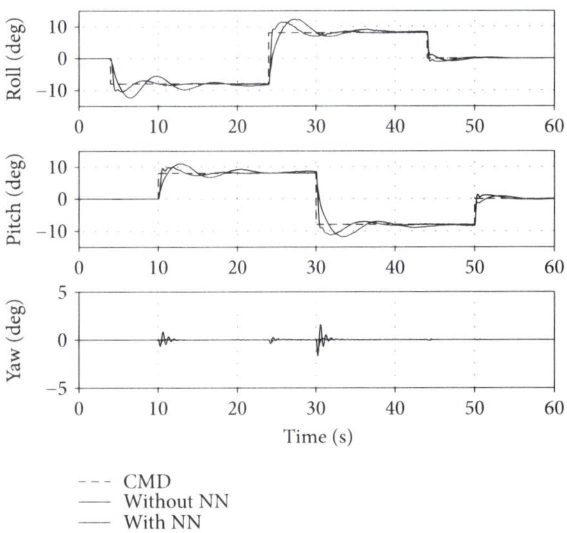

FIGURE 8: Euler angle response in scenario 1.

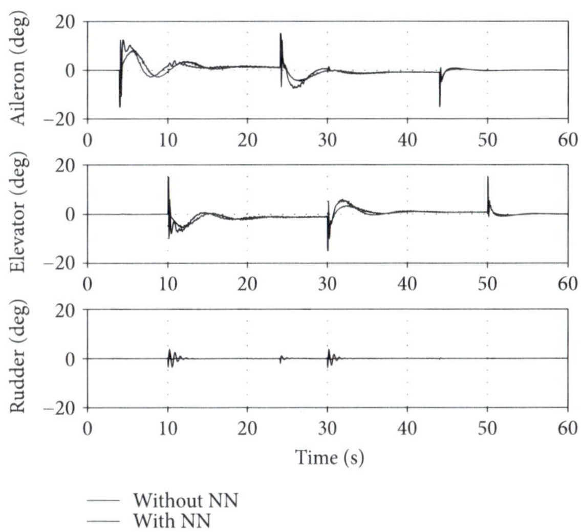

FIGURE 10: Control surface deflection angle in scenario 1.

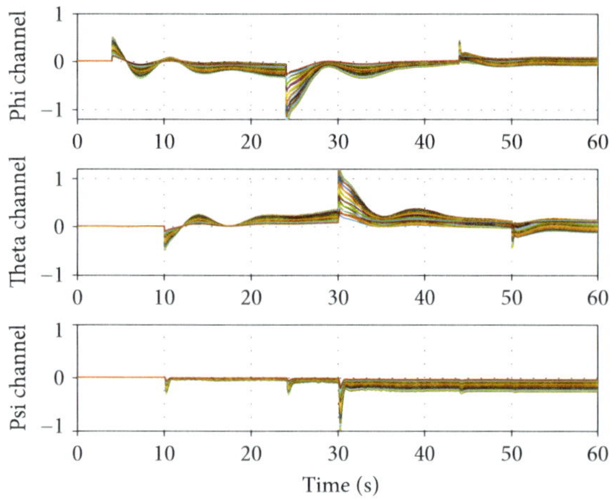

FIGURE 11: Weight value of the neural network in scenario 1.

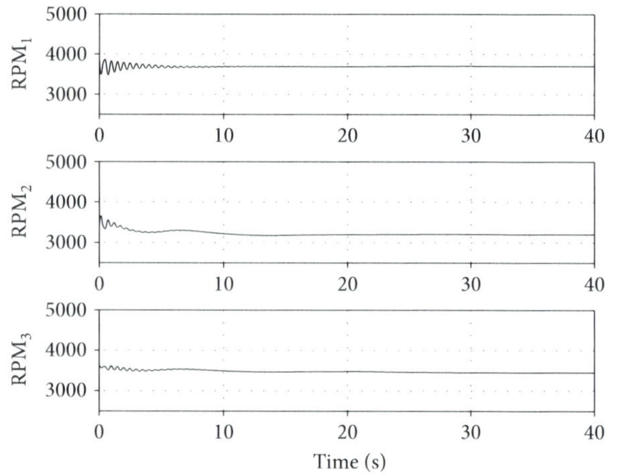

FIGURE 14: RPM control input when considering the neural network in scenario 2.

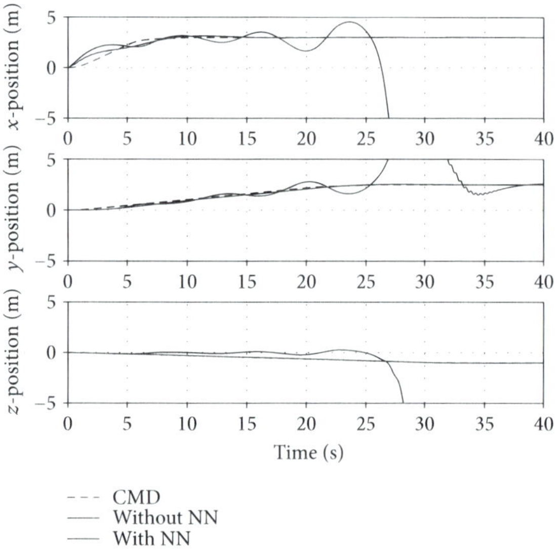

FIGURE 12: Position of the OFA in scenario 2.

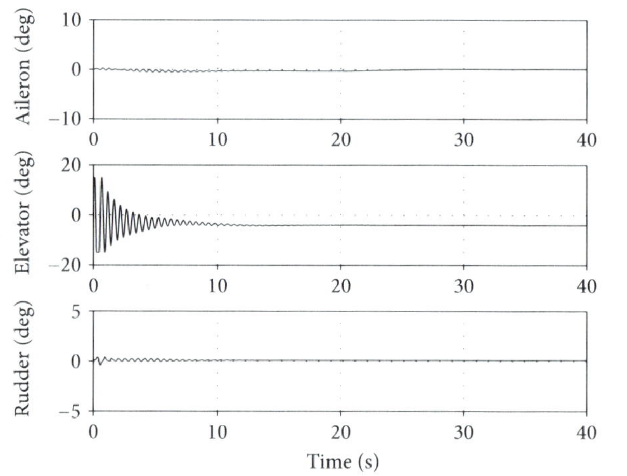

FIGURE 15: Control surface deflection angle when considering the neural network in scenario 2.

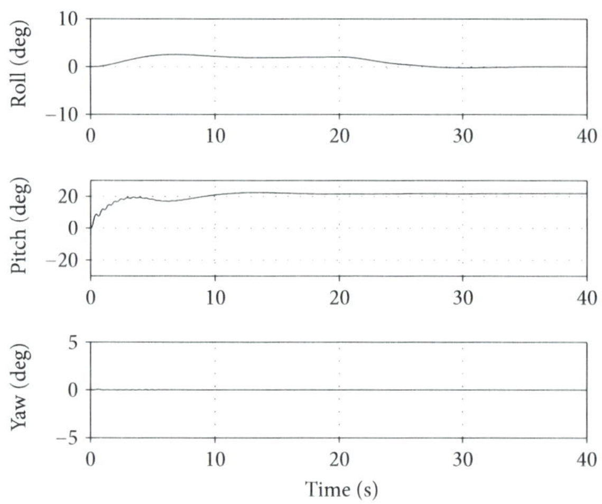

FIGURE 13: Attitude response when considering the neural network in scenario 2.

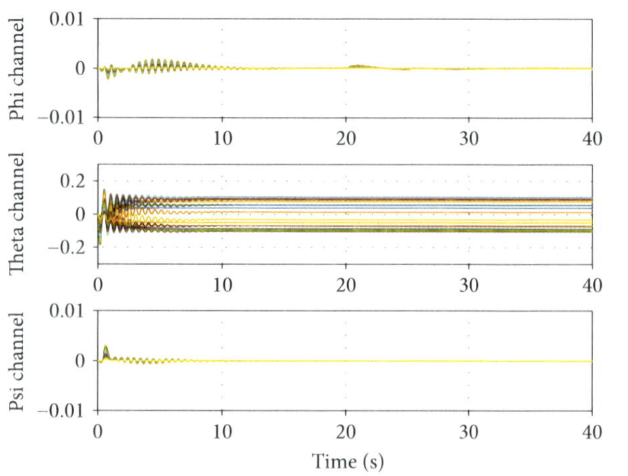

FIGURE 16: Weight value of the neural network in scenario 2.

the control stability can be guaranteed when the neural network is used as an adaptive controller in a situation where the model uncertainty is large or the disturbance is strong.

Conflicts of Interest

The authors declare that they have no conflicts of interest.

Acknowledgments

This work is partly supported by the research project (no. 10062497) funded by the Ministry of Trade, Industry & Energy, Korean Government.

References

[1] S. Bouabdallah, P. Murrieri, and R. Siegwart, "Towards autonomous indoor micro VTOL," *Autonomous Robots*, vol. 18, no. 2, pp. 171–183, 2005.

[2] C. Huerzeler, K. Alexis, and R. Siegwart, "Configurable real-time simulation suite for coaxial rotor UAVS," in *2013 IEEE International Conference on Robotics and Automation*, pp. 309–316, Karlsruhe, Germany, May 2013.

[3] J. T. K. Ping, C. S. Tan, and V. O. T. Tat, "Coanda Effect Test Bench (CoETB) — design enhancement of the Coanda^{JLT} craft," in *2011 IEEE Conference on Sustainable Utilization and Development in Engineering and Technology (STUDENT)*, pp. 25–30, Semenyih, Malaysia, October 2011.

[4] E. N. Johnson and M. A. Turbe, "Modeling, control, and flight testing of a small-ducted fan aircraft," *Journal of Guidance, Control, and Dynamics*, vol. 29, no. 4, pp. 769–779, 2006.

[5] C. Nicol, C. J. B. Macnab, and A. Ramirez-Serrano, "Robust adaptive control of a quadrotor helicopter," *Mechatronics*, vol. 21, no. 6, pp. 927–938, 2011.

[6] J. M. Pflimlin, P. Soueres, and T. Hamel, "Position control of a ducted fan VTOL UAV in crosswind," *International Journal of Control*, vol. 80, no. 5, pp. 666–683, 2007.

[7] D. Mellinger, M. Shomin, N. Michael, and V. Kumar, "Cooperative grasping and transport using multiple quadrotors," in *Distributed Autonomous Robotic Systems*, Springer, 2013.

[8] R. Oung, A. Ramezani, and R. D'Andrea, "Feasibility of a distributed flight array," in *Proceedings of the 48h IEEE Conference on Decision and Control (CDC) held jointly with 2009 28th Chinese Control Conference*, pp. 3038–3044, Shanghai, China, December 2009.

[9] R. Oung and R. D'Andrea, "The distributed flight array," *Mechatronics*, vol. 21, no. 6, pp. 908–917, 2011.

[10] J. Abedin and R. Akmeliawati, "Distributed flight array of autonomous flying vehicles," in *Proceedings of international conference on intelligent unmanned systems*, vol. 11, Bali, Indonesia, August 2015.

[11] B. Oh, "*Design of control system for organic flight array based on neural network controller*, M.S. thesis," ChungNam National University, 2017.

[12] B.-M. Kim, B. S. Kim, and N.-W. Kim, "Trajectory tracking controller design using neural networks for a tiltrotor unmanned aerial vehicle," *Proceedings of the Institution of Mechanical Engineers, Part G: Journal of Aerospace Engineering*, vol. 224, no. 8, pp. 881–896, 2010.

[13] C. H. Hong, K. C. Choi, and B. S. Kim, "Applications of adaptive neural network control to an unmanned airship," *International Journal of Control, Automation and Systems*, vol. 7, no. 6, p. 911, 2009.

[14] B. S. Kim and A. J. Calise, "Nonlinear flight control using neural networks," *Journal of Guidance, Control, and Dynamics*, vol. 20, no. 1, pp. 26–33, 1997.

[15] M. Bahita and K. Belarbi, "Neural feedback linearization adaptive control for affine nonlinear systems based on neural network estimator," *Serbian Journal of Electrical Engineering*, vol. 8, no. 3, pp. 307–323, 2011.

[16] H. W. Zhao, "*Development of a dynamic model of a ducted fan VTOL UAV*, Ph.D. thesis," RMIT University, 2009.

[17] B. Kim, "*The design of guidance and control for tilt rotor aircraft using neural networks*, Ph.D. thesis," Gyeongsang National University, 2008.

2

Azimuth and Elevation Dynamic Tracking of UAVs via 3-Axial ULA and Particle Filtering

Andrea Papaiz[1] **and Andrea M. Tonello**[2]

[1] *WiPLi Lab, University of Udine, Via delle Scienze 206, 33100 Udine, Italy*
[2] *EcoSys Lab, Alpen-Adria-Universität Klagenfurt, Universitätsstraße 65-67, 9020 Klagenfurt, Austria*

Correspondence should be addressed to Andrea Papaiz; andrea.papaiz@uniud.it

Academic Editor: Yanbo Zhu

Unmanned Aerial Vehicles (UAVs) localization has become crucial in recent years, mainly for navigation or self-positioning and for UAV based security monitoring and surveillance. In this paper, azimuth and elevation radio positioning of UAVs are considered. The localization is based on multiple differential phase-of-arrival measures exploiting a 3-Axial Uniform Linear Array of antennas. An ad hoc particle filtering algorithm is applied to improve the positioning performance using a dynamic motion model. A novel adaptive algorithm, namely, Particles Swarm Adaptive Scattering (PSAS), is proposed to increment the algorithm stability and precision. To assess performance a Confined Area Random Aerial Trajectory Emulator (CARATE) algorithm has been developed to generate actual paths of flying UAVs. The algorithm performance is compared with the baseline method and with the average trajectory Cramér Rao lower bound to show the effectiveness of the proposed algorithm.

1. Introduction

Unmanned Aerial Vehicles (UAVs) are attracting considerable attention since they can be used for a number of consumer, industrial, and military applications ranging, for instance, from sport video making to environmental monitoring and parcel delivery [1, 2]. A key component is the technology that allows to locate and navigate the UAVs. Presently, global navigation satellite systems (GNSS) and inertial sensors are used to provide information on position, speed, and direction of movement. Reliable localization is very important also in view of new regulations that aim at better controlling the use and the status of the UAVs for higher safety and security [3]. Furthermore, recently, UAVs technology has started to be under the spotlight of police audits because of possible security threats [4].

In this paper, a radio localization approach is considered and it is based on azimuth and elevation positioning using a transmitting source as a reference. Azimuth and elevation are determined by processing with particle filtering (PF) the signals that impinge on a 3-Axial Uniform Linear Array (3A-ULA). The 3A-ULA can be mounted either on a ground base station or on the UAVs. In the first case, namely, Ground-Localization Scenario (GL-S), the base station passively eavesdrops the signals emitted by the UAVs to determine their angular coordinates. In the latter case, namely, Self-Localization Scenario (SL-S), the ground node acts as a radio anchor allowing UAVs self-localization. The system can be used in a standalone way or to complement existing GNSS/inertial or inertial sensors employing data fusion techniques [5].

In Section 2, the analytical model for the signals received by the 3A-ULA is described. The baseline method (BM) coordinates estimates obtained from the 3A-ULA in Section 3 are iteratively processed with an ad hoc particle filtering algorithm (PF) described in Section 4. The PF technique uses a novel dynamic particles swarm management routine, namely, Particles Swarm Adaptive Scattering (PSAS), introduced in Section 5 to improve both convergence and precision performance.

In order to quantify the performance of the proposed tracking algorithm an emulator for the UAV behaviours is specifically developed in Section 6. This ad hoc numerical model, namely, Confined Area Random Aerial Trajectory

Emulator (CARATE), randomly generates trajectories in the 3-dimensional space following a well defined set of rules in order to emulate real UAV behaviours.

Finally, in Section 8, numerical results are reported to assess performance as a function of the UAVs motion model, the PF algorithm parameters, and external detrimental impairments, such as phase noise (PN) and carrier frequency Doppler shift (DS). A comparison with the average trajectory Cramér Rao lower bound (AT-CRLB) for the angular estimation of the elevation θ and azimuth ϕ is also provided.

2. Positioning System Description

A system that determines the position of flying UAVs through the analysis of a signal impinging on the 3 branches of a 3A-ULA is considered.

The GL-S is built as follows. UAVs send narrow band radio signals that are captured by a ground base station equipped with a 3A-ULA. The radio signals are properly frequency/time duplexed to allow multiple UAVs tracking. The ground node can then compute locally the UAVs azimuth and elevation coordinates [6]. By sharing the data collected from more ground nodes, it is possible to achieve a full 3D positioning for the UAVs trajectories.

The SL-S is analogously arranged. Each 3A-ULA is mounted on the UAVs so that the ground device takes the role of a radio beacon anchor [7], allowing the UAVs to be self-aware of their position. Using more anchors transmitting at different frequency or in a time division multiplexing fashion, the UAVs can accomplish a full 3D positioning. Clearly, the implementation of the SL-S is dependent on the UAV and array sizes.

The 3A-ULA is constituted, for each one of the 3 branches, by N_a antennas. Each branch is labelled with $a \in [x, y, z]$ to indicate along which one of the axes it is displaced. An antenna of the same branch a is indexed with $i \in [1, 2, \ldots, N_a]$ to indicate its position along the branch. Thus, the coordinates of the antennas can be written as

$$
\begin{aligned}
x^{(a,i)} &= (i-1)\, D\delta_{xa},\\
y^{(a,i)} &= (i-1)\, D\delta_{ya},\\
z^{(a,i)} &= (i-1)\, D\delta_{za},
\end{aligned} \tag{1}
$$

where δ_{ij} is the Kronecker delta, that is, 1 for $i = j$ and 0 otherwise, and D is the constant distance between antennas.

The signal model considered in the following describes for simplicity the scenario with only one UAV; this assumption can be easily extended to a multi-UAVs application. Both GL-S and SL-S scenarios, due to symmetry, can be modelled in the same way. Hence, in general, the downconverted signal received by the ith antenna of the 3A-ULA branch a can be written as

$$
s_n^{(a,i)} = A_n e^{j[\psi_n^{(a,i)} + \gamma_n^{(a,i)} + \varphi_n + \zeta]} + w_n^{(a,i)}, \tag{2}
$$

where n indexes the time, sampled with period T. The quantity $\psi_n^{(a,i)}$ is the phase of the demodulated signal impinging on the ith sensor of the branch a, namely, the phase of arrival (PoA), in the ideal case without impairing effects. $\gamma_n^{(a,i)}$ represents the phase Doppler shift (DS) caused by the nonzero speed of the radio source with respect to each receiving array element. The parameter φ_n is the phase noise (PN) of the oscillators. It is considered white for this application [8]. The component ζ is the phase offset (PO) between modulating and demodulating oscillators. The phase components of (2) are described in more detail in the following:

$$
\begin{aligned}
\psi_n^{(a,i)} &= 2\pi f_0 \tau_n^{(a,i)},\\
\gamma_n^{(a,i)} &= 2\pi f_0 C_{\mathrm{DS}} \left(nT - \tau_n^{(a,i)}\right) \tau_n^{(a,i)},
\end{aligned} \tag{3}
$$

where f_0 is the carrier frequency of the received modulated signal and where $\tau_n^{(a,i)}$ is the time of arrival (ToA) of the plane wave from the source to the sensor. $C_{\mathrm{DS}}(t) = v_r(t)/c_0$ is the DS coefficient due to the movement of the UAV. The parameter $v_r(t)$ is the UAV speed in the direction of the center of the 3A-ULA and c_0 is the speed of light. Finally, $w_n^{(a,i)}$ is circular additive white Gaussian noise (C-AWGN) with standard deviation σ_w. The receiving architecture comprises N_a downconverting branches for each axis of the 3A-ULA all fed by the same oscillator. For this reason PN and PO components are identical for all antennas.

Considering the scenario where the transmitted signals carry also information, the component A_n of (2) is representative of the unknown informative part of the signal, specifically the modulated complex data symbols. For the application considered in this paper A_n is assumed to be an equiprobable symbol of an unknown order PSK modulation scheme. For this reason the symbol is modelled as complex uniform ring shaped noise (RSN) $A_n = |A|e^{j\xi_n}$ where $\xi_n \sim U(0, 2\pi)$ is a random uniform variable between 0 and 2π and $|A|$ is the constant amplitude of the modulated signal.

In this model, the presence of multipath propagation is not considered; a line of sight environment is assumed. This can hold true when the ground and UAV antennas have a wide vertical lobe, respectively, directed upward and downward.

3. Azimuth and Elevation Estimation with 3A-ULA

A radio source moving in the 3D Cartesian space is considered. Its angular spherical coordinates θ_n and ϕ_n are determined using the same coordinate system of the 3A-ULA defined in (1). A trajectory example is visible in Figure 2. The positioning technique used to locate the radio source is based on the estimation of the phase difference of the received signals between different antenna couples belonging to the same 3A-ULA branch a, namely, the differential phase of arrival (D-PoA) $\widehat{\psi}_n^{(a)}$. The D-PoA is defined as $\widehat{\psi}_n^{(a)} = \psi_n^{(a,2i-1)} - \psi_n^{(a,2i)}$ $\forall i \in [1, 2, \ldots, N_a/2]$.

In [8] a technique to estimate $\widehat{\psi}_n^{(a)}$ in a scenario with impaired signals similar to (2) is described. Applying this technique for the model considered in this paper, it allows exploiting the D-PoAs from the signal model introduced in

(2). $\widehat{\psi}_n^{(a)}$ are estimated as $\angle u_n^{(a)}$, where $\angle$ stands for the phase operator, where

$$u_n^{(a)} = \frac{2}{N_a} \sum_{i=1}^{N_a/2} s_n^{(a,2i-1)} s_n^{(a,2i)^*}, \quad (4)$$

where an average over $N_a/2$ distinct antenna pairs is performed to mitigate noise and compensate other zero mean impairments. Using $u_n^{(a)}$, the angular spherical coordinates θ_n and ϕ_n are estimated, respectively, with $\widetilde{\theta}_n$ and $\widetilde{\phi}_n$ as follows:

$$\widetilde{\theta}_n = \arctan\left(\sqrt{\left(\angle u_n^{(x)}\right)^2 + \left(\angle u_n^{(y)}\right)^2}, \angle u_n^{(z)}\right),$$
$$\widetilde{\phi}_n = \arctan\left(\angle u_n^{(y)}, \angle u_n^{(x)}\right), \quad (5)$$

where $\arctan(\cdot,\cdot)$ is the arctangent function extended to the $[0, 2\pi)$ domain.

The estimations computed in (5) will be referred to as baseline method (BM) estimations to differ them from the particle filter (PF) estimations that will be introduced in Section 4.

4. PF Algorithm Estimation

In this section, the application of the PF algorithms [9] to the localization problem introduced in Section 3 is briefly described. PF algorithm techniques have been chosen because of their simple, scalable, and flexible numerical implementation with respect to other methods like Extended Kalman Filter [10] that approach the problem linearising the system and approximating the probability density functions as Gaussians. PF is specifically designed to fit complex system models as well as nonlinear and non-Gaussian configurations. The target of PF methods is to estimate the dynamic evolution of the hidden states in a system that are the coordinates θ_n and ϕ_n of the radio source, using the sequential discrete noisy measures $u_n^{(a)}$ introduced in (4). In PF, it is necessary to hypothesize a model for the observations and for the states temporal evolution [9]. The evolution model for the hidden states θ_n and ϕ_n is

$$\theta_n = g_\theta\left(\theta_{[n-N_{mm} \div n-1]}, p_{\theta,n}\right),$$
$$\phi_n = g_\phi\left(\phi_{[n-N_{mm} \div n-1]}, p_{\phi,n}\right), \quad (6)$$

while the model for the observations related to the hidden states introduced in (4) is

$$u_n^{(a)} = h\left(\left[\theta_n, \phi_n\right], q_n\right). \quad (7)$$

The relations (6) show how the current hidden state $[\theta_n, \phi_n]$ evolves thanks to the previous N_{mm} states and the random uncertain variables $p_{\theta,n}$ and $p_{\phi,n}$ through the motion model function g. Relation (7) shows how the measurements $u_n^{(a)}$ depend only on the current state $[\theta_n, \phi_n]$ and on a random uncertain variable q_n through the measure model function h.

For notational simplicity, both θ_n and ϕ_n will be denoted with the generic angle α_n. When needed this generalization will be removed.

PF has been chosen because of its simplicity and versatility. PF algorithms are a numeric implementation of Bayesian estimation [10], a statistical approach to iteratively increase the knowledge of the states to be estimated by narrowing their pdf conditioned by the sequential acquisition of new measures during time. The pdf of α_n, defined as $f_{\alpha_n}[\alpha]$, is approximated in a discrete way [11] with a set of M particles $\{\alpha_n^{(m)}\}_{m=1,\dots,M}$ and a set of weights $\{w_{\alpha,n}^{(m)}\}_{m=1,\dots,M}$. Particles and weights are generated from the partial knowledge given from the measure model h and the state evolution model g_α. This knowledge is more and more incremented by the continuous time acquisition of the measures $u_n^{(a)}$. Clearly, the UAV moves so that its coordinates change over time n.

As shown in (6), in the PF algorithm an assumption about the model that controls the evolution of the hidden state of the system g_α is made. For this purpose the motion model used is the Drift Motion Model (DMM) proposed in [12]. The DMM generates the new particles cloud $\{\alpha_n^{(m)}\}$ at step n shifting the previous step resampled cloud [9] $\{\alpha_{n-1,\mathrm{RES}}^{(m)}\}$ using the regression constant $\mathbf{m}_{n-1}$:

$$\alpha_n^{(m)} = \alpha_{n-1,\mathrm{RES}}^{(m)} + \mathbf{m}_{n-1}T + \eta, \quad (8)$$

where $\alpha_{n-1,\mathrm{RES}}^{(m)}$ is the mth particle of $n-1$ step cloud after the resampling process [11]. $\eta \sim \mathcal{N}(\sigma_s, m_s = 0)$ is a Gaussian random variable that represents the fundamental statistical scattering part of the motion model. The parameter $\mathbf{m}_{n-1}$ is the slope of the linear regression of N_{mm} previous values of $\widehat{\alpha}_n$ that are the PF estimates of α_n. It is important to emphasize that in this regression model, due to the periodic nature of α_n, it is necessary to use the previously saved N_{mm} unwrapped values of $\widehat{\alpha}_n$. This is necessary to correctly track the azimuthal coordinate ϕ_n during 2π angular shifts from $-\pi$ to $+\pi$ going from the third to the second Cartesian quadrant.

Particles weights $w_{\alpha,n}^{(m)}$ are calculated by the PF algorithm using in each cycle the temporary estimates $\widetilde{\alpha}_n$ obtained from the real acquired measures in (5). This temporary estimations are compared with the cloud of generated particles $\alpha_n^{(m)}$ through an exponential weighting function to obtain the set of weights $w_{\alpha,n}^{(m)}$:

$$w_{\alpha,n}^{(m)} = \exp\left(-\left|\alpha_n^{(m)} - \widetilde{\alpha}_n\right|\right), \quad (9)$$

where, to correct unnecessary 2π offsets of $|\alpha_n^{(m)} - \widetilde{\alpha}_n|$ due to the periodicity of α_n, the minimum absolute value of the distance of the two angles $\alpha_n^{(m)}$ and $\widetilde{\alpha}_n$ in 2π-modulo space is computed. A reshaping technique [12] is then applied, before estimation, to regularize the particle cloud weights.

The PF estimation $\widehat{\alpha}_n$ of the hidden state α_n is then computed through a weighted average of the reshaped particles swarm:

$$\widehat{\alpha}_n = \sum_{m=1}^{M} w_{\alpha,n,\mathrm{RESH}}^{(m)} \alpha_n^{(m)}, \quad (10)$$

where $w_{\alpha,n,\mathrm{RESH}}^{(m)}$ is the weight after the reshaping process. Finally, after the estimation stage, the particles cloud is resampled, generating the set $\alpha_{n,\mathrm{RES}}^{(m)}$, to eliminate spurious particles and avoid fragmentation.

5. Novel Particles Swarm Adaptive Scattering Algorithm (PSAS)

The parameter σ_s is the standard deviation of the random variable η introduced in (8). In the previous section σ_s, it was considered constant over time. However, depending on the situation of the tracking algorithm, it will be shown that it is better to have a dynamic $\sigma_{s,n}$. It is possible to define two parameters that describe the particle cloud: Ω_n, namely, the Particles Cloud Width (PC-W) that describes how much wide the set of particles is and, ε_n, the Particles Cloud Granularity (PC-G) that indicates the interparticle distance:

$$\Omega_n = \max_m \left\{\alpha_n^{(m)}\right\} - \min_m \left\{\alpha_n^{(m)}\right\} \le 2\pi,$$
$$\varepsilon_n = \frac{\Omega_n}{(M-1)}. \tag{11}$$

The value of Ω_n influences the convergence speed of the algorithm. In fact, if the particles cloud is positioned over the value to be estimated the convergence is fast because the the exponential weighting function introduced in (9) gives high weights to the nearest particles and negligible weights to the further ones. Instead, if the particles cloud is located far from the value to be estimated, convergence is slower because the exponential weighting function algorithm gives almost the same weights to all particles. Thus, it will need more iterations to move the particles swarm, cycle by cycle, near the convergence point. For these reasons a bigger value of Ω_n will lead to wider clouds, encouraging the convergence. On the other hand, smaller values of Ω_n will make the algorithm slower and more unstable to fast state changes.

The value of ε_n influences the precision of convergence. In fact, hypothesising that it has reached the convergence point, the estimation precision is upper bounded by the average interparticle distance, that is, the PC-G.

Considering the spreading variable η introduced in (8) as a Gaussian variable with zero mean and standard deviation σ_s, it is possible to estimate the PC-W using the Tchebysheff's inequality as $\Omega_n \approx 6\sigma_s \propto \sigma_s$. Likewise the PC-G can be estimated as $\varepsilon_n \approx 6\sigma_s/(M-1) \propto \sigma_s$. These relations show that σ_s is directly proportional to PC-W and PC-G. Because of the previous observation it is possible to conclude that, given a swarm of M particles, during the convergence phase a bigger value of σ_s would be more appropriate because it would lead to faster results. On the other hand, after the convergence phase, smaller values of σ_s would be more appropriate to increase the estimation precision.

For these reasons in this paper a novel algorithm that manages the dynamic evolution of σ_s across the iterations, namely, Particle Swarm Adaptive Scattering (PSAS), is proposed.

The distance metric $d_n^{(m)}$ is defined as a baseline error cloud for the particles and is used by the PSAS to select the convergence status of the PF algorithm. The error cloud $d_n^{(m)}$ is calculated between the baseline estimation $\tilde{\alpha}_n$ introduced in (5) and the resampled particle swarm $\alpha_{n,\mathrm{RES}}^{(m)}$ as follows:

$$d_n^+ = \operatorname{avg}\left(d_n^{(m)}\right) + \operatorname{std}\left(d_n^{(m)}\right),$$
$$d_n^- = \operatorname{avg}\left(d_n^{(m)}\right) - \operatorname{std}\left(d_n^{(m)}\right), \tag{12}$$
$$d_n^{(m)} = \left|\alpha_{n,\mathrm{RES}}^{(m)} - \tilde{\alpha}_n\right|,$$

where the operators avg and std denote, respectively, the numeric operations of average and standard deviation and the absolute value operates in the periodic $[-\pi, \pi)$ domain. The parameters d_n^- and d_n^+ represent, respectively, the lower and higher bounds of the error cloud $d_n^{(m)}$ and are used as thresholds by the PSAS algorithm to trigger the increment or the decrement of $\sigma_{s,n}$ as follows:

$$\sigma_{s,n} = \sigma_{s,n-1}\left(1 \pm \delta\sigma_s\right), \tag{13}$$

where $\delta\sigma_s > 0$ is the proportional feedback parameter of the PSAS. The value of $\delta\sigma_s$, depending on the two feedback parameters d_n^+ and d_n^- and the threshold value γ_{th}, is set positive to iterative increment $\sigma_{s,n}$ and negative to decrement $\sigma_{s,n}$ and is set to zero to keep it constant.

The value of $\sigma_{s,n}$ is incremented with respect to its previous value when the condition of nonconvergence is detected. This condition holds true when the error cloud $d_n^{(m)}$ is reasonably out of $[-\gamma_{\mathrm{th}}, \gamma_{\mathrm{th}}]$ convergence interval, that is, when $d_n^- > \gamma_{\mathrm{th}}$ or $d_n^+ < -\gamma_{\mathrm{th}}$. Then, the PC-W is incremented to encourage convergence. If the previous condition is not satisfied, the value of $\sigma_{s,n}$ is decremented w.r.t. its previous value when the convergence state is detected. This condition holds true when the error cloud $d_n^{(m)}$ is on the edges of $[-\gamma_{\mathrm{th}}, \gamma_{\mathrm{th}}]$ convergence interval, that is, when $d_n^+ > \gamma_{\mathrm{th}}$ or $d_n^- < -\gamma_{\mathrm{th}}$. Then, the PC-G is reduced to increment the precision. For the remaining values of d_n^+ and d_n^- the condition of tight convergence is reached. In this state the value of $\sigma_{s,n}$ is kept the same w.r.t. its previous value. This PSAS case keeps the cloud at the same size to avoid instability conditions due to rapid changes of the convergence point. The values of the feedback parameter $\delta\sigma_s$ and the threshold parameter γ_{th} are related to the dynamic behaviour of the convergence point that is related to the UAV motion model. Optimal values for $\delta\sigma_s$ and γ_{th} were numerically calculated for the trajectories generated by CARATE in this paper.

6. UAVs Trajectory Emulator

To assess performance of BM and PF algorithms estimations and to evaluate the effects of different parameters on the estimation process, it is necessary to emulate UAVs trajectories using a proper coordinates evolution algorithm. Herein, a possible trajectory generation method is proposed.

The proposed algorithm, named Confined Area Random Aerial Trajectory Emulator (CARATE), generates iteratively a 3D path obtained from a variable length previous history of the trajectory and a tunable set of random variables. CARATE is specifically designed to emulate UAVs trajectories inside a limited flight area, in the GL-S, around the receiving array.

Thus, the generated trajectories allow testing the tracking algorithms for a broad range of arrival angles. An UAV located far from the receiving array instead would be localized using a limited range of angles. The Cartesian position $[x_n, y_n, z_n]$ at time n of the UAV evolves as

$$\begin{aligned} x_n &= x_{n-1} + T v_n \cos\left(\phi_n'\right) \sin\left(\theta_n'\right), \\ y_n &= y_{n-1} + T v_n \sin\left(\phi_n'\right) \sin\left(\theta_n'\right), \\ z_n &= z_{n-1} + T v_n \cos\left(\theta_n'\right). \end{aligned} \tag{14}$$

The angle ϕ_n', depicted in Figure 1, is the local azimuthal angle at time n of the actual trajectory coordinates $[x_n, y_n, z_n]$ seen from the previous path point $[x_{n-1}, y_{n-1}, z_{n-1}]$. The local elevation angle θ_n' can be defined analogously. v_n is the absolute speed of the UAV. At each step, the trajectory increments generated by θ_n', ϕ_n', and v_n are related to the "seed" components $\theta_{\text{seed},n}$, $\phi_{\text{seed},n}$, and $v_{\text{seed},n}$ and to the random components $\delta\theta_n$, $\delta\phi_n$, and δv_n as follows:

$$\begin{aligned} \theta_n' &= \theta_{\text{seed},n} + \delta\theta_n, \\ \phi_n' &= \phi_{\text{seed},n} + \delta\phi_n, \\ v_n &= v_{\text{seed},n} + \delta v_n. \end{aligned} \tag{15}$$

The random components are generated following, respectively, the normal distributions $\delta\theta \sim N(m_{\delta\theta}, \sigma_{\delta\theta})$, $\delta\phi \sim N(m_{\delta\phi}, \sigma_{\delta\phi})$, $\delta v \sim N(m_{\delta v}, \sigma_{\delta v})$. The parameters of the random component distributions affect the UAV path behaviour. For example, for $m_\delta v > 0$, the UAV will have an accelerating trend and for $m_{\delta\phi} > 0$ the UAV will tend to turn anticlockwise. The seed values introduced in (15) are the history parameters that express how much the evolution of the path is related to its past and are given by

$$\begin{aligned} \theta_{\text{seed},n} &= \arccos\left[\Delta z_n\left(N_{\text{ang}}\right), \Delta r_n\left(N_{\text{ang}}\right)\right], \\ \phi_{\text{seed},n} &= \arctan\left[\Delta y_n\left(N_{\text{ang}}\right), \Delta x_n\left(N_{\text{ang}}\right)\right], \\ v_{\text{seed},n} &= \sum_{k=1}^{N_{\text{sp}}} \frac{v_{n-k}}{N_{\text{sp}}}, \end{aligned} \tag{16}$$

where $\arctan(\cdot,\cdot)$ is the arctangent function defined within $[0, 2\pi)$ and where

$$\begin{aligned} \Delta x_n\left(N_{\text{ang}}\right) &= x_{n-1} - x_{n-N_{\text{ang}}}, \\ \Delta y_n\left(N_{\text{ang}}\right) &= y_{n-1} - y_{n-N_{\text{ang}}}, \\ \Delta z_n\left(N_{\text{ang}}\right) &= z_{n-1} - z_{n-N_{\text{ang}}}, \\ \Delta r_n\left(N_{\text{ang}}\right) & \\ &= \sqrt{\Delta x_n\left(N_{\text{ang}}\right)^2 + \Delta y_n\left(N_{\text{ang}}\right)^2 + \Delta z_n\left(N_{\text{ang}}\right)^2}. \end{aligned} \tag{17}$$

In Figure 1, the relations among the CARATE components for the variable ϕ_n are represented. As shown in (16), the smaller

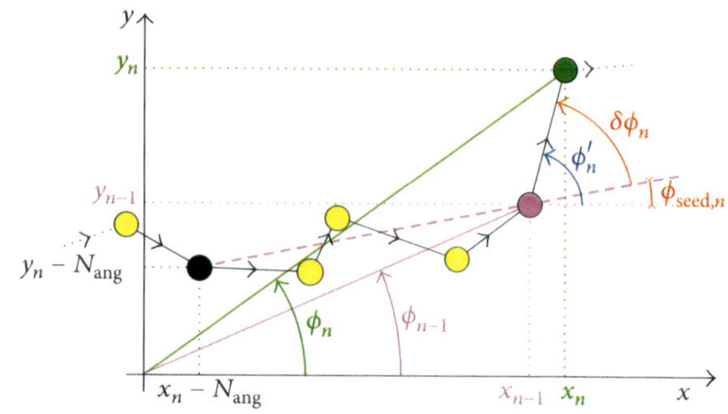

FIGURE 1: Diagram of CARATE components for variable ϕ ($N_{\text{ang}} = 5$).

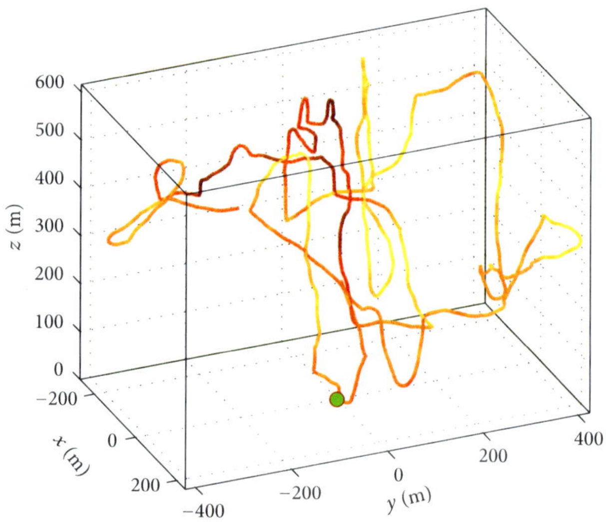

FIGURE 2: Example of UAV 3D trajectory generated with CARATE.

the window width of the seed parameters N_{ang} and N_{sp}, the higher the influence of the path randomization with respect to the past history of the path itself. The same effect will occur for higher values of the standard deviation $\sigma_{\delta\theta}$, $\sigma_{\delta\phi}$, and $\sigma_{\delta v}$ of the random components $\delta\theta$, $\delta\phi$, and δv introduced in (15).

The CARATE algorithm, as visible in the sample path in Figure 2, generates a 3D trajectory inside a limited flight area, formed by the horizontal bound region S_{xy} and the vertical bound region S_z. S_{xy} is defined as $\{|x| < L_{xy} \cap |y| < L_{xy}\}$ while S_z is defined as $\{z > L_z^{\min} \cap z < L_z^{\max}\}$. A specific routine of the algorithm has been developed to keep the trajectory inside the limited flight area, preserving it smooth and without sharp edges on region bounds. If x_n or y_n exceed S_{xy} the value of $\phi_{\text{seed},n}$ introduced in (16) is forced for the subsequent N_{bend} steps to follow a set of equispaced values between $\phi_{\text{seed},n}$ and $\phi_{\text{seed,END}}$. The value of $\phi_{\text{seed,END}}$ is appropriately defined depending on the Cartesian quadrant where the trajectory crosses the surface of S_{xy}. Analogously, if z_n exceeds S_z the value of $\theta_{\text{seed},n}$ introduced in (16) is forced for the next N_{bend} steps to a set of equally spaced values between $\theta_{\text{seed},n}$ and

$\theta_{\text{seed,END}}$. The value of $\theta_{\text{seed,END}}$ is also defined depending on the Cartesian quadrant where the surface of S_z has been crossed.

7. Cramér-Rao Lower Bound

In statistics the Cramér-Rao Lower Bound (CRLB) expresses the lowest value of the Root Mean Square Error (RMSE) of an optimal estimator given a certain set of data [13]. In this paper the estimations of the coordinates $[\theta_n, \phi_n]$ are carried out from the information given by the measurements $u_n^{(x)}$, $u_n^{(y)}$, and $u_n^{(z)}$.

Now, we define $\mathbf{b}_n = [b_n^{(a)}]_{n\in[x,y,z]}$ as the vector of ideal measured values, namely, $u_n^{(a)}$, in (4) but without the detrimental effects due to the impairments introduced in (2) such as PN, DS, and C-AWGN:

$$b_n^{(a)} = |A|^2 e^{j\widehat{\psi}_n^{(a)}}, \quad (18)$$

where, as introduced in Section 2, $\widehat{\psi}_n^{(a)}$ is the D-PoA for ath 3A-ULA antenna branch and $|A|$ is the constant amplitude of the complex downconverted constellation. The vector of real measurements $\mathbf{u}_n = [u_n^{(x)}, u_n^{(y)}, u_n^{(z)}]$, defined in (4), can be written as

$$u_n^{(a)} = \frac{2}{N_a}\sum_{i=1}^{N_a/2}\left(|A|^2 e^{j\widehat{\psi}_n^{(a)}} + W_n^{(a,i)}\right) = \widehat{A}e^{j\widehat{\psi}_n^{(a)}} + \widehat{W}_n^{(a)}, \quad (19)$$

where the intermediate signal $|A|^2 e^{j\widehat{\psi}_n^{(a)}} + W_n^{(a,i)}$ is averaged through antenna pairs and $\widehat{A} = (2/N_a)\sum_{i=1}^{N_a/2}|A|^2 = |A|^2$. The noise components in (19) are

$$\begin{aligned} W_n^{(a,i)} &= |A|\, e^{j\psi_n^{(a,2i-1)}} w_n^{(a,2i)^*} + |A|^*\, e^{-j\psi_n^{(a,2i)}} w_n^{(a,2i-1)} \\ &\quad + w_n^{(a,2i-1)} w_n^{(a,2i)^*}, \\ \widehat{W}_n^{(a)} &= \frac{2}{N_a}\sum_{i=1}^{N_a/2} W_n^{(a,i)}, \end{aligned} \quad (20)$$

where the notation $(\cdot)^*$ denotes the complex conjugate operator. $w_n^{(a,i)}$ is the C-AWGN with zero mean and variance σ_w^2. $\psi_n^{(a,i)}$ is the PoA as defined in (3). The expectation of the differential noise $W_n^{(a,i)}$ among the antennas of the same branch is $m_W = 0$, while its variance $\sigma_W^2 = 2|A|^2\sigma_w^2 + \sigma_w^4$. Analogously, the average differential noise $\widehat{W}_n^{(a)}$ is described by its mean $m_{\widehat{W}} = 0$ and its variance is $\sigma_{\widehat{W}}^2 = (2/N_a)\sigma_W^2$.

To calculate the CRLB it is necessary to evaluate the log-likelihood $L(\phi,\theta) = \ln P[\mathbf{u}_n \mid \mathbf{b}_n]$ of the acquired data given the parameters to be estimated. $\ln(\cdot)$ represents the natural logarithm. In this context, for the calculation of CRLB, the measurements vector $\mathbf{u}_n$ is considered affected only by C-AWGN noise without the influence of the other impairments introduced in (2). This assumption yields to the likelihood Gaussianity and is verified via numerical fitting similarly to [14], so that

$$L(\phi,\theta) = -3\ln\left(\pi\sigma_{\widehat{W}}^2\right) - \frac{1}{\sigma_{\widehat{W}}^2}\|\mathbf{b}-\mathbf{u}\|^2, \quad (21)$$

where the time index n is omitted until the end of the CRLB derivation for notational simplicity. The CRLB of the variable α_r, defined as rth element of the unknowns vector $\boldsymbol{\alpha} = [\phi,\theta]$, is

$$\text{CRLB}_{\alpha_r}(\phi,\theta) = \left\{F(\phi,\theta)^{-1}\right\}_{(r,r)}, \quad (22)$$

where $F(\phi,\theta)$ is the Fischer information matrix:

$$F(\phi,\theta) = -E\left\{\left[\frac{\partial^2}{\partial\alpha_h\partial\alpha_k}L(\phi,\theta)\right]_{(h,k)\in[1,2]}\right\}. \quad (23)$$

For the log-likelihood function described in (21), we obtain

$$\begin{aligned} F(\phi,\theta) &= \begin{bmatrix} \left\|\dfrac{\partial\boldsymbol{\beta}}{\partial\phi}\right\|^2 & \dfrac{\partial\boldsymbol{\beta}}{\partial\phi}\dfrac{\partial\boldsymbol{\beta}^{\mathrm{T}}}{\partial\theta} \\ \dfrac{\partial\boldsymbol{\beta}}{\partial\theta}\dfrac{\partial\boldsymbol{\beta}^{\mathrm{T}}}{\partial\phi} & \left\|\dfrac{\partial\boldsymbol{\beta}}{\partial\theta}\right\|^2 \end{bmatrix} \\ &= 2\left(\frac{2\pi|A|^2 D_{\text{norm}}}{\sigma_{\widehat{W}}}\right)^2 \begin{bmatrix} (\sin\theta)^2 & 0 \\ 0 & 1 \end{bmatrix}, \end{aligned} \quad (24)$$

where $\boldsymbol{\beta} = [\cos\phi\sin\theta, \sin\phi\sin\theta, \cos\theta]$ is the unit direction vector and $D_{\text{norm}} = D/\lambda_0$ is the normalized interantenna distance (Figure 4). The notation $(\cdot)^{\mathrm{T}}$ denotes the transpose operator. Finally, the CRLB expressions for ϕ and θ are

$$\begin{aligned} \text{CRLB}_\phi(\phi,\theta) &= \frac{1}{2}\left(\frac{\sigma_{\widehat{W}}}{2\pi|A|^2 D_{\text{norm}}}\right)^2\left(\frac{1}{\sin\theta}\right)^2, \\ \text{CRLB}_\theta(\phi,\theta) &= \frac{1}{2}\left(\frac{\sigma_{\widehat{W}}}{2\pi|A|^2 D_{\text{norm}}}\right)^2. \end{aligned} \quad (25)$$

The RMSE of every estimator of the unknown α_r from the measurements $\mathbf{u}$ is lower bounded by $\sqrt{\text{CRLB}_{\alpha_r}}$. It is important to state again that this C-AWGN-only CRLB calculated in (25) takes into account C-AWGN but not the other previously introduced impairments, in particular DS and PN. Given the previous assumptions, the AWGN-only CRLB is lower or equal to the fully impaired CRLB. Another relevant point is that the computed CRLB does not take into account the motion model and the trajectory history differently from the PF algorithm. The CRLB is simply computed from the 3 observations $u_n^{(a)}$ introduced in (4) at time nT, as the BM does. Instead, the proposed PF algorithm exploits the motion model jointly with PSAS, using information coming from the trajectory history. For this reason, it may occur that the PF estimation offers better performance than the CRLB.

8. Performance Analysis

The performance of the PF algorithm is now assessed. A comparison with the BM estimation introduced in Section 3 and the CRLB calculated in Section 7 is also made. Different UAV trajectories are generated with the CARATE to assess performance. Each UAV trajectory is N time samples long. In order to have a common performance evaluation metric, in the following we define the average of the Root Mean Square Error (RMSE) calculated through every trajectory sample, namely, the Average Trajectory RMSE (AT-RMSE).

n_{traj}th generated trajectory is described with the set $\{\theta_n^{(n_{\text{traj}})}, \phi_n^{(n_{\text{traj}})}\}$, its BM estimation is $\{\tilde{\theta}_n^{(n_{\text{traj}})}, \tilde{\phi}_n^{(n_{\text{traj}})}\}$, and its PF estimation $\{\hat{\theta}_n^{(n_{\text{traj}})}, \hat{\phi}_n^{(n_{\text{traj}})}\}$. Their AT-RMSE are, respectively, $\overline{\text{RMSE}}_{\alpha,\text{BM}} = (1/N_{\text{traj}})\sum_{n_{\text{traj}}=1}^{N_{\text{traj}}} \text{RMSE}_{\alpha,\text{BM}}^{(n_{\text{traj}})}$ and $\overline{\text{RMSE}}_{\alpha,\text{PF}} = (1/N_{\text{traj}})\sum_{n_{\text{traj}}=1}^{N_{\text{traj}}} \text{RMSE}_{\alpha,\text{PF}}^{(n_{\text{traj}})}$ for $\alpha \in [\theta, \phi]$ where

$$\text{RMSE}_{\alpha,\text{BM}}^{(n_{\text{traj}})} = \sqrt{\frac{\sum_{n=1}^{N}\left(\alpha_n^{(n_{\text{traj}})} - \tilde{\alpha}_n^{(n_{\text{traj}})}\right)^2}{N}},$$
$$\text{RMSE}_{\alpha,\text{PF}}^{(n_{\text{traj}})} = \sqrt{\frac{\sum_{n=1}^{N}\left(\alpha_n^{(n_{\text{traj}})} - \hat{\alpha}_n^{(n_{\text{traj}})}\right)^2}{N}} \quad (26)$$

are the RMSEs calculated for each trajectory n_{traj}. N_{traj} is the number of generated trajectories.

Analogously, with the purpose to allow a performance comparison with the CRLB, we define the average trajectory CRLB (AT-CRLB), as $\overline{\text{CRLB}}_\alpha = (1/N_{\text{traj}})\sum_{n_{\text{traj}}=1}^{N_{\text{traj}}} \text{CRLB}_\alpha^{(n_{\text{traj}})}$ for $\alpha \in [\theta, \phi]$ and where

$$\text{CRLB}_\alpha^{(n_{\text{traj}})} = \frac{1}{N}\sum_{n=1}^{N} \text{CRLB}_\alpha\left(\phi_n^{(n_{\text{traj}})}, \theta_n^{(n_{\text{traj}})}\right). \quad (27)$$

The default parameters for the performance evaluation are (a) angles $\sigma_{\delta\theta} = 7^\circ$, $\sigma_{\delta\phi} = 7^\circ$, and $N_{\text{ang}} = 20$; (b) speeds $v_{\min} = 0.5\,\text{km/h}$, $v_{\max} = 10\,\text{km/h}$, and $N_{\text{sp}} = 2$. The array parameters are $N_a = 6$ and $D_{\text{norm}} = 1/3$. The PF algorithm basic settings are $M = 20$ and $N_{mm} = 20$. PSAS parameters $\delta\sigma_s = 25\%$ and $\gamma_{\text{th}} = 10^\circ$ were numerically optimized for the CARATE algorithm. In the signal model introduced in (2), the PN is considered white with standard deviation $\sigma_{\text{PN}} = 1^\circ$, the PO is considered uniform in $[0, 2\pi)$ for each array element, and the signal-to-noise ratio is $\text{SNR} = [|A|^2/\sigma_w^2]$ dB =10 dB. The number of simulated trajectories N_{traj} and their length N have been properly dimensioned to guarantee the statistical confidence of the results.

In Figure 2 an example of 3D-trajectory generated for performance evaluation is shown. The 3A-ULA is positioned in the center of the axis.

8.1. Performance with Static and Dynamic σ_s. We now consider the performance as a function of the signal-to-noise ratio $\text{SNR} = |A|^2/\sigma_w^2$. In Figure 3, the AT-RMSE of the azimuthal coordinate ϕ and the elevation coordinate θ for both the PF algorithm and the BM are reported. Furthermore the impact on performance of the PSAS algorithm introduced in Section 5 is analysed. It is possible to observe how, due to the geometrical asymmetry, the performance in terms of AT-RMSE is significantly different between θ and ϕ. Greater SNR values lead to lower AT-RMSE for both PF and BM estimations. It is important to state that, for a static value of σ_s, PF performs better than BM in a broad range of SNR values. For example, at 2 dB of SNR, it outperforms the BM algorithm by about 17° for ϕ and 10° for θ. The analysis shows that the introduction of the PSAS algorithm leads to a further improvement of performance. For example, at 2 dB of SNR the PSAS leads to a performance improvement of 5° for ϕ and 3° for θ with respect to the use of a static σ_s. It is visible in Figure 3 how the introduction of the PF algorithm allows overcoming the AT-CRLB. This is possible thanks to the a priori knowledge of the UAV path extracted by the motion model from the previous steps of the trajectory. For higher SNR the performance improvement of PF with respect to the BM algorithm is less prominent than for lower SNR.

8.2. Performance for Different Interantenna Distances D_{norm}. Now, the behaviour of the AT-RMSE for the BM and PF algorithms is analysed varying the normalized interantenna distance D_{norm}. To unambiguously extract the phase from $u_n^{(a)}$ the interelement distance must be less than half of the impinging signal wavelength [15]: $D_{\text{norm}} \le 0.5$. Furthermore in (25) it is visible that for higher values of D_{norm} the CRLBs increase. This leads to $D_{\text{norm}} = 0.5$ as the optimal value for the estimation. However, in lower SNR scenarios and for D_{norm} values near 0.5 the CRLB is found to be overly optimistic. In fact, high C-AWGN and the other impairments in (2) severely impair performance corrupting the value of $\angle u_n^{(x)}$ in (4) and making it exceeding 180°. The PF algorithm, thanks to its hypothesised a priori knowledge of the UAV path given by the motion model, attenuates this impairing effect. In Figure 3, we can see that the proposed PF algorithm is less dependent from $D_{\text{norm}} = Df_0/c_0$ than the BM. Thus, the same 3A-ULA is usable for a wider set of carrier frequency f_0 with respect to the BM that exhibits its optimal performance for a narrow interval of D_{norm} around 0.4. For this reason, the PF algorithm can offer good performance also in a multiple UAVs scenario where multiplexing is implemented in a frequency division fashion.

9. Conclusions

We have discussed the application of an appropriately designed PF algorithm for self-localization (SL-S) and ground localization (GL-S) of UAVs using a 3A-ULA of antennas, showing an overall increase of performance with respect to the BM. A novel algorithm to manage the amplitude of the particle swarm, namely, Particles Swarm Adaptive Scattering (PSAS), has been developed and tested, showing a further increase of precision. A complete, fully adjustable, and effective 3D UAVs trajectory emulator, namely, CARATE, has been proposed and used to assess performance. Strong impairing effects like Doppler spread, phase offset, and phase noise have been considered in the performance evaluation. The effects on the proposed localization algorithm of the PF model parameters as well as the SNR and the 3A-ULA characteristics have been studied. Numerical results show that the proposed PF algorithm is able of dynamically track the UAVs angular position better that the BM. A critical point of this approach is that, similarly to PSAS behaviour, a dynamic calibration procedure has to be implemented to optimize the PF parameters. Future work will also investigate

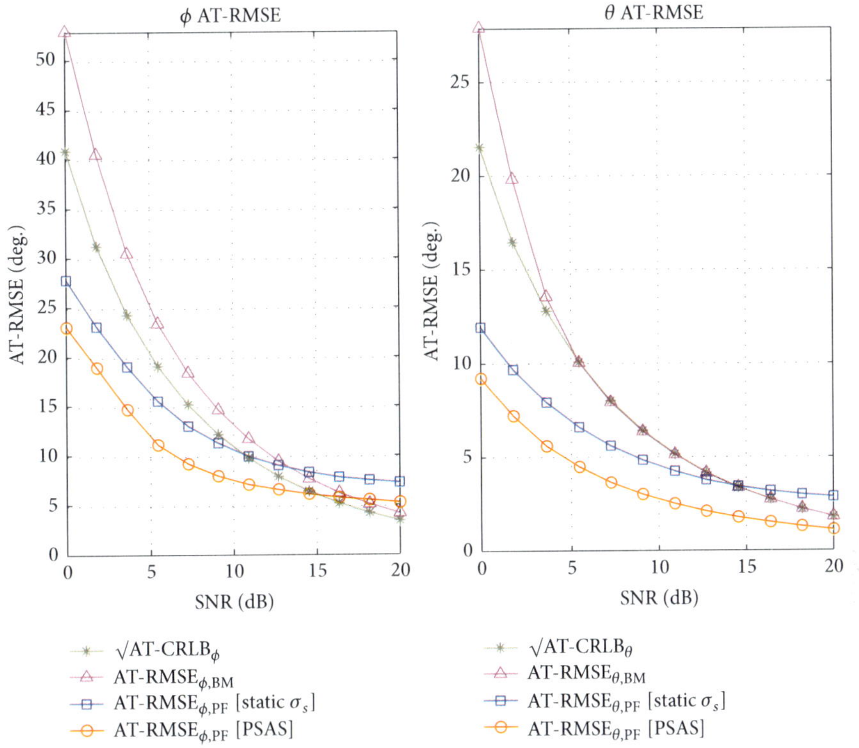

FIGURE 3: AT-RMSE for different $\mathrm{SNR}_{\mathrm{dB}}$ conditions with dynamic and static σ_s.

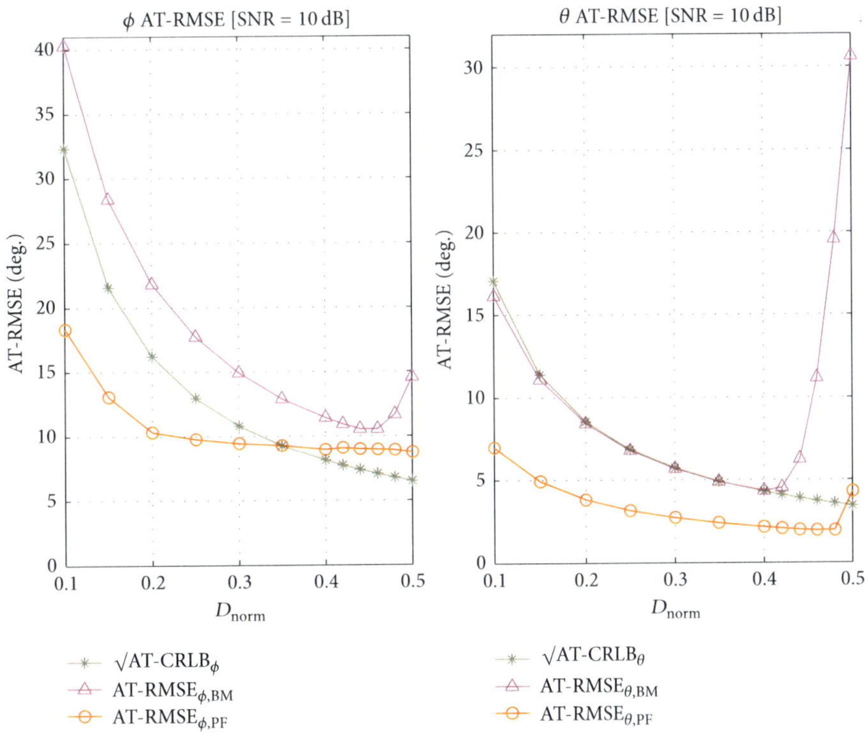

FIGURE 4: AT-RMSE for different normalized 3A-ULA interantenna distance $D_{\mathrm{norm}} = D/\lambda_0$.

the integration of ranging algorithms like Received Signal Strength Estimation (RSSE) and ToA to PF to provide full 3D localization.

Competing Interests

The authors declare that they have no competing interests.

References

[1] P. B. Quater, F. Grimaccia, S. Leva, M. Mussetta, and M. Aghaei, "Light Unmanned Aerial Vehicles (UAVs) for cooperative inspection of PV plants," *IEEE Journal of Photovoltaics*, vol. 4, no. 4, pp. 1107–1113, 2014.

[2] M. R. Haque, M. Muhammad, D. Swarnaker, and M. Arifuzzaman, "Autonomous Quadcopter for product home delivery," in *Proceedings of the 1st International Conference on Electrical Engineering and Information & Communication Technology (ICEEICT '14)*, pp. 1–5, Dhaka, Bangladesh, April 2014.

[3] E. Ackerman, "FAA unveils drone rules: autonomy is in, drone delivery is out," IEEE Spectrum, February 2015, http://spectrum.ieee.org/automaton/robotics/drones/faa-proposed-commercial-drone-rules/.

[4] M. Peacock and M. N. Johnstone, "Towards detection and control of civilian unmanned aerial vehicles," in *Proceedings of the Australian Information Warfare Conference (WCICA '13)*, SRI Security Research Institute, Edith Cowan University, Perth, Western Australia, 2013.

[5] S. Panzieri, F. Pascucci, and G. Ulivi, "An outdoor navigation system using GPS and inertial platform," *IEEE/ASME Transactions on Mechatronics*, vol. 7, no. 2, pp. 134–142, 2002.

[6] G. Mao, B. Fidan, and B. D. O. Anderson, "Wireless sensor network localization techniques," *Computer Networks*, vol. 51, no. 10, pp. 2529–2553, 2007.

[7] K. Dogancay, "UAV path planning for passive emitter localization," *IEEE Transactions on Aerospace and Electronic Systems*, vol. 48, no. 2, pp. 1150–1166, 2012.

[8] D. Inserra and A. M. Tonello, "Characterization of hardware impairments in multiple antenna systems for doa estimation," *Journal of Electrical and Computer Engineering*, vol. 2011, Article ID 908234, 10 pages, 2011.

[9] M. S. Arulampalam, S. Maskell, N. Gordon, and T. Clapp, "A tutorial on particle filters for online nonlinear/non-Gaussian Bayesian tracking," *IEEE Transactions on Signal Processing*, vol. 50, no. 2, pp. 174–188, 2002.

[10] S. Maskell and N. Gordon, "A tutorial on particle filters for on-line nonlinear/non-gaussian bayesian tracking," in *Target Tracking: Algorithms and Applications*, vol. 2, Ref. no. 2001/174, pp. 2/1–2/15, IEE, 2001.

[11] Y. Liu, B. Wang, W. He, J. Zhao, and Z. Ding, "Fundamental principles and applications of particle filters," in *Proceedings of the 6th World Congress on Intelligent Control and Automation (WCICA '06)*, vol. 2, pp. 5327–5331, June 2006.

[12] A. Papaiz and A. M. Tonello, "Particle filtering with weight reshaping for opportunistic angle of arrival estimation in a vehicular scenario," in *Proceedings of the IEEE 5th International Conference on Consumer Electronics (ICCE '15)*, pp. 145–149, Berlin, Germany, September 2015.

[13] S. Kay, *Fundamentals of Statistical Signal Processing: Estimation Theory*, Prentice Hall, 1993.

[14] D. Inserra and A. M. Tonello, "DoA estimation with compensation of hardware impairments," in *Proceedings of the IEEE 72nd Vehicular Technology Conference Fall (VTC '10)*, pp. 1–5, Ottawa, Canada, September 2010.

[15] Z.-M. Liu and F.-C. Guo, "Azimuth and elevation estimation with rotating long-baseline interferometers," *IEEE Transactions on Signal Processing*, vol. 63, no. 9, pp. 2405–2419, 2015.

3

A Moving Frame Trajectory Tracking Method of a Flying-Wing UAV using the Differential Geometry

Yi Zhu, Xin Chen, and Chuntao Li

College of Automation Engineering, Nanjing University of Aeronautics and Astronautics, Nanjing 210016, China

Correspondence should be addressed to Yi Zhu; zhuyi73@126.com

Academic Editor: Mahmut Reyhanoglu

The problem of UAV trajectory tracking is a difficult issue for scholars and engineers, especially when the target curve is a complex curve in the three-dimensional space. In this paper, the coordinate frames during the tracking process are transformed to improve the tracking result. Firstly, the basic concepts of the moving frame are given. Secondly the transfer principles of various moving frames are formulated and the Bishop frame is selected as a final choice for its flexibility. Thirdly, the detailed dynamic equations of the moving frame tracking method are formulated. In simulation, a moving frame of an elliptic cylinder helix is formulated precisely. Then, the devised tracking method on the basis of the dynamic equations is tested in a complete flight control system with 6 DOF nonlinear equations of the UAV. The simulation result shows a satisfactory trajectory tracking performance so that the effectiveness and efficiency of the devised tracking method is proved.

1. Introduction

The main difficulties of a UAV trajectory tracking issue differing from the ordinary tracking problems lie in the strict real-time requirement and the complexity of 3-dimensional curves. Some literature only attempts to deal with the reduced situations such as the "path following" problem without a constrained time parameter [1–3] or the simplified target curves in 2D plane [4–6]. What is more, different types of curves usually call for different guidance strategies. In contrast, the moving frame tracking method is employed to give a more universal solution for various kinds of complex curves in the 3D space.

There are two procedures of the moving frame tracking method.

The first procedure of the moving frame tracking method is to form a moving frame of the given target curve. A moving frame is a moving coordinate axis system with its three axes being the curve's tangent vector, principal normal vector, and secondary normal vector. There are many ways to obtain a moving frame of a given curve [7, 8]. The Bishop frame which is also called the parallel frame is the most famous one [9, 10]. In recent years, papers tend to concentrate on the improvements of the Bishop frame [11, 12]. In this paper by comparisons of the Frenet frame and the type 1 Bishop frames, we can see some insights of forming a moving frame and find out the reason why the Bishop frame is suitable for solving a UAV trajectory problem.

The second procedure of the moving frame tracking method is to formulate the dynamic error equations of the whole tracking process. In this procedure a quite unusual tracking process is presented. We no longer need to care about the individual shape of the target curve [13, 14]. That is to say, whether the target curve is a straight line, a circle, or an arc is unimportant. Instead, the moving frame of the curve with its tangent vector, principal normal vector, and secondary normal vector is the objectives to follow. Some knowledge of the theoretical mechanics about the motions of rigid bodies is necessary. In essence, the objective dynamic error equations are the result of a series of coordinate transformations and vector operations [15, 16]. In this procedure, the final aim is to obtain the kinematics errors in the moving frame of the target curve rather than the inertial coordinate frame.

The previous papers treating the trajectory tracking problem with the moving frame method are not sufficient. The reason is that the moving frame tracking method is very interdisciplinary for the theorists of guidance, navigation, and control. However, the analysis of geometrical

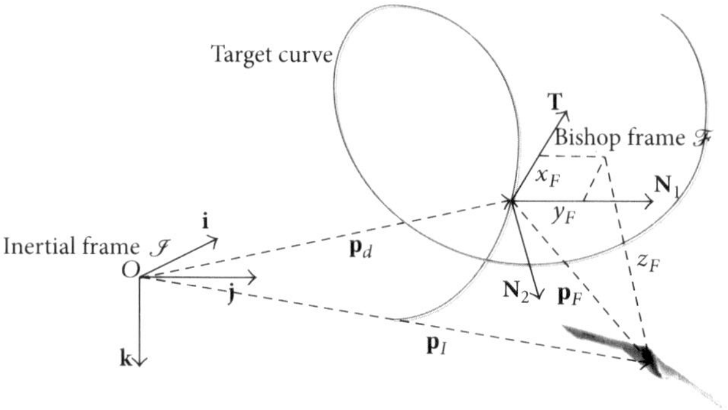

FIGURE 1: Illustration of the moving frame tracking method.

properties of curves actually plays an important role in practical applications. Thus, the moving frame tracking is an effective and practical method for the UAV 3D trajectory tracking problem. In the previous literature some trials have been made to apply this tracking method to the tiny UAVs or quadrotors [15, 17]. Some other sources of literature are concerned about the mathematical features of this tracking method [18, 19]. All these sources of literature tackle the particle models of the microaircrafts with very low flight speed. For example, in literature [15], the velocity of the aircraft is 0.4 m/s. In this paper a much larger and heavier UAV is used and its velocity is about 200 m/s so that the situation is different. Also, it is well known that when a particle model is used, the angular motions inside the rigid body are omitted. So another distinction with the previous literature is that in this paper a 6 DOF UAV model is employed instead of a particle model.

2. Formulation of the Moving Frame Tracking Method

2.1. General Illustration. The core idea of the moving frame tracking method lies in the conception that the position errors are resolved in the moving coordinate frame of the target curve instead of the inertial frame in ordinary tracking methods. In this method the time constraint is well satisfied since the target curve is formulated with a time parameter t. The primary aim is aligning the heading direction to the tangent vector of the target curve with almost no position deviation. The principle of the moving frame tracking method is illustrated in Figure 1.

It can be seen from Figure 1 that the aircraft used in this paper is a UAV with flying-wing configuration. The target curve, that is, the target trajectory to track, is an elliptic cylinder helix. There are five coordinate frames that need to be introduced which are the inertial frame $\{\mathcal{I}\}$, the body frame $\{\mathcal{B}\}$, the wind frame $\{\mathcal{W}\}$, and the curve frame $\{\mathcal{F}\}$. The inertial frame describes the physical positions on the earth. The body frame is a frame attached to the body of the aircraft. The wind frame is a speed frame with its x-axis aligned with the true airspeed of the aircraft, while the other two axes are orthogonally decomposed in the inertial frame. The curve frame is a moving frame (as the Frenet frame or Bishop frame) of the target trajectory.

2.2. Forming a Moving Frame. By mathematical ways, it is implementable to form a moving frame of a 3D Curve. In the early stages of classical differential geometry, the Frenet frame was adopted to formulate a curve in the three-dimensional Euclidean space E^3. Let $\mathbf{T}$, $\mathbf{N}$, and $\mathbf{B}$ represent the tangent vector, principal normal vector, and secondary normal vector, respectively. For all the curves in E^3, by transforming the time parameter t into an arc length parameter s, the Frenet frame of a curve is defined by

$$\frac{d\mathbf{r}}{ds} = \mathbf{T},$$
$$\begin{bmatrix} \frac{d\mathbf{T}}{ds} \\ \frac{d\mathbf{N}}{ds} \\ \frac{d\mathbf{B}}{ds} \end{bmatrix} = \begin{bmatrix} \kappa\mathbf{N} \\ -\kappa\mathbf{T} + \tau\mathbf{B} \\ -\tau\mathbf{N} \end{bmatrix} = \begin{bmatrix} 0 & \kappa & 0 \\ -\kappa & 0 & \tau \\ 0 & -\tau & 0 \end{bmatrix} \begin{bmatrix} \mathbf{T} \\ \mathbf{N} \\ \mathbf{B} \end{bmatrix}, \quad (1)$$

where κ and τ are curvature and torsion of the Frenet frame, defined by

$$\kappa = \left| \frac{d\mathbf{T}}{ds} \right|,$$
$$\tau = -\mathbf{B}' \cdot \mathbf{N}. \quad (2)$$

Today the Frenet frame has been improved to become the Bishop frame. The type 1 Bishop frame is defined by

$$\begin{bmatrix} \frac{d\mathbf{T}}{ds} \\ \frac{d\mathbf{N}_1}{ds} \\ \frac{d\mathbf{N}_2}{ds} \end{bmatrix} = \begin{bmatrix} k_1(s)\mathbf{N}_1 + k_2(s)\mathbf{N}_2 \\ -k_1(s)\mathbf{T} \\ -k_2(s)\mathbf{T} \end{bmatrix}$$
$$= \begin{bmatrix} 0 & k_1(s) & k_2(s) \\ -k_1(s) & 0 & 0 \\ -k_2(s) & 0 & 0 \end{bmatrix} \begin{bmatrix} \mathbf{T} \\ \mathbf{N}_1 \\ \mathbf{N}_2 \end{bmatrix}. \quad (3)$$

The relationship between the Frenet frame and type 1 Bishop frame is described as follows:

$$\mathbf{T} = \mathbf{T},$$
$$\mathbf{N} = \cos\theta_F \mathbf{N}_1 + \sin\theta_F \mathbf{N}_2, \quad (4)$$
$$\mathbf{B} = -\sin\theta_F \mathbf{N}_1 + \cos\theta_F \mathbf{N}_2,$$

where

$$\theta_F(s) = \arctan\frac{k_2}{k_1},$$
$$\tau(s) = \frac{d\theta(s)}{ds}, \quad (5)$$
$$\kappa(s) = \sqrt{k_1^2 + k_2^2}.$$

The type 1 Bishop curvatures are defined by

$$k_1 = \kappa \cos\theta_F, \quad k_2 = \kappa \sin\theta_F. \tag{6}$$

From the definition of the Bishop frame and the relationship between it and the Frenet frame we can see that, roughly speaking, the Bishop frame is a frame which has the same original point with the Frenet frame and aligns one of its axes with one of the axes of the Frenet frame. Then the rest of the two axes of the Bishop frame rotate at a certain angle θ_F in the normal plane of the first axis. Obviously, a new generated Bishop frame is always isomorphic to the Frenet frame in E^3.

2.3. Dynamic Errors Equations of Moving Frame Tracking. Let the inertial frame be defined by $\{\mathcal{I}\} = \{O : \mathbf{i}, \mathbf{j}, \mathbf{k}\}$ with axis $\mathbf{i}$ pointing to the north, axis $\mathbf{j}$ pointing to the east, and axis $\mathbf{k}$ pointing vertically to the center of the earth. So the frame $\{\mathcal{I}\}$ is a NED (North-Easy-Down) frame. Suppose the target curve with an arc length parameter is

$$\mathbf{P}_d(s) = \left[P_{dx}(s) \;\; P_{dy}(s) \;\; P_{dz}(s)\right]^{\mathrm{T}}. \tag{7}$$

Then the trajectory tracking problem is presented as follows: for a target curve $\mathbf{P}_d$ with its moving frame and an aircraft model, a controller needs to be designed to force the aircraft to fly along the target curve by means of minimizing the position error vector which is resolved in the moving frame.

Assuming the Bishop frame of the target curve is defined by $\{\mathcal{F}\} = \{\mathbf{T}, \mathbf{N}_1, \mathbf{N}_2\}$, according to the definition of the frame $\{\mathcal{W}\}$, the airspeed V equals ground speed when there is no wind disturbance. Thus the speed vector from frame $\{\mathcal{W}\}$ to $\{\mathcal{I}\}$ resolved in frame $\{\mathcal{W}\}$ is represented as $\mathbf{v}_{W2I}|_W = [V \;\; 0 \;\; 0]$.

Let $\mathbf{p}_I = [x \;\; y \;\; z]^{\mathrm{T}}$ be the vector from the origin of the frame $\{\mathcal{I}\}$ pointing to the aircraft's center of mass resolved in $\{\mathcal{I}\}$ and $\mathbf{p}_F = [x_F \;\; y_F \;\; z_F]^{\mathrm{T}}$ be the vector from the origin of the frame $\{\mathcal{F}\}$ pointing to the aircraft's center of mass resolved in $\{\mathcal{F}\}$. The angular velocity vector of the body frame $\{\mathcal{B}\}$ rotating around $\{\mathcal{I}\}$ and resolved in frame $\{\mathcal{F}\}$ is presented as

$$\boldsymbol{\omega}_d|_F = \mathbf{R}_{I2F}\boldsymbol{\omega}_d = \left[\omega_{dx}|_F \;\; \omega_{dy}|_F \;\; \omega_{dz}|_F\right]^{\mathrm{T}}, \tag{8}$$

where $\mathbf{R}_{I2F} \in SO(3)$ is the rotation matrix from $\{\mathcal{I}\}$ to $\{\mathcal{F}\}$. SO(3) is the 3-dimensional special orthogonal group. Also, the matrix $\mathbf{R}_{F2I}$ represents the rotation matrix from the frame $\{\mathcal{F}\}$ to the inertial frame. Then, in the light of the kinematics of rigid bodies, (9) holds.

$$\mathbf{R}_{F2I} = [\mathbf{T}, \mathbf{N}_1, \mathbf{N}_2]. \tag{9}$$

The condition is just the same with the other frames, which is no longer repeated here. So formula (9) is an instance of showing a general rule of the coordinate transformation with unitary frames in E^3.

To be more precise about the meaning and resolving method of the angular velocity vector just mentioned above, some discussions of the Darboux vector are given in the following. The Darboux vector is named after Gaston Darboux who discovered it. It is also called angular momentum vector, because it is directly proportional to angular momentum. For a rigid body moving absolutely along a general curve which is expressed by a moving frame, its motions can be divided into two kinds of motions as the translational motion and the rotational motion. The rotational motion is formulated by the Darboux vector $\boldsymbol{\omega}$ which is a composition of three axes' areal velocity components.

$$\boldsymbol{\omega} = \boldsymbol{\omega}_{\tilde{\mathbf{i}}} + \boldsymbol{\omega}_{\tilde{\mathbf{j}}} + \boldsymbol{\omega}_{\tilde{\mathbf{k}}}, \tag{10}$$

where $\tilde{\mathbf{i}}, \tilde{\mathbf{j}}, \tilde{\mathbf{k}}$ represent axes vectors of a moving frame. Each axis vector rotates around a point. When this moving frame is a unitary frame, each axis areal velocity is defined by

$$\begin{aligned}
\boldsymbol{\omega}_{\tilde{\mathbf{i}}} &= \lim_{\Delta t \to 0} \frac{\tilde{\mathbf{i}}(t) \times \tilde{\mathbf{i}}(t+\Delta t)}{2\Delta t} = \frac{\tilde{\mathbf{i}}(t) \times \tilde{\mathbf{i}}'(t)}{2}, \\
\boldsymbol{\omega}_{\tilde{\mathbf{j}}} &= \lim_{\Delta t \to 0} \frac{\tilde{\mathbf{j}}(t) \times \tilde{\mathbf{j}}(t+\Delta t)}{2\Delta t} = \frac{\tilde{\mathbf{j}}(t) \times \tilde{\mathbf{j}}'(t)}{2}, \\
\boldsymbol{\omega}_{\tilde{\mathbf{k}}} &= \lim_{\Delta t \to 0} \frac{\tilde{\mathbf{k}}(t) \times \tilde{\mathbf{k}}(t+\Delta t)}{2\Delta t} = \frac{\tilde{\mathbf{k}}(t) \times \tilde{\mathbf{k}}'(t)}{2}.
\end{aligned} \tag{11}$$

Taking the Frenet frame, for instance, with $\tilde{\mathbf{i}}, \tilde{\mathbf{j}}$, and $\tilde{\mathbf{k}}$ replaced by $\mathbf{T}$, $\mathbf{N}$, and $\mathbf{B}$, one has

$$\begin{aligned}
\boldsymbol{\omega}_{\mathbf{T}} &= \frac{\mathbf{T}(t) \times \mathbf{T}'(t)}{2} = \frac{\kappa\mathbf{B}}{2}, \\
\boldsymbol{\omega}_{\mathbf{N}} &= \frac{\mathbf{N}(t) \times \mathbf{N}'(t)}{2} = \frac{\tau\mathbf{T} + \kappa\mathbf{B}}{2}, \\
\boldsymbol{\omega}_{\mathbf{B}} &= \frac{\mathbf{B}(t) \times \mathbf{B}'(t)}{2} = \frac{\tau\mathbf{T}}{2}, \\
\boldsymbol{\omega} &= \boldsymbol{\omega}_{\mathbf{T}} + \boldsymbol{\omega}_{\mathbf{N}} + \boldsymbol{\omega}_{\mathbf{B}} = \tau\mathbf{T} + \kappa\mathbf{B}.
\end{aligned} \tag{12}$$

It is easy to see that the vector $\boldsymbol{\omega}$ satisfies the equations

$$\begin{bmatrix} \dfrac{d\mathbf{T}}{dt} \\ \dfrac{d\mathbf{N}}{dt} \\ \dfrac{d\mathbf{B}}{dt} \end{bmatrix} = \begin{bmatrix} \boldsymbol{\omega} \times \mathbf{T} \\ \boldsymbol{\omega} \times \mathbf{N} \\ \boldsymbol{\omega} \times \mathbf{B} \end{bmatrix}. \tag{13}$$

Then concerning the Bishop frame, its Darboux vector is

$$\begin{aligned}
\boldsymbol{\omega}_{\mathbf{T}} &= \frac{\mathbf{T} \times \mathbf{T}}{2} = \frac{1}{2}\mathbf{T} \times (k_1\mathbf{N}_1 + k_2\mathbf{N}_2) \\
&= \frac{1}{2}k_1\mathbf{N}_2 - \frac{1}{2}k_2\mathbf{N}_1, \\
\boldsymbol{\omega}_{\mathbf{N}_1} &= \frac{\mathbf{N}_1 \times \mathbf{N}_1'}{2} = \mathbf{N}_1 \times \left(-\frac{1}{2}k_1\right)\mathbf{T} = \frac{1}{2}k_1\mathbf{N}_2, \\
\boldsymbol{\omega}_{\mathbf{N}_2} &= \frac{\mathbf{N}_2 \times \mathbf{N}_2'}{2} = \mathbf{N}_2 \times \left(-\frac{1}{2}k_2\right)\mathbf{T} = -\frac{1}{2}k_2\mathbf{N}_1, \\
\boldsymbol{\omega} &= \boldsymbol{\omega}_{\mathbf{T}} + \boldsymbol{\omega}_{\mathbf{N}_1} + \boldsymbol{\omega}_{\mathbf{N}_2} = k_1\mathbf{N}_2 - k_2\mathbf{N}_1.
\end{aligned} \tag{14}$$

When the moving frame is parameterized by a time parameter t, the Darboux vector of the Bishop frame and resolved in the Bishop frame is

$$\boldsymbol{\omega}_{\text{Darboux|Bishop}} = \left[0 \;\; -k_2(s)\dot{s} \;\; k_1(s)\dot{s}\right]^{\text{T}}, \tag{15}$$

where s is the arc length parameter. For a parameterized curve $\mathbf{r}(t)$ with time t, by the chain rule for finding the derivatives of composite functions, we get

$$\frac{d\mathbf{r}(t)}{dt} = \frac{d\mathbf{r}(s)}{ds}\frac{ds}{dt} = \frac{d\mathbf{r}(s)}{ds}\dot{s}. \tag{16}$$

Also, by the definition of the arc differentiation,

$$ds = \left|\mathbf{r}'(t)\right| dt = \sqrt{\left(\mathbf{r}_1'(t)\right)^2 + \left(\mathbf{r}_2'(t)\right)^2 + \left(\mathbf{r}_3'(t)\right)^2} dt. \tag{17}$$

If the frame of the target curve rotates with an extra angular velocity $\boldsymbol{\omega}_d|_F$, then the angular velocity of $\{\mathscr{F}\}$ around inertial frame $\{\mathscr{I}\}$ resolved in frame $\{\mathscr{F}\}$ is defined by

$$\begin{aligned} \boldsymbol{\omega}_{F2I}\big|_F &= \boldsymbol{\omega}_d\big|_F + \boldsymbol{\omega}_{\text{Darboux|Bishop}} \\ &= \left[\boldsymbol{\omega}_{dx}\big|_F \;\; -k_2(s)\dot{s} + \boldsymbol{\omega}_{dy}\big|_F \;\; k_1(s)\dot{s} + \boldsymbol{\omega}_{dz}\big|_F\right]^{\text{T}}. \end{aligned} \tag{18}$$

The translational speed vector of $\{\mathscr{W}\}$ towards $\{\mathscr{I}\}$ resolved in $\{\mathscr{I}\}$ is $\mathbf{v}_W|_I = \left[\dot{x} \;\; \dot{y} \;\; \dot{z}\right] = \mathbf{R}_{W2I}\mathbf{v}_W|_W$, where $\mathbf{R}_{W2I}$ is the rotation matrix from $\{\mathscr{W}\}$ to $\{\mathscr{I}\}$. From Figure 2, the position vector of the UAV resolved in $\{\mathscr{I}\}$ can be presented as

$$\mathbf{p}_I = \mathbf{p}_d + \mathbf{R}_{F2I}\mathbf{p}_F. \tag{19}$$

According to the rules of finding derivatives of speed vectors in different frames in the kinematics of rigid bodies, we can find the derivative of (19)

$$\dot{\mathbf{p}}_I = \dot{\mathbf{p}}_d + \mathbf{R}_{F2I}\dot{\mathbf{p}}_F + \mathbf{R}_{F2I}\left(\boldsymbol{\omega}_{F2I}\big|_F\right)^\wedge \mathbf{p}_F, \tag{20}$$

where $(\,)^\wedge$ is the so-called hat mapping, which means, for any two vector, $\mathbf{a}$ and $\mathbf{b}$, $(\mathbf{a})^\wedge\mathbf{b} = \mathbf{a}\times\mathbf{b}$ is satisfied. Particularly, for a 3-dimension case, the form of a skew-symmetric matrix of vector $\mathbf{a} = (a_1, a_2, a_3)^{\text{T}}$ in accordance with a hat mapping is presented as

$$(\mathbf{a})^\wedge = \begin{bmatrix} 0 & -a_3 & a_2 \\ a_3 & 0 & -a_1 \\ -a_2 & a_1 & 0 \end{bmatrix}. \tag{21}$$

Then we state that $(\mathbf{a})^\wedge$ belongs to so(3). so(3) is the Lie algebra of 3-dimensional special orthogonal group SO(3). By premultiplying both sides of (20) with $\mathbf{R}_{I2F}$, we get

$$\mathbf{R}_{I2F}\dot{\mathbf{p}} = \mathbf{R}_{I2F}\dot{\mathbf{p}}_d + \dot{\mathbf{p}}_F + \left(\boldsymbol{\omega}_{F2I}\big|_F\right)^\wedge \mathbf{p}_F. \tag{22}$$

Of (22) $\mathbf{R}_{I2F}\dot{\mathbf{p}}_d$ represents the speed of a point in the target curve towards the frame $\{\mathscr{I}\}$ and is resolved in the frame $\{\mathscr{F}\}$. It is a composition of two parts. The first is the speed resolved in frame $\{\mathscr{F}\}$, that is, $\mathbf{v}_F|_F = \left(\dot{s} \;\; 0 \;\; 0\right)^{\text{T}}$. The second is the speed from the frame $\{\mathscr{F}\}$ towards $\{\mathscr{I}\}$, which is also resolved in $\{\mathscr{F}\}$ denoted as $\mathbf{v}_{F2I}|_F$. To get an analytic expression of $\mathbf{R}_{I2F}\dot{\mathbf{p}}_d$, some knowledge about theoretical mechanics is necessary to be introduced. When a rigid body makes a general motion which is a combination of translation and rotation, the speed of any point $\mathbf{P}$ in the rigid body can be formulated as

$$\mathbf{v} = \mathbf{v}_c + \boldsymbol{\omega}\times\mathbf{r}, \tag{23}$$

where $\mathbf{v}_c$ is the speed of the basic point and $\mathbf{r}$ is the position vector from the basic point to the point $\mathbf{P}$. It is arbitrary to choose the position of the basic point, for $\boldsymbol{\omega}$ never changes with the basic point.

Assume there is an initial position $\mathbf{P}_0$ moving with a speed of $\dot{\mathbf{P}}_0 = \mathbf{v}_d$ and rotating with an angular velocity of $\mathbf{P}_d$. Then the point $\mathbf{P}_0$ moves with the target curve meanwhile the relative position between $\mathbf{P}_0$ and the point of the target curve remains unchanged. By (23), one has

$$\mathbf{v}_{F2I}\big|_F = \mathbf{R}_{I2F}\left(\mathbf{v}_d + (\omega_d)^\wedge(\mathbf{p}_d - \mathbf{p}_0)\right). \tag{24}$$

Thus

$$\begin{aligned} \mathbf{R}_{I2F}\dot{\mathbf{p}}_d &= \mathbf{v}_F\big|_F + \mathbf{v}_{F2I}\big|_F \\ &= \mathbf{v}_F\big|_F + \mathbf{R}_{I2F}\left(\mathbf{v}_d + (\omega_d)^\wedge(\mathbf{p}_d - \mathbf{p}_0)\right). \end{aligned} \tag{25}$$

On the other hand, the left side of (22) is expressed as

$$\mathbf{R}_{I2F}\dot{\mathbf{p}}_I = \mathbf{R}_{W2F}\mathbf{v}_{W2I}\big|_W. \tag{26}$$

With regard to (25) and (26), (22) can be rewritten as

$$\begin{aligned} \dot{\mathbf{p}}_F = \;& \mathbf{R}_{W2F}\mathbf{v}_{W2I}\big|_W - \mathbf{v}_F\big|_F \\ &- \mathbf{R}_{I2F}\left(\mathbf{v}_d + (\omega_d)^\wedge(\mathbf{p}_d - \mathbf{p}_0)\right) - \left(\boldsymbol{\omega}_{F2I}\big|_F\right)^\wedge \mathbf{p}_F. \end{aligned} \tag{27}$$

Let $\mathbf{R}_{B2N}$ be the rotation matrix from the body frame $\{\mathscr{B}\}$ to a navigation frame $\{\mathscr{N}\}$. $\Omega_b = \left[\omega_x, \omega_y, \omega_z\right]^{\text{T}}$ represents the angular velocity of $\{\mathscr{B}\}$ around the frame $\{\widetilde{\mathscr{N}}\}$ resolved in $\{\mathscr{B}\}$. According to the general rule of finding the derivatives of the rotation matrices, the derivative of $\mathbf{R}_{B2N}$ with respect to time is

$$\dot{\mathbf{R}}_{B2N} = \mathbf{R}_{B2N}\Omega_b^\wedge. \tag{28}$$

Since in most time the navigation frame $\{\widetilde{\mathscr{N}}\}$ is considered as an inertial frame, the conclusion is still suitable for the UAV systems. Also, the angular velocity from $\{\mathscr{F}\}$ to $\{\mathscr{W}\}$ resolved in $\{\mathscr{W}\}$ is

$$\boldsymbol{\omega}_{W2F}\big|_W = \boldsymbol{\omega}_{W2I}\big|_W - \boldsymbol{\omega}_{F2I}\big|_W. \tag{29}$$

So one has

$$\dot{\mathbf{R}}_{W2F} = \mathbf{R}_{W2F}\left(\boldsymbol{\omega}_{W2F}\big|_W\right)^\wedge. \tag{30}$$

Then (27) and (30) construct the dynamic error equations of the moving frame tracking method.

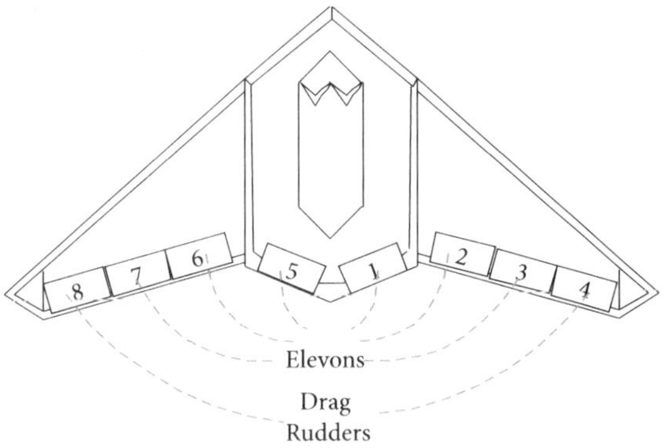

FIGURE 2: The planform and the actuators distribution of the flying-wing UAV.

3. Simulation

3.1. Structure of the System. The mass of the UAV, which is used in the simulation, is 2300 kg and its wingspan equals 8.76 m with a typical flying-wing configuration. This kind of UAV has no vertical or horizontal tails, so its aerodynamic characteristic is unordinary in various aspects. That leads to a worse static heading stability. The actuators of a flying-wing UAV are not the conventional elevators, ailerons, or rudders. Thus the allocation has to be made to transform the signals of actuators of a flying-wing UAV into the signals of the virtually ordinary actuators, such as δ_e, δ_a, and δ_r, so that the universality of the designing methodology devised in this paper is guaranteed. Figure 2 shows a planform and the actuators distribution of a flying-wing UAV.

Figure 3 shows the structure of the devised UAV flight control systems in the simulation. The blocks of the target curve, tracking errors of the moving frame, reverse calculation, the UAV model, and the controllers of the innerloops are all included, which can be seen from Figure 3.

A UAV nonlinear state-space system can be obtained with state variables defined by $\mathbf{X}^{\mathrm{T}} = [u, v, w, \phi, \theta, \psi, p, q, r, x_g, y_g, h_g]$ and control input vector $\mathbf{U}^{\mathrm{T}} = [\delta_e, \delta_\alpha, \delta_r, \delta_T]$. According to the traditional definition of a flight control system, u, v, and w are speed components of body frame, θ, ψ, and ϕ represent pitch angle, yaw angle, and bank angle, respectively, p, q, and r are angular velocity from body frame to inertial frame resolved in body frame, x_g, y_g, and h_g represent the positions of the aircraft in the ground frame which is almost the same as the inertial frame, V is true airspeed, and γ and φ are flight-path angles between the first/second axis of wind frame and inertial frame, respectively.

Particularly, the controllers of the attitude and trajectory loops are all the ordinary controllers for simplicity. Also when a better tracking result is obtained, the success can be clearly attributed to the devised method rather than the other advanced and complex controllers of the innerloops.

3.2. Moving Frame of an Elliptic Cylinder Helix. The target curve is chosen as an elliptic cylinder helix extending along the horizontal direction. The expression of the target curve with regard to a time parameter is defined as

$$\mathbf{r}(t) = (a_1 t + c_1, a_2 \cos(b_1 t) + c_2, a_3 \sin(b_2 t) + c_3)^{\mathrm{T}}, \quad (31)$$

where a_1, a_2, a_3, b_1, b_2, c_1, c_2, and c_3 are all coefficients. By finding the first-order derivative of it with regard to time we have

$$\mathbf{r}'(t) = (a_1, -a_2 b_1 \sin b_1 t, a_3 b_2 \cos b_2 t)^{\mathrm{T}}. \quad (32)$$

Its magnitude is

$$|\mathbf{r}'(t)| = \sqrt{a_1^2 + (-a_2 b_1 \sin(b_1 t))^2 + (a_3 b_2 \cos(b_2 t))^2}. \quad (33)$$

By the definition of the Frenet frame and (32), one has

$$\begin{aligned}
\mathbf{T}(t) &= \frac{\mathbf{r}'(t)}{|\mathbf{r}'(t)|} \\
&= \frac{(a_1, -a_2 b_1 \sin b_1 t, a_3 b_2 \cos b_2 t)^{\mathrm{T}}}{\sqrt{a_1^2 + (-a_2 b_1 \sin(b_1 t))^2 + (a_3 b_2 \cos(b_2 t))^2}}, \\
\frac{d\mathbf{T}}{ds} &= \frac{d\mathbf{T}(t)}{dt}\frac{dt}{ds} = \frac{d\mathbf{T}(t)}{dt}\frac{1}{|\mathbf{r}'(t)|} \\
&= \frac{1}{|\mathbf{r}'(t)|}\frac{d}{dt}\left(\frac{\mathbf{r}'(t)}{|\mathbf{r}'(t)|}\right) \\
&= \frac{1}{|\mathbf{r}'(t)|}\left(\frac{\mathbf{r}''(t)|\mathbf{r}'(t)| - \mathbf{r}'(t)(|\mathbf{r}'(t)|)'}{|\mathbf{r}'(t)|^2}\right).
\end{aligned} \quad (34)$$

For the simplicity of the expression, denote $m = a_1^2 + (a_2 b_1)^2$, $n = (a_3 b_2)^2 - (a_2 b_1)^2$, and den $= |\mathbf{r}'(t)|$. Let $b_1 = b_2$; we have

$$\begin{aligned}
\text{den} &= |\mathbf{r}'(t)| = \sqrt{a_1^2 + (-a_2 b_1 \sin(b_1 t))^2 + (a_3 b_2 \cos(b_2 t))^2} = \sqrt{a_1^2 + (a_2 b_1)^2 + ((a_3 b_2)^2 - (a_2 b_1)^2)\cos^2(b_1 t)} \\
&= \sqrt{m + n\cos^2(b_1 t)}, \\
\text{den}' &= \frac{-n \cdot b_1 \cdot \sin(b_1 t) \cdot \cos(b_1 t)}{\text{den}}, \\
\frac{d\mathbf{T}(t)}{dt} &= \frac{(-a_1 \text{den}', -a_2 b_1^2 \cos(b_1 t) \cdot \text{den} + a_2 b_1 \sin(b_1 t) \cdot \text{den}', -a_3 b_2^2 \sin(b_2 t) \cdot \text{den} - a_3 b_2 \cos(b_2 t) \cdot \text{den}')^{\mathrm{T}}}{\text{den}^2}.
\end{aligned} \quad (35)$$

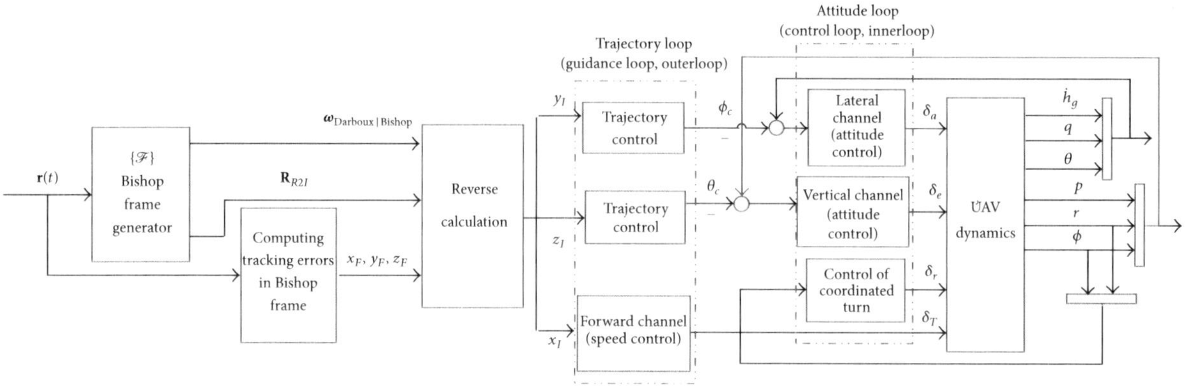

FIGURE 3: Simulation structure of the UAV flight control system.

By (34) and (35)

$$\begin{aligned}\frac{d\mathbf{T}}{ds} &= \frac{d\mathbf{T}(t)}{dt}\frac{1}{|\mathbf{r}'(t)|}\\ &= \frac{\left(-a_1\mathrm{den}', -a_2b_1^{\,2}\cos(b_1t)\cdot\mathrm{den} + a_2b_1\sin(b_1t)\cdot\mathrm{den}', -a_3b_2^{\,2}\sin(b_2t)\cdot\mathrm{den} - a_3b_2\cos(b_2t)\cdot\mathrm{den}'\right)^{\mathrm{T}}}{\mathrm{den}^2|\mathbf{r}'(t)|}.\end{aligned} \tag{36}$$

Then $\mathbf{N}(t)$ and $\mathbf{B}(t)$ are obtained

$$\begin{aligned}\mathbf{N}(t) &= \frac{d\mathbf{T}}{ds}\left|\frac{d\mathbf{T}}{ds}\right|^{-1},\\ \mathbf{B}(t) &= \mathbf{T}(t)\times\mathbf{N}(t).\end{aligned} \tag{37}$$

According to (4)~(5), that is, the transformation rule of the Frenet frame and the Bishop-1 frame, one has

$$\begin{aligned}\mathbf{T} &= \mathbf{T},\\ \mathbf{N}_1 &= \cos\theta_F\mathbf{N} - \sin\theta_F\mathbf{B},\\ \mathbf{N}_2 &= \sin\theta_F\mathbf{N} + \cos\theta_F\mathbf{B},\\ \theta_F(t) &= \int\tau(s)\,ds + \theta_0 = \int\tau(t)\,|r(t)|'\,dt + \theta_0,\\ \tan\theta_F &= \frac{k_2}{k_1},\\ \kappa &= \sqrt{k_1^2 + k_2^2} = \sqrt{k_1^2 + k_1^2\tan^2\theta_F}\\ &= \sqrt{1+\tan^2\theta_F}\,|k_1|.\end{aligned} \tag{38}$$

Thus

$$\begin{aligned}k_1 &= \pm\frac{\kappa}{\sqrt{1+\tan^2\theta_F}},\\ k_2 &= \tan\theta_F\cdot k_1.\end{aligned} \tag{39}$$

By (2), the expressions of the Frenet curvature and torsion, we get

$$\begin{aligned}\kappa &= \left|\frac{d\mathbf{T}}{ds}\right| = \left|\frac{d\mathbf{T}}{dt}\frac{dt}{ds}\right| = \left|\mathbf{T}'(t)\,|r'(t)|^{-1}\right|,\\ \tau &= -\frac{d\mathbf{B}}{ds}\cdot\mathbf{N} = -\frac{dt}{ds}\frac{d\mathbf{B}}{dt}\cdot\mathbf{N} = -|r'(t)|^{-1}\frac{d\mathbf{B}}{dt}\cdot\mathbf{N}.\end{aligned} \tag{40}$$

By (31), (32), (34), (35), (38), and (39), the Bishop frame of the target curve $\mathbf{r}(t)$ used in the simulation is formulated. For other complex curves of 3D trajectory, this methodology is still feasible as long as their analytic expressions are available (either formulated by their definitions or curve fitting methods) with regard to the time parameter t. Actually if a curve is fitted by a linear or quadratic function, the computation of its moving frame may be easier as its higher derivatives equal zero with orders being more than 3.

In simulation, according to (31), the values of the coefficients are chosen by

$$\begin{aligned}a_1 &= 200,\\ a_2 &= 300,\\ a_3 &= -250,\\ b_1 &= 0.1,\end{aligned}$$

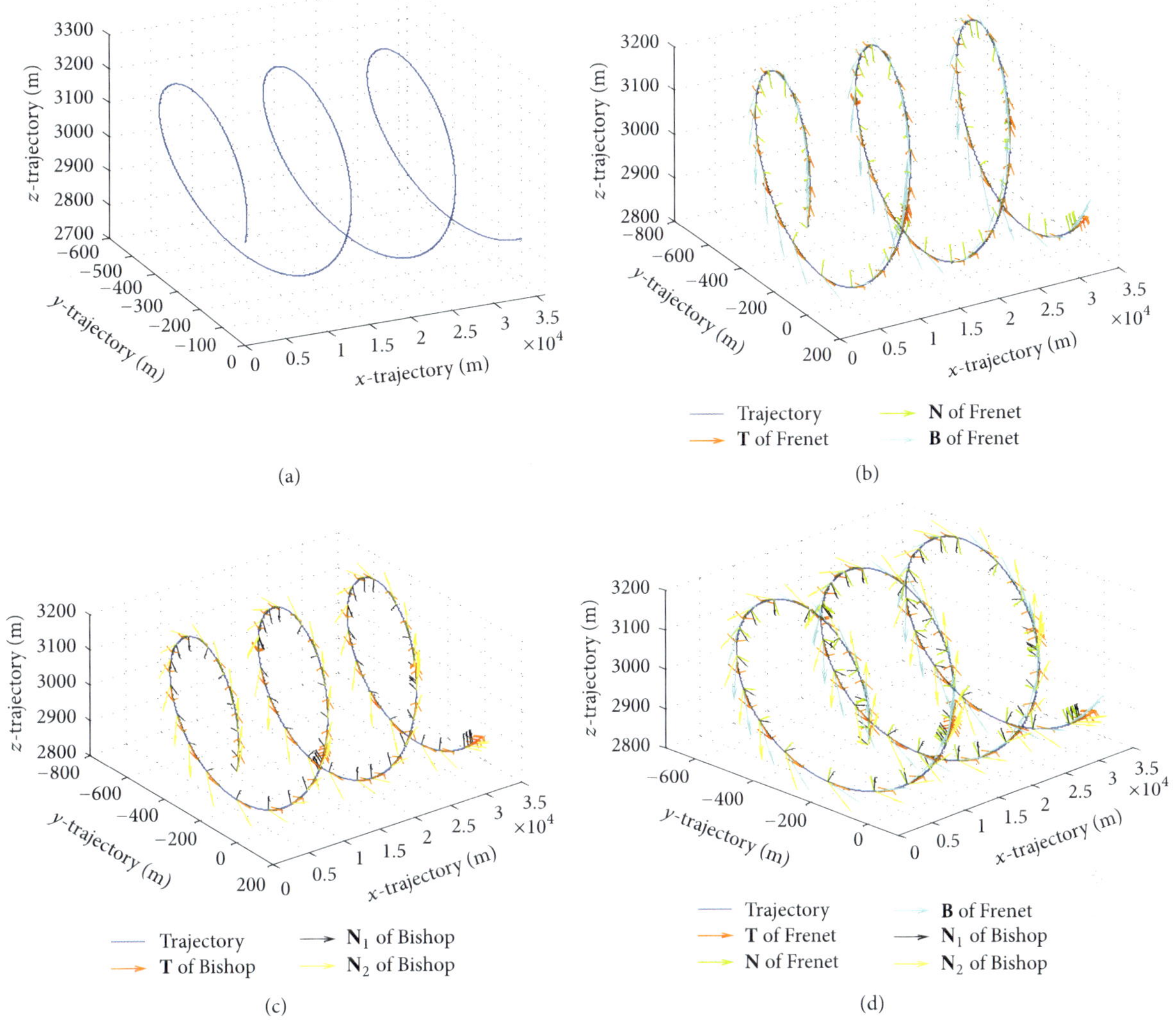

FIGURE 4: (a) Trajectory of target curve, (b) the target curve and its Frenet frame, (c) the target curve and its Bishop frame, and (d) the target curve and its moving frames.

$$
\begin{aligned}
b_2 &= 0.1, \\
c_1 &= 0, \\
c_2 &= -300, \\
c_3 &= -3000.
\end{aligned}
\tag{41}
$$

The curves of the target curve $\mathbf{r}(t)$ and its moving frames are shown in Figures 4(a)–4(d).

In Figure 4 the target curve is the blue curve which is the target to track. **T**, **N**, and **B**, respectively, represent the tangent vector, principal normal vector, and secondary normal vector and they are the three orthonormal axes of the Frenet frame. Meanwhile **T**, $\mathbf{N}_1$, and $\mathbf{N}_2$ represent the three orthonormal axes of the Bishop frame.

For the lengths of the intervals of three axes of the moving frames are different, there are some zoom in and out of these axes in Figure 4 for the purpose of clear identification. In Figures 4(b)–4(d) it is easy to identify the tangent vector **T** of the curve as the red arrow. The other two axes such as **N** and **B** or $\mathbf{N}_1$ and $\mathbf{N}_2$ construct the orthonormal coordinate frames with **T**. In Figures 4(c)-4(d) the initial parameter θ_F of the Bishop frame is chosen as 50°. According to the definitions of the two moving frames, there is a superposition of them when that parameter equals zero.

3.3. Tracking Result. The beginning of the tracking action starts from the moment the UAV arrives in a small neighborhood of an arbitrary point of the target curve. The initial value of the speed vector of the body frame is chosen at $\mathbf{V}_{b0} = [200, 0, -2]$ and the attitude angles vector is $\phi, \theta, \psi|_0 = [0, 2, 0]$, where the units are, respectively, "m/s" and "°". The high speed is needed to provide enough lift. Also a larger drag force, which is proportional to the square of the airspeed, is needed to keep the maneuverability in the forward channel. The opening range of the throttle is $[0, 1]$, which means the throttle cannot provide a force in the opposite direction. So the total force of the forward channel

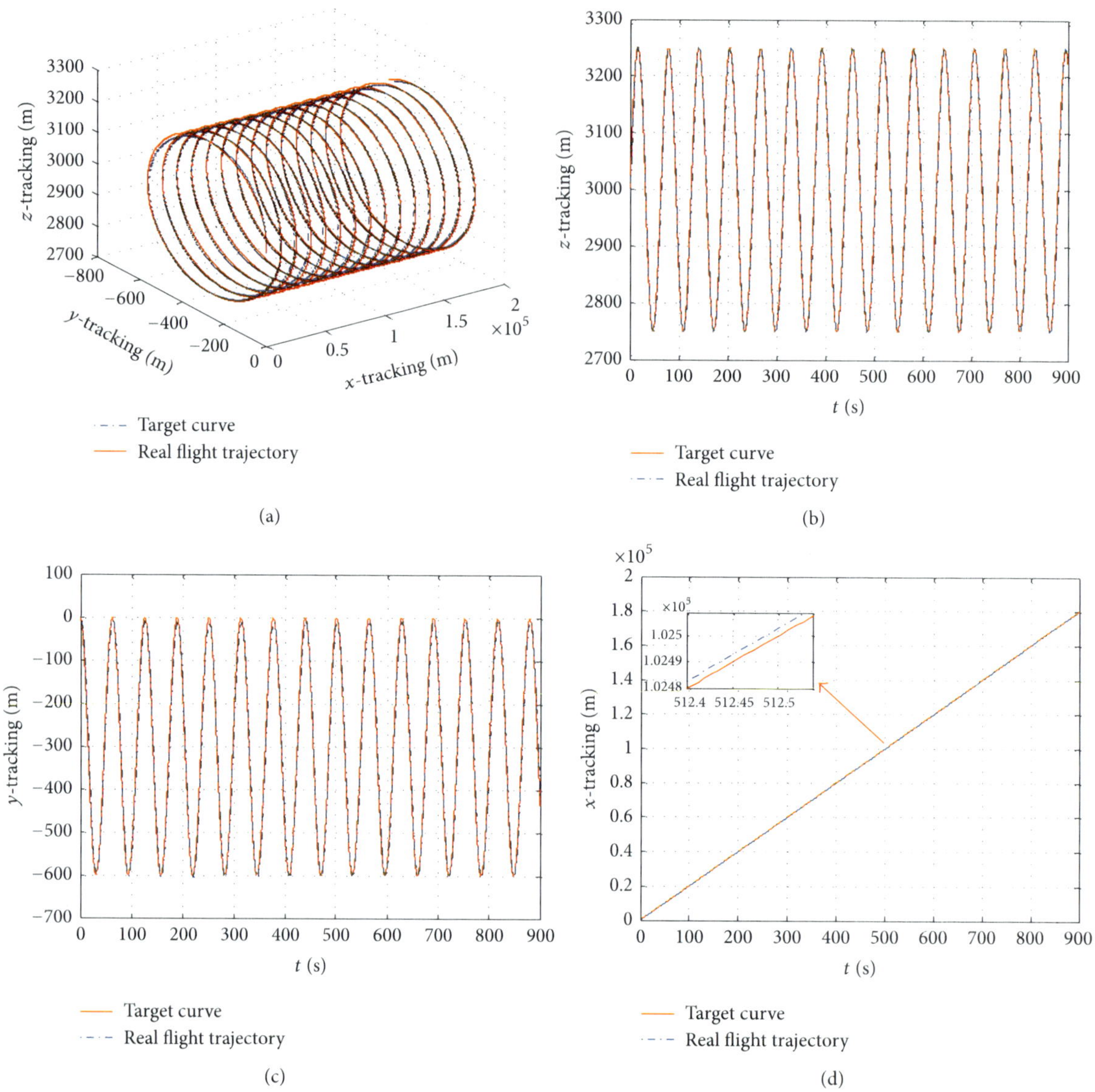

FIGURE 5: (a) 3D trajectory tracking result, (b) tracking curves in the vertical direction, (c) tracking curves in the lateral direction, and (d) tracking curves in the forward direction.

will not change its direction if the drag is not larger enough and finally it will result in an increasing forward speed. The result of trajectory tracking is shown in Figures 5(a)–5(d).

Figure 5(a) shows the 3D trajectory tracking result. From Figure 5(a) we can see that, during the simulation time of 900 s, UAV keeps on following the target curve with hardly any time delay. The mission of trajectory tracking is satisfactorily completed. Figures 5(b)–5(d), respectively, show the tracking curves with respect to time in the vertical, lateral, and forward directions of the inertial frame. As the airspeed of the UAV is about 200 m/s, the deviation could be quite large if the tracking method is not good enough. The vertical steady-state tracking error is less than 2 m within 0.8 s time delay, the lateral error is less than 7 m within 1.1 s time delay, and the forward error is less than 5 m, which are shown in these figures.

4. Conclusions

The moving frame tracking method devised in this paper based on the classical differential geometry is interdisciplinary. It is a trial to solve the trajectory tracking problems in a new direction avoiding the complex designing of the innerloop controllers. Although the innerloop controllers are simple, the whole flight control system in the simulation is much more complicated since a 6 DOF model of a large flying-wing UAV is employed. From the simulation it can be seen that a rapid, accurate, and safe trajectory tracking result is obtained.

Competing Interests

The authors declare that there are no competing interests regarding the publication of this paper.

References

[1] R. Curry, M. Lizarraga, B. Mairs, G. Elkaim, and G. H. Elkaim, "L_2^+ an improved line of sight guidance law for UAVs," in *Proceedings of the American Control Conference (ACC '13)*, June 2013.

[2] J. J. Deyst, J. P. How, and S. Park, "Lyapunov stability of a nonlinear guidance law for UAVs," in *Proceedings of the AIAA Atmospheric Flight Mechanics Conference*, pp. 1192–1208, August 2005.

[3] H. Chen, K.-C. Chang, and C. S. Agate, "Tracking with UAV using tangent-plus-Lyapunov vector field guidance," in *Proceedings of the 12th International Conference on Information Fusion (FUSION '09)*, pp. 363–372, Seattle, Wash, USA, July 2009.

[4] A. M. Shinji, "Lateral guidance control of UAV using feedback error learning," in *Proceedings of the AIAA Infotech@Aerospace 2007 Conference and Exhibit*, pp. 2007–2727, May 2007.

[5] S. Park, J. Deyst, and J. P. How, "Performance and Lyapunov stability of a nonlinear path-following guidance method," *Journal of Guidance, Control, and Dynamics*, vol. 30, no. 6, pp. 1718–1728, 2007.

[6] R. Yanushevsky, *Guidance of Unmanned Aerial Vehicles*, CRC Press, 2011.

[7] R. L. Bishop, "There is more than one way to frame a curve," *The American Mathematical Monthly*, vol. 82, pp. 246–251, 1975.

[8] A. J. Hanson and H. Ma, "Parallel transport approach to curve framing," Tech. Rep., Indiana University Compute Science Department, 1995.

[9] D. Carroll, E. Köse, and I. Sterling, "Improving Frenet's frame using Bishop's frame," *Journal of Mathematics Research*, vol. 5, no. 4, pp. 97–106, 2013.

[10] S. Kiziltug, S. Kaya, and O. Tarakci, "Tube surfaces with type-2 bishop frame of weingarten types in E^3," *International Journal of Mathematical Analysis*, vol. 7, no. 1-4, pp. 9–18, 2013.

[11] Ş. Kiliçoğlu and H. H. Hacisalihoğlu, "On the ruled surfaces whose frame is the Bishop frame in the Euclidean 3-space," *International Electronic Journal of Geometry*, vol. 6, no. 2, pp. 110–117, 2013.

[12] T. Körpınar and E. Turhan, "Biharmonic curves according to parallel transport frame in E4," *Boletim da Sociedade Paranaense de Matemática*, vol. 31, no. 2, pp. 213–217, 2013.

[13] T. Oliveira, P. Encarnacao, and A. P. Aguiar, "Moving path following for autonomous robotic vehicles," in *Proceedings of the 12th European Control Conference (ECC '13)*, pp. 3320–3325, Zurich, Switzerland, July 2013.

[14] I. Kaminer, O. Yakimenko, A. Pascoal, and R. Ghabcheloo, "Path generation, path following and coordinated control for time-critical missions of multiple UAVs," in *Proceedings of the American Control Conference*, pp. 4906–4913, Minneapolis, Minn, USA, June 2006.

[15] V. Cichella, V. Dobrokhodov, I. Kaminer, L. Marconi, and R. Naldi, "On 3D path following control of a ducted-fan UAV on SO(3)," in *Proceedings of the 50th IEEE Conference on Decision and Control and European Control Conference (CDC-ECC '11)*, pp. 3578–3583, IEEE, Orlando, Fla, USA, December 2011.

[16] V. Cichella, E. Xargay, V. Dobrokhodov, I. Kaminer, A. M. Pascoal, and N. Hovakimyan, "Geometric 3D path-following control for a fixed-wing UAV on SO(3)," in *Proceedings of the AIAA Guidance, Navigation and Control Conference*, Portland, Ore, USA, August 2011.

[17] T. Lee, M. Leok, and N. McClamroch, "Geometric tracking control of a quadrotor UAV on SE(3)," https://arxiv.org/abs/1003.2005v1.

[18] T. Lee, "Robust global exponential attitude tracking controls on SO(3)," in *Proceedings of the 1st American Control Conference (ACC '13)*, pp. 2103–2108, June 2013.

[19] T. Lee, "Geometric tracking control of the attitude dynamics of a rigid body on SO(3)," in *Proceedings of the American Control Conference*, pp. 1200–1205, San Francisco, Calif, USA, 2011.

4

Design of Robust Systems for Stabilization of Unmanned Aerial Vehicle Equipment

Olha Sushchenko[1] and Andriy Goncharenko[2]

[1] *Aircraft Control Systems Department, Educational and Research Institute of Air Navigation, National Aviation University, Cosmonaut Komarova Avenue 1, Kyiv 03058, Ukraine*
[2] *Aircraft Airworthiness Department, Aircraft Faculty, Educational and Research Airspace Institute, National Aviation University, Cosmonaut Komarova Avenue 1, Kyiv 03058, Ukraine*

Correspondence should be addressed to Olha Sushchenko; sushoa@ukr.net

Academic Editor: Paul Williams

The paper deals with the structural synthesis of robust system for stabilization of observation equipment operated on unmanned aerial vehicles. The model of the triaxial stabilization system taking into consideration necessary kinematic transformations is developed. The matrix weighting transfer functions ensuring design of the system with the desired amplitude-frequency characteristics of the system are chosen. The features of the robust structural synthesis for the researched system are considered. The structure and parameters of the robust controller, based on robust structural synthesis including the methods of the mixed sensitivity and loop-shaping, are obtained. The results of the synthesized system simulation are represented. The obtained results allow implementing stabilization of observation equipment in difficult conditions of real operation. This improves the quality of photography, mapping, survey, and so forth and gives advantages of accuracy for images representations of the territory flown. The obtained results are significant for stabilization of equipment operated at a moving base.

1. Introduction

Nowadays unmanned aerial vehicles (UAVs) carry out different functions including photography, mapping, monitoring, and survey. Cameras, laser scanners, and other information and measuring devices can be used for solving these important tasks. The problems of cartography can be solved by means of the digital photogrammetry and aerial survey using the high resolution TV cameras. A laser scanner can be used for aerial survey. The high quality of image could be provided by means of stabilization of the information and measuring devices in the inertial space.

All these applications require solving such problems as stabilization and control by orientation of the device line-of-sight. The appropriate control systems may differ according to the performed tasks, but main functions of these systems are stabilization and pointing of the above listed devices. The main components of the systems are the platform with the payload (observation equipment), inertial measuring instruments (rate gyros, accelerometers), gimbals, and drives [1–4].

As a rule, such systems implement stabilization by two axes [2]. In the most responsible cases, it is necessary to use three-axial stabilization. The paper proposes a method of design of the robust system providing stabilization of UAV payload by three axes. Such choice improves functionality of the system. In this case, it is possible both to stabilize platform and to control by platform angular position providing the necessary orientation of the payload line-of-sight.

Control in conditions of uncertainty is very important for design problems of the modern UAVs. The robust stabilization provides the required quality of operation in conditions of changed plant parameters and external disturbances. It is worth noticing that implementation of the robust systems is less expensive in comparison with the adaptive systems. Therefore the robust systems have some advantages in comparison with adaptive ones.

The paper deals with design of a robust system of the researched type. Such system provides the robust stability and robust performance in conditions of parametrical and coordinate (external) disturbance. It is based on minimization of H_∞-norm of the system sensitivity and complementary sensitivity functions.

Operation of the observation equipment at UAV is implemented in conditions of the external disturbances caused, first of all, by wind influence. Required accuracy of stabilization, tracking, and pointing in such conditions could be achieved by means of the robust control.

Features of the robust control systems design are represented in many papers and textbooks [5–8]. It should be noticed that the modern scientific-technical literature gives sufficient attention to the design of the robust systems for control of aircrafts motion [9, 10]. At the same time problems of creation of the robust systems for stabilization of equipment operated at aircrafts have no appropriate development. The method described in the paper allows solving these problems. Papers [11, 12] are dedicated to the applicable issues. Some basic concepts and representations of the considered approach are briefly given in [12].

The paper deals with the robust stabilization of the observation equipment operated at UAV. This is a new problem in the framework of the robust systems application.

Such approach provides the needed accuracy of stabilization in difficult conditions of UAV real operation. At that, costs for design works are much lower than those for adaptive systems. The lower cost of the design works is an advantage of the system of the researched type.

Expediency of alternative aircraft or their equipment type use can be modeled in the paradigm described and discussed in [13].

The paper proposes mathematical model of the stabilization plant and the weighting transfer functions for the plant augmentation and represents features of the robust structural synthesis for the specified systems.

2. Mathematical Model of Stabilization Plant

A stabilization plant represents a platform with observation equipment and measuring instruments. In order to provide the high accuracy of the observation processes, stabilization must be implemented by all three UAV axes. The stabilized platform is installed into gimbals with three degrees of freedom. Such approach makes it possible to implement the platform rotations in a sufficiently wide range of angles such as 360 degrees of the yaw and ±90 degrees of angles of the pitch and roll. (In practical situations rotations at angles of the pitch and the roll are bounded by values ±70 degrees.) In the general case, stabilization system may operate in modes of prestabilization and precision stabilization. Control in the prestabilization mode may be realized by means of accelerometers (in fact deviation from stabilization plane may be determined using accelerometers signals because $\alpha = w/g$; here α is angle of a platform deviation, w is acceleration, and g is the gravity acceleration).

These modes require using of the inertial sensors such as accelerometers and rate gyros. The pendulous accelerometers and MEMS-gyros can be used as measuring instruments for systems of the researched kind.

A mathematical model of such a plant must include equations of dynamics and kinematics. The dynamics of the platform could be described by Euler equations [14]:

$$\begin{aligned}
&\dot{\omega}_X J_X + \omega_Y\omega_Z\left(J_Z - J_Y\right) - \left(\omega_Y^2 - \omega_Z^2\right) J_{YZ} \\
&\qquad - \left(\omega_X\omega_Y + \dot{\omega}_Z\right) J_{XZ} + \left(\omega_X\omega_Z - \dot{\omega}_Y\right) J_{XY} = M_X; \\
&\dot{\omega}_Y J_Y + \omega_X\omega_Z\left(J_X - J_Z\right) - \left(\omega_Z^2 - \omega_X^2\right) J_{XZ} \\
&\qquad - \left(\omega_Z\omega_Y + \dot{\omega}_X\right) J_{XY} + \left(\omega_X\omega_Y - \dot{\omega}_Z\right) J_{YZ} = M_Y; \\
&\dot{\omega}_Z J_Z + \omega_X\omega_Y\left(J_Y - J_X\right) - \left(\omega_X^2 - \omega_Y^2\right) J_{XY} \\
&\qquad - \left(\omega_X\omega_Z + \dot{\omega}_Y\right) J_{YZ} + \left(\omega_X\omega_Z - \dot{\omega}_X\right) J_{XZ} = M_Z,
\end{aligned} \tag{1}$$

where $\omega_X, \omega_Y, \omega_Z$ represent projections of the angular rates on the platform axes; J_X, J_Y, J_Z represent inertia moments of the platform with the payload and the measuring instruments relative to the gimbals axes; J_{YZ}, J_{XZ}, J_{XY} represent the centrifugal moments of inertia relative to the gimbals axes; $\dot{\omega}_X, \dot{\omega}_Y, \dot{\omega}_Z$ represent projections of angular accelerations on the platform axes; M_X, M_Y, M_Z are moments acting along the gimbals axes. It should be noted that constituents of expressions (1) J_X, J_Y, J_Z represent the equivalent inertia moments, which include the inertia moments of the platform J_p and the motors J_m. They can be described in the following way:

$$J_i = J_{pi} + n_r^2 J_m, \quad i = X, Y, Z, \tag{2}$$

where J_{pi}, J_m represent the inertia moments of the platform with the payload and measuring instruments and the inertia moment of the motor, respectively; n_r represents a ratio of the reduction gear. Relationships for the centrifugal inertia moments of the platform can be derived in a similar way.

The moments acting on the platform include such components as the moments of the dry friction in bearings of the gimbals, moments created with motors, and disturbing moments.

For the system, working in the precision stabilization mode, it is necessary to take into consideration that every contour channel includes the platform with payload, motor, pulse-width-modulator (PWM), and rate gyro [12, 15]. Measuring range, sensitivity, and resistance to shocks are main characteristics of the rate gyro choice. Bandwidth of the chosen gyro is 100 Hz. Angular random walk of the gyro is 0, 3 deg/$\sqrt{h}$. Stability of the scale factor is of great importance for the considered application.

Components of moments acting by the gimbals axes can be defined in the following way [16]:

$$\begin{aligned}
M_{1i} &= M_{\text{fr}} \operatorname{sign} \omega_i; \\
M_{2i} &= \frac{\tilde{n}_m U_{ai}}{R_a}; \\
M_{3i} &= M_{\text{dist}\, i}; \\
&\qquad i = X, Y, Z,
\end{aligned} \tag{3}$$

where M_{fr} represents nominal moment of the dry friction in bearings mounted in the gimbals axes; $\tilde{n}_m$ represents coefficient of loading at the motor shaft; U_{ai} represent voltages of the motor control windings; R_a represents resistance of the motor armature winding; $M_{\text{dist}\,i}$ represent disturbance moments.

The voltages created in the motor armature control windings are described by the following expressions [16, 17]:

$$T_a \dot{U}_{ai} + U_{ai} = k_{\text{PWM}} U_{\text{PWM}i} - n_r c_e \omega_i, \quad i = X, Y, Z, \qquad (4)$$

where T_a represents the time constant of the armature circuit; $\tilde{n}_e$ represents the coefficient of proportionality between the electromotive force and the angular rate of the motor; k_{PWM} represents the transfer constant of the linearized PWM; $U_{\text{PWM}i}$ represent voltages at PWM input.

Control voltages at the output of the angular rate gyro sensors $U_{\omega i}$ can be described in the following way [12, 15]:

$$T_g^2 \ddot{U}_{\omega i} + 2\xi T_g \dot{U}_{\omega i} + U_{\omega i} = k_g \omega_i, \quad i = X, Y, Z, \qquad (5)$$

where T_g represents the time constant of the rate gyro; ξ represents the damping coefficient; k_g represents the gain of the rate gyro.

In order to describe kinematics of the stabilization system of the researched kind it is necessary to take into consideration the reference frames describing a position of the stabilized platform in the inertial space.

The position of the body-axes reference frame of the platform $O\ XYZ$ in the inertial space relative to the initial reference frame $O\ X_A Y_A Z_A$ is defined by means of the sequence of rotations on angles ψ, ϑ, γ as it is shown in Figure 1.

The appropriate matrices of the direction cosines are represented in the form

$$A_1 = \begin{array}{cccc} & X_1 & Y_1 & Z_1 \\ X_A & \cos\psi & 0 & \sin\psi \\ Y_A & 0 & 1 & 0 \\ Z_A & -\sin\psi & 0 & \cos\psi \end{array},$$

$$A_2 = \begin{array}{cccc} & X_2 & Y_2 & Z_2 \\ X_1 & 1 & 0 & 0 \\ Y_1 & 0 & \cos\vartheta & -\sin\vartheta \\ Z_1 & 0 & \sin\vartheta & \cos\vartheta \end{array}, \qquad (6)$$

$$A_3 = \begin{array}{cccc} & X & Y & Z \\ X_2 & \cos\gamma & -\sin\gamma & 0 \\ Y_2 & \sin\gamma & \cos\gamma & 0 \\ Z_2 & 0 & 0 & 1 \end{array}.$$

Based on relationships (6) the matrices of direction cosines between the axes XYZ, $X_A Y_A Z_A$ and the axes $X_A Y_A Z_A$, XYZ will look like

$$A = \begin{array}{cccc} & X & Y & Z \\ X_A & \cos\psi\cos\gamma + \sin\psi\sin\vartheta\sin\gamma & -\cos\psi\sin\gamma + \sin\psi\sin\vartheta\cos\gamma & \sin\psi\cos\vartheta \\ Y_A & \cos\vartheta\sin\gamma & \cos\vartheta\cos\gamma & -\sin\vartheta \\ Z_A & -\sin\psi\cos\gamma + \cos\psi\sin\vartheta\sin\gamma & \sin\psi\sin\gamma + \cos\psi\sin\vartheta\sin\gamma & \cos\psi\cos\vartheta \end{array},$$

$$A^T = \begin{array}{cccc} & X_A & Y_A & Z_A \\ X & \cos\psi\cos\gamma + \sin\psi\sin\vartheta\sin\gamma & \cos\vartheta\sin\gamma & -\sin\psi\cos\gamma + \cos\psi\sin\vartheta\sin\gamma \\ Y & -\cos\psi\sin\gamma + \sin\psi\sin\vartheta\cos\gamma & \cos\vartheta\cos\gamma & \sin\psi\sin\gamma + \cos\psi\sin\vartheta\sin\gamma \\ Z & \sin\psi\cos\vartheta & -\sin\vartheta & \cos\psi\cos\vartheta \end{array}. \qquad (7)$$

Determination of the stabilization angular rates can be implemented in the following way. If ω represents an angular rate of a platform, on which observation equipment and instruments for measurement of the platform angular rate in the inertial space are mounted, and Ω represents the platform angular rate conditioned by control moments, the condition of the precision stabilization becomes [18]

$$\Omega + \omega = 0. \qquad (8)$$

If transition to a new position of the platform in the inertial space is implemented by means of rotations on angles ψ, ϑ, γ in accordance with Figure 1, angular rates of the gimbals axes, taking into consideration matrices of the directional cosines (6), will be determined by the relationships [12]

$$\begin{bmatrix} \omega_{X_1} \\ \omega_{Y_1} \\ \omega_{Z_1} \end{bmatrix} = A_1^T \begin{bmatrix} \omega_{X_A} \\ \omega_{Y_A} \\ \omega_{Z_A} \end{bmatrix} + A_1^T \begin{bmatrix} 0 \\ \dot{\psi} \\ 0 \end{bmatrix}; \qquad (9)$$

$$\begin{bmatrix} \omega_{X_2} \\ \omega_{Y_2} \\ \omega_{Z_2} \end{bmatrix} = A_2^T A_1^T \begin{bmatrix} \omega_{X_A} \\ \omega_{Y_A} \\ \omega_{Z_A} \end{bmatrix} + A_2^T \left(A_1^T \begin{bmatrix} 0 \\ \dot{\psi} \\ 0 \end{bmatrix} + \begin{bmatrix} \dot{\vartheta} \\ 0 \\ 0 \end{bmatrix} \right); \qquad (10)$$

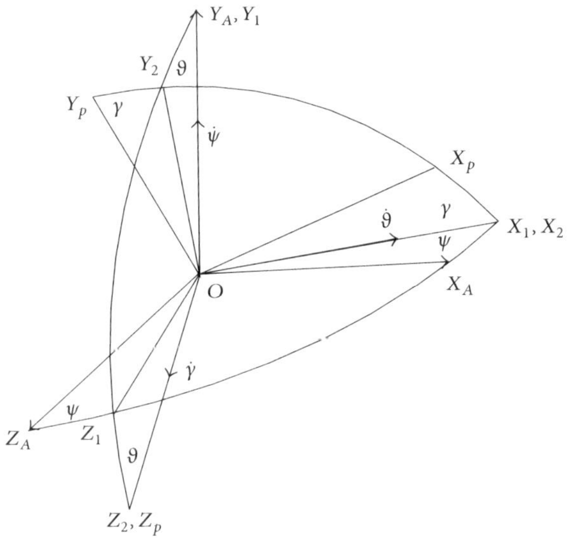

FIGURE 1: Sequence of rotations determining the platform position.

$$\begin{bmatrix} \omega_X \\ \omega_Y \\ \omega_Z \end{bmatrix} = A_3^T A_2^T A_1^T \begin{bmatrix} \omega_{X_A} \\ \omega_{Y_A} \\ \omega_{Z_A} \end{bmatrix} + A_3^T \left(A_2^T A_1^T \begin{bmatrix} 0 \\ \dot{\psi} \\ 0 \end{bmatrix} + A_2^T \begin{bmatrix} \dot{\vartheta} \\ 0 \\ 0 \end{bmatrix} + \begin{bmatrix} 0 \\ 0 \\ \dot{\gamma} \end{bmatrix} \right) = A^T \begin{bmatrix} \omega_{X_A} \\ \omega_{Y_A} \\ \omega_{Z_A} \end{bmatrix} + \begin{bmatrix} \dot{\vartheta}\cos\gamma + \dot{\psi}\cos\vartheta\sin\gamma \\ -\dot{\vartheta}\sin\gamma + \dot{\psi}\cos\vartheta\cos\gamma \\ \dot{\gamma} - \dot{\psi}\sin\vartheta \end{bmatrix}. \quad (11)$$

The first components of the obtained formula (11) are the projections of UAV angular rate onto the platform axes. In other words, these constituents represent projections of the rate of translation. The second constituents define appropriate projections of the platform relative angular rate.

While creating the mathematical model of the stabilization system it is necessary to take into account that the control moments are formed relatively to the gimbals of the yaw, pitch, and roll ψ, ϑ, γ and the constituents of relationships (1) are projections onto the proper axes of the platform. Therefore the expressions for the control moments determination must be based on the following transformations:

$$\begin{bmatrix} M_{Xp} \\ M_{Yp} \\ M_{Zp} \end{bmatrix} = \mathbf{A}_3^T \begin{bmatrix} 0 \\ 0 \\ M_{Z\gamma} \end{bmatrix} = \begin{bmatrix} 0 \\ 0 \\ M_{Z\gamma} \end{bmatrix};$$

$$\begin{bmatrix} M_{Xp} \\ M_{Yp} \\ M_{Zp} \end{bmatrix} = \mathbf{A}_3^T \mathbf{A}_2^T \begin{bmatrix} M_{X\vartheta} \\ 0 \\ 0 \end{bmatrix} = \begin{bmatrix} \cos\gamma M_{X\vartheta} \\ -\sin\gamma M_{X\vartheta} \\ 0 \end{bmatrix}; \quad (12)$$

$$\begin{bmatrix} M_{Xp} \\ M_{Yp} \\ M_{Zp} \end{bmatrix} = \mathbf{A}_3^T \mathbf{A}_2^T \mathbf{A}_1^T \begin{bmatrix} 0 \\ M_{Y\psi} \\ 0 \end{bmatrix} = \begin{bmatrix} \cos\vartheta\sin\gamma M_{Y\psi} \\ \cos\vartheta\cos\gamma M_{Y\psi} \\ -\sin\vartheta M_{Y\psi} \end{bmatrix}.$$

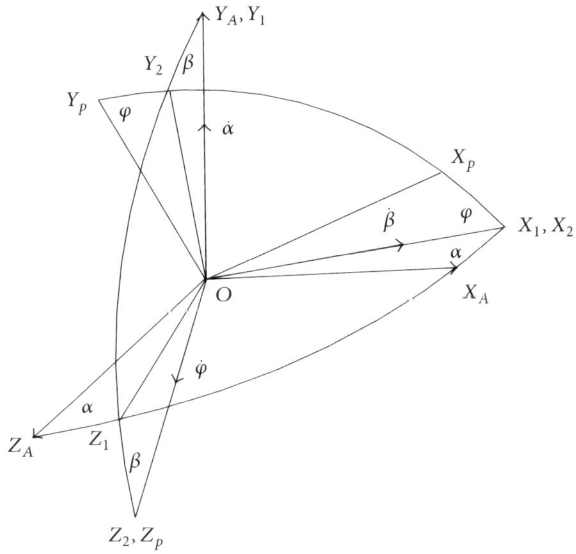

FIGURE 2: Sequence of rotations of the gimbals axes during platform stabilization.

Based on expressions (12) the projections of the moments on the platform axes are represented as

$$M_{Xp} = \cos\gamma M_{X\vartheta} + \cos\vartheta\sin\gamma M_{Y\psi};$$
$$M_{Yp} = -\sin\gamma M_{X\vartheta} + \cos\vartheta\cos\gamma M_{Y\psi}; \quad (13)$$
$$M_{Zp} = M_{Z\gamma} - \sin\vartheta M_{Y\psi}.$$

It should be noticed that angular rates of the platform are defined by expressions (9)–(11) if the gimbals axes coincide with the UAV axes at the initial instant of time. Taking into consideration this supposition, the sequence of rotations of the platform with the installed equipment and measuring instruments during stabilization will correspond to the sequence of rotations shown in Figure 2.

In accordance with Figure 2 the vector of the stabilization rate can be determined by the expression

$$\Omega = \dot{\alpha} Y_p + \dot{\beta} X_1 + \dot{\varphi} Z_2, \quad (14)$$

where $\dot{\alpha}$, $\dot{\beta}$, $\dot{\varphi}$ are angular rates of the rotations of the external and internal gimbals.

Conditions of the precision stabilization (8) expressed in the projections on the platform axes can be represented in the following form:

$$\Omega_X + \omega_X = 0;$$
$$\Omega_Y + \omega_Y = 0; \quad (15)$$
$$\Omega_Z + \omega_Z = 0.$$

Projections of stabilization rates of the platform on proper axes in accordance with Figure 2 could be defined by the following relationships:

$$\begin{aligned}\Omega_X &= \dot{\beta}\cos\varphi + \dot{\alpha}\cos\beta\sin\varphi;\\ \Omega_Y &= -\dot{\beta}\sin\varphi + \dot{\alpha}\cos\beta\cos\varphi;\\ \Omega_Z &= \dot{\varphi} - \dot{\alpha}\sin\beta.\end{aligned} \tag{16}$$

Based on relationships (16) and taking into consideration (15) it is possible to derive expressions for angular rates of the platform stabilization

$$\begin{aligned}\dot{\alpha} &= -\frac{1}{\cos\beta}(\omega_X\sin\varphi + \omega_Y\cos\varphi);\\ -\omega_Y &= -\dot{\beta}\sin\varphi + \dot{\alpha}\cos\beta\cos\varphi;\\ -\omega_Z &= \dot{\varphi} - \dot{\alpha}\sin\beta.\end{aligned} \tag{17}$$

After some transformations, using expressions (17), it is possible to obtain the differential equations for determination of the angular rates of the gimbals stabilization

$$\begin{aligned}\dot{\alpha} &= -\frac{1}{\cos\beta}(\omega_X\sin\varphi + \omega_Y\cos\varphi);\\ \dot{\beta} &= -\omega_X\cos\varphi + \omega_Y\sin\varphi;\\ \dot{\varphi} &= \frac{\sin\beta}{\cos\beta}(\omega_X\sin\varphi + \omega_Y\cos\varphi) - \omega_Z.\end{aligned} \tag{18}$$

Relationships (1)–(5), (7), and (13) and (18) represent the mathematical model of the system for stabilization of the observation equipment operated on UAV.

In accordance with concepts of the modern control theory, design of robust systems is implemented in two stages. At the first stage the robust synthesis on the basis of the use of the linearized model in the state space is determined. At the second stage the check of the synthesized system by means of simulation is carried out. Depending on the obtained results the first stage may be implemented repeatedly after the change of the initial conditions and weighting transfer functions of the optimization criterion.

The robust optimization requires an application of the linearized model in the state space. Linearization of the derived model can be carried out using the following steps:

(1) neglecting the centrifugal moments and differences of axial moments of the platform that allows simplifying expressions (1),

(2) taking into consideration only small angles of the platform rotations which allows simplifying expressions (7), (13), and (18),

(3) linearization of nonlinear dry friction moments [17],

(4) using of the simplified linearized model of PWM,

(5) neglecting the disturbing moments acting on the platform.

Using expressions (1)–(6), (7), and (13) and (18) and the above listed steps the linear model of the stabilization plant for the vector of states x can be described by the quadruple of the matrices A, B, C, D in the state space

$$x^T = [\alpha\ \beta\ \gamma\ U_{\omega X}\ U_{\omega Y}\ U_{\omega Z}\ U_{aX}\ U_{aY}\ U_{aZ}\ \omega_X\ \omega_Y\ \omega_Z\ U_{\omega dX}\ U_{\omega dY}\ U_{\omega dZ}],$$

$$A = \begin{bmatrix} 0&0&0&0&0&0&0&0&0&1&0&0&0&0&0\\ 0&0&0&0&0&0&0&0&0&0&1&0&0&0&0\\ 0&0&0&0&0&0&0&0&0&0&0&1&0&0&0\\ 0&0&0&0&0&0&0&0&0&0&0&0&1&0&0\\ 0&0&0&0&0&0&0&0&0&0&0&0&0&1&0\\ 0&0&0&0&0&0&0&0&0&0&0&0&0&0&1\\ 0&0&0&0&0&0&-\frac{1}{T_a}&0&0&-\frac{n_r c_e}{T_a}&0&0&0&0&0\\ 0&0&0&0&0&0&0&-1/T_a&0&0&-\frac{n_r c_e}{T_a}&0&0&0&0\\ 0&0&0&0&0&0&0&0&-\frac{1}{T_a}&0&0&-\frac{n_r c_e}{T_a}&0&0&0\\ 0&0&0&0&0&0&\frac{c_m}{(R_a J_X)}&0&0&-\frac{f_X}{J_X}&0&0&0&0&0\\ 0&0&0&0&0&0&0&\frac{c_m}{(R_a J_Y)}&0&0&-\frac{f_Y}{J_Y}&0&0&0&0\\ 0&0&0&0&0&0&0&0&\frac{c_m}{(R_a J_Z)}&0&0&-\frac{f_Z}{J_Z}&0&0&0\\ 0&0&0&-\frac{1}{T_g^2}&0&0&0&0&0&\frac{k_g}{T_g^2}&0&0&-\frac{2\xi}{T_g}&0&0\\ 0&0&0&0&-\frac{1}{T_g^2}&0&0&0&0&0&\frac{k_g}{T_g^2}&0&0&-\frac{2\xi}{T_g}&0\\ 0&0&0&0&0&-\frac{1}{T_g^2}&0&0&0&0&0&\frac{k_g}{T_g^2}&0&0&-\frac{2\xi}{T_g} \end{bmatrix},$$

$$B^T = \begin{bmatrix} 0 & 0 & 0 & 0 & 0 & 0 & k_{\text{PWM}} & 0 & 0 & 0 & 0 & 0 & 0 & 0 & 0 \\ 0 & 0 & 0 & 0 & 0 & 0 & 0 & k_{\text{PWM}} & 0 & 0 & 0 & 0 & 0 & 0 & 0 \\ 0 & 0 & 0 & 0 & 0 & 0 & 0 & 0 & k_{\text{PWM}} & 0 & 0 & 0 & 0 & 0 & 0 \end{bmatrix};$$

$$C = \begin{bmatrix} 0 & 0 & 0 & 1 & 0 & 0 & 0 & 0 & 0 & 0 & 0 & 0 & 0 & 0 & 0 \\ 0 & 0 & 0 & 0 & 1 & 0 & 0 & 0 & 0 & 0 & 0 & 0 & 0 & 0 & 0 \\ 0 & 0 & 0 & 0 & 0 & 1 & 0 & 0 & 0 & 0 & 0 & 0 & 0 & 0 & 0 \end{bmatrix};$$

$$D = \begin{bmatrix} 0 & 0 & 0 \\ 0 & 0 & 0 \\ 0 & 0 & 0 \end{bmatrix}. \tag{19}$$

Elements $U_{\omega dX}$, $U_{\omega dY}$, $U_{\omega dZ}$ of the matrix A represent derivatives of the voltages $U_{\omega X}, U_{\omega Y}, U_{\omega Z}$; f_X, f_Y, f_Z represent coefficients of linearized moments of the dry friction [17].

The mathematical models (19) can be used in the procedure of the robust structural optimization.

3. Algorithm of Structural Synthesis of Robust Stabilization System

One of the modern approaches to the structural synthesis of the robust stabilization systems is H_∞-synthesis. Its basic principles are represented in many papers and textbooks, for example, [7, 8]. The standard configuration of the system designed by means of the H_∞-synthesis is shown in Figure 3.

The system consists of the plant G and the controller K and is characterized with the vectors of output parameters z to be optimized, external input signals w, control signals u, and measured signals y entering the input of the controller [7, 8].

The researched system can be described by the scheme represented in Figure 4 [8].

Interconnection between signals of such system can be described by the following expressions [8]:

$$w = r;$$

$$z = \begin{bmatrix} e \\ u \\ y \end{bmatrix};$$

$$P = \begin{bmatrix} P_{11} & P_{12} \\ P_{21} & P_{22} \end{bmatrix} = \begin{bmatrix} I & -G \\ 0 & I \\ 0 & G \\ I & -G \end{bmatrix}, \tag{20}$$

where P represents the transfer function of the interconnected system.

The transfer function from the input w to the output z becomes

$$T_w^z = P_{11} + P_{12}K\left(I - P_{22}K\right)^{-1} P_{21}. \tag{21}$$

The principle of the guaranteed result may be applied to robust systems. In accordance with this principle [7]

$$\frac{\|P(s)\, w(s)\|_\infty}{\|w(s)\|_\infty} < \gamma, \tag{22}$$

where γ is a small value; $w(s)$ is representation of the external input signals including disturbances.

Optimization problem of the robust structural synthesis can be represented in the following form [7, 8]:

$$K_{\text{opt}} = \arg \inf_{K_{\text{opt}} \in K_{\text{per}}} J(G, K), \tag{23}$$

where

$$J(G, K) = \left\| \begin{bmatrix} (I + GK)^{-1} \\ K(I + GK)^{-1} \\ GK(I + GK)^{-1} \end{bmatrix} \right\|_\infty . \tag{24}$$

The optimization problem (23) may be solved by means of the method of the mixed sensitivity [7, 8]. The modern approach to solving the problem of the robust structural optimization is based on forming the desired amplitude-frequency characteristics of the system (loop-shaping) by means of the plant augmentation due to introducing the weighting transfer functions as it is shown in Figure 5 [7].

Optimization criterion of the method of the mixed sensitivity can be represented based on formula (24) in the form of the expression representing the H_∞-norm of the function of the mixed sensitivity of the augmented system [7, 8]

$$J(G, K) = \left\| \begin{bmatrix} W_1 (I + GK)^{-1} \\ W_2 K(I + GK)^{-1} \\ W_3 GK(I + GK)^{-1} \end{bmatrix} \right\|_\infty = \left\| \begin{bmatrix} W_1 S \\ W_2 R \\ W_3 T \end{bmatrix} \right\|_\infty , \tag{25}$$

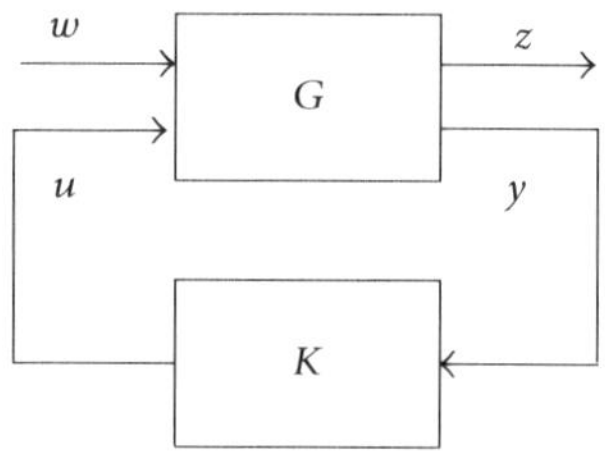

FIGURE 3: Standard H_∞-configuration.

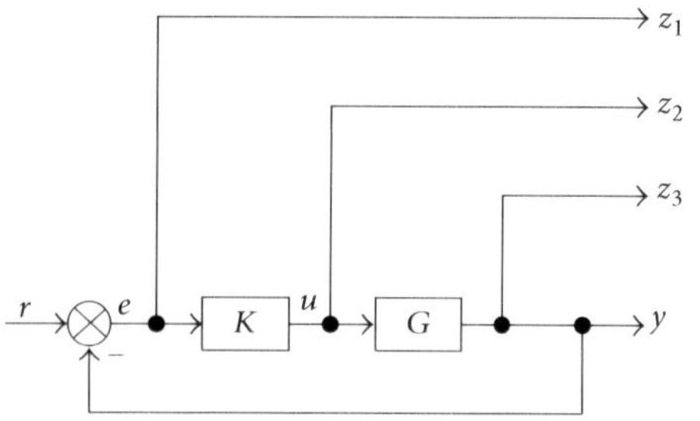

FIGURE 4: Structural scheme of the designed system.

where W_1, W_2, W_3 represent the weighting transfer functions; S, R, T are functions of sensitivity with respect to the reference and control signals and complementary sensitivity function.

Implementation of the H_∞-synthesis by the method of the mixed sensitivity is based on solutions of two Riccati equations, checking some conditions and minimization of the H_∞-norm of the function of the mixed sensitivity (25) [7, 8]. It should be noticed that there are automated means of this problem solution based on MatLab software. The appropriate optimization procedure requires using the mathematical model of the interconnected system (20).

The procedure of the H_∞-synthesis includes such stages as

(1) creation of the mathematical models including the linear model and the model which takes into consideration the nonlinearities inherent to real systems,

(2) choice of the weighting transfer functions,

(3) augmentation of the plant,

(4) structural robust synthesis.

These stages can be implemented by means of Robust Control Toolbox of MatLab software.

One of the most important stages of the robust structural synthesis is a choice of the weighting transfer functions, which provide forming of the augmented plant and loop-shaping approach. There are the following general recommendations for determination of weighting transfer functions for augmented plant forming [7]:

$$W_1 = \frac{s/M + \omega_0}{s + \omega_0 A};$$
$$W_2 = \text{const}; \qquad (26)$$
$$W_3 = \frac{s + \omega_0/M}{As + \omega_0},$$

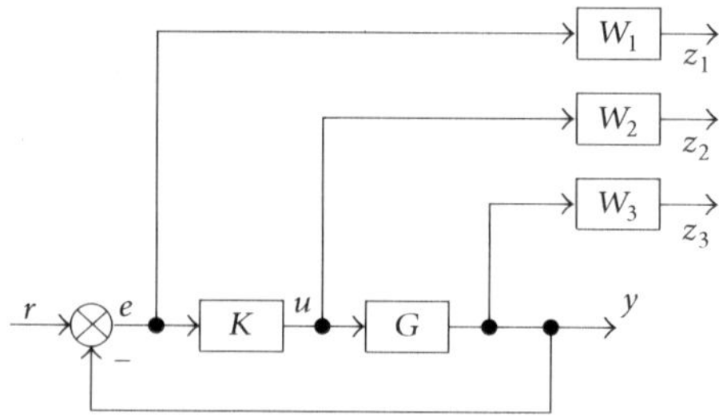

FIGURE 5: Structural scheme of the system augmented by means of the weighting transfer functions.

where A is the desired stable error in the steady mode; ω_0 is the desired bandwidth; M is sensitivity peak.

An experience of the rate drives design showed the necessity of using integrator in the expression of the weighting transfer function W_1. But in real applications it is necessary to specify weighting transfer functions by methods of trials and errors.

In accordance with Figure 5 the weighting transfer functions impose a penalty on the error, control and output signals. Changing parameters of the weighting transfer functions A, ω_0, M it is possible to achieve the desired frequency characteristics of the system [7].

Usage of the weighting transfer functions provides loop-shaping of the designed system. Such approach allows forming the designed system with the desired frequency characteristics.

During synthesis of the researched system the following expressions for the weighting transfer functions determination were obtained:

$$W_1 = \begin{bmatrix} \frac{1}{s}\frac{0.1s+20}{s+0.005} & 0 & 0 \\ 0 & \frac{1}{s}\frac{0.02s+10}{s+0.002} & 0 \\ 0 & 0 & \frac{1}{s}\frac{0.01s+10}{s+0.001} \end{bmatrix};$$
$$W_2 = \begin{bmatrix} 0.04 & 0 & 0 \\ 0 & 0.04 & 0 \\ 0 & 0 & 0.04 \end{bmatrix}; \qquad (27)$$
$$W_3 = \begin{bmatrix} \frac{s}{0.005s+50} & 0 & 0 \\ 0 & \frac{s}{0.002s+20} & 0 \\ 0 & 0 & \frac{s}{0.001s+10} \end{bmatrix}.$$

Synthesis of the robust controller was implemented by means of MatLab function *hinfopt*. First, an order of the synthesized controller was equalled to 8. Decrease of controller order was implemented by means of function *schmr*. This allows decreasing the controller order to 5. Discretization of controller was implemented by means of function *c2d*.

Manufacturing experience showed that controller of such order is realizable in the form of digital controller. In this case

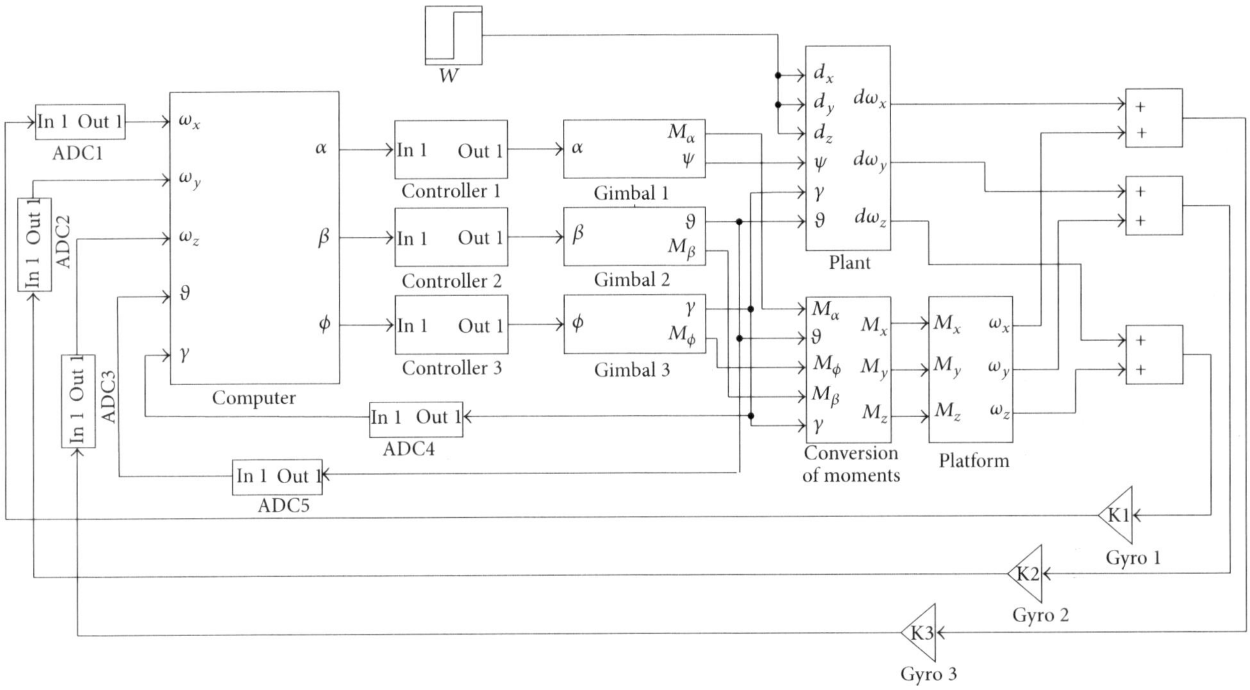

FIGURE 6: Simulink-model of the stabilization system.

the difference equations are realized by means of C-language assigned for microcontrollers.

The structure of the synthesized robust controller is described in the state space by means of the following quadruple of matrices:

$$A_p = \begin{bmatrix} 186.3 & 352.6 & 45.83 & 11.33 & 43.7 \\ 711.8 & -966.3 & -179.7 & -12.17 & -11.6 \\ -117.75 & 126.7 & -18.41 & -15.84 & -15.62 \\ -343.7 & 517.3 & 136.8 & 19.39 & -23.43 \\ -218.3 & 268.8 & 20.53 & 18.75 & -25.89 \end{bmatrix};$$

$$B_p^T = \begin{bmatrix} 20.98 & -37.1 & -8.3 & -5.36 & -7.25 \\ 17.6 & -32.1 & -7.3 & -4.2 & 6.23 \\ 15.6 & -30.5 & 5.4 & -3.1 & -5.21 \end{bmatrix}; \quad (28)$$

$$C_p = \begin{bmatrix} 8.31 & -9.8 & -4.78 & -1.43 & 2.35 \\ 9.6 & -6.4 & 12.2 & -1.2 & 16.3 \\ 11.2 & -7.5 & 9.2 & -11.1 & -15.21 \end{bmatrix};$$

$$D_p = \begin{bmatrix} -0.9 & -0.543 & -0.217 \\ -0.6 & -0.312 & -0.123 \\ -0.4 & -0.213 & -0.223 \end{bmatrix}.$$

The Simulink-model for system simulation is represented in Figure 6. Results of the synthesized system simulation are represented in Figures 7 and 8.

The angular rates of UAV, taking into consideration influence of the wind, were considered as disturbances during simulation. Forming the wind influence was carried out by means of the Dryden filter. Represented results prove the possibility of achieving the acceptable quality of the stabilization accuracy and speed of operation in conditions of influence of the external disturbances.

Advantages in comparison with other controllers may be described in the following way. Such systems keep required performances in conditions of changed parameters and external disturbances. Controllers of other types do not guarantee even stability of the system in such conditions. This can lead to inability to perform functions of the system not to mention providing of its high characteristics.

Adaptive systems take into consideration changes of plant characteristics during operation but they are too expensive in comparison with the robust systems.

4. Conclusions

(a) The kinematics expressions for three gimbal configurations were derived.

(b) The mathematical model of the stabilization plant of the considered system was obtained.

(c) The weighting transfer functions, which provide loop-shaping of the designed system, were determined.

(d) The robust controller was synthesized. Representation in the form of the quadruple of the state space matrices is convenient for digital controller programming.

(e) In order to provide the high accuracy of observation processes it is necessary to use the three-axial stabilization systems.

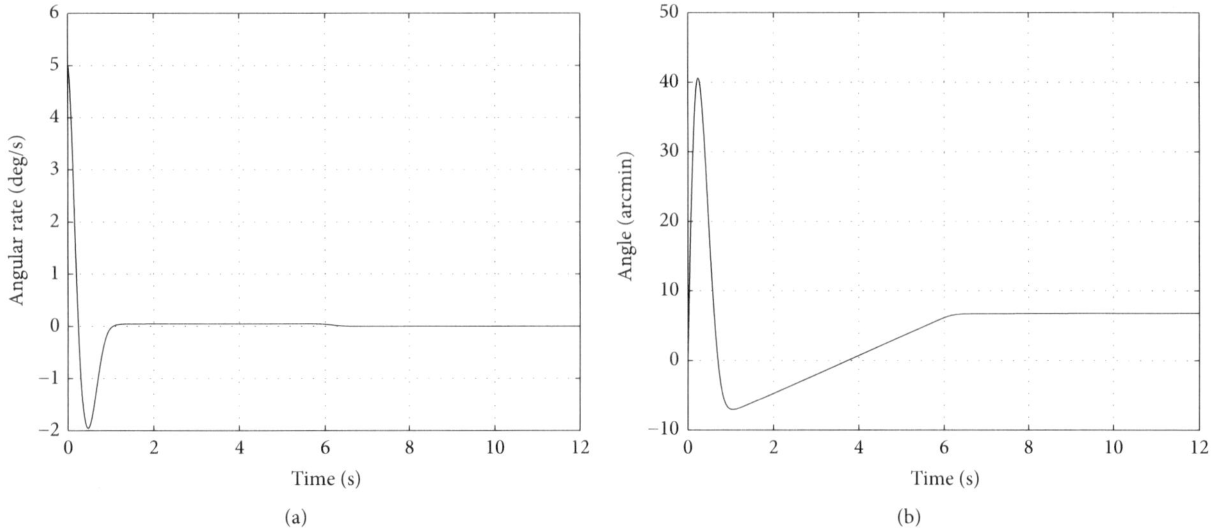

FIGURE 7: Results of simulation of the platform motion in the mode of prestabilization: (a) is transient of the given angular rate; (b) is error of the angular position determination.

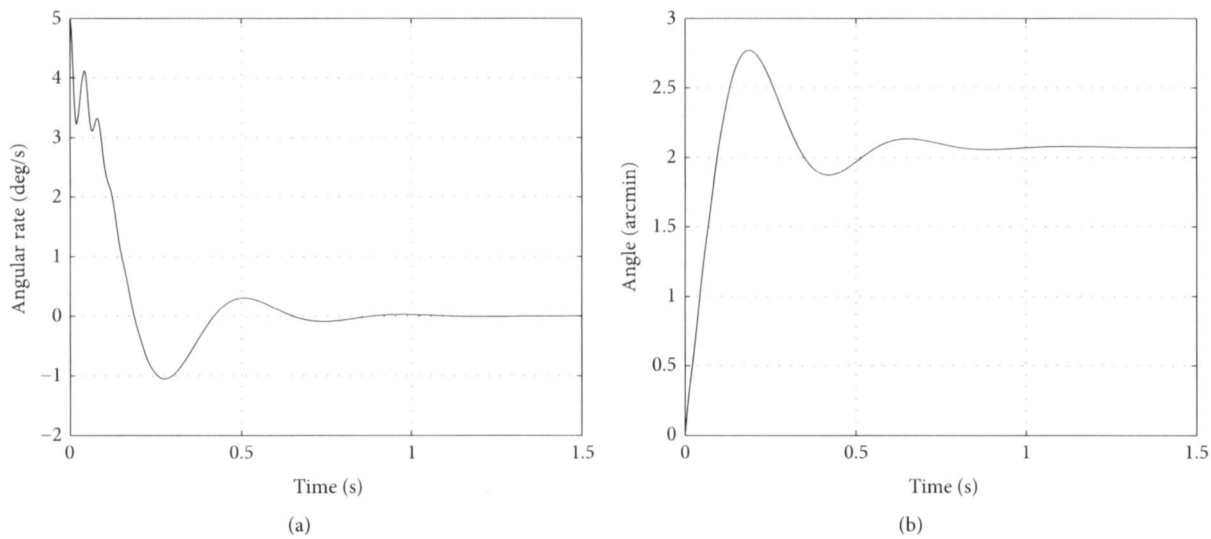

FIGURE 8: Results of simulation of the platform motion in the mode of precision stabilization: (a) is transient of the given angular rate; (b) is error of the angular position determination.

(f) Creation of the mathematical model of the three-axial stabilization system requires using the coordinate transformations.

(g) Efficiency of the suggested approaches is proved by the results of simulation of the synthesized system, while simulation influence of the wind was taken into consideration.

(h) The obtained results will be useful perspective for stabilization of the observation equipment operated on the vehicles of the wide class.

Competing Interests

The authors declare that there is no conflict of interests regarding the publication of this paper.

References

[1] J. M. Hilkert, "Inertially stabilized platform technology: concepts and principles," *IEEE Control Systems Magazine*, vol. 28, no. 1, pp. 26–46, 2008.

[2] M. K. Masten, "Inertially stabilized platforms for optical imaging systems," *IEEE Control Systems Magazine*, vol. 28, no. 1, pp. 47–64, 2008.

[3] D. Debruin, "Control systems for mobile satcom antennas," *IEEE Control Systems Magazine*, vol. 28, no. 1, pp. 86–101, 2008.

[4] H. G. Wang and T. C. Williams, "Strategic inertial navigation systems—high-accuracy inertially stabilized platforms for hostile environments," *IEEE Control Systems Magazine*, vol. 28, no. 1, pp. 65–85, 2008.

[5] B. A. Francis, J. W. Helton, and G. Zames, "H^{∞}-optimal feedback controllers for linear multivariable systems," *IEEE*

Transactions on Automatic Control, vol. 29, no. 10, pp. 888–900, 1984.

[6] K. Zhou and J. Doyle, *Essentials of Robust Control*, Prentice Hall, New Jersey, NJ, USA, 1999.

[7] S. Skogestad and I. Postlethwaite, *Multivariable Feedback Control*, John Wiley & Sons, New York, NY, USA, 2001.

[8] D. W. Gu, P. H. Petkov, and M. M. Konstantinov, *Robust Control Design with MATLAB*, Springer, London, UK, 2005.

[9] A. A. Tunik, J. C. Kim, and C. S. Yoo, "The parameter optimization of aircrafts control law from the viewpoint of some airworthiness requirements," in *Proceedings of the 12th Korea Automatic Control Conference (KACC '97)*, pp. 1651–1654, Seoul, South Korea, 1997.

[10] A. A. Tunik, H. Rye, and H. C. Lee, "Parametric optimization procedure for robust flight control system design," *KSAS International Journal*, vol. 2, no. 2, pp. 95–107, 2001.

[11] O. A. Sushchenko and A. A. Tunik, "Robust stabilization of UAV observation equipment," in *Proceedings of the IEEE 2nd International Conference on Actual Problems of Unmanned Air Vehicles Developments (APUAVD '13)*, pp. 176–180, IEEE, Kiev, Ukraine, October 2013.

[12] O. A. Sushchenko, "Approach to design of robust systems for stabilization of unmanned aerial vehicles equipment," in *Proceedings of the 6th World Congress 'Aviation in XX-st Century' 'Safety in Aviation and Space Technologies'*, vol. 2, pp. 3.5.1–3.5.15, Kyiv, Ukraine, September 2014.

[13] A. V. Goncharenko, "Expediency of unmanned air vehicles application in the framework of subjective analysis," in *Proceedings of the IEEE 2nd International Conference on Actual Problems of Unmanned Air Vehicles Developments (APUAVD '13)*, pp. 129–133, IEEE, Kyiv, Ukraine, October 2013.

[14] M. A. Aizerman, *Classical Mechanics*, Nauka, Moscow, Russia, 1980 (Russian).

[15] O. A. Sushchenko, "Mathematical model of inertially stabilized platform for aircraft observation equipment," in *Proceedings of the 4th World Congress Aviation in the 21st Century. Safety in Aviation and Space Technology*, vol. 1, pp. 21.43–21.46, Kyiv, Ukraine, September 2010.

[16] D. S. Pelpor, *Gyroscopic Systems*, part 1, Vishaya Shkola, Moscow, Russia, 1976 (Russian).

[17] V. V. Kochergin, *Servo-Systems with Direct Current Motors*, Ehergoatomizdat, St-Petersburg, Russia, 1988 (Russian).

[18] S. S. Rivkin, *Stabilization of Information and Measuring Devices on Swinging Base*, Nauka, Moscow, Russia, 1978 (Russian).

VTOL UAV Transition Maneuver using Incremental Nonlinear Dynamic Inversion

Zhenchang Liu, Jie Guo, Mengting Li, Shengjing Tang, and Xiao Wang

School of Aerospace Engineering, Beijing Institute of Technology, Beijing 100081, China

Correspondence should be addressed to Jie Guo; guojie1981@bit.edu.cn

Academic Editor: Paul Williams

The paper seeks to study the control system design of a novel unmanned aerial vehicle (UAV). The UAV is capable of vertical takeoff and landing (VTOL), transition flight and cruising via the technique of direct force control. The incremental nonlinear dynamic inversion (INDI) approach is adopted for the 6-DOF nonlinear and nonaffine control of the UAV. Based on the INDI control law, a method of two-layer cascaded optimal control allocation is proposed to handle the redundant and coupled control variables. For the weight selection in optimal control allocation, a dynamic weight strategy is proposed. This strategy can adjust the weight of the objective function according to the flight states and mission requirements, thus determining the optimizing direction and ensuring the rationality of the allocation results. Simulation results indicate that the UAV can track the target trajectory accurately and exhibit continuous maneuverability in transition flight.

1. Introduction

UAVs have increasing applications including surveillance, communications, search and rescue operations, and other military tasks. Among different flight conditions of UAVs, the 0–1000 m height area in cities is one of the most significant applications, where complex terrain and significant gusts arising from atmospheric turbulence exist.

This research studies a novel fixed-wing VTOL UAV with thrust vector engines, which can be applied in urban areas. The UAV can get rid of the restrictions imposed by takeoff and landing conditions and be accurately recovered by hover function; the UAV owns a larger combat range and higher flight speed by forward flight capability. A new concept of low-speed cruising is studied in this paper. For the most current VTOL aircrafts, the transition from hover to forward flight is short and stable. However, for low-speed cruising, the transition is prolonged as a normal flight state. The vehicle can maintain transition flight for long time cruising by adopting direct force control and has favorable maneuverability. This flight mode is appropriate for vehicles flying in low-altitude complex conditions. In the meantime, under the low speed of UAV and the inefficient aerodynamic surface control during the transition flight, the incorporated control strategy of vectoring nozzles and aerodynamic surface should be adopted to control the attitude. In view of this, the thrust vector direct force control is of critical significance for transition maneuver. However, large nonlinearities, redundancies, and coupling effects arise when this technique is adopted.

In recent years, the research on VTOL UAVs is increasingly prosperous with the advancements in automatic control and the increasing popularity of UAV platforms [1]. The problem of transition maneuvers has been studied for different types of UAVs, including a fixed-wing aircraft equipped with a thrust vector engine and lift fan [1–3], tilt-rotor aircraft [4–7], tail-sitter aircraft [8, 9], ducted-fan VTOL aircraft [10, 11], and tilt-wing aircraft [12, 13]. A pilot auxiliary control system was designed by Francesco and Mattei [1] for a fixed-wing tiltrotor UAV, and the control logics for different flight states were specified and synthesized through adopting INDI method. The daisy-chaining logic was employed to handle the control redundancy. The autonomous transition control of two V/STOL aircrafts was studied by Xili et al. [2]. The nonlinear trajectory control strategies in longitudinal direction have been designed specifically for aircrafts in different types. The maneuver of a tilt quadcopter was researched by Ryll et al. [7]. The control design was basically dependent on the exact linearization of the motion equations, and the actuation redundancy was calculated through employing pseudoinverse matrices. The previous works generally decouple the longitudinal and lateral control, design control logics for hover, transition flight, and forward flight, respectively, and regard some redundant control variables as constants to solve the transition maneuver control problem. The UAV's maneuver potential with vectored thrust cannot be fully utilized.

The INDI method is adopted in this paper to control UAV's position/attitude during transition maneuver. The INDI method, which originates from the nonlinear dynamic inversion (NDI), solves the incremental form of equations of motion and generates a control law substantially reducing the dependence on aerodynamic model and other vehicle models. INDI was firstly adopted to control UAV's attitude control by Sieberling et al. [14]. The INDI attitude control law of a quadcopter was proposed, and the momentum of the propellers was incorporated in the controller by Smeur et al. [15]. They generalized this method based on the previous work to the outer control loop (position control loop) of a quadcopter under severe gust loads [16, 17]. Lu et al. [18] applied the INDI method to the fixed-wing aircraft trajectory controller. They compared the performance of the INDI controller with that of NDI approach and proved that the INDI method can reduce the model uncertainties effectively.

An INDI control system is designed in this paper to address the 6-DOF nonlinear control of the UAV in transition flight, and the main contributions are listed below:

(1) Given the problems of strong nonlinearities and multi-axis coupling characteristics, a unified 6-DOF nonlinear control strategy is proposed to control position/attitude and there is no need to switch the control logic according to different flight states. The INDI method is introduced to address the model uncertainty and control coupling problem. Different from the work conducted by Lu et al. [18], the sideslip angle β is not assumed to be zero, and the vectored thrust in 3 directions of the body axis is considered

(2) A two-layer cascaded optimal control allocation method is proposed to address the control redundancy based on the INDI control law. The first-layer optimal allocation is conducted to allocate the increment of flight attitude and vectored thrust in the translational dynamics control loop. The solution of the engine thrust, vectoring nozzle deflections, and aerodynamic surface deflections are calculated in the second-layer control allocation

(3) A dynamic weight selection strategy is designed for the objective function of the two-layer cascaded optimal control allocation. In dynamic weight selection strategy, a weight generator and a weight regulator are designed, which calculate weight through an analytic hierarchy process (AHP) and adjust weight according to flight states and mission requirement, respectively. The dynamic weight selection strategy can allocate control variables according to mission requirement and ensure the optimal results to be feasible

This paper is constructed as follows: the configuration and aerodynamics characteristics of the VTOL UAV researched in this paper are described in Section 2. A 6-DOF mathematical model and an INDI control system of the vehicle are given in Section 3. The two-layer cascaded optimal control allocation method is presented in Section 4. Section 5 provides the simulations results. Eventually, conclusions are drawn in Section 6.

2. UAV Configuration and Aerodynamics

The UAV is designed in a tandem-wing plus lift-body configuration. This aerodynamic configuration can provide more lift under the limitation imposed by wing span, and it makes UAV applicable to fly in low-altitude complex flight condition. The power system consists of a lift fan in the front part of fuselage and two thrust vector engines in each side of the rear part of fuselage, as presented in Figures 1(a) and 1(b). At the bottom of the lift fan, a control rudder is equipped with the shaft. The control rudder can be deflected left and right with 12 degrees and provides lateral vectored thrust T_y and yaw moments n_T, as illustrated in Figures 1(c) and 1(d). The engine's vectoring nozzle can be swung up with 15 degrees and down with 90 degrees. It provides 3-axis vectored thrust for a direct force control, as shown in Figures 1(e) and 1(f). Through the cooperation of the lift fan and thrust vector engine, the UAV can realize VTOL, transition flight, and cruise.

In the case of no wind tunnel experiment, the UAV's aerodynamic coefficients were obtained by the method of aerodynamic estimation and CFD calculation together. In aerodynamic coefficients, the static pitch moment derivative $C_{m\alpha}$ ranges from −0.04 to 0.056. When the angle of attack is from 0 to 18 degrees, $C_{m\alpha} > 0$, which indicates that the UAV is statically instable in pitch channel. The reason of instability design is that for VTOL UAV, the alignment of the mass center should be considered with power system distribution and the location of aerodynamic center, which is in the front of the mass center, facilitates the vehicle to pitch up in transition maneuver. The static yaw moment derivative $C_{n\beta}$ ranges from −0.017 to −0.03. The static stability in yaw channel is caused by the rear wing dihedral effect. The vehicle's main dynamic moment derivatives are $C_{lp} = -0.0084$, $C_{mq} = -0.01739$, and $C_{nr} = -0.0011$. They are all less than zero, and it indicates that the damping moments can reduce the angular velocities and keep the vehicle dynamics stable. The mapping relationship between the actuators, aerodynamic forces, and moments are usually nonlinear. In this paper, to meet the requirement of the control allocation, polynomial approximation is used to fit the aerodynamic data curve between the actuator deflections and aerodynamic coefficients. This will alleviate the need of large lookup tables

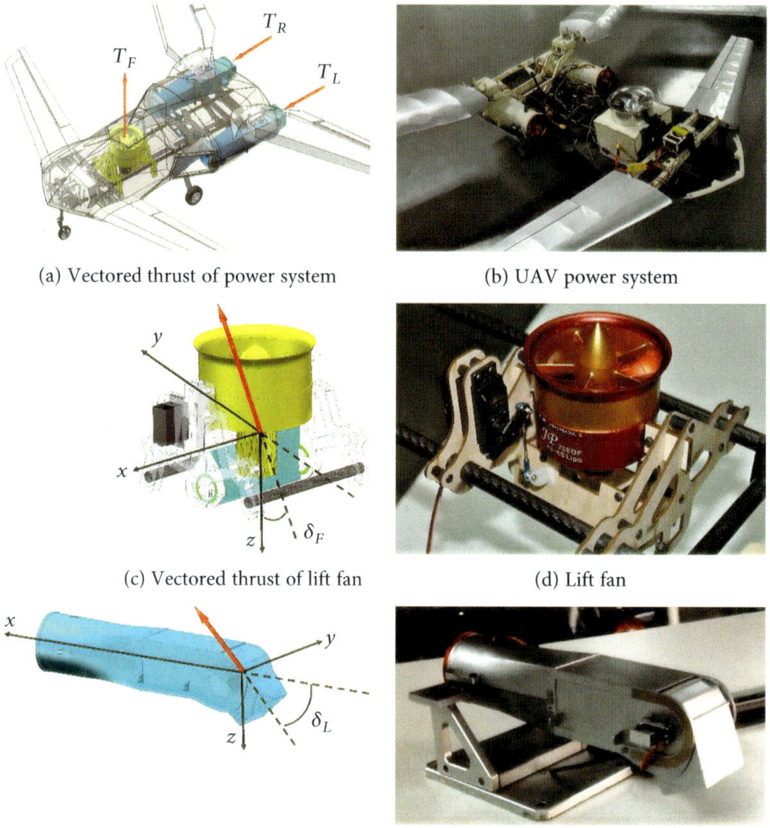

(a) Vectored thrust of power system

(b) UAV power system

(c) Vectored thrust of lift fan

(d) Lift fan

(e) Vectored thrust of engine

(f) Thrust vector engine

FIGURE 1: UAV aerodynamic and power configurations.

and speed up the control allocation computational process. The basic data of UAV is presented in Table 1.

TABLE 1: Prototype basic data.

Item	Data
Running takeoff speed	15 m/s
Cruise speed	20 m/s
Vehicle length	1 m
Rear wing span	1.57 m
Vehicle weight	5 kg
Maximum engine thrust (single)	2.6 kg
Maximum lift fan thrust	3.8 kg
Flight range	6 km

3. INDI Control System Design

This section presents the design process of the control system. The trajectory tracking controller, by adopting time-scaled method, is split into four control loops, i.e., translational kinematics (position) control loop, translational dynamics (flight path) control loop, rotational kinematics (attitude) control loop, and rotational dynamics (angular rate) control loop. Given the differences of the inner and outer loops in the time constants, the control laws for inner and outer loops can be designed independently [19, 20]. Considering that the model uncertainties caused by aerodynamic force and moments merely exist in translational dynamics and rotational dynamics control loops, these two control loops are designed by the INDI method while the other two loops are designed by the NDI method. The equations of the 6-DOF flight dynamic model and thrust vector engine model adopted to design the control system are described in this section.

3.1. Translational Kinematic Control Loop. The reference flight path vector is calculated in translational kinematic control loop based on the target trajectory. The position vector and flight path vector of the aircraft are defined as

$$\mathbf{x}_0 = [x \;\; y \;\; z]^{\mathrm{T}},$$
$$\mathbf{x}_1 = [V \quad \chi \quad \gamma]^{\mathrm{T}}, \tag{1}$$

where x, y, and z represent the positions of the aircraft in the North, East, and down directions, respectively. V denotes the total velocity of the aircraft, χ indicates the kinematic azimuth angle, and γ represents the flight path angle. The equation of translational kinematics is expressed as

$$\dot{\mathbf{x}}_0 = \mathbf{g}_0(\mathbf{x}_1) = \begin{bmatrix} V\cos\chi\cos\gamma \\ V\sin\chi\cos\gamma \\ -V\sin\gamma \end{bmatrix}. \tag{2}$$

The desired derivatives of position vector can be designed by a classical linear controller according to the reference trajectory and the feedback of the vehicle's position.

$$\dot{\mathbf{x}}_0^{\text{des}} = \dot{\mathbf{x}}_0^{\text{ref}} + \mathbf{K}_0\left(\mathbf{x}_0 - \mathbf{x}_0^{\text{ref}}\right). \tag{3}$$

In (3), $\mathbf{K}_0 = \begin{bmatrix} K_x & K_y & K_z \end{bmatrix}^{\text{T}}$ are the gains of the position linear controller. For a better explanation, the definitions are given below: the variables with superscript "des" denote the desired commands generated by the linear controller, and the variables with superscript "ref" denote the reference commands given for the controller to follow. The translational kinematic control loop contains no model uncertainties, and the NDI method is adopted to calculate the control input. The reference flight path command for the next control loop is calculated in

$$\mathbf{x}_1^{\text{ref}} = \begin{bmatrix} \dfrac{-\dot{Z}_{\text{des}}}{\sin\left(\arctan\left(-\left(\dot{Z}_{\text{des}}/\dot{Y}_{\text{des}}\right)\sin\left(\arctan\left(\dot{Y}_{\text{des}}/\dot{X}_{\text{des}}\right)\right)\right)\right)} \\ \arctan\dfrac{\dot{Y}_{\text{des}}}{\dot{X}_{\text{des}}} \\ \arctan\left(-\dfrac{\dot{Z}_{\text{des}}}{\dot{Y}_{\text{des}}}\sin\left(\arctan\dfrac{\dot{Y}_{\text{des}}}{\dot{X}_{\text{des}}}\right)\right) \end{bmatrix}. \tag{4}$$

3.2. Translational Dynamic Control Loop

3.2.1. INDI Control Law Design. In translational dynamic control loop, the vehicle's flight path status $\mathbf{x}_1$ is controlled by aerodynamic force and vectored thrust together to track the reference command $\mathbf{x}_1^{\text{ref}}$ obtained by the last control loop. Define the control variables as

$$\begin{aligned} \mathbf{x}_2 &= \begin{bmatrix} \alpha & \beta & \mu & T_x & T_y & T_z \end{bmatrix}^{\text{T}}, \\ \mathbf{x}_{2a} &= \begin{bmatrix} \alpha & \beta & \mu \end{bmatrix}^{\text{T}}, \\ \mathbf{x}_{2t} &= \begin{bmatrix} T_x & T_y & T_z \end{bmatrix}^{\text{T}}. \end{aligned} \tag{5}$$

In this loop, there are totally 6 control inputs. Among them, $\mathbf{x}_{2a}$ represent the aerodynamic control variables, where μ is the bank angle, and α and β are the angle of attack and angle of sideslip, respectively. $\mathbf{x}_{2t}$ represent the vectored thrusts, and T_x, T_y, and T_z are the components of the thrust in the body axis. The equation of translational dynamics is expressed as

$$\dot{\mathbf{x}}_1 = \mathbf{f}_1(\mathbf{x}_1) + \mathbf{g}_1(\mathbf{x}_1, \mathbf{x}_2), \tag{6}$$

$$\mathbf{f}_1(\mathbf{x}_1) = \begin{bmatrix} \dfrac{1}{m} & 0 & 0 \\ 0 & \dfrac{1}{mV\cos\gamma} & 0 \\ 0 & 0 & \dfrac{-1}{mV} \end{bmatrix} \mathbf{L}_{kg} \begin{bmatrix} 0 \\ 0 \\ mg \end{bmatrix}, \tag{7}$$

$$\mathbf{g}_1(\mathbf{x}_1, \mathbf{x}_2) = \begin{bmatrix} \dfrac{1}{m} & 0 & 0 \\ 0 & \dfrac{1}{mV\cos\gamma} & 0 \\ 0 & 0 & \dfrac{-1}{mV} \end{bmatrix} \mathbf{L}_{kb}(\alpha, \beta, \mu)(\mathbf{x}_{2t} + \mathbf{L}_{ka}(\mu)\mathbf{F}_A). \tag{8}$$

In (7) and (8), $\mathbf{L}_{kg}$ is the direction cosine matrix from earth axis to flight path axis, $\mathbf{L}_{kb}(\alpha, \beta, \mu)$ is the direction cosine matrix from body axis to flight path axis, and $\mathbf{L}_{ka}(\mu)$ is the direction cosine matrix from wind axis to flight path axis. $\mathbf{F}_A = \left[-D(\alpha) \quad C(\beta) \quad -L(\alpha)\right]^{\text{T}}$ is the aerodynamic force in which D represents drag, C represents aerodynamic side force, and L represents lift. $\mathbf{F}_A$ and $\mathbf{L}_{kb}(\alpha, \beta, \mu)$ and $\mathbf{L}_{ka}(\mu)$ are assumed as the functions of $\mathbf{x}_{2a}$. The multiplication of direction cosine matrixes causes severe coupling in control variables. Furthermore, in translational dynamic control loop, the error caused by the estimation of $\mathbf{F}_A$ will bring uncertainties in the control system. Therefore, the INDI method is adopted to address the control problem. To rewrite (6) in incremental form, it is defined that $\mathbf{x}_{1n}$ and $\mathbf{x}_{2n}$ represent the values of $\mathbf{x}_1$ and $\mathbf{x}_2$ in the next time step, and their relationships are described below:

$$\begin{aligned} \mathbf{x}_{1n} &= \mathbf{x}_1 + \Delta\mathbf{x}_1, \\ \mathbf{x}_{2n} &= \mathbf{x}_2 + \Delta\mathbf{x}_2. \end{aligned} \tag{9}$$

Applying Taylor expansion to $\dot{\mathbf{x}}_{1n}$ at $(\mathbf{x}_1, \mathbf{x}_2)$ and with higher-order terms neglected, the result can be expressed as

$$\begin{aligned} \dot{\mathbf{x}}_{1n} = \dot{\mathbf{x}}_1 &+ \frac{\partial \mathbf{f}_1(\mathbf{x}_1)}{\partial \mathbf{x}_1}(\mathbf{x}_{1n} - \mathbf{x}_1) + \frac{\partial \mathbf{g}_1(\mathbf{x}_1, \mathbf{x}_2)}{\partial \mathbf{x}_1}(\mathbf{x}_{1n} - \mathbf{x}_1) \\ &+ \frac{\partial \mathbf{g}_1(\mathbf{x}_1, \mathbf{x}_2)}{\partial \mathbf{x}_2}(\mathbf{x}_{2n} - \mathbf{x}_2). \end{aligned} \tag{10}$$

In (10), the second and third terms partial to $\mathbf{x}_1$ are assumed much smaller than the forth term, partial to $\mathbf{x}_2$. This commonly arises from the principle of time scale separation [20]. For simplification, it is approximated that

$$\frac{\partial \mathbf{g}_1(\mathbf{x}_1,\mathbf{x}_2)}{\partial \mathbf{x}_2}(\mathbf{x}_{2n}-\mathbf{x}_2)=\frac{\mathbf{g}_1(\mathbf{x}_1,\mathbf{x}_2+\Delta\mathbf{x}_2)-\mathbf{g}_1(\mathbf{x}_1,\mathbf{x}_2)}{\Delta\mathbf{x}_2}(\mathbf{x}_{2n}-\mathbf{x}_2)$$
$$=\mathbf{g}_1(\mathbf{x}_1,\mathbf{x}_2+\Delta\mathbf{x}_2)-\mathbf{g}_1(\mathbf{x}_1,\mathbf{x}_2). \tag{11}$$

Replace $\dot{\mathbf{x}}_{1n}$ with ideal derivatives of flight path vector $\dot{\mathbf{x}}_1^{\mathrm{des}}$, which is calculated by the linear controller. The incremental control equation can be denoted as

$$\dot{\mathbf{x}}_1^{\mathrm{des}}-\dot{\mathbf{x}}_1=\mathbf{g}_1(\mathbf{x}_1,\mathbf{x}_2+\Delta\mathbf{x}_2)-\mathbf{g}_1(\mathbf{x}_1,\mathbf{x}_2). \tag{12}$$

To simplify the computation process, it is assumed that

$$\Delta x_{2i}\Delta x_{2j}=0, \tag{13}$$

where x_{2i} and x_{2j} represent the elements in vector $\mathbf{x}_2$. Accordingly, the equation can be rewritten into the affine-in-control form

$$\dot{\mathbf{x}}_1^{\mathrm{des}}-\dot{\mathbf{x}}_1=\begin{bmatrix}\mathbf{g}_{1a} & 0\\ 0 & \mathbf{g}_{1t}\end{bmatrix}\begin{bmatrix}\Delta\mathbf{x}_{2a}\\ \Delta\mathbf{x}_{2t}\end{bmatrix}, \tag{14}$$

where $\mathbf{g}_{1a}$ and $\mathbf{g}_{1t}$ are the 3×3 matrixes, representing the aerodynamic control matrix and thrust vector control matrix. The detail information of $\mathbf{g}_{1a}$ and $\mathbf{g}_{1t}$ is shown in Appendix. Compared with a traditional aircraft in incremental approach, the thrust vector control matrix $\mathbf{g}_{1t}$ is additional. The elements in $\mathbf{g}_{1t}$ are polynomials of control variables in $\mathbf{x}_{2a}$. Thus, the control system is transferred into a linear and time variant system, and the increment of control variables is decoupled. Then the ideal incremental virtual command $\Delta\mathbf{x}_2$ can be calculated by

$$\begin{bmatrix}\Delta\mathbf{x}_{2a}\\ \Delta\mathbf{x}_{2t}\end{bmatrix}=\begin{bmatrix}\mathbf{g}_{1a} & 0\\ 0 & \mathbf{g}_{1t}\end{bmatrix}^{\ddagger}\left(\dot{\mathbf{x}}_1^{\mathrm{des}}-\dot{\mathbf{x}}_1\right). \tag{15}$$

In (15), the superscript "‡" represents the dynamic weight pseudoinverse method, which will be introduced in the next section. The reference command for the next control loop is obtained by

$$\mathbf{x}_2^{\mathrm{ref}}=\begin{bmatrix}\mathbf{x}_{2a}\\ \mathbf{x}_{2t}\end{bmatrix}+\begin{bmatrix}\Delta\mathbf{x}_{2a}\\ \Delta\mathbf{x}_{2t}\end{bmatrix}. \tag{16}$$

3.2.2. Flight Path Vector Derivative Acquisition. In (12), (14), and (15), $\dot{\mathbf{x}}_1$ can be derived by V, χ, and γ, which are measured by onboard sensors. With the $\dot{\mathbf{x}}_1^{\mathrm{des}}-\dot{\mathbf{x}}_1$ introduced, the term $\mathbf{f}_1(\mathbf{x}_1)$ is cancelled, which is the reason why INDI is referred to a sensor-based approach. This approach transfers the dependence on the model accuracy into the dependence on the sensor accuracy. In most cases, the signals were measured by sensor contain noise, and the differentiation of noisy signal amplifies the noise. To make results accurate, a filter can be adopted to abate the noise in sensor data. In this paper, a second-order filter is employed. As stated in literatures [16, 18, 21], the washout filter can be expressed in Laplace domain as

$$H(s)=\frac{\omega_n^2}{s^2+2\zeta\omega_n s+\omega_n^2}. \tag{17}$$

However, the filter leads to a delay which should be compensated. In the Taylor expansion shown in (10), $\dot{\mathbf{x}}_1$, $\mathbf{x}_1$, and $\mathbf{x}_2$ should be from the same moment. In this regard, a second-order filter is also applied for $\mathbf{x}_2$ to counteract the impact caused by time delay in $\mathbf{x}_1$ and $\dot{\mathbf{x}}_1$, and (12) can be rewritten in a time-synchronized form

$$\dot{\mathbf{x}}_1^{\mathrm{des}}-\dot{\mathbf{x}}_{1f}=\mathbf{g}_1\left(\mathbf{x}_{1f},\mathbf{x}_{2f}+\Delta\mathbf{x}_2\right)-\mathbf{g}_1\left(\mathbf{x}_{1f},\mathbf{x}_{2f}\right), \tag{18}$$

where subscript "f" represents the filtered variable. Alternatively, other methods can also deal with the measurement delay problem, such as predictive filtering proposed by Sieberling et al. [14]. However, the prediction requires additional modeling and cannot predict disturbances. The final reference command for $\mathbf{x}_2^{\mathrm{ref}}$ can be given by

$$\mathbf{x}_2^{\mathrm{ref}}=\mathbf{x}_{2f}+\Delta\mathbf{x}_2. \tag{19}$$

3.2.3. Uncertainty Analysis. The model uncertainty stemmed from aerodynamic force in translational dynamic control loop is analyzed in this subsection. The change of aerodynamic coefficients is assumed to be primarily caused by the angle of attack (α) and sideslip angle (β) for simplification. The aerodynamic coefficient can be expressed as

$$\begin{cases}C_D=\bar{C}_{D0}+C_{D\alpha}\alpha,\\ C_C=\bar{C}_{C0}+C_{C\beta}\beta,\\ C_L=\bar{C}_{L0}+C_{L\alpha}\alpha.\end{cases} \tag{20}$$

In (20), C_D, C_C, and C_L represent the total drag coefficient, side force coefficient, and lift coefficient, respectively; $\bar{C}_{D0}$ represents the sum of drag coefficients without the part contributed by α; $\bar{C}_{C0}$ represents the sum of side force coefficients without the part contributed by β; $\bar{C}_{L0}$ represents the sum of lift coefficients without the part contributed by α; and $C_{D\alpha}$, $C_{C\beta}$, and $C_{L\alpha}$ represent derivative coefficients about α and β. These coefficients vary with the flight condition and can be assumed as constants in the calculation of each control period. The methods to obtain these parameters involve mathematic estimation and CFD analysis, as described in Section 2. The errors between the estimate coefficients and accurate coefficients make the model uncertain and assume that the uncertainty is mainly caused by $\bar{C}_{D0}$, $\bar{C}_{C0}$, and $\bar{C}_{L0}$. As the work by Lu et al. [18] indicates, the kinematic roll angle (μ) can be calculated directly under the assumption of $\beta = 0$, which largely simplifies the control equation. Additionally, coefficients $\bar{C}_{D0}$, $\bar{C}_{C0}$, and $\bar{C}_{L0}$ can be eliminated by adopting INDI, which evidently reduces model uncertainty. However, the UAV in this paper with direct force control owns great lateral maneuverability. Therefore, β cannot be approximated as zero, and μ is calculated through solving the control equation. As a result, coefficients $\bar{C}_{D0}$, $\bar{C}_{C0}$, and $\bar{C}_{L0}$ arise in the third column of $\mathbf{g}_{1a}$ and cannot be eliminated. Nevertheless, no uncertainty exists in the thrust vector control matrix $\mathbf{g}_{1t}$.

The application of INDI in the translational dynamics control loop has two primary advantages. First, INDI control law restrains the model uncertainty caused by aerodynamic force in control matrix $\mathbf{g}_{1a}$ and isolates its impact on direct force incremental control. Second, INDI can decouple the control variables in $\mathbf{g}_1(\mathbf{x}_1, \mathbf{x}_2)$ and transfer the control equation into a linear form. Accordingly, the aerodynamic force and vectored thrust control allocation are simplified, the complex numerical solution of nonlinear coupling control is avoided, and computational load of onboard control system is reduced.

3.3. Rotational Kinematic Control Loop. The objective in this control loop is to track $\mathbf{x}_{2a}^{\text{ref}}$ given in the translational dynamics control loop. The angular rate vector of the vehicle is defined as

$$\mathbf{x}_3 = \begin{bmatrix} p & q & r \end{bmatrix}^{\mathrm{T}}. \tag{21}$$

In (21), p, q, and r represent the roll, pitch, and yaw rates in body axis, respectively. No model uncertainty exists in this control loop, and the control law based on the standard NDI approach is established below:

$$\begin{aligned} \dot{\mathbf{x}}_{2a} &= \mathbf{f}_2(\mathbf{x}_{2a}) + \mathbf{g}_2(\mathbf{x}_{2a})\mathbf{x}_3, \\ \mathbf{f}_2(\mathbf{x}_{2a}) &= \frac{1}{\cos\beta}\begin{bmatrix} -\cos\gamma & -\sin\mu\cos\gamma \\ -\sin\mu\cos\beta & \cos\mu\cos\gamma\cos\beta \\ \cos\mu\sin\beta & \sin\gamma\cos\beta + \sin\mu\cos\gamma\sin\beta \end{bmatrix}\begin{bmatrix} \dot{\gamma} \\ \dot{\chi} \end{bmatrix}, \\ \mathbf{g}_2(\mathbf{x}_{2a}) &= \frac{1}{\cos\beta}\begin{bmatrix} -\cos\alpha\sin\beta & \cos\beta & -\sin\alpha\sin\beta \\ \sin\alpha\cos\beta & 0 & -\cos\alpha\cos\beta \\ \cos\alpha & 0 & \sin\alpha \end{bmatrix}. \end{aligned} \tag{22}$$

It is noteworthy that $\dot{\gamma}$, $\dot{\chi}$, and μ are not directly measured onboard in this control system. They are calculated by (23) derived from (6).

$$\begin{aligned} \dot{\gamma} &= \frac{(\cos\alpha\sin\beta\sin\mu + \sin\alpha\cos\mu)A_x - \cos\beta\sin\mu A_y + (\sin\alpha\sin\beta\sin\mu - \cos\alpha\cos\mu)A_z}{V}, \\ \dot{\chi} &= \frac{(-\cos\alpha\sin\beta\cos\mu + \sin\alpha\sin\mu)A_x + \cos\beta\cos\mu A_y + (-\sin\alpha\sin\beta\cos\mu - \cos\alpha\sin\mu)A_z}{V\cos\gamma}, \end{aligned} \tag{23}$$

where A_x, A_y, and A_z are directly measured by the accelerometers of the aircraft. The reason why $\dot{\gamma}$ and $\dot{\chi}$ are not acquired by the same method adopted in translational dynamic control loop is that the filter will delay the measurement and cause errors from accurate results. The reference angular rates can be obtained by

$$\mathbf{x}_3^{\text{ref}} = \mathbf{g}_2(\mathbf{x}_{2a})^{-1}\left(\dot{\mathbf{x}}_{2a}^{\text{des}} - \mathbf{f}_2(\mathbf{x}_{2a})\right), \tag{24}$$

where $\dot{\mathbf{x}}_{2a}^{\text{des}}$ is the desired command and designed by linear controller, which is similar to the one adopted to design $\dot{\mathbf{x}}_1^{\text{des}}$.

3.4. Rotational Dynamic Control Loop. $\mathbf{x}_4$ is defined as the control moments acting on the vehicle, and it can be denoted as

$$\mathbf{x}_4 = \mathbf{x}_{4s} + \mathbf{x}_{4t} = \begin{bmatrix} l_c & m_c & n_c \end{bmatrix}^{\mathrm{T}}, \tag{25}$$

where $\mathbf{x}_{4s}$ denote the moments generated by aerodynamic control surface; $\mathbf{x}_{4t}$ indicate the moments generated by vectored thrust; and l_c, m_c, and n_c represent the rolling, pitching, and yawing control moments, respectively. The dynamics of the angular rates of the vehicle can be expressed into the following affine-in-control form

$$\begin{aligned} \dot{\mathbf{x}}_3 &= \mathbf{f}_3(\mathbf{x}_3) + \mathbf{J}^{-1}\mathbf{x}_4, \\ \mathbf{f}_3(\mathbf{x}_3) &= \mathbf{J}^{-1}(\mathbf{M}_a - \mathbf{x}_3 \times \mathbf{J}\mathbf{x}_3), \end{aligned} \tag{26}$$

where $\mathbf{J}$ denotes the inertia matrix and $\mathbf{M}_a$ refers to the aerodynamic moments generated by derivatives unrelated to control surface deflections.

$$\mathbf{M}_a = \bar{q}S\begin{bmatrix} \bar{C}_{l0} \\ \bar{C}_{m0} \\ \bar{C}_{n0} \end{bmatrix}. \tag{27}$$

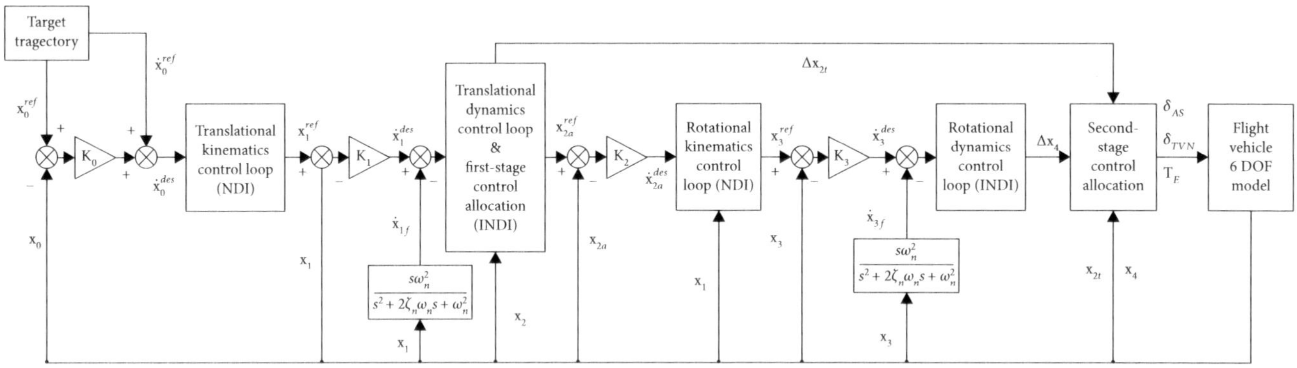

FIGURE 2: Conceptual block diagram of the control system.

Like aerodynamic force, the aerodynamic coefficients can also be denoted as

$$\begin{cases} C_l = \bar{C}_{l0} + C_{l\delta_a}\delta_a + C_{l\delta_r}\delta_r, \\ C_m = \bar{C}_{m0} + C_{m\delta_e}\delta_e, \\ C_n = \bar{C}_{n0} + C_{n\delta_a}\delta_a + C_{n\delta_r}\delta_r. \end{cases} \quad (28)$$

In (27) and (28), C_l, C_m, and C_n represent the total 3-axis aerodynamic moment coefficients, respectively; $\bar{C}_{l0}$ represents the sum of aerodynamic rolling moment coefficients without the part contributed by δ_a and δ_r; $\bar{C}_{m0}$ represents the sum of aerodynamic pitching moment coefficients without the part contributed by δ_e; $\bar{C}_{n0}$ represents the sum of aerodynamic yawing moment without the part contributed by δ_a and δ_r; and $C_{l\delta_a}$, $C_{l\delta_r}$, $C_{m\delta_e}$, $C_{n\delta_a}$, and $C_{n\delta_r}$ represent derivative aerodynamic moment coefficients about δ_a, δ_e, and δ_r.

The aerodynamic control moments are denoted as

$$\mathbf{x}_{4s} = \bar{q}S\mathbf{b}_1\mathbf{C}_{AS}\boldsymbol{\delta}_{AS},$$
$$\mathbf{C}_{AS} = \begin{bmatrix} C_{l\delta_a} & 0 & C_{l\delta_r} \\ 0 & C_{m\delta_e} & 0 \\ C_{n\delta_a} & 0 & C_{n\delta_r} \end{bmatrix}, \quad (29)$$
$$\mathbf{b}_1 = \mathrm{diag}\,(b, \bar{c}, b),$$

where $\bar{q}$ denotes the dynamic pressure, S denotes the reference wing area, b denotes the wing span, and $\bar{c}$ denotes the mean aerodynamic chord. $\boldsymbol{\delta}_{AS} = [\delta_a \quad \delta_e \quad \delta_r]^{\mathrm{T}}$ are control surface deflections. The control moment coefficients $\mathbf{C}_{AS}$ are assumed to be accurate in aerodynamic estimation, and the uncertainties are considered to be caused principally by $\bar{C}_{l0}$, $\bar{C}_{m0}$, and $\bar{C}_{n0}$. Rewriting (26) into the incremental form, as presented in (30), the term $\mathbf{f}_3(\mathbf{x}_3)$ can be cleared up.

$$\Delta\mathbf{x}_4 = \mathbf{J}\left(\dot{\mathbf{x}}_3^{\mathrm{des}} - \dot{\mathbf{x}}_{3f}\right). \quad (30)$$

In (30), $\dot{\mathbf{x}}_{3f}$ represents the derivative of $\mathbf{x}_3$, which is the filtered signal collected by the sensor. In the absence of $\mathbf{f}_3(\mathbf{x}_3)$, the uncertainty caused by $\mathbf{M}_a$ is eliminated and the nonlinear cross couplings of the angular rate term is also cancelled. The reference control moments can be calculated as

$$\mathbf{x}_4^{\mathrm{ref}} = \mathbf{x}_{4f} + \Delta\mathbf{x}_4. \quad (31)$$

After the calculation of four control loops, $\Delta\mathbf{x}_{2t}$ and $\Delta\mathbf{x}_4$ will be output as virtual commands to the second-layer control allocation. The control allocation of the engine thrust, vectoring nozzles, and aerodynamic control surface deflections will be introduced in the next section. The block diagram of the control system is illustrated in Figure 2.

4. Two-Layer Cascaded Optimal Control Allocation

The allocation method of redundant control variables (elements in $\mathbf{x}_2$ and $\mathbf{x}_4$) in translational dynamics and rotational dynamics control loops is introduced in this section. Given this, the allocation result of $\mathbf{x}_2$ will affect $\mathbf{x}_4$, and the allocation results of power system actuators T_L, T_R, T_F δ_R, δ_L, and δ_F are determined by $\mathbf{x}_{2t}$ and $\mathbf{x}_{4t}$ synthetically. A two-layer cascade optimal control allocation method is designed to address UAV's control allocation based on INDI control law.

4.1. First-Layer Trajectory Incremental Control Allocation. The first-layer control allocation is conducted to allocate the increment of flight attitude $\Delta\mathbf{x}_{2a}$ and vectored thrust $\Delta\mathbf{x}_{2t}$ in trajectory control.

4.1.1. Incremental Pseudoinverse Method. In translational dynamics control loop, the control equation is linearized by the INDI method. Accordingly, the dynamic weight pseudoinverse method can be used in control allocation. Based on (14), it is defined as

$$\mathbf{B} = \begin{bmatrix} \mathbf{g}_{1a} & 0 \\ 0 & \mathbf{g}_{1t} \end{bmatrix},$$
$$\mathbf{v} = \dot{\mathbf{x}}_1^{\mathrm{des}} - \dot{\mathbf{x}}_{1f}, \quad (32)$$

where $\dot{\mathbf{x}}_{1f}$ represents the derivative of $\mathbf{x}_1$, which is the filtered signal collected by the sensor. As the filter will delay the time in $\dot{\mathbf{x}}_{1f}$, to keep all signals synchronized in control equations, the signal of $\mathbf{x}_2$ is also required to be filtered to counteract the error generated by the delay of $\dot{\mathbf{x}}_{1f}$. Regardless of the control variable constrains, the optimal control allocation can be denoted as

$$\min_{\Delta\mathbf{x}_2} \quad \left\|\mathbf{W}_1\left(\Delta\mathbf{x}_2 + \mathbf{x}_{2f}\right)\right\|^2 + \left\|\mathbf{W}_2\Delta\mathbf{x}_2\right\|^2 \tag{33}$$

$$\text{s.t.} \quad \mathbf{B}\Delta\mathbf{x}_2 = \mathbf{v}, \tag{34}$$

where $\Delta\mathbf{x}_2$ denotes the optimization variable and $\mathbf{x}_{2f}$ and $\mathbf{v}$ are considered constants, they are all 3×1 vectors, and $\mathbf{W}_1$ and $\mathbf{W}_2$ indicate the diagonal matrixes of weight, which are generated by dynamic weight strategy, and will be discussed later. The first term of objective function, as shown in (33), is adopted to control the scale of control variables in $\mathbf{x}_2$, while the second term is employed to limit the change rates of the control variables. Equation (33) can be expanded as

$$\begin{aligned}\left\|\mathbf{W}_1\left(\Delta\mathbf{x}_2 + \mathbf{x}_{2f}\right)\right\|^2 + \left\|\mathbf{W}_2\Delta\mathbf{x}_2\right\|^2 &= \left(\Delta\mathbf{x}_2 + \mathbf{x}_{2f}\right)^T\mathbf{W}_1^2\left(\Delta\mathbf{x}_2 + \mathbf{x}_{2f}\right) + \mathbf{x}^T\mathbf{W}_2^2\Delta\mathbf{x}_2\\ &= \Delta\mathbf{x}_2^T\mathbf{W}_1^2\Delta\mathbf{x}_2 + \Delta\mathbf{x}_2^T\mathbf{W}_1^2\mathbf{x}_{2f} + \mathbf{x}_{2f}^T\mathbf{W}_1^2\Delta\mathbf{x}_2\\ &\quad + \mathbf{x}_{2f}^T\mathbf{W}_1^2\mathbf{x}_{2f} + \Delta\mathbf{x}_2^T\mathbf{W}_2^2\Delta\mathbf{x}_2.\end{aligned} \tag{35}$$

In (35), $\Delta{\mathbf{x}_2}^T\mathbf{W}_1^2\mathbf{x}_{2f} = \mathbf{x}_{2f}^T\mathbf{W}_1^2\Delta\mathbf{x}_2$ and $\mathbf{x}_{2f}^T\mathbf{W}_1^2\mathbf{x}_{2f}$ are constants which can be calculated directly. Accordingly, it is deduced that (33) has the same minimizing argument as

$$\Delta\mathbf{x}_2^T\mathbf{W}_1^2\Delta\mathbf{x}_2 + 2\Delta\mathbf{x}_2^T\mathbf{W}_1^2\mathbf{x}_{2f} + \Delta\mathbf{x}_2^T\mathbf{W}_2^2\Delta\mathbf{x}_2. \tag{36}$$

Define

$$\begin{aligned}\mathbf{W} &= \sqrt{\mathbf{W}_1^2 + \mathbf{W}_2^2},\\ \mathbf{x}_{20} &= \mathbf{W}^{-2}\left(\mathbf{W}_1^2\mathbf{x}_{2f}\right).\end{aligned} \tag{37}$$

Substitute (37) into (36) and incorporate with constant $\mathbf{x}_{20}^T\mathbf{W}^2\mathbf{x}_{20}$.

$$\begin{aligned}\Delta\mathbf{x}_2^T\mathbf{W}^2\Delta\mathbf{x}_2 + 2\Delta\mathbf{x}_2^T\mathbf{W}^2\mathbf{x}_{20} + \mathbf{x}_{20}^T\mathbf{W}^2\mathbf{x}_{20} &= \left(\Delta\mathbf{x}_2 + \mathbf{x}_{20}\right)^T\mathbf{W}^2\left(\Delta\mathbf{x}_2 + \mathbf{x}_{20}\right)\\ &= \left\|\mathbf{W}\left(\Delta\mathbf{x}_2 + \mathbf{x}_{20}\right)\right\|^2.\end{aligned} \tag{38}$$

Based on the foregoing deduction, the optimal problem presented in (33) is equivalence

$$\begin{aligned}&\min_{\Delta\mathbf{x}_2} \quad \left\|\mathbf{W}\left(\Delta\mathbf{x}_2 + \mathbf{x}_{20}\right)\right\|^2\\ &\text{s.t.} \quad \mathbf{B}\Delta\mathbf{x}_2 = \mathbf{v}.\end{aligned} \tag{39}$$

On that basis, the minimum norm solution of the control allocation problem is obtained as

$$\Delta\mathbf{x}_2 = \mathbf{G}\mathbf{v} + \left(\mathbf{G}\mathbf{B} - \mathbf{I}\right)\mathbf{x}_{20}, \tag{40}$$

$$\mathbf{G} = \mathbf{W}^{-1}\left(\mathbf{B}\mathbf{W}^{-1}\right)^{\dagger}. \tag{41}$$

In (41), superscript "†" represents the pseudoinverse operation for the matrix. In order to make the first-layer control allocation an equality constrained optimization problem, the range of $\Delta\mathbf{x}_2$ is not limited. Then an analytical solution is obtained, and numerical iterations are avoided, which is suitable for onboard use [22]. With the allocation results of $\Delta\mathbf{x}_{2a}$, the $\mathbf{x}_{2a}^{\text{ref}}$ can be calculated by (19). After adjusting $\mathbf{x}_{2a}^{\text{ref}}$ into the right quadrant, as shown in (42), $\mathbf{x}_{2a}^{\text{ref}}$ is transmitted into the rotational kinematics control loop.

$$\mathbf{x}_{2a}^{\text{ref}} \in \left[-0.5\pi \quad 0.5\pi\right]. \tag{42}$$

The $\Delta\mathbf{x}_{2t}$ is considered the virtual command to transmit into the second-layer control allocation, in which $\Delta\mathbf{x}_{2t}$ and $\Delta\mathbf{x}_{4t}$ are used to solve the values of power system actuators T_L, T_R, T_F δ_R, δ_L, and δ_F.

4.1.2. Dynamic Weight Strategy. In the course of control allocation, the weight is adopted to describe the differences of control variables' significance. On the basis of the significance of control variables changing with flight states and mission requirements, a dynamic weight strategy is proposed to allocate control variables optimally and properly.

The traditional determination of control variable's weight largely depends on human experience. In this regard, there is no absolute criteria for weight determination. It is commonly difficult to judge a control variable's significance globally, while the significance between every two control variables can be easily compared. In dynamic weight strategy, the analytic hierarchy process (AHP) is adopted to synthesize comparison results of every two control variables and calculate each control variable's weight by AHP-judgment matrix. The weight matrixes in the objective function can be denoted as

$$\begin{aligned}\mathbf{W}_1 &= \operatorname{diag}\left(W_\alpha, W_\beta, W_\mu, W_{Tx}, W_{Ty}, W_{Tz}\right),\\ \mathbf{W}_2 &= \operatorname{diag}\left(W_{\Delta\alpha}, W_{\Delta\beta}, W_{\Delta\mu}, W_{\Delta Tx}, W_{\Delta Ty}, W_{\Delta Tz}\right).\end{aligned} \tag{43}$$

In (43), each element in $\mathbf{W}_1$ and $\mathbf{W}_2$ represents the weight of the control variable corresponding to its subscript. According to different physical properties, control variables are divided into different sets. The

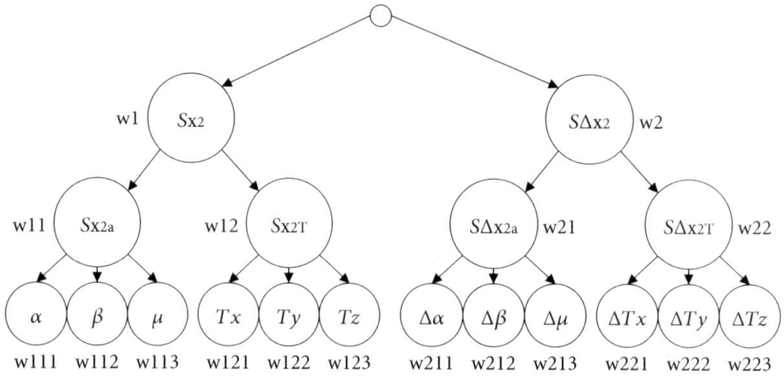

FIGURE 3: Illustration of weight structure.

definitions and subordinations, respectively, of the sets are expressed in

$$\begin{cases} S_{\mathbf{x}_2} = \{\alpha, \beta, \mu, T_x, T_y, T_z\}, \\ S_{\mathbf{x}_{2a}} = \{\alpha, \beta, \mu\}, \\ S_{\mathbf{x}_{2t}} = \{T_x, T_y, T_z\}, \\ S_{\Delta\mathbf{x}_2} = \{\Delta\alpha, \Delta\beta, \Delta\mu, \Delta T_x, \Delta T_y, \Delta T_z\}, \\ S_{\Delta\mathbf{x}_{2a}} = \{\Delta\alpha, \Delta\beta, \Delta\mu\}, \\ S_{\Delta\mathbf{x}_{2t}} = \{\Delta T_x, \Delta T_y, \Delta T_z\}, \end{cases} \quad (44)$$

$$\begin{cases} S_{\mathbf{x}_{2a}} \subseteq S_{\mathbf{x}_2}, \\ S_{\mathbf{x}_{2t}} \subseteq S_{\mathbf{x}_2}, \\ S_{\Delta\mathbf{x}_{2a}} \subseteq S_{\Delta\mathbf{x}_2}, \\ S_{\Delta\mathbf{x}_{2t}} \subseteq S_{\Delta\mathbf{x}_2}. \end{cases} \quad (45)$$

Based on the subordinations, the sets and control variables are classified into three hierarchies, known as the weight structure, as presented in Figure 3.

The weights in different hierarchies are determined, respectively, and they all obey the rules listed below:

(1) Weights in the same set are required to be normalized

$$\begin{cases} \sum_{i=1}^{2} w_i = 1, \\ \sum_{j=1}^{2} w_{ij} = 1 \quad (i = 1, 2), \\ \sum_{k=1}^{3} w_{ijk} = 1 \quad (i, j = 1, 2) \end{cases} \quad (46)$$

(2) Weights in the first and second hierarchies satisfy

$$w_1, w_2, w_{11}, w_{12}, w_{21}, w_{22} \in \{0 : 0.01 : 1\} \quad (47)$$

(3) In the third hierarchy, w_{ijk} is acquired by the APH-judgment matrix. The judgment matrix is established for each set, with elements representing the priorities of every two control variables in the same set. The processes of APH method are introduced in [23, 24]. According to the weights in each hierarchy, the final. weights in the objective function can be calculated by

$$\begin{cases} W_{1(k,k)} = w_1 \times w_{11} \times w_{11k} (k = 1, 2, 3), \\ W_{1(k+3,k+3)} = w_1 \times w_{12} \times w_{12k} (k = 1, 2, 3), \\ W_{2(k,k)} = w_2 \times w_{21} \times w_{21k} (k = 1, 2, 3), \\ W_{2(k+3,k+3)} = w_2 \times w_{22} \times w_{22k} (k = 1, 2, 3), \end{cases} \quad (48)$$

where $W_{i(k,k)}$ represents the element in weight matrix $\mathbf{W}_i$ row k column k. In practical flight, the controller is required to generate the initial value of the weight structure according to the mission requirement and dynamically adjust the value of the weight structure according to the conditions of actuator saturation. In dynamic weight strategy, the weight generator and weight regulator are designed to realize the foregoing functions.

The weight generator is designed to generate the initial value of the weight structure, in every control period before optimization. Its working principle is introduced below:

Step 1. Establish weight structures which consist of w_i, w_{ij}, and w_{ijk}.

Step 2. Artificially design and test several weight structures according to some typical flight states and mission requirements. On this basis, index weight structures with their relative flight states and mission requirements save weight structures into repository.

Step 3. Acquire weight structure corresponding to the current flight state and mission requirement through traversing the repository.

A simplified repository is established in this paper. For flight states, merely the impact of velocity is factored in, and velocities 5 m/s, 10 m/s, 15 m/s, 20 m/s, and 25 m/s count as typical flight states. In mission requirement, only the impact of direct force control level ($DFC_{level} \in [0, 1]$) is factored in, and DFC_{level} equals 0, 0.5, and 1 count as typical mission requirements. Larger DFC_{level} represents less attitude changes and more direct force control, while smaller DFC_{level} which represents less vectored thrust usage and more attitude maneuver during trajectory tracking. The repository of 3×5 weight structures employed in this paper is acquired by artificial design and computer-based simulation. In control allocation, the weight structure for the current demand is interpolated by velocity and DFC_{level}.

The weight regulator is designed to adjust the weight when control variables exceed their limitation and ensure the rationality of allocation results. The working process of weight regulator is shown below.

Step 1. Extract control variables saturated in the last control allocation according to the feedback information. The saturated control variable can be single or multiple.

Step 2. Update saturation counter. Every control variable has a corresponding saturation counter $n_{ijk} \in N\ (i, j = 1, 2,\ k = 1, 2, 3)$, which is adopted to record saturation times. If the control variable is not saturated in the last allocation, its saturation counter will be established at zero.

Step 3. Adjust the weight structure in accordance with the saturation counter results. For different hierarchies of weight structure, the adjustment strategies are different, as illustrated below:

Strategy 1. For $n_{ijk} \in [1,150]$ with the interval of every 5 count times, increase the significance of the saturated control variables relative to other control variables in the identical minimum set. On that basis, update the APH-judgment matrix and recalculate w_{ijk} in set. If more than one control variable gets saturated, the comparison between the saturated control variables in significance remains unchanged, and their significance is improved compared with those unsaturated.

Strategy 2. For $n_{ijk} \in [151,300]$, the weight (w_{ij}) in the middle hierarchy is required to be adjusted as Strategy 1 is exercised. Take the angle of attack α as an example. If α continues to be saturated with the interval of every 5 count times, increase the weight (w_{11}) of set $S_{\mathbf{x}_{2a}}$. In (47), w_{ij} is divided into 101 levels, and w_{ij} is improved to a new level for each weight adjustment. Assume $w_{11} = 0.6$ and $w_{12} = 0.4$ before the adjustment, and with the improvement of w_{11}, the values turn out to be $w_{11} = 0.61$ and $w_{12} = 0.39$.

Strategy 3. For $n_{ijk} \in [301, +\infty]$ with an interval of every 10 count times, the weight (w_i) of set ($S_{\mathbf{x}_2}$ or $S_{\Delta\mathbf{x}_2}$) is required to be adjusted, as Strategy 1 and Strategy 2 are exercised. If the weight reaches the upper limit with $w_i, w_{ij} = 1$, the weight stops increasing; if the weight reaches the lower limit with $w_i, w_{ij} = 0$, the decrement stops.

The range of saturation counter n_{ijk} and the adjustment of interval in each strategy are determined by control frequency in the weight regulator. For example, for 100 Hz control frequency, the control is allocated for every 0.01 second. Following the foregoing strategy, if a control variable is saturated for more than 3 seconds, the controller will adjust the weight for every hierarchy in the weight structure to get optimal control allocation to satisfy the current flight requirement. For the excessively large range of n_{ijk}, the weight adjustment turns out to be oversluggish. The control variables will be saturated continuously, and the vehicle will deviate from target trajectory seriously. For the excessively small range of n_{ijk}, the weight adjustment will be excessively sensitive. The fast-changing weight in objective function will cause vehicle oscillation and control divergence. The working process of the first-layer control allocation and dynamic weight strategy is illustrated in Figure 4.

4.2. Second-Layer Actuator Control Allocation. In the second-layer control allocation, the allocation of aerodynamic control surfaces (δ_a, δ_e, δ_r) and power system actuators (T_L, T_R, T_F, δ_L, δ_R, δ_F) are conducted according to the information of vectored thrust and control moments calculated in translational dynamics control loop and rotational dynamics control loops, respectively.

The dynamic model of the engine system can be denoted as

$$\mathbf{x}_{2t} = \begin{bmatrix} T_x \\ T_y \\ T_z \end{bmatrix} = \begin{bmatrix} T_L \cos\delta_L + T_R \cos\delta_R \\ T_F \sin\delta_F \\ T_L \sin\delta_L + T_R \sin\delta_R - T_F \cos\delta_F \end{bmatrix},$$
$$\mathbf{x}_{4t} = \begin{bmatrix} l_T \\ m_T \\ n_T \end{bmatrix} = \begin{bmatrix} -T_L \sin\delta_L d_{yt} + T_R \sin\delta_R d_{yt} \\ (T_L \sin\delta_L + T_L \sin\delta_L) d_{xt} + T_F \cos\delta_F d_{xf} \\ (T_L \cos\delta_L - T_R \cos\delta_R) d_{yt} + T_F \sin\delta_F d_{xf} \end{bmatrix}. \tag{49}$$

Define

$$\begin{cases} \mathbf{T}_E = [T_L \quad T_R \quad T_F]^T, \\ \boldsymbol{\delta}_{TVN} = [\delta_L \quad \delta_R \quad \delta_F]^T, \\ \boldsymbol{\delta}_{AS} = [\delta_a \quad \delta_e \quad \delta_r]^T. \end{cases} \tag{50}$$

In (49) and (50), T_F denotes lift fan thrust; T_L and T_R denote the left and right engine thrusts, respectively; δ_F

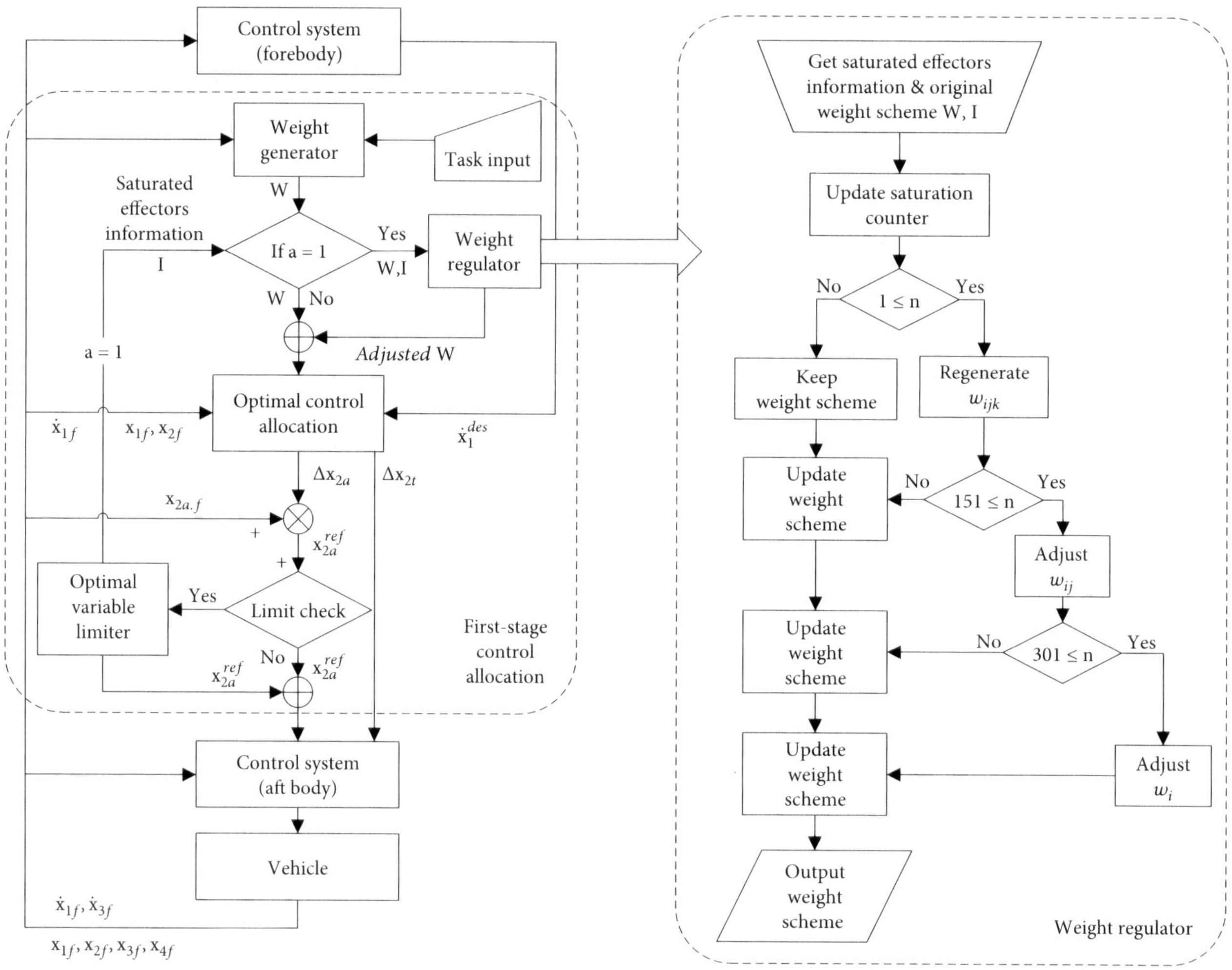

FIGURE 4: Working process of first-layer control allocation and dynamic weight strategy.

denotes the lift fan rudder deflection angle; δ_L and δ_R denote the left and right engine thrust vector nozzle deflections, respectively; d_{xf} denotes the distance from the lift fan center to the vehicle mass center in body x -axis; d_{xt} and d_{yt} denote the distance from the engine nozzle to the vehicle mass center in body x-axis and y-axis, respectively; and δ_a, δ_e, and δ_r represent the alerion, elevator, and rudder, respectively. To keep time synchronized, the output of the actuator model will go through a second-order filter to get $\mathbf{T}_{E.f}, \boldsymbol{\delta}_{\mathrm{TVN}.f}, \boldsymbol{\delta}_{AS.f}$, where subscript "$f$" denotes the filtered control output. In the second-layer control allocation, the daisy chaining method is adopted to allocate aerodynamic control moment and thrust vector moment. On that basis, the control output of the power system is calculated by $\mathbf{x}_{2t}$ and $\mathbf{x}_{4t}$. The second-layer control allocation can be conducted in the incremental linear form or normal nonlinear form, and the two methods are introduced below.

4.2.1. Incremental Linear Allocation Method. The incremental linear allocation method allocates $\Delta\mathbf{T}_E$ $\Delta\boldsymbol{\delta}_{\mathrm{TVN}}$ and $\Delta\boldsymbol{\delta}_{AS}$ according to $\Delta\mathbf{x}_4$ and $\Delta\mathbf{x}_{2t}$. The incremental daisy chaining is adopted to allocate $\Delta\mathbf{x}_{4s}$ and $\Delta\mathbf{x}_{4t}$ based on (51). In this method, the priority of $\Delta\mathbf{x}_{4s}$ outstrips $\Delta\mathbf{x}_{4t}$, and thus, the thrust vector moment is adopted only when the aerodynamic control moment gets saturated.

$$\Delta\mathbf{x}_4 = \Delta\mathbf{x}_{4s} + \Delta\mathbf{x}_{4t}. \tag{51}$$

The increment of aerodynamic control surface is calculated by

$$\Delta\boldsymbol{\delta}_{AS} = \frac{\Delta\mathbf{x}_{4s}}{\bar{q}S\mathbf{b}_1\mathbf{C}_{AS}}. \tag{52}$$

The reference value of aerodynamic control surface can be expressed as

$$\boldsymbol{\delta}_{AS}^{ref} = \boldsymbol{\delta}_{AS.f} + \Delta\boldsymbol{\delta}_{AS}. \tag{53}$$

On that basis, introduce (49) into the incremental form, as shown in (54), in which the incremental

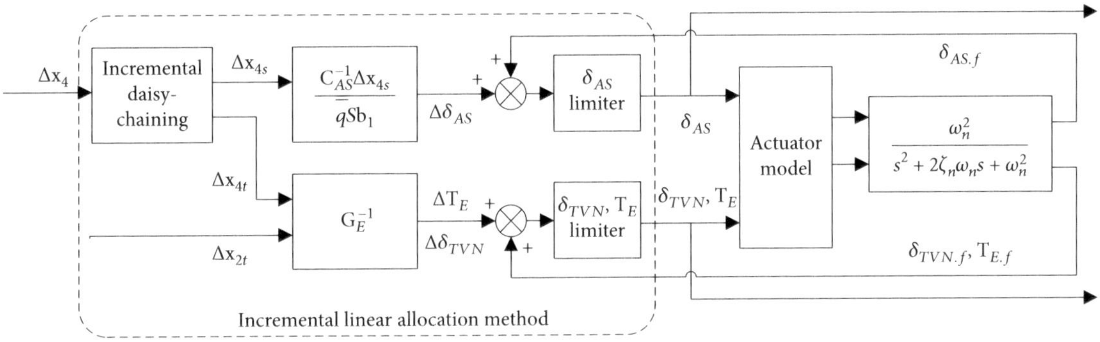

FIGURE 5: Second-layer incremental linear allocation method.

multiplication is considered a high-order small quantity and can be ignored.

$$\begin{bmatrix} \Delta\mathbf{x}_{2t} \\ \Delta\mathbf{x}_{4t} \end{bmatrix} = \mathbf{G}_E \begin{bmatrix} \Delta\mathbf{T}_E \\ \Delta\boldsymbol{\delta}_{\mathrm{TVN}} \end{bmatrix}, \tag{54}$$

$$\mathbf{G}_E = \begin{bmatrix} \cos\delta_L & \cos\delta_R & 0 & -T_L\sin\delta_L & -T_R\sin\delta_R & 0 \\ 0 & 0 & \sin\delta_F & 0 & 0 & T_F\cos\delta_F \\ \sin\delta_L & \sin\delta_R & -\cos\delta_F & T_L\cos\delta_L & T_R\cos\delta_R & T_F\sin\delta_F \\ -\sin\delta_L d_{yt} & \sin\delta_R d_{yt} & 0 & -T_L\cos\delta_L d_{yt} & T_R\cos\delta_R d_{yt} & 0 \\ \sin\delta_L d_{xt} & \sin\delta_R d_{xt} & \cos\delta_F d_{xf} & T_L\cos\delta_L d_{xt} & T_R\cos\delta_R d_{xt} & -T_F\sin\delta_F d_{xf} \\ \cos\delta_L d_{yt} & -\cos\delta_R d_{yt} & \sin\delta_F d_{xf} & -T_L\sin\delta_L d_{yt} & T_R\sin\delta_R d_{yt} & T_F\cos\delta_F d_{xf} \end{bmatrix}. \tag{55}$$

The thrust vector engine model is transformed from a nonlinear system into a linear system, and $\Delta\mathbf{T}_E$ and $\Delta\boldsymbol{\delta}_{TVN}$ can be calculated directly by matrix inversion through the incremental linear allocation, as shown in (56). This method can evidently expedite allocation without iterative solution of nonlinear equations. However, like (14), when UAV is in high maneuver flight, the increment of control variables will be large, and the neglection of incremental multiplication will cause approximation errors. In this regard, this method is appropriate for minor or moderate maneuverable flights.

$$\begin{bmatrix} \Delta\mathbf{T}_E \\ \Delta\boldsymbol{\delta}_{\mathrm{TVN}} \end{bmatrix} = \mathbf{G}_E^{-1} \begin{bmatrix} \Delta\mathbf{x}_{2t} \\ \Delta\mathbf{x}_{4t} \end{bmatrix}. \tag{56}$$

The reference thrust vector engine actuators are calculated by (57).

$$\begin{aligned} \mathbf{T}_E^{\mathrm{ref}} &= \mathbf{T}_{E.f} + \Delta\mathbf{T}_E, \\ \boldsymbol{\delta}_{\mathrm{TVN}}^{\mathrm{ref}} &= \boldsymbol{\delta}_{\mathrm{TVN}.f} + \Delta\boldsymbol{\delta}_{\mathrm{TVN}}. \end{aligned} \tag{57}$$

The range of T_L, T_R, T_F and δ_L, δ_R, δ_F is limited on the basis of power system, as shown in (58). Eventually, the control signals are output to the aerodynamic control surfaces and power system.

$$\begin{cases} T_i = T_{i\,\max}, & \text{if } T_i \geq T_{i\,\max}, \\ T_i = T_{i\,\min}, & \text{if } T_i \leq T_{i\,\min}, \\ \delta_i = \min\left(\delta_i(t_{n-1}) + \dot{\delta}_{i\,\max}\Delta t, \delta_i\,\max\right), & \text{if } \delta_i(t_n) \geq \min\left(\delta_i(t_{n-1}) + \dot{\delta}_{i\,\max}\Delta t, \delta_i\,\max\right), \\ \delta_i = \max\left(\delta_i(t_{n-1}) - \dot{\delta}_{i\,\max}\Delta t, \delta_i\,\min\right), & \text{if } \delta_i(t_n) \leq \max\left(\delta_i(t_{n-1}) - \dot{\delta}_{i\,\max}\Delta t, \delta_i\,\min\right). \end{cases} \tag{58}$$

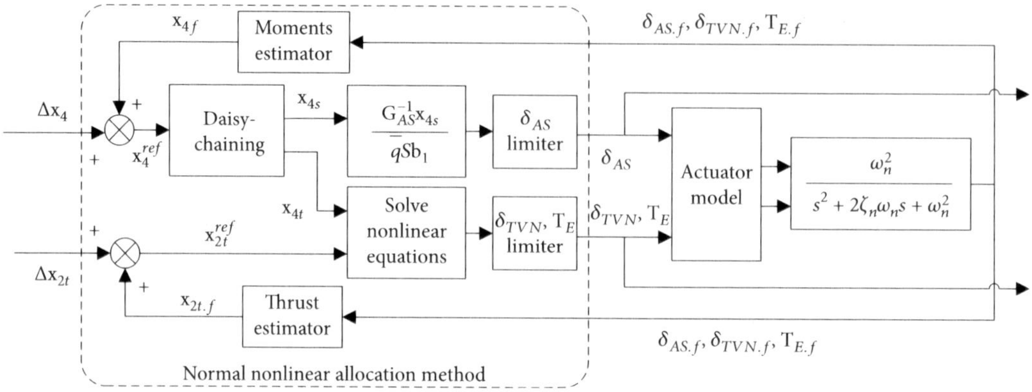

FIGURE 6: Second-layer normal nonlinear allocation method.

The flow chart of the incremental linear allocation is presented in Figure 5.

4.2.2. Normal Nonlinear Allocation Method. In nonlinear allocation, other than calculating $\Delta \mathbf{T}_E$ and $\Delta \boldsymbol{\delta}_{\mathrm{TVN}}$, the outputs of thrust vector engine system $\mathbf{T}_E$ and $\boldsymbol{\delta}_{\mathrm{TVN}}$ are calculated directly through solving nonlinear equations. This method is suitable for all maneuver flights and will not cause approximation error. The total time consumption remains rational and applicable for onboard use though the calculation time is longer than the incremental linear allocation method. The flow chart of the allocation process is exhibited in Figure 6.

5. Simulation

An example is designed in this section to test the control method which is discussed in the previous section. The target trajectory is designed as follows:

$$
\begin{aligned}
v_x &= \begin{cases} 0.5t, & 0 \le t < 20, \\ 10, & 20 \le t < 60, \\ 10 + \dfrac{(t-60)}{2}, & 60 \le t < 80, \end{cases} \\
v_y &= \begin{cases} 0, & 0 \le t < 20, \\ 7\sin\left(\dfrac{2\pi(t-20)}{40}\right), & 20 \le t < 60, \\ 0, & 60 \le t < 80, \end{cases} \\
v_z &= \begin{cases} -2\sin\left(\dfrac{\pi t}{20}\right), & 0 \le t < 20, \\ 0, & 20 \le t < 80. \end{cases}
\end{aligned}
\tag{59}
$$

The initial state of UAV is static. From 0 to 20 seconds, the vehicle is climbing up and accelerating to 10 m/s in ground x-axis. From 20 to 60 seconds, the UAV is in transition flight and maneuvering laterally, showing UAV's flight ability in low-altitude complex flight conditions, including cities and mountainous areas. After 60 seconds, the vehicle makes uniformly accelerated flight in ground x-axis and transforms from transition flight to cruise.

The example involves three different flight states, i.e., VTOL, transition flight, and cruise. These states can adequately show UAV's longitudinal and lateral maneuverability required for an urban flight vehicle. The controller designed in this paper does not need to switch as flight states change. The trajectory can be tracked merely through adjusting the weights in the controller. As the simulation results prove, the control method can solve nonlinear, nonaffine, and coupled control problems effectively and allocate redundant control variables appropriately.

To show the effectiveness of the two-layer cascaded control allocation for different flight states and mission requirements, two flight strategies are designed. In the first strategy, the UAV will track the trajectory with small attitude angle magnitude and fluctuation. In the second strategy, the vehicle is required to track the target trajectory through employing small direct force control and primarily adopting attitude control. For all strategies, the incremental linear allocation method is adopted in the second-layer control allocation, and 0.01 s is adopted as the calculation step.

5.1. Strategy 1: Trajectory Tracking. For Strategy 1, the initial values of the weights are presented in (60). To avoid the errors caused by the small magnitudes of weighted control variables, the weights have been expanded by 100 times.

$$
\begin{aligned}
\mathbf{W}_1 &= \mathrm{diag}\,[4.41 \quad 22.90 \quad 22.90 \quad 0.11 \quad 1.08 \quad 0.05], \\
\mathbf{W}_2 &= \mathrm{diag}\,[9.56 \quad 19.13 \quad 19.13 \quad 0.24 \quad 0.24 \quad 0.24].
\end{aligned}
\tag{60}
$$

The weight structure is generated and regulated all through the trajectory tracking, and at last, the weight changes into (61).

$$
\begin{aligned}
\mathbf{W}_1 &= \mathrm{diag}\,[5.78 \quad 30.20 \quad 30.20 \quad 0.09 \quad 0.83 \quad 0.04], \\
\mathbf{W}_2 &= \mathrm{diag}\,[6.49 \quad 12.95 \quad 12.95 \quad 0.19 \quad 0.19 \quad 0.19].
\end{aligned}
\tag{61}
$$

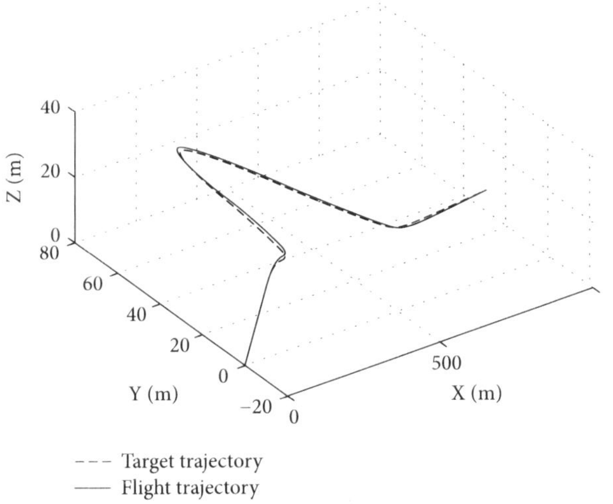

FIGURE 7: Trajectory tracking results.

Additionally, the weight change would be better consecutive for different flight states and mission requirements, since inconsecutive weight change will cause the saltation of flight states. The trajectory tracking result is presented in Figure 7.

The flight states of Strategy 1 are exhibited in Figure 8.

The wind angle and body angle of the UAV are presented in Figures 8(c) and 8(d). As indicated, the angle of attack α is negative in climbing phase, and α increases with the increasing of velocity and pitch angle. The UAV first keeps the minus pitch angle, and the body z-axis thrust is adopted to assist x-axis acceleration. With the rise of lift and decline of climb rate, the pitch angle (θ) and vectoring nozzle deflection angle increase, and more body x-axis thrust is adopted to accelerate the vehicle in ground x-axis. These maneuvering strategies are calculated by the controller automatically according to current flight states and weight structure.

Figures 8(i) and 8(j) show the changes of control surface deflections and aerodynamic control moments, respectively. In Figure 8(i), the black curve (delta-a) represents the aileron deflection angle (δ_a), and the blue curve (delta-e) represents the elevator deflection angle (δ_e). In Figure 8(j), the black curve (PTC-roll) represents the percentage of aerodynamic roll control moment in total roll control moments, and the blue curve (PTC-pitch) represents the percentage of aerodynamic pitch control moment in total pitch control moments. As the daisy chaining method is adopted in the second layer control allocation, the thrust vector control moments will be used only when aerodynamic control moments get saturated. Also, the effect of aerodynamic control surface is affected by the magnitude of wind angle. With large magnitude of angle of attack, the effect of control surface is less and vice versa. Therefore, in the takeoff phase, the control surfaces are in maximum deflection angle ($\pm 32^\circ$). From 0 to 20 seconds, with the increase of velocity, the percentage of aerodynamic control moments increases, and control surfaces are no longer saturated.

From 20 to 60 seconds, the UAV is in transition flight and making lateral maneuvering, the flight velocity is 10 m/s, and α ranges from 0 degrees to 17 degrees. In Figure 8(f), T_z is negative as the body x-axis is pointing down, indicating that UAV employs vectored thrust to compensate the lift. The change tendencies of T_z and α are similar with the absolute value of velocity. As indicated, the larger lift is generated with the rise of velocity and smaller T_z is required to compensate the lift. The change of engine thrust and nozzle deflection angle after control allocation are presented in Figures 8(g) and 8(h). In Figures 8(i) and 8(j), the aerodynamic control moments get saturated again when the large control moments are required, and their usage percentage decreases as the thrust vector control moments are added in. During lateral maneuver, roll angle ϕ (Figure 8(d)) and sideslip angle β (Figure 8(c)) are changed jointly, meaning that STT and BTT controls are adopted in the UAV simultaneously. Additionally, different values of T_L and T_R change similarly with β. As indicated, the UAV uses engine differential thrust to realize the STT control. The change of these flight states is corresponded to kinematic azimuth angle χ, as presented in Figure 8(b). Overall, when χ is increasing, ϕ, β, and n_c keep positive; when χ is decreasing, ϕ, β, and n_c keep negative, as presented in Figures 8(b), 8(c), 8(d), and 8(e).

When $t > 60$s, with the increase of velocity, α and T_z decrease, the usage percentage of aerodynamic roll control moment rises to 100%, and the usage of aerodynamic pitch control moments increases. The UAV is transformed from transition flight to cruise.

5.2. Strategy 2: Trajectory Tracking. For Strategy 2, the initial values of the weights are presented in

$$\mathbf{W}_1 = \text{diag}\left[2.28 \quad 22.78 \quad 22.78 \quad 0.13 \quad 1.28 \quad 0.12\right],$$
$$\mathbf{W}_1 = \text{diag}\left[9.96 \quad 19.92 \quad 19.92 \quad 0.28 \quad 0.28 \quad 0.28\right]. \tag{62}$$

With the generation and regulation of weight structure during trajectory tracking, at the end of simulation, the weight changes into (63).

$$\mathbf{W}_1 = \text{diag}\left[1.81 \quad 32.53 \quad 32.53 \quad 0.08 \quad 0.81 \quad 0.08\right],$$
$$\mathbf{W}_1 = \text{diag}\left[6.32 \quad 12.65 \quad 12.65 \quad 0.18 \quad 0.18 \quad 0.18\right]. \tag{63}$$

The flight states of Strategy 2 are presented in Figure 9.

Due to different weight selections, the tracking maneuver in Strategy 2 differs from that in Strategy 1 evidently. In Strategy 2, α and θ increase quickly (Figures 9(a) and 9(b)) during takeoff, and the vehicle climbs and accelerates in forward with the thrust in body x-axis. In the lateral maneuver process, the ranges of attitude angle and control moments (Figures 9(a) and 9(c)) in Strategy 2 are larger than the one in Strategy 1 evidently. However, the vectored thrust T_Z and power system actuators T_F, δ_L, and δ_R adopted in Strategy 2 (Figures 9(d), 9(e), and 9(f)) are less than those in Strategy 1 (Figures 8(f), 8(g), and 8(h)). Compared with Strategy 1, Strategy 2 adopts less direct force control, and the maneuver is dominated by attitude control. In the meantime, similar with Strategy 1, BTT and STT controls are both adopted in Strategy 2. In

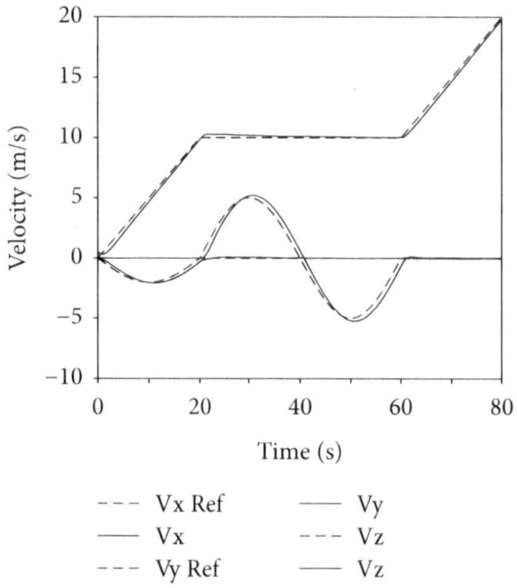

(a) Three-axis velocity

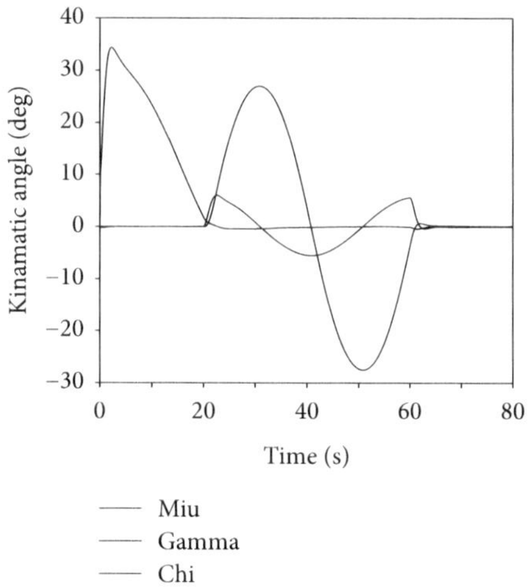

(b) Kinematic angle

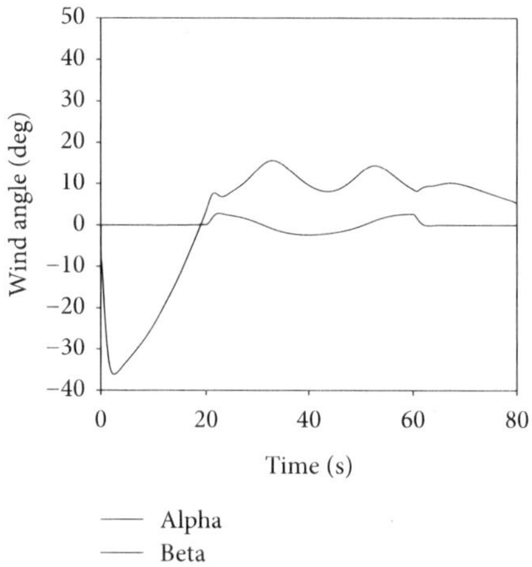

(c) Wind angle

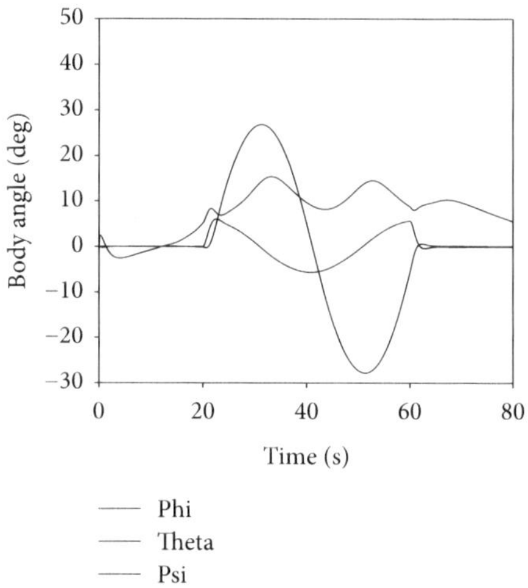

(d) Body angle

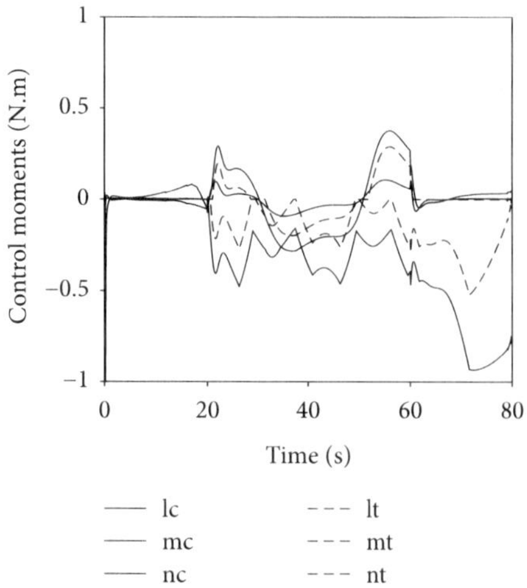

(e) Control moments

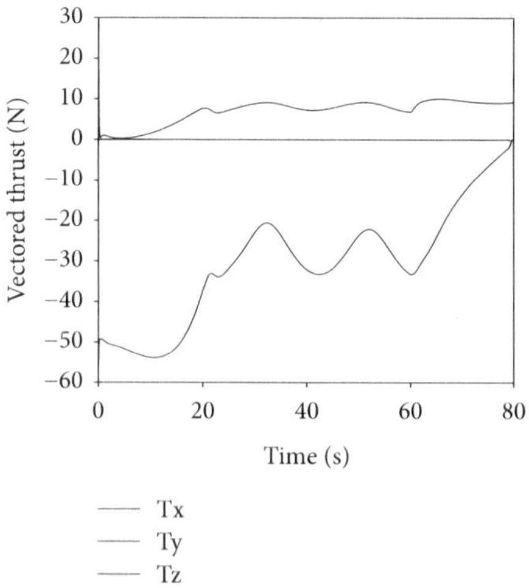

(f) Vectored thrust in body axis

Figure 8: Continued.

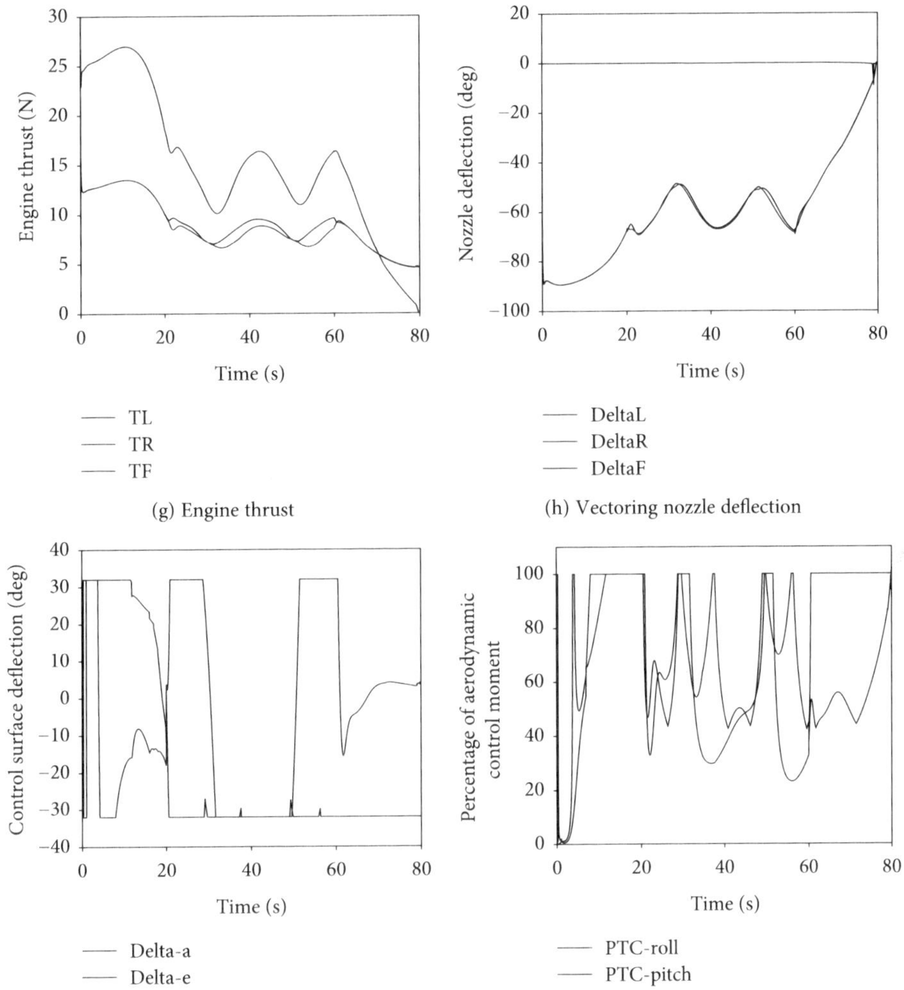

(g) Engine thrust

(h) Vectoring nozzle deflection

(i) Control surface deflection

(j) Percentage of aerodynamic control moments

FIGURE 8: Flight states of Strategy 1.

Figure 9(c), from 60 seconds to 80 seconds, the saltation of control moment m_c exists, which is caused by the aerodynamic interference between the tandem wings. During this period, α decreases from 36 degrees to 5 degrees. When α is around 18 degrees, the first saltation of pitch moment coefficient C_m is caused by the different stall angles between the front wing and the rear wing. When α is around 9 degrees, the wake flow of the front wing beats on the rear wing directly, which leads to the second saltation of C_m. As indicated, in the INDI control system, the control moment m_c is used to counteract the aerodynamic disturbances and keep the stability of flight attitude.

Comparing the change of α in Figures 8(c) and 9(a), we can find that when UAV makes lateral maneuver, α in Strategy 1 increases with the increase of V, while Strategy 2 is on the contrary. This is because the average α approaching to 12 degrees is small in Strategy 1 and does not reach the stall angle. With the rise of velocity, the increment of α can largely increase lift and reduce T_Z. This maneuver strategy based on weight selection is an optimal result, which makes the objective function get minor results. In Strategy 2, the average value of α is around 34 degrees, which exceeds the stall angle evidently. In this condition, increasing α will result in lift loss. Accordingly, with the increase of velocity, more lift is generated, and reducing α is the best solution for optimal control allocation.

The application of the INDI control method has a requirement for the frequency of the control system. High control frequency is especially required for fiercer maneuver. In INDI control, the multiplication of increment caused by coupled control variables counts as a high-order small quantity and is neglected, which will cause large errors under fierce maneuver and large sampling time. Also, in the discrete differentiator, which is used to calculate the derivative of control vectors, the larger sampling time will result in less derivation accuracy. These errors can be restrained through reducing the sampling time, which is equivalent to increase the control frequency. As the simulation results indicate, UAV can track the target trajectory accurately. The application of the INDI method and two-layer cascaded optimal control allocation in a controller of 100 Hz can satisfy the need of maneuver in this case. Furthermore, the simulation validates the effectiveness of the control method developed in this paper.

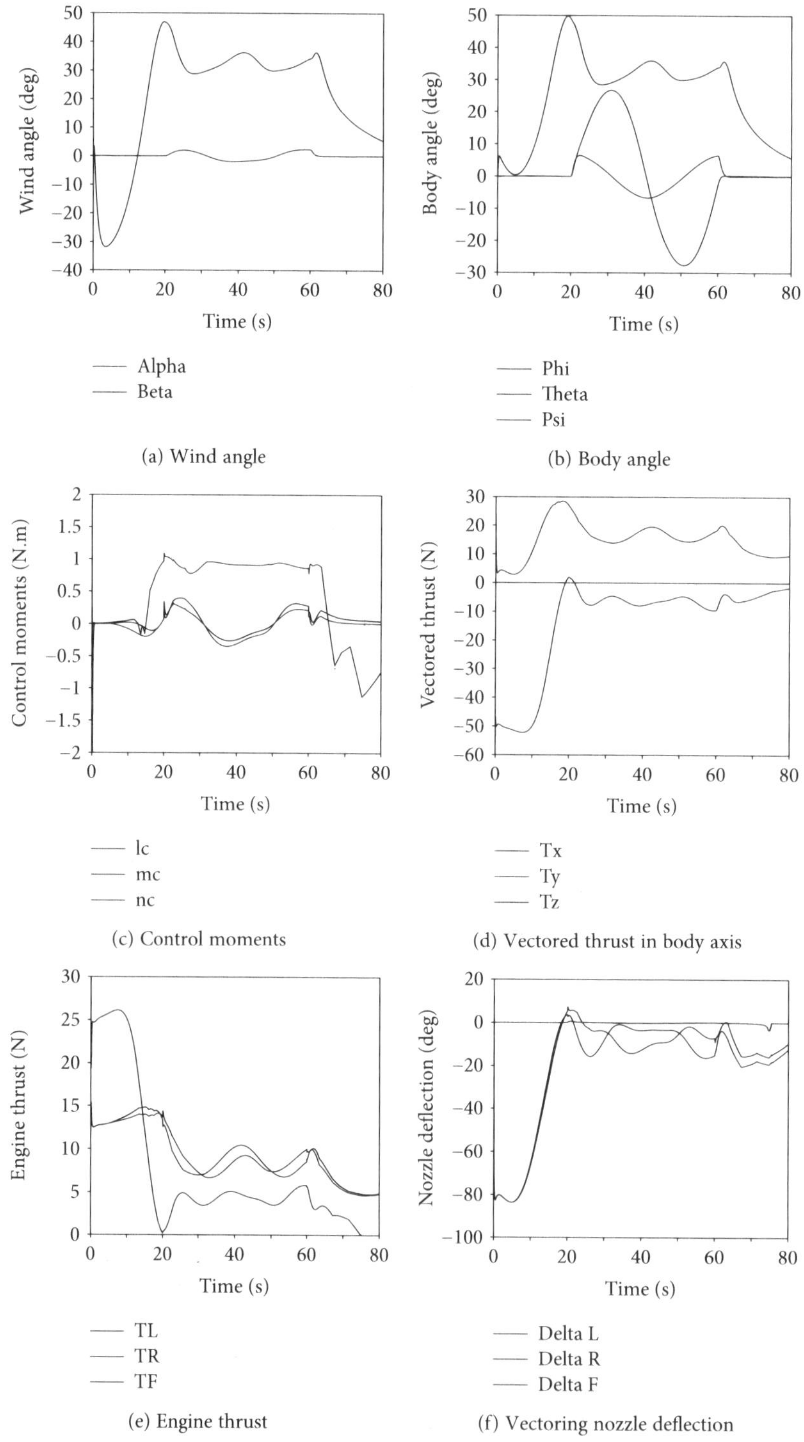

FIGURE 9: Flight states of Strategy 2.

6. Conclusions

A trajectory control system is proposed in this paper to solve the nonlinear, nonaffine, redundant, and coupled control problems existing in a UAV, which adopts direct force control technique. The research results are concluded below: (1) as the simulation indicates, the INDI controller is applicable for different flight modes, and the UAV with the INDI controller exhibits continuous maneuverability in transition flight. (2) The two-layer cascaded optimal control allocation is designed based on INDI control law; with rationally designed constraints and cascaded solving strategy, this method transforms the nonlinear coupled allocation problem into an incremental linear-equality constraint optimization problem and largely improves the allocation speed. (3) The dynamic weight strategy can adjust the weights according to different flight states and mission requirements, which ensures the rationality of control allocation. In the future, the control system presented in this paper will be incorporated with the controller of a prototype. A flight test is required to be further conducted to verify the effectiveness of the proposed control system.

Appendix

$$\mathbf{g}_{1a} = \begin{bmatrix} g_{1a(11)} & g_{1a(12)} & g_{1a(13)} \\ g_{1a(21)} & g_{1a(22)} & g_{1a(23)} \\ g_{1a(31)} & g_{1a(32)} & g_{1a(33)} \end{bmatrix},$$

$$\begin{aligned}
g_{1a(11)} &= -T_x \sin\alpha\cos\beta + T_z\cos\alpha\cos\beta - \frac{1}{2}\rho V^2 S C_{L\alpha},\\
g_{1a(12)} &= -T_x\sin\beta\cos\alpha + T_y\cos\beta - T_z\sin\beta\sin\alpha,\\
g_{1a(13)} &= 0,\\
g_{1a(21)} &= T_x(\sin\alpha\sin\beta\cos\mu + \cos\alpha\sin\mu) + T_z(-\cos\alpha\sin\beta\cos\mu_* + \sin\alpha\sin\mu) + \sin\mu\frac{1}{2}\rho V^2 S C_{L\alpha},\\
g_{1a(22)} &= -T_x\cos\alpha\cos\beta\cos\mu - T_y\sin\beta\cos\mu - T_z\sin\alpha\cos\beta\cos\mu + cos\mu\frac{1}{2}\rho V^2 S C_{C\beta},\\
g_{1a(23)} &= T_x(\cos\alpha\sin\beta\sin\mu + \sin\alpha\cos\mu) - T_y\sin\mu\cos\beta + T_z(\sin\alpha\sin\beta\sin\mu - \cos\alpha\cos\mu)\\
&\quad - \sin\mu\frac{1}{2}\rho V^2 S(\bar{C}_{C0} + C_{C\beta}\beta) + \cos\mu\frac{1}{2}\rho V^2 S(\bar{C}_{L0} + C_{L\alpha}\alpha),\\
g_{1a(31)} &= T_x(-\cos\alpha\sin\beta\cos\mu + \sin\alpha\sin\beta\sin\mu - \cos\alpha\cos\mu) + T_z(-\cos\alpha\sin\beta\sin\mu - \sin\alpha\cos\mu) - \cos\mu\frac{1}{2}\rho V^2 S C_{L\alpha},\\
g_{1a(32)} &= -T_x\cos\alpha\cos\beta\sin\mu - T_y\sin\beta\sin\mu - T_z\sin\alpha\cos\beta\sin\mu + \sin\mu\frac{1}{2}\rho V^2 S C_{C\beta},\\
g_{1a(33)} &= T_x\sin\alpha\sin\mu + T_y\sin\alpha\sin\mu + T_z(-\sin\alpha\sin\beta\cos\mu - \cos\alpha\sin\mu)\\
&\quad + \cos\mu\frac{1}{2}\rho V^2 S(\bar{C}_{C0} + C_{C\beta}\beta) + \sin\mu\frac{1}{2}\rho V^2 S(\bar{C}_{L0} + C_{L\alpha}\alpha),
\end{aligned} \tag{A.1}$$

$$\mathbf{g}_{1t} = \begin{bmatrix} \cos\alpha\cos\beta & \sin\beta & \sin\alpha\cos\beta \\ -\cos\alpha\sin\beta\cos\mu + \sin\alpha\sin\mu & \cos\mu\cos\beta & -\sin\alpha\sin\beta\cos\mu - \cos\alpha\sin\mu \\ -\cos\alpha\sin\beta\sin\mu - \sin\alpha\cos\mu & \sin\mu\cos\beta & -\sin\alpha\sin\beta\sin\mu + \cos\alpha\cos\mu \end{bmatrix}.$$

Conflicts of Interest

The authors declare that there is no conflict of interest regarding the publication of this paper.

Acknowledgments

This work is supported by the Aeronautical Science Foundation of China Grant 2016ZA72001 and the National Natural Science Foundation of China Grant 11572036. The authors would like to thank all members of the Microraptor Works, Key Lab of Dynamics and Control of Flight Vehicle, Ministry of Education, School of Aerospace Engineering, Beijing Institute of Technology.

References

[1] G. Di Francesco and M. Mattei, "Modeling and incremental nonlinear dynamic inversion control of a novel unmanned tiltrotor," *Journal of Aircraft*, vol. 53, no. 1, pp. 73–86, 2016.

[2] Y. Xili, F. Yong, and Z. Jihong, "Transition flight control of two vertical/short takeoff and landing aircraft," *Journal of Guidance, Control, and Dynamics*, vol. 31, no. 2, pp. 371–385, 2008.

[3] Y. O. Aktas, U. Ozdemir, Y. Dereli et al., "Rapid prototyping of a fixed-wing VTOL UAV for design testing," *Journal of Intelligent & Robotic Systems*, vol. 84, no. 1–4, pp. 639–664, 2016.

[4] M. D. Hua, T. Hamel, P. Morin, and C. Samson, "Control of VTOL vehicles with thrust-tilting augmentation," *Automatica*, vol. 52, pp. 1–7, 2015.

[5] P. Hartmann, C. Meyer, and D. Moormann, "Unified velocity

control and flight state transition of unmanned tilt-wing aircraft," *Journal of Guidance, Control, and Dynamics*, vol. 40, no. 6, pp. 1348–1359, 2017.

[6] E. Cetinsoy, C. Hancer, K. T. Oner, E. Sirimoglu, and M. Unel, "Aerodynamic design and characterization of a quad tilt-wing UAV via wind tunnel tests," *Journal of Aerospace Engineering*, vol. 25, no. 4, pp. 574–587, 2012.

[7] M. Ryll, H. H. Bülthoff, and P. R. Giordano, "A novel overactuated quadrotor unmanned aerial vehicle: modeling, control, and experimental validation," *IEEE Transactions on Control Systems Technology*, vol. 23, no. 2, pp. 540–556, 2015.

[8] D. Kubo and S. Suzuki, "Tail-sitter vertical takeoff and landing unmanned aerial vehicle: transitional flight analysis," *Journal of Aircraft*, vol. 45, no. 1, pp. 292–297, 2008.

[9] A. Banazadeh and N. Taymourtash, "Optimal control of an aerial tail sitter in transition flight phases," *Journal of Aircraft*, vol. 53, no. 4, pp. 914–921, 2016.

[10] R. Naldi and L. Marconi, "Optimal transition maneuvers for a class of V/STOL aircraft," *Automatica*, vol. 47, no. 5, pp. 870–879, 2011.

[11] R. Naldi and L. Marconi, "Robust control of transition maneuvers for a class of V/STOL aircraft," *Automatica*, vol. 49, no. 6, pp. 1693–1704, 2013.

[12] A. Maqsood and T. H. Go, "Optimization of hover-to-cruise transition maneuver using variable-incidence wing," *Journal of Aircraft*, vol. 47, no. 3, pp. 1060–1064, 2010.

[13] A. Maqsood and T. H. Go, "Optimization of transition maneuvers through aerodynamic vectoring," *Aerospace Science and Technology*, vol. 23, no. 1, pp. 363–371, 2012.

[14] S. Sieberling, Q. P. Chu, and J. A. Mulder, "Robust flight control using incremental nonlinear dynamic inversion and angular acceleration prediction," *Journal of Guidance, Control, and Dynamics*, vol. 33, no. 6, pp. 1732–1742, 2010.

[15] E. J. J. Smeur, G. C. H. E. D. Croon, and Q. Chu, "Gust disturbance alleviation with incremental nonlinear dynamic inversion," in *2016 IEEE/RSJ International Conference on Intelligent Robots and Systems (IROS)*, Daejeon, Korea, October 2016.

[16] E. J. J. Smeur, Q. Chu, and G. C. H. E. de Croon, "Adaptive incremental nonlinear dynamic inversion for attitude control of micro air vehicles," *Journal of Guidance, Control, and Dynamics*, vol. 39, no. 3, pp. 450–461, 2016.

[17] E. J. J. Smeur, G. C. H. E. de Croon, and Q. Chu, "Cascaded incremental nonlinear dynamic inversion for MAV disturbance rejection," *Control Engineering Practice*, vol. 73, pp. 79–90, 2018.

[18] P. Lu, E. J. van Kampen, C. de Visser, and Q. Chu, "Aircraft fault-tolerant trajectory control using incremental nonlinear dynamic inversion," *Control Engineering Practice*, vol. 57, pp. 126–141, 2016.

[19] J. Reiner, G. J. Balas, and W. L. Garrard, "Robust dynamic inversion for control of highly maneuverable aircraft," *Journal of Guidance, Control, and Dynamics*, vol. 18, no. 1, pp. 18–24, 1995.

[20] P. Simplício, M. D. Pavel, E. van Kampen, and Q. P. Chu, "An acceleration measurements-based approach for helicopter nonlinear flight control using incremental nonlinear dynamic inversion," *Control Engineering Practice*, vol. 21, no. 8, pp. 1065–1077, 2013.

[21] B. J. Bacon, A. J. Ostroff, and S. M. Joshi, "Reconfigurable NDI controller using inertial sensor failure detection & isolation," *IEEE Transactions on Aerospace and Electronic Systems*, vol. 37, no. 4, pp. 1373–1383, 2001.

[22] J. Jin, "Modified pseudoinverse redistribution methods for redundant controls allocation," *Journal of Guidance, Control, and Dynamics*, vol. 28, no. 5, pp. 1076–1079, 2005.

[23] T. L. Saaty, "Decision making with the analytic hierarchy process," *International Journal of Services Sciences*, vol. 1, no. 1, pp. 83–98, 2008.

[24] O. S. Vaidya and S. Kumar, "Analytic hierarchy process: an overview of applications," *European Journal of Operational Research*, vol. 169, no. 1, pp. 1–29, 2006.

6

Design of Meteorological Element Detection Platform for Atmospheric Boundary Layer based on UAV

Yonghong Zhang,[1] Tiantian Dong,[1] and Yunping Liu[1,2]

[1]*B-DAT, C-MEIC, CICAEET, School of Information and Control, Nanjing University of Information Science & Technology, Nanjing, China*
[2]*Department of Mechanical and Industrial Engineering, Ryerson University, Toronto, ON, Canada*

Correspondence should be addressed to Yunping Liu; liuyunping@nuist.edu.cn

Academic Editor: Hikmat Asadov

Among current detection methods of the atmospheric boundary layer, sounding balloon has disadvantages such as low recovery and low reuse rate, anemometer tower has disadvantages such as fixed location and high cost, and remote sensing detection has disadvantages such as low data accuracy. In this paper, a meteorological element sensor was carried on a six-rotor UAV platform to achieve detection of meteorological elements of the atmospheric boundary layer, and the influence of different installation positions of the meteorological element sensor on the detection accuracy of the meteorological element sensor was analyzed through many experiments. Firstly, a six-rotor UAV platform was built through mechanical structure design and control system design. Secondly, data such as temperature, relative humidity, pressure, elevation, and latitude and longitude were collected by designing a meteorological element detection system. Thirdly, data management of the collected data was conducted, including local storage and real-time display on ground host computer. Finally, combined with the comprehensive analysis of the data of automatic weather station, the validity of the data was verified. This six-rotor UAV platform carrying a meteorological element sensor can effectively realize the direct measurement of the atmospheric boundary layer and in some cases can make up for the deficiency of sounding balloon, anemometer tower, and remote sensing detection.

1. Introduction

The current detection methods of the atmospheric boundary layer include sounding balloon, anemometer tower, radar detection, and so on [1, 2]. Among which, conventional sounding balloon and captive balloon have some limitations in the detection of the atmospheric boundary layer due to disadvantages such as low recovery and low reuse rate. The observational approaches of radar detection and remote sensing detection can only obtain relatively low accuracy and reliability of data [3]. Anemometer tower has difficulty in giving consideration to the purpose of research on the atmospheric boundary layer due to disadvantages such as fixed location and high cost. Above all, to seek a high stability and low cost, continuously used detection instrument for the atmospheric boundary layer is an urgent problem to be solved. By virtue of features such as easy landing, automatic cruise, constant height, and constant point [4], it is possible that a meteorological element sensor is carried on a six-rotor UAV (unmanned aerial vehicles) platform to realize the direct measurement of atmospheric sounding. Therefore, developing a meteorological element detection platform for the atmospheric boundary layer based on UAV has important scientific significance.

Among current detection methods of atmospheric sounding, sounding balloon carrying a radiosonde for direct measurement is an important means; however, conventional sounding balloon and captive balloon have low detection accuracy of the atmospheric boundary layer due to disadvantages such as low recovery and high cost along with low altitude airflow and terrain environment, so they cannot meet the needs of modern meteorological applications. Therefore, by virtue of features such as easy vertical takeoff and landing, high positioning accuracy, and reusability, it is possible that a

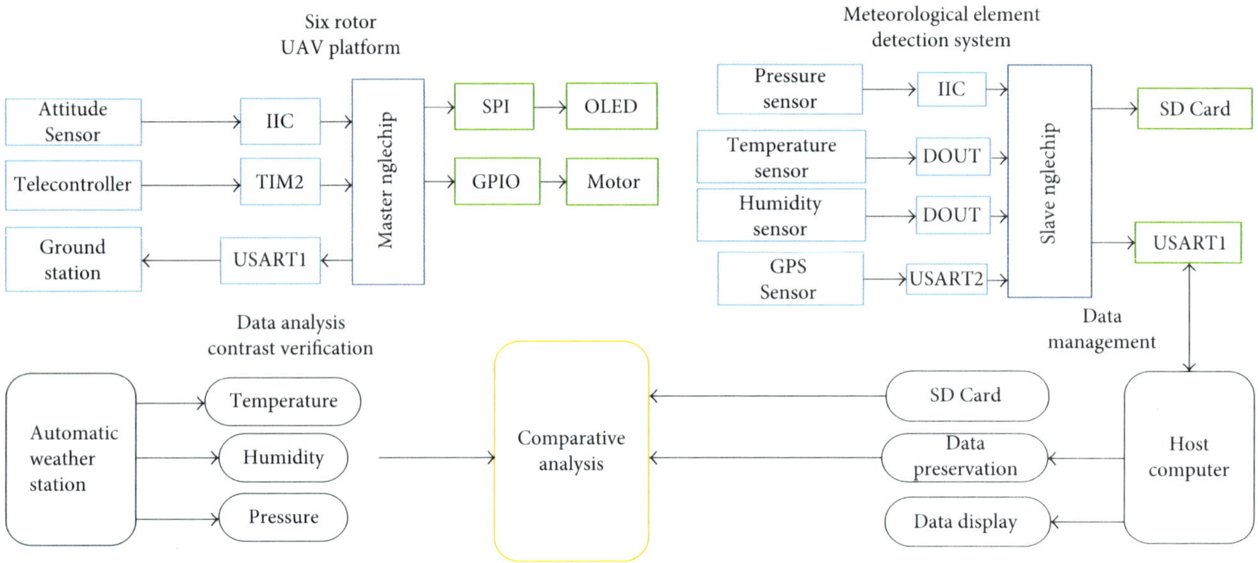

FIGURE 1: System diagram.

meteorological element sensor is carried on a six-rotor UAV platform to make up for the deficiency of sounding balloon in some cases. The observational approaches of radar detection and remote sensing detection obtain meteorological parameters through inversion and observation of indirect variables, and it can only obtain relatively low accuracy and reliability of data relative to physical measurement; therefore, it is possible that a meteorological element sensor is carried on a six-rotor UAV platform to realize the direct measurement of atmospheric boundary layer. Anemometer tower is a facility to observe the vertical distribution of meteorological elements of the atmospheric boundary layer but has difficulty in giving consideration to the purpose of research on the atmospheric boundary layer due to disadvantages such as fixed location and high cost. Therefore, detecting the atmospheric boundary layer with an UAV has advantages of low cost and good mobility [5–10].

At present, many research achievements have been made in this field: the United States, Australia, France, and China have developed UAV meteorological remote sensing systems, such as Perseus, Theseus, and Aerosonde. Aerosonde Ltd. is an Australian-based developer and manufacturer of unmanned aerial vehicles; in 1995, it started to provide products and services related to UAV meteorological detection system, taking the leading position in this field. Reuder et al. observed the meteorological elements of the atmospheric boundary layer such as temperature, humidity, and pressure by using SUMO, a small fixed-wing UAV, and achieved good results [11]. China has done a lot of work in this field and has achieved preliminary results. However, performing atmospheric sounding with a fixed-wing UAV has deviation in the measurement of meteorological data because the fixed-wing UAV will shift substantially on windy days during uniform straight up. In contrast, a multirotor UAV can fly straight up and down on windy days and has better applicability than a fixed-wing UAV, so it can perform meteorological detection of atmospheric layer of different height [12, 13]. The commonly used multirotor UAVs include four-rotor UAV and six-rotor UAV, among which, the six-rotor UAV has two additional rotors compared with four-rotor UAV; thus, it can exhibit better stability when it experiences strong external disturbance or part of the rotor is disturbed [14]. Therefore, in this paper, firstly, a six-rotor UAV platform was built through mechanical structure design and control system design; secondly, data such as temperature, relative humidity, pressure, elevation, and latitude and longitude were collected by designing a meteorological element detection system; thirdly, data management of the collected data was conducted, including local storage and real-time display on ground host computer [15]; and finally, combined with the comprehensive analysis of the data of automatic weather station, the validity of the data was verified.

2. Design of Overall System Scheme

The overall system included four parts, namely, six-rotor UAV platform design, meteorological element detection system design, data management and system testing, and data analysis and contrast verification, and the relationships between the parts are shown in Figure 1. Firstly, a six-rotor UAV, as the flight carrier of detection system, carried a meteorological element detection system to achieve detection of meteorological elements of the atmospheric boundary layer. Secondly, the collected data were saved in the SD card and sent to the host computer for real-time display through the wireless data transmission module at the same time. Finally, the comprehensive analysis of the data of automatic weather station was conducted, and contrast verification by experiment was conducted. The parts are described below:

In this paper, combining the design method of six-rotor UAV flight control system with the design method of meteorological element detection system, the meteorological sounding of atmospheric boundary layer was studied, on this basis, combined with the comprehensive analysis of the data of automatic weather station; the research methods and key technologies involved in the parts are as follows:

(1) Six-rotor UAV technology mainly adopted mechanical structure design, Mahony complementary filter algorithm design [16], and hardware and software design.

The aerodynamic layout of six-rotor UAV had a great influence on the control characteristics and flight quality. Therefore, firstly, the mechanical structure design was conducted using Solid Works software. Secondly, the master control conducted programming of flight control system software using STM32 series chips under MDK (Microcontroller Development kit) development environment through Mahony complementary filter algorithm design and other designs, and the design of the main program completed the overall function by means of interrupt nesting. Finally, the hardware circuit and PCB (printed circuit board) layout were designed using Altium Designer software, including board processing, welding and debugging, and model six-rotor UAV building.

(2) Meteorological element detection system technology included hardware design, program design, host computer design, and data management.

The master control of the meteorological element detection system collected the data of temperature and humidity sensor, pressure sensor, and GPS sensor using STM32 series chips, among which, the hardware design and software program design were conducted using Altium Designer software under KEIL environment. As for data management, the meteorological data were displayed by designing the host computer software using Visual Basic and saved in the SD card in the form of file system.

(3) As for data analysis and experimental comparison, the system test and contrast test were conducted in the simulation environment, and the influence of different installation positions of the meteorological element detection system on the data was analyzed.

In this experiment, under the standard environment simulated by a split-type precision humidity generator, firstly, the temperature and humidity of the meteorological element detection system board were tested and analyzed under different conditions. Secondly, in light of the influence of the rotor wind of six-rotor UAV on the detection accuracy of the meteorological element sensor during the detection of atmospheric boundary layer, the influence of different installation positions of the meteorological element sensor on the detection accuracy of the meteorological element sensor was analyzed through many experiments. Finally, in the simulation environment, the temperature and humidity of the meteorological element detection system board and DAVIS weather station collection board were tested; in the actual test, the meteorological elements such as temperature, relative humidity, pressure, elevation, and latitude and longitude were detected by a six-rotor UAV platform carrying a meteorological element sensor and DAVIS weather station collection board, and the validity of the data was verified through many experiments.

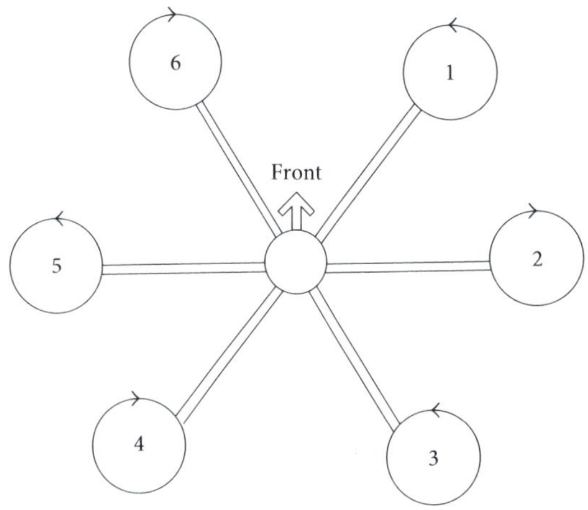

Figure 2: Aerodynamic layout diagram.

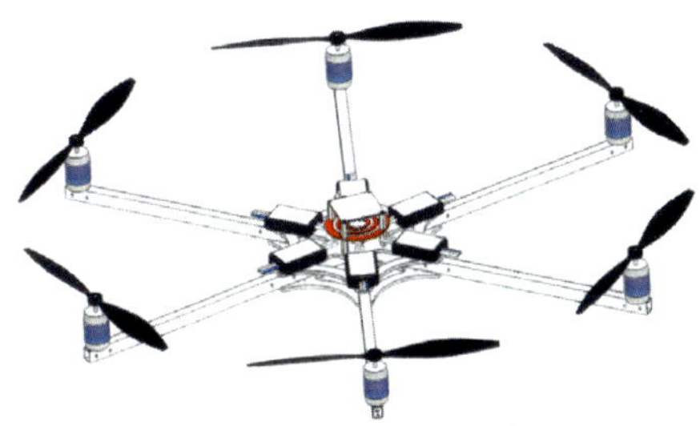

Figure 3: Six-rotor UAV platform assembly drawing.

3. Design of Six-Rotor UAV Platform

In light of the influence of meteorological element detection system carried on a six-rotor UAV and other loads on the stability of six-rotor UAV, on the basis of open-source platform, a six-rotor UAV platform of high reliability was built through the optimization of mechanical structure design. The main research contents included the design of six-rotor UAV mechanical structure, design of control system hardware, software design, and design of the host computer of the ground station.

3.1. Design of Six-Rotor UAV Mechanical Structure. The design of six-rotor UAV mechanical structure included aerodynamic layout design and material selection. The six-rotor UAV was a fully actuated system with six outputs and six freedoms of motion. The aerodynamic layout design was the premise for good flight performance of six-rotor UAV; therefore, the aerodynamic layout of the six-rotor UAV designed in this paper chose X6 structure. The aerodynamic layout and assembly of six-rotor UAV platform are shown in Figures 2 and 3.

The design of six-rotor UAV platform required the assembly of multiple parts, such as center plate of six-rotor

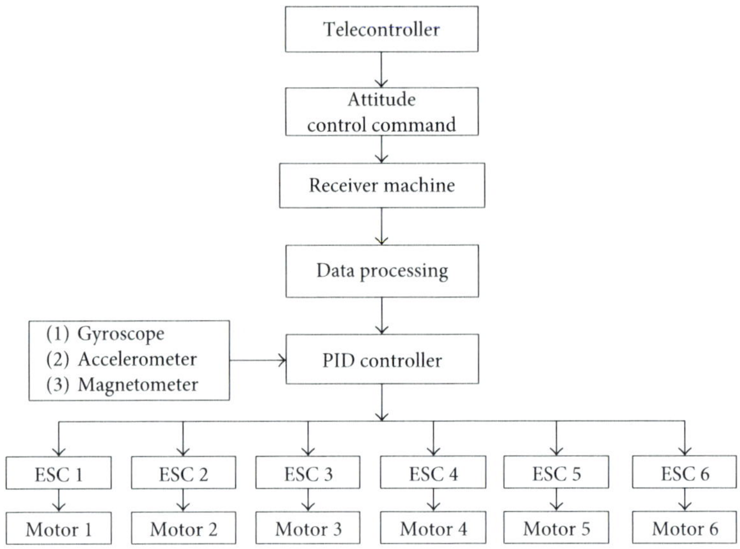

FIGURE 4: Six-rotor UAV system structure diagram.

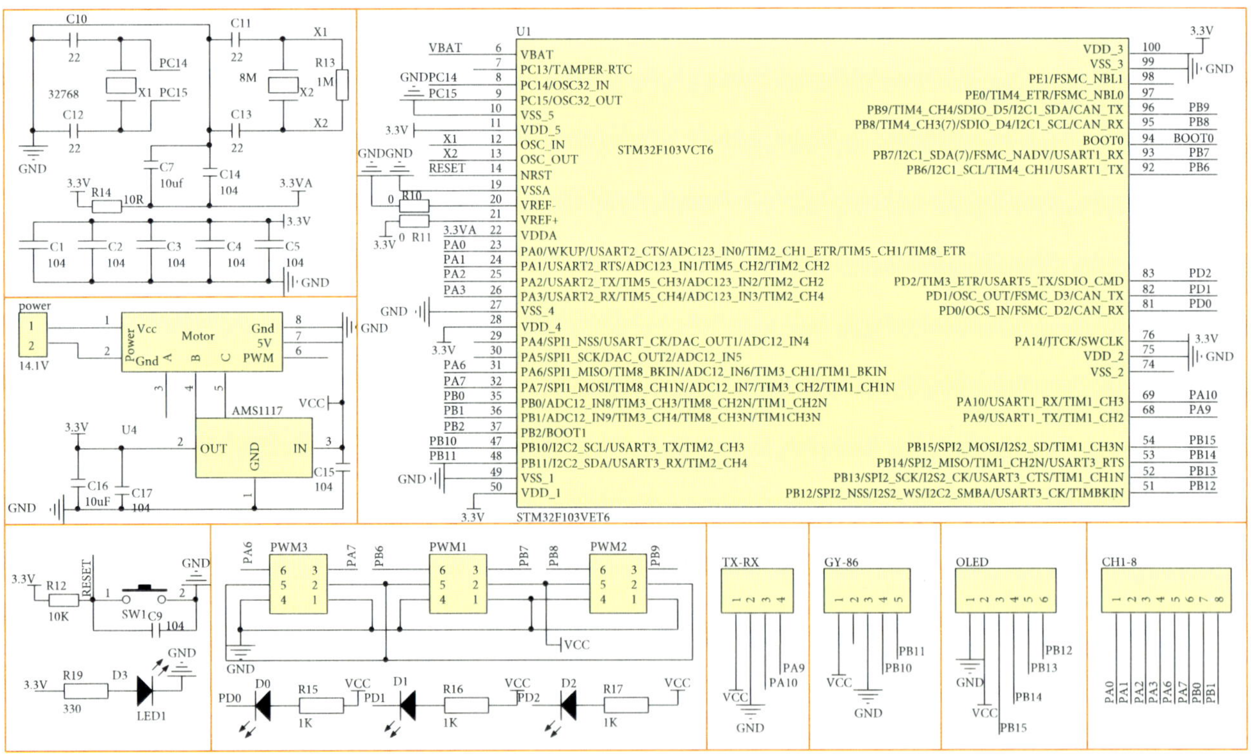

FIGURE 5: Flight control system circuit.

UAV frame, shaft arm, flight control placing board, screw, and nut, as shown in the following figure. The length of the wheelbase was 650 cm, the shaft arm was designed as a square of 15 cm × 15 cm, and the center plate of six-rotor UAV frame and shaft arm were fixed with screws and nuts. Through the assembly of multiple parts, a six-rotor UAV platform was built; on this basis, brushless motor, ESC (electronic speed control), and propeller were assembled, and the X6 structure of six-rotor UAV had better reliability than other structures.

3.2. Design of Flight Control Circuit. The correct design of hardware circuit was the premise for stable flight of six-rotor UAV. With modular design method being adopted, a meteorological element detection system was carried on the six-rotor UAV flight control system studied in this paper to perform atmospheric sounding on the premise of ensuring reliable flight of six-rotor UAV. The six-rotor UAV system structure is shown in Figure 4.

The six-rotor UAV flight control system circuit included main controller circuit, filter circuit, reset circuit, crystal

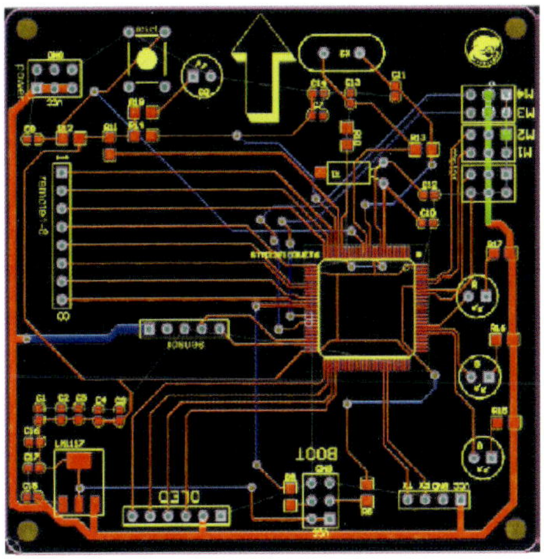

Figure 6: PCB wiring diagram.

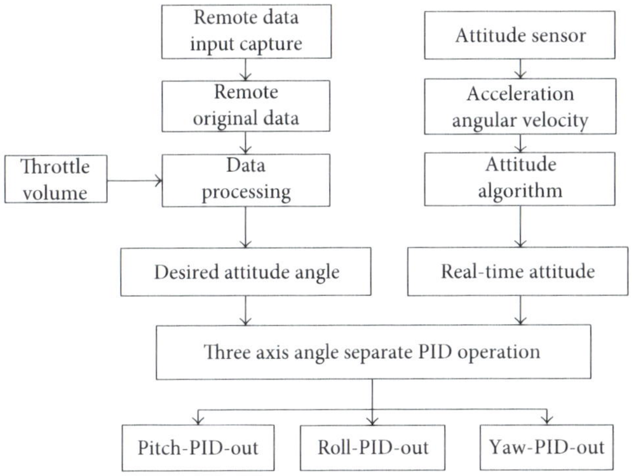

Figure 7: Flight control system flow chart.

oscillator circuit, power interface circuit, serial port circuit, GY-86 attitude sensor circuit, OLED display interface circuit, remote control interface circuit, motor output circuit, and status indicator circuit. The normal operation of each module circuit was the premise for stable flight of six-rotor UAV. The six-rotor UAV system circuit and PCB wiring are shown in Figures 5 and 6.

By adopting the above circuit design, the PCB circuit was designed using Altium Designer10 software, and the flight control board was designed as a square. The main control chip and attitude sensor were located at the center of the flight control board; thus, the six-rotor UAV had easy access to real attitude information, and the coordinate axis of attitude sensor was coincident with that of six-rotor UAV frame.

3.3. Software Design. Based on the above design, the master control conducted programming of UAV flight control system software using STM32 series chips under MDK development environment, and the design of the main program completed the overall function by means of interrupt nesting. The core of the program was that three interruptions (200 Hz, 50 Hz, and 10 Hz) were set up by timers in the main cycle, namely, scanning the remote control command once, updating the attitude sensor data once, and updating the motor control once to realize control.

After initialization of flight control system, first, the original data of remote control were obtained, and the throttle was added to data processing to obtain the desired attitude angle. Meanwhile, the real-time attitude angle of UAV was obtained using attitude solution algorithm for three axis acceleration and angular velocity acquired by the attitude sensor; then, the desired attitude angle and real-time attitude angle were input into a PID controller for operation to obtain the PID outputs of the three attitude angles. The control flow of control motor after ESC was input is shown in Figure 7.

The flight control system designed in this paper obtained accurate attitude information through the attitude calculation of six-rotor UAV software using Mahony

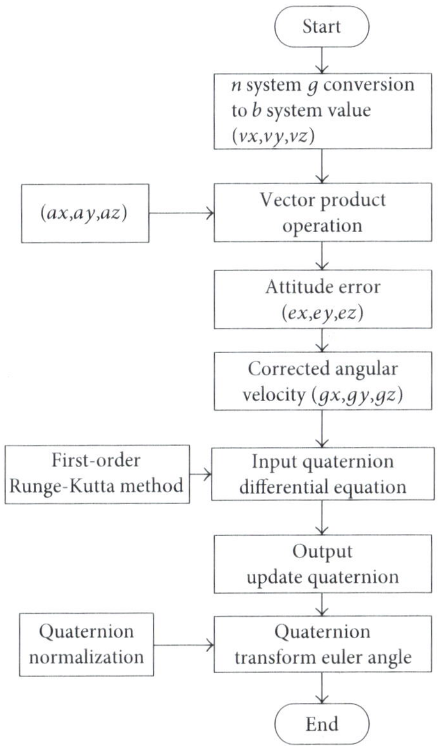

Figure 8: Attitude calculation.

complementary filter algorithm. The attitude calculation is shown in Figure 8.

First, low-pass filtering and normalization of acceleration output by MUP6050 gyroscope were conducted to obtain the unit acceleration. Second, four elements ranging from geographic coordinate system to body coordinate system were converted into direction cosine matrix; then, vector product operation between gravity vector (ax,ay,az) measured by the accelerometer on the body coordinate system and gravity vector (vx,vy,vz) calculated before was conducted to obtain the attitude error (ex,ey,ez) between the two, which was used as the PI-modified gyro bias to obtain the modified angular

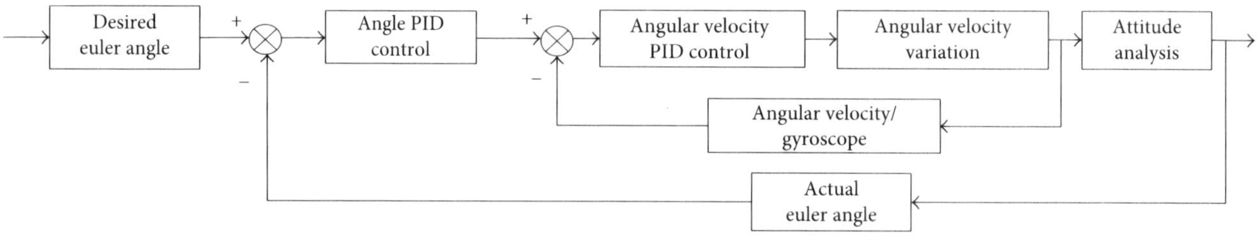

FIGURE 9: Attitude controller.

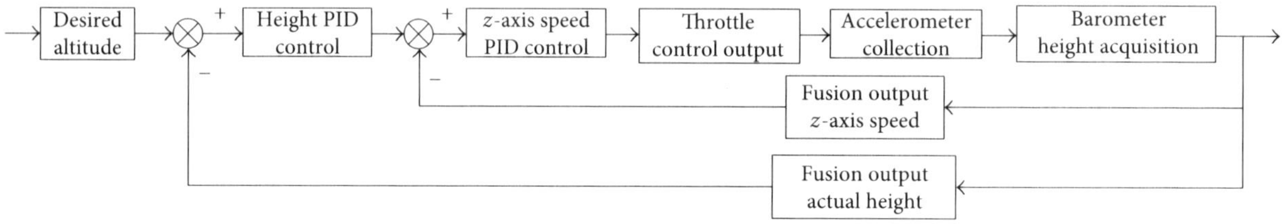

FIGURE 10: Position controller.

velocity (gx,gy,gz). Finally, the quaternion differential equation was solved using the first-order Runge-Kutta method to obtain the quaternion and then the quaternion was converted into Euler angle.

In this paper, the attitude control algorithm adopted a double closed loop PID attitude controller, as shown in Figure 9. The main function of this part mainly read attitude sensor data for solution and control. This design adopted the soft solution posture, and the data read were AD (analog to digital conversion) values of accelerometer and gyroscope; after calibration, filtering, and correction of the data, the three axis Euler angle was obtained by fusing the four elements. However, the data acquired by the acceleration sensor were susceptible to distortion, resulting in wrong attitude angle through attitude calculation. It was difficult for the system to operate stably when only using a single loop, so the angular velocity was added as the inner loop, which was output through data acquisition by the gyroscope. The data acquired by the gyroscope were generally free from external influence, with strong anti-interference ability; in addition, the angular velocity was sensitive to variation, and its rapid recovery from external disturbance enhanced the robustness of the system.

The six-rotor UAV was a three-dimensional coordinate (x,y,z) in the air, and it had requirements for two dimensions to stay in the air. The first dimension was horizontal direction $(x,y,0)$ in which the six-rotor UAV cannot move back and forth or left and right, and the second dimension was vertical direction $(0,0,z)$ in which the six-rotor UAV cannot run out of altitude. There were different technical solutions to these two dimensions. In the horizontal direction, to determine the position of the six-rotor UAV, GPS positioning was generally used; in the vertical direction, height was determined using a pressure gauge. The six-rotor UAV required self-stability and absolute coordinates in the air to achieve positioning, and the barometer adopted in this paper was MS5611, a high precision barometer, which can achieve z-axis positioning in the air. On the z-axis, firstly, the pressure collected by the pressure sensor was calculated and corrected; secondly, the accurate pressure was obtained through second-order temperature compensation; and finally, the absolute height of relative takeoff point was obtained using the transformation formula. Because the accuracy of pressure sensor MS5611 was 10 cm, it was necessary to fuse the accelerometer complementary filter to obtain the appropriate height, z-axis speed, and acceleration. In this design, a height double-loop PID controller was formed with height as the outer ring and speed as the inner ring, and the altitude hold of z-axis was achieved by adjusting the output throttle. The position controller is shown in Figure 10.

4. Design of Meteorological Element Detection System

At present, the basic meteorological elements in the detection of atmospheric boundary layer included temperature, humidity, pressure, wind speed, wind direction, and precipitation. The precipitation was generally measured using a rain measuring glass and measuring cup, which was a static measurement, while the measurement of wind speed and wind direction was greatly influenced by the rotor wind of six-rotor UAV. Therefore, in this paper, a meteorological element detection system was designed based on the actual demand of meteorological detection, and the three elements (temperature, humidity, and pressure) as well as elevation and latitude and longitude were measured. The system structure is shown in Figure 11.

4.1. Circuit Design. The hardware circuit was the basis of the overall detection platform, and its design needed to consider many aspects, including working environment and selection of component models, and only the reasonable design of the overall hardware circuit can complete the work of hardware system of this detection platform. With modular circuit design method being adopted, the circuits were connected through Place Net Label, and the circuits as a whole were

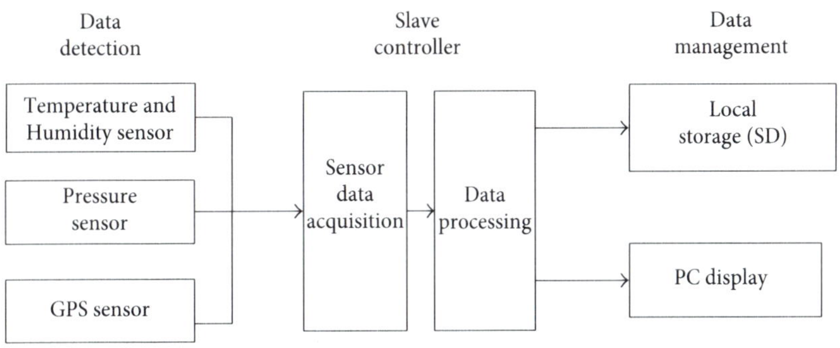

FIGURE 11: Meteorological element detection system structure diagram.

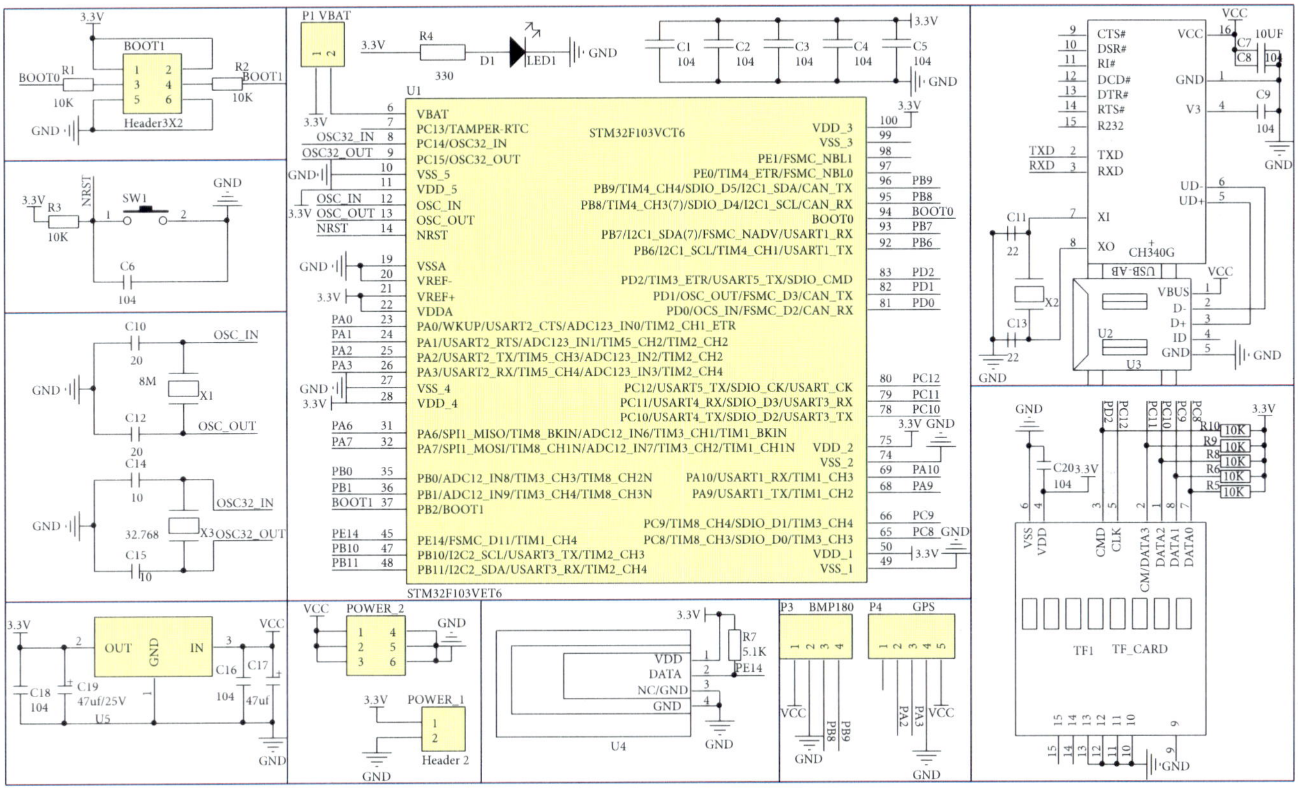

FIGURE 12: Detection system circuit.

simple, intuitive, and readable, as shown in Figure 12. The circuit board of the meteorological element detection system designed the position of the meteorological element sensor to ensure a secure position of the meteorological element sensor, and the square design in the bottommost corner of PCB circuit board was mainly the reserved slot position of memory card, as shown in Figure 13.

The meteorological element detection system circuits included controller circuit, reset circuit, crystal oscillation circuit, BOOT setting circuit, power supply circuit, AM2320 circuit, BMP180 circuit, GPS interface circuit, SD card storage circuit, and USB serial port conversion circuit. Among which, the USB serial port conversion circuit had multiplex functions, which can be used for program flash port or wireless data transmission module interface.

4.2. Program Design. The meteorological element detection system circuits included temperature and humidity collection, pressure acquisition, and geographic coordinate acquisition. The communication mode of the meteorological element detection sensor was as follows: temperature and humidity were transmitted through single bus, pressure was transmitted through I2C, and elevation, latitude and longitude, and Beijing time were transmitted through serial port. The program design included pin configuration, communication interface initialization, data format conversion, data processing, and data storage. The program flow is shown in Figure 14.

The temperature and humidity were acquired with a AM2320 sensor. In the communication process, the data bus was pulled down 18 ms from the controller, and

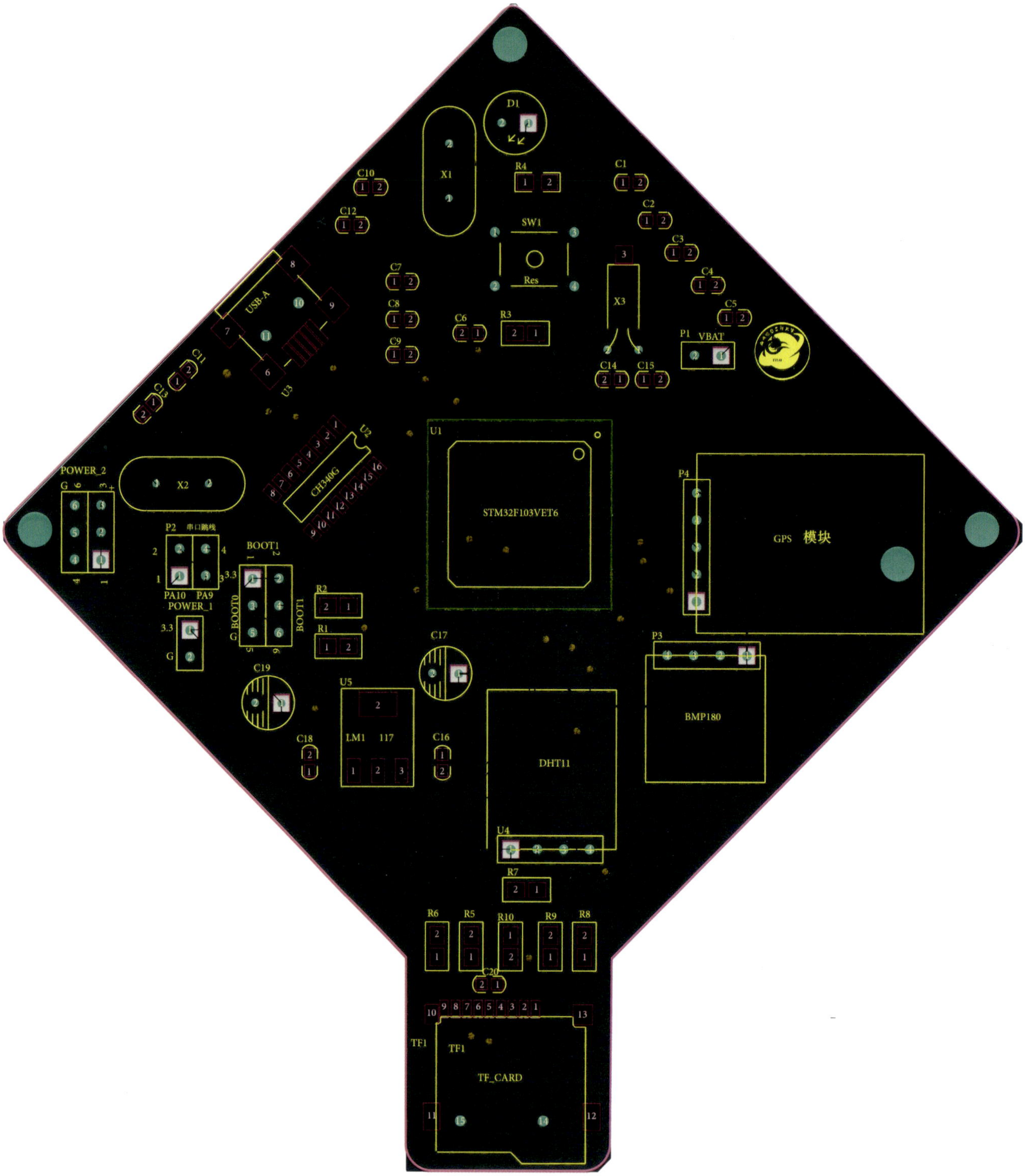

Figure 13: PCB wiring diagram.

AM2320 sensor was converted from sleep mode to high-speed mode and sent response signal after waiting the start signal from the controller to end; it sent 40-bit data from the data bus, after which the information acquisition was triggered once, and after the information collection ended, AM2320 sensor automatically entered sleep mode to wait for the next communication. The communication process is shown in Figure 15(a).

The pressure was acquired with a BMP180 pressure sensor and communicated via I2C. When the BMP180 pressure sensor started working, it read the data of pressure sensor from the I2C interface of the controller, then conducted

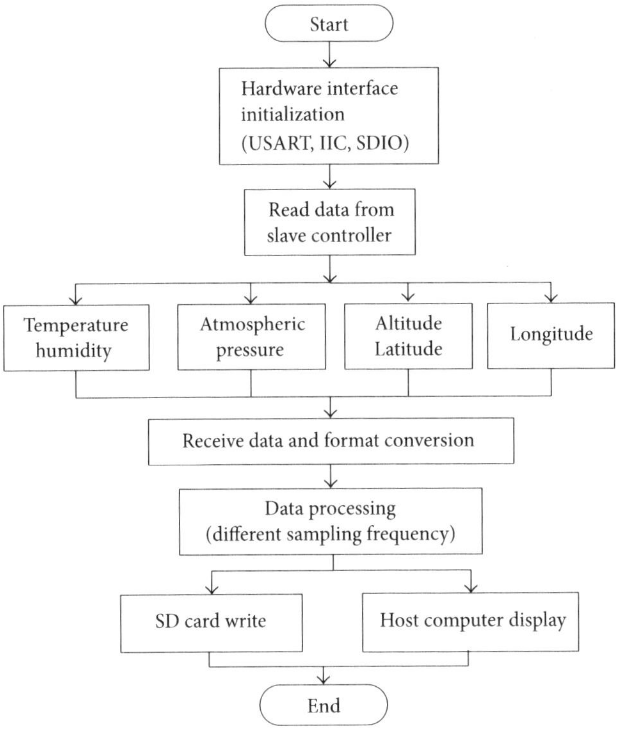

FIGURE 14: Program design flow chart.

temperature compensation for such data, and finally obtained high accuracy pressure. The communication process is shown in Figure 15(b).

As for geographic coordinate acquisition, the elevation, latitude, and longitude were acquired with a NEO-6M GPS sensor and then decoded through a NMEA base. The data were communicated from serial port 2 (USART2) of STM32F103VET6 controller through serial port and sent to memory from peripheral (serial port 2) of STM32F103VET6 controller through direct memory access (DMA). The communication process is shown in Figure 15(c).

In this paper, a meteorological element detection system was designed, and the three elements (temperature, humidity, and pressure) and elevation, latitude, and longitude were collected and saved in the SD card from the controller. Meanwhile, the data were sent to the host computer of the meteorological element detection system through the wireless data transmission module for real-time display.

4.3. Design of Host Computer Software. The core task of the host computer software of the meteorological element detection system designed in this paper was to display, process, and store data, as shown in Figure 16.

The host computer software communicated with the meteorological element detection system through the wireless data transmission module, which was convenient for the data transmission of meteorological elements. In addition to the operating system, there was no need to configure other software environment. This host computer was written using Visual Basic, and the interfaces included serial port parameter setting, real-time data display, time data, historical data query, and various function keys, which can realize the function of receiving and displaying meteorological elements.

The data of the host computer software and meteorological element detection system were transmitted through serial port; in the connection mode, the transmitting terminal of wireless data transmission module was connected with the meteorological element detection system, and the receiving terminal was connected with the ground PC terminal (computer), and the data were sent to the host computer for real-time display from the meteorological element detection system through the serial port communication between wireless data transmission modules. The meteorological element detection system conducted design of communication protocols for the transmitted data, including frame header, data length, data block, check, and frame tail, so as to ensure the integrity of the data. The protocol rules are shown in Table 1.

5. Building of Complete Six-Rotor UAV

The building of model six-rotor UAV was the premise for normal operation, and function requirements can be realized through welding, debugging, final checking function, and various indicators of the complete six-rotor UAV. The flight control board of six-rotor UAV is shown in Figure 17(a), and the meteorological element detection board of six-rotor UAV is shown in Figure 17(b).

In this paper, through hardware and software debugging, a meteorological element detection system was carried on a six-rotor UAV platform to achieve collection of meteorological elements of atmospheric boundary layer, such as temperature, relative humidity, pressure, elevation, latitude and longitude, and current time, and the collected data were saved in the SD card; meanwhile, the data were sent to the host computer of the meteorological element detection system through the wireless data transmission module for real-time display. In the test flight experiment, the six-rotor UAV platform maintained stable flight, and the collection and storage of meteorological elements by the meteorological element detection system as well as real-time display on the host computer were realized, as shown in Figure 18.

As can be seen from flight waveform in Figure 19, the six-rotor UAV platform carrying a meteorological element detection system showed small fluctuation of attitude angle during flight, which was around 0.4°, proving that the six-rotor UAV flight control platform designed in this paper had good self-stability effect.

6. Analysis of Experimental Data

To ensure the correctness and validity of data analysis, first, the influence of different installation positions of the meteorological element sensor on the data analysis result was analyzed, then, the meteorological elements were detected and the data were analyzed by a six-rotor UAV platform carrying a meteorological element detection system and DAVIS weather station collection board, and the validity of the system was verified.

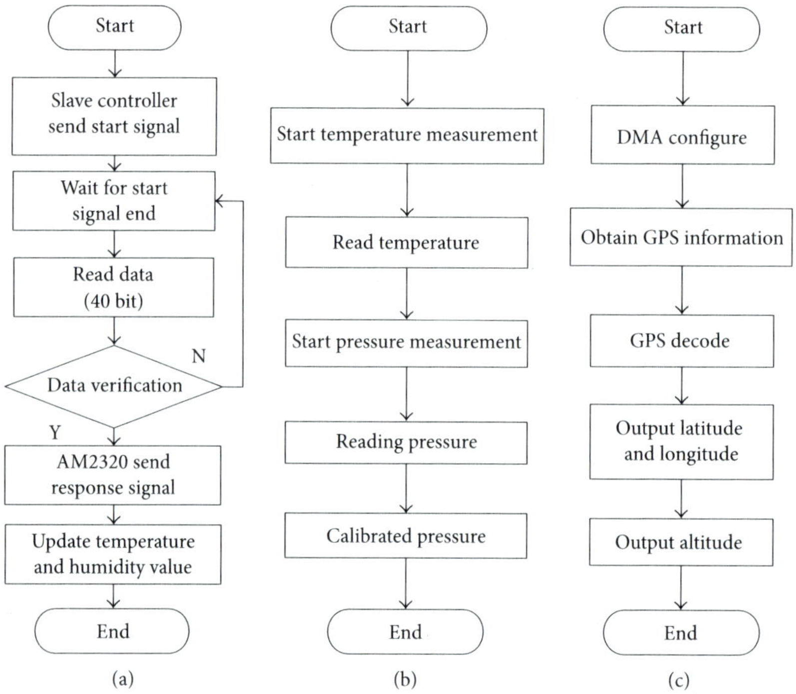

Figure 15: (a) Temperature and humidity collection. (b) Pressure collection. (c) Elevation, latitude, and longitude collection.

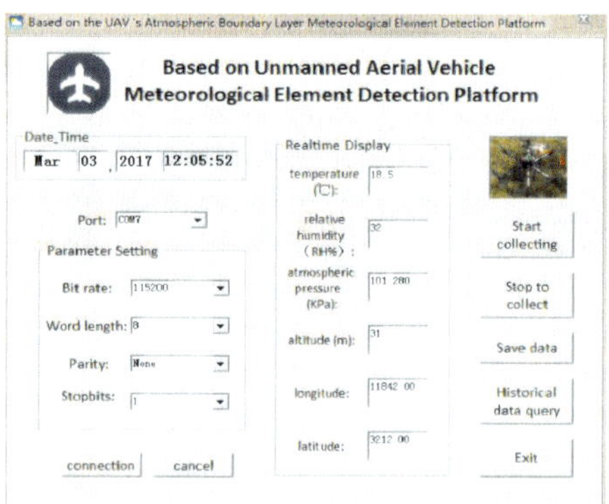

Figure 16: Host computer software.

Table 1: Format frame.

Explain	Code	Byte
Frame header	0x45, 0xBB, 0x7E	3
Data length	0x02	2
Data block	0x23, 0x45	0~255
Check	SUM	1
Frame tail	0x0D	1

6.1. Test of Position Installation. At present, it has been a trend that a meteorological element sensor is carried on a UAV platform to achieve detection of meteorological elements of the atmospheric boundary layer, which can realize detection and research of different regions, but the rotor wind of UAV has some influence on the detection accuracy of the meteorological element sensor. The integrated and coordinated detection by a UAV-borne multisensor has become the mainstream of the current and future work model, which is the future development trend. Therefore, in this chapter, based on the future integrated detection by a UAV-borne multisensor, different installation positions of the UAV-borne multisensor were studied, and the influence of different installation positions of the meteorological element sensor on the accuracy of detection data was analyzed through many experiments.

According to the survey, the experiment mainly conducted an analysis of detection data of three different installation positions of the meteorological element detection system. Through hydrodynamic analysis, the meteorological element detection system was installed below, above, and 10 cm above the center plate of six-rotor UAV frame. The test data of this experiment are shown in Figure 20.

The experimental data selected the test data in 10 minutes in the experimental test, and analysis of temperature, humidity, and pressure was conducted, respectively. As can be seen from Figure 20, when the meteorological element detection system was installed above and below the center plate of six-rotor UAV frame, the temperature and humidity curves showed large fluctuation, and the pressure curve showed small fluctuation; when the meteorological element detection system was installed 10 cm above the center plate of six-rotor UAV frame, the temperature and humidity curves showed smooth fluctuation, and the pressure curve showed small fluctuation.

In the experiment, when the meteorological element detection system was installed above and below the center plate of six-rotor UAV frame, the temperature and humidity curves showed large fluctuation, because of influence of the

(a) (b)

Figure 17: (a) Flight control board. (b) Detection board.

(a) (b)

Figure 18: (a) Model six-rotor UAV. (b) Test flight experiment.

downward rotor wind of six-rotor UAV as the positions above and below the center plate of six-rotor UAV frame were below the propeller; when the meteorological element detection system was installed 10 cm above the center plate of six-rotor UAV frame, the temperature and humidity curves showed smooth fluctuation, because of little influence of the rotor wind of six-rotor UAV as the position 10 cm above the center plate of six-rotor UAV frame was above the propeller. Therefore, the meteorological element detection system was installed 10 cm above the center plate of UAV frame.

6.2. Contrast Test. The meteorological elements such as temperature, humidity, and pressure were detected by a six-rotor UAV platform carrying a meteorological element sensor and DAVIS weather station collection board, and the experimental data selected two sets of test data in 10 minutes in the experimental test, as shown in Figure 21.

As can be seen from Figure 21, the temperature and humidity curves and pressure curve were consistent with the trend of curve of the data collected by the DAVIS weather station, which met the design requirements. As can be seen from Figure 21(a), the error between the temperature curve of the meteorological element detection system and the temperature curve of the DAVIS weather station was about 0.5°C, which was in line with the margin of error of temperature accuracy. Figures 21(c) and 21(d) show that the humidity curves of the meteorological element detection system and the DAVIS weather station were basically coincident with each other. Figures 21(e) and 21(f) show that the pressure curves of the meteorological element detection system and the DAVIS weather station showed certain fluctuation and the overall trends were the same due to the small fluctuation, so the data were valid.

7. Conclusions

In this paper, a meteorological element detection system was carried on a six-rotor UAV platform to achieve collection and analysis of meteorological elements of the atmospheric boundary layer. The main works were summarized as follows:

(1) In light of the influence of meteorological element detection system carried on a six-rotor UAV and other loads on the stability of six-rotor UAV, the

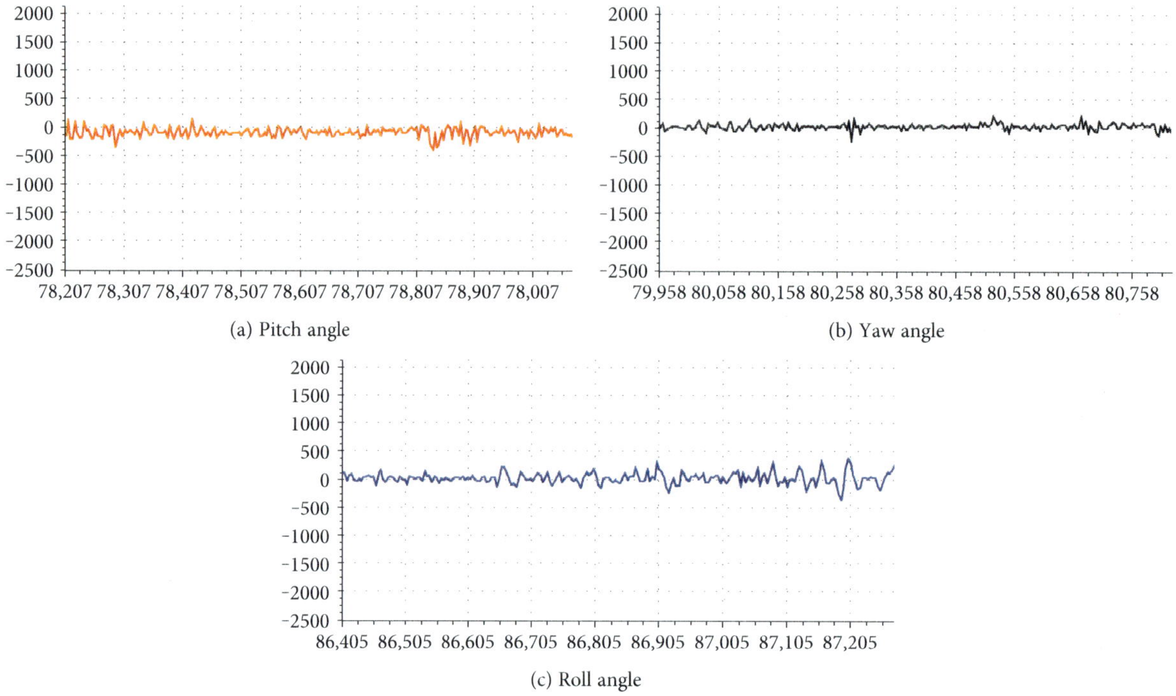

FIGURE 19: Flight attitude angle. Note: x-axis was time (t/s), and y-axis was angle (°/1000).

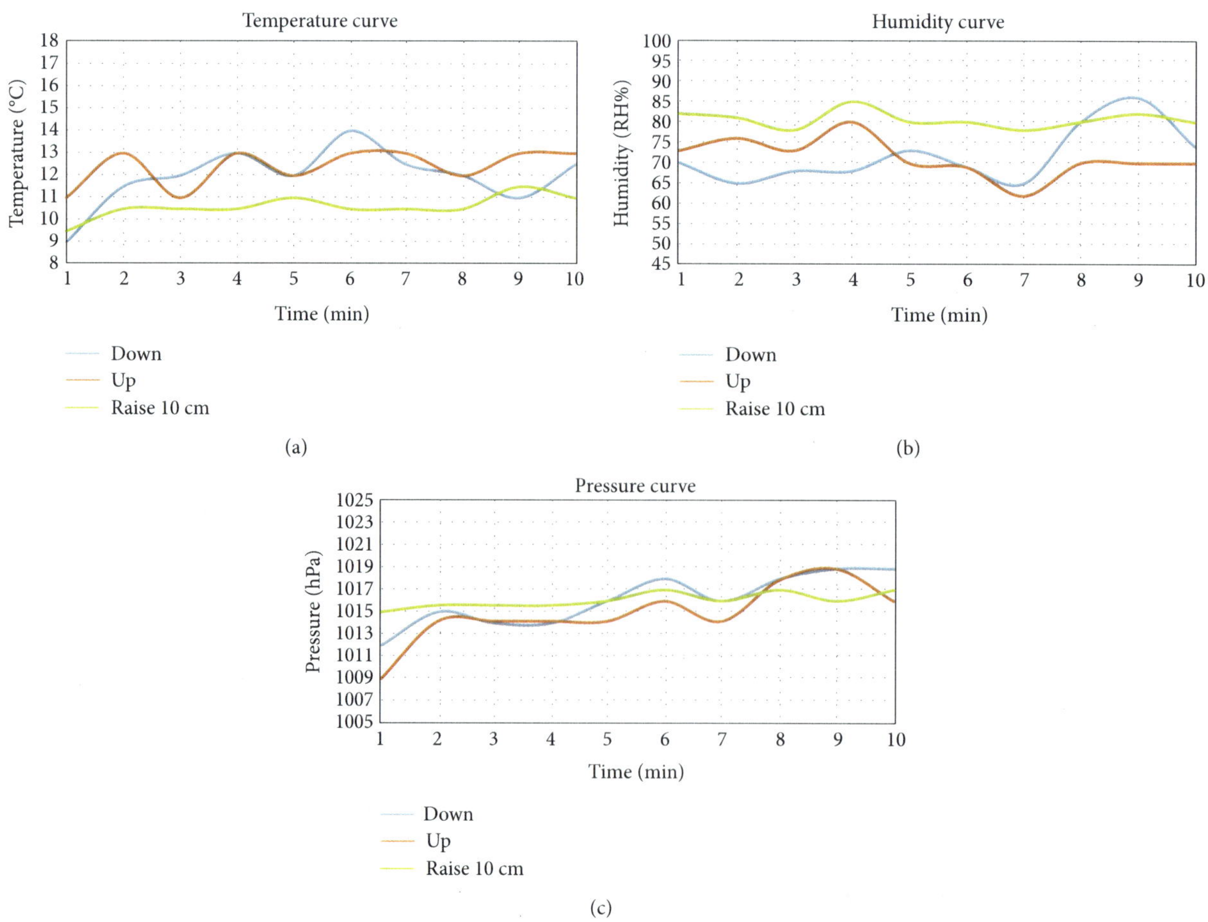

FIGURE 20: Test curves.

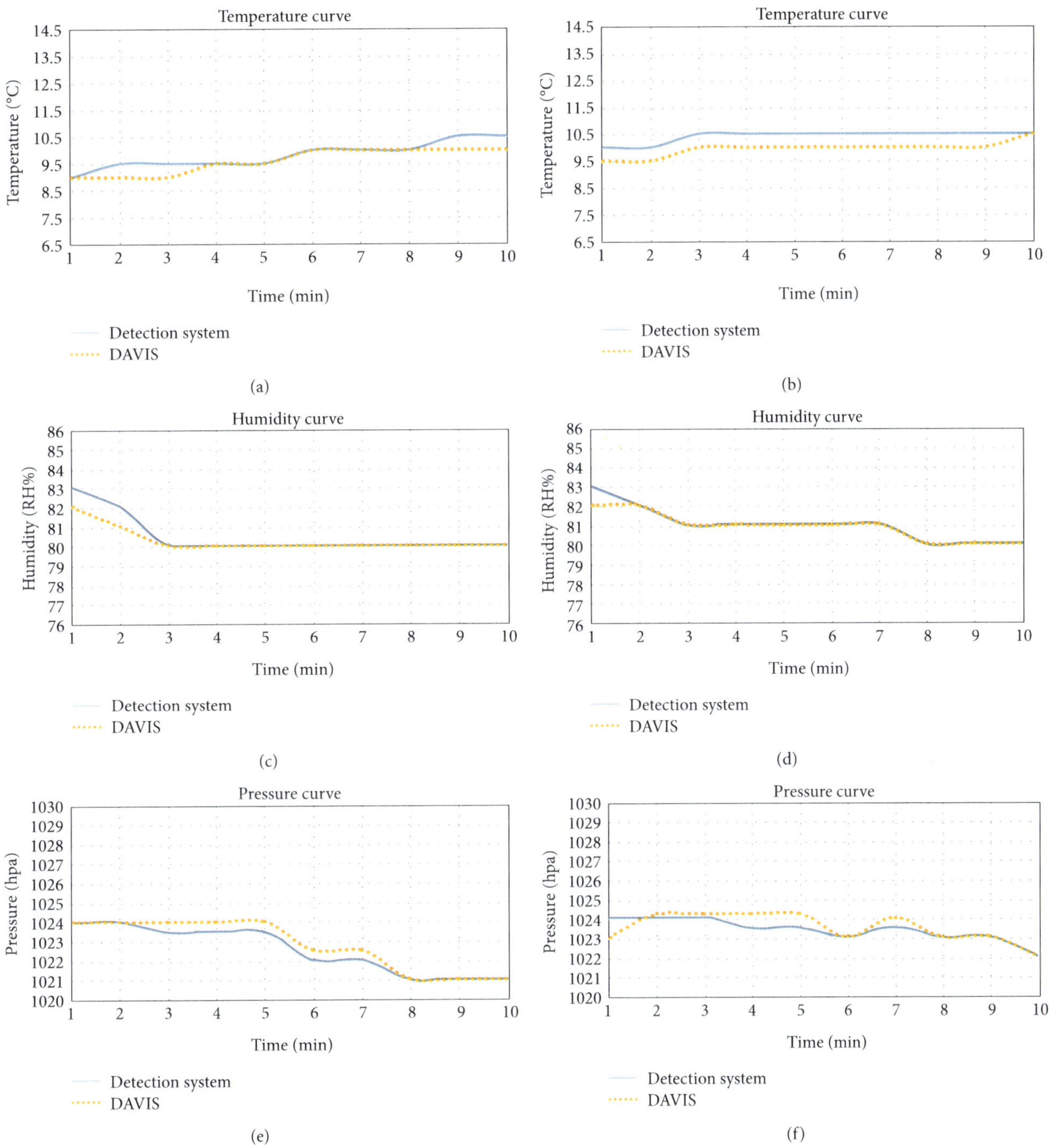

FIGURE 21: Test curves.

flight stability of six-rotor UAV was improved through the optimization of attitude controller and position controller.

(2) In light of the detection requirements of the atmospheric boundary layer, a meteorological element detection system was designed to achieve detection and storage of meteorological elements such as temperature, relative humidity, pressure, elevation, and latitude and longitude. In view of the influence of the rotor wind on the detection accuracy of the meteorological element detection sensor, the installation location is spatially higher than the propeller plane, and the influence of the rotor wind is the smallest. Through a large number of experimental studies, it is found that the installation position is 10 cm above the frame, and sensor accuracy can be achieved: temperature accuracy: $\triangle T \leq 0.5°C$; humidity accuracy: $\triangle U \leq 6RH\%$; air pressure accuracy: 0.5 hPa; altitude: 3 M, to ensure the accuracy of the detection results. It reveals the position relation between unmanned aerial vehicle and meteorological sensor, which is of great significance to guide the system structure design.

(3) In light of the meteorological elements collected, the temperature, humidity, and air pressure were analyzed by automatic weather station data. The temperature curve, humidity curve, and air pressure curve were consistent with the trend of data curve collected by DAVIS weather station, and the validity of the design was clarified.

Conflicts of Interest

The authors declare that there is no conflict of interests regarding the publication of this paper.

Acknowledgments

This work was supported by project supported by the National Natural Science Foundation of China (51575283 and 51405243).

References

[1] L. I. D. G. J. S. Zhenfeng, "From digital earth to smart earth," *Geomatics and Information Science of Wuhan University*, vol. 2, no. 002, 2010.

[2] D. R. Li, "Development prospect of photogrammetry and remote sensing," *Geomatics and Information Science of Wuhan University*, vol. 33, pp. 1211–1215, 2008.

[3] J. Kim, D. woo Lee, K. Cho et al., "Development of an electro-optical system for small UAV," *Aerospace Science and Technology*, vol. 14, no. 7, pp. 505–511, 2010.

[4] A. C. Watts, V. G. Ambrosia, and E. A. Hinkley, "Unmanned aircraft systems in remote sensing and scientific research: classification and considerations of use," *Remote Sensing*, vol. 4, no. 6, pp. 1671–1692, 2012.

[5] J. A. J. Berni, P. J. Zarco-Tejada, L. Suárez, and E. Fereres, "Thermal and narrowband multispectral remote sensing for vegetation monitoring from an unmanned aerial vehicle," *IEEE Transactions on Geoscience and Remote Sensing*, vol. 47, no. 3, pp. 722–738, 2009.

[6] H. Xiang and L. Tian, "Development of a low-cost agricultural remote sensing system based on an autonomous unmanned aerial vehicle (UAV)," *Biosystems Engineering*, vol. 108, no. 2, pp. 174–190, 2011.

[7] A. Khan, D. Schaefer, L. Tao et al., "Low power greenhouse gas sensors for unmanned aerial vehicles," *Remote Sensing*, vol. 4, no. 5, pp. 1355–1368, 2012.

[8] J. Kelcey and A. Lucieer, "Sensor correction of a 6-band multispectral imaging sensor for UAV remote sensing," *Remote Sensing*, vol. 4, no. 5, pp. 1462–1493, 2012.

[9] A. C. Watts, J. H. Perry, S. E. Smith et al., "Small unmanned aircraft systems for low-altitude aerial surveys," *Journal of Wildlife Management*, vol. 74, no. 7, pp. 1614–1619, 2010.

[10] M. Li, D. Li, and D. Fan, *A study on automatic UAV image mosaic method for paroxysmal disaster*, vol. 25, Proceedings of the International Society of Photogrammetry and Remote Sensing Congress, Melbourne, Australia, 2012.

[11] J. Reuder, P. Brisset, M. Jonassen, M. Müller, and S. Mayer, "SUMO: a small unmanned meteorological observer for atmospheric boundary layer research," *IOP Conference Series: Earth and Environmental Science*, vol. 1, no. 1, article 012014, 2008.

[12] L. Yangjun, *Control circuit design in digital aerial camera shooting of unmanned aerial vehicles*, Beijing University of Posts and Telecommunications, Beijing, 2008.

[13] J. Wei, G. Hongli, D. U. Hua-qiang, and X. J. Xu, "A review on unmanned aerial vehicle remote sensing and its application," *Remote Sensing Information*, vol. 1, pp. 88–92, 2009.

[14] H. Wu, *Unmanned Aircraft System Introduction (Second Edition)*, Publishing House of Electronices Industry, Beijing, 2003.

[15] A. Ameti, R. J. Fontana, E. J. Knight, and E. Richley, "Ultra wideband technology for aircraft wireless intercommunications systems (AWICS) design," *IEEE Aerospace and Electronic Systems Magazine*, vol. 19, no. 7, pp. 14–18, 2004.

[16] Y. Watanabe and P. Fabiani, "Optimal guidance design for UAV visual target tracking in an urban environment," *IFAC Proceedings Volumes*, vol. 43, no. 15, pp. 69–74, 2010.

7

Formation Control for Unmanned Aerial Vehicles with Directed and Switching Topologies

Yahui Qi, Shaolei Zhou, Yuhang Kang, and Shi Yan

Department of Control Engineering, Naval Aeronautical and Astronautical University, Yantai 264000, China

Correspondence should be addressed to Yahui Qi; luckyqiyahui@163.com

Academic Editor: Mahmut Reyhanoglu

Formation control problems for unmanned aerial vehicle (UAV) swarm systems with directed and switching topologies are investigated. A general formation control protocol is proposed firstly. Then, by variable transformation, the formation problem is transformed into a consensus problem, which can be solved by a novel matrix decomposition method. Sufficient conditions to achieve formation with directed and switching topologies are provided and an explicit expression of the formation reference function is given. Furthermore, an algorithm to design the gain matrices of the protocol is presented. Finally, numerical simulations are provided to illustrate the effectiveness of the theoretical results.

1. Introduction

In the past decades, unmanned aerial vehicles (UAVs) have been widely used in civilian and military areas, such as surveillance and reconnaissance [1, 2] and target search and localization [3]. Since the performance of a team of UAVs working cooperatively exceeds the performance of individual UAVs, formation control of UAVs is of importance and has received a lot of attention.

The formation control of UAVs has been studied with many different methods, such as leader–follower [4], behavior [5], and virtual structure-based [6] approaches. Recently, with the development of consensus theory [7–15], some related methods are also used to deal with the formation control problems of UAVs. Consensus means that all agents reach a common state. The results in [16] show that consensus approaches can be used to deal with formation control problems, and leader–follower, behavior, and virtual structure-based formation control approaches are special cases of consensus-based approaches.

Based on consensus method, Abdessameud and Tayebi [17] proposed controllers for UAV swarm systems to achieve formation in the presence of communication delays. A consensus protocol together with an output feedback linearization method is presented in [18] such that the UAV swarm systems can achieve partially time-varying formation. Besides, indoor and outdoor flight experiments for quadrotor swarm systems to achieve formation by consensus approaches are carried out in [19] and [20], respectively. Based on consensus theory, we know that the achievement of formation depends on not only the individual UAV dynamics but also the structure of the networks between UAVs which can be modeled by directed and undirected graphs. However, the interaction topologies between UAVs in [19, 20] are assumed to be fixed. In practical applications, the interaction topologies of UAV swarm systems may be switching due to the fact that the communication channel may fail and new channels may be created during flight. Time-varying formation control for UAV swarm systems and high-order LTI systems with switching interaction topologies are studied by Dong [20, 21], but the topologies are assumed to be undirected. To the best of our knowledge, there is still work to do on formation control of UAV swarm systems with directed and switching topologies.

In this paper, we aim to solve the formation problem of UAV swarm systems with directed and switching topologies. Compared with the existing results, the assumptions of the communication topology are quite general. The remainder of this paper is organized as follows. In Section 2, some necessary concepts and useful results on graph theory are summarized and the problem formulation is given. Main theoretical results are proposed in Section 3. In Section 4, a numerical simulation is presented. Section 5 is the conclusion.

2. Preliminaries and Problem Description

2.1. Notations and Graph Theory. In this paper, the following notations will be used. $R^{n\times n}$ and $C^{n\times n}$ denote the set of $n \times n$ real and complex matrices, respectively. For $\mu \in C$, the real part is $\mathrm{Re}(\mu)$. $\otimes$ denotes the Kronecker product. I_n is the identity matrix of order n. For a square matrix A, $\lambda(A)$ denotes the eigenvalues of matrix A. $A > 0$ ($A \geq 0$) means that A is positive definite (positive semidefinite). $\max\{\lambda(A)\}$ ($\min\{\lambda(A)\}$) denotes the largest (smallest) eigenvalue of the matrix A.

A directed graph $G = (\mathcal{V}, \mathcal{E}, \mathcal{A})$ contains the vertex set $\mathcal{V} = \{1, 2, \ldots, N\}$, the directed edges set $\mathcal{E} \subseteq \mathcal{V} \times \mathcal{V}$, and the adjacency matrix $\mathcal{A} = [a_{ij}]_{N\times N}$ with nonnegative elements a_{ij}. $a_{ij} = 1$ if there is a directed edge from vertex j to i; $a_{ij} = 0$, otherwise. The Laplacian matrix of the graph G is defined as $L = [L_{ij}]_{N\times N}$, where $L_{ii} = \sum_{j\neq i} a_{ij}$ and $L_{ij} = -a_{ij}$ $(i \neq j)$. Zero is an eigenvalue of L with the eigenvector $\mathbf{1}_N$. A directed graph is said to have a spanning tree if there is a vertex such that there is a directed path from this vertex to every other vertex.

Lemma 1 (see [8]). *Zero is a simple eigenvalue of L and all the other nonzero eigenvalues have positive real parts if and only if the graph has a directed spanning tree.*

2.2. Problem Description. Consider UAV swarm systems with N UAVs. The interaction topology of the UAV swarm systems can be described by a directed graph G, in which UAV i can be denoted by a vertex and the interaction channel from UAV i to UAV j can be denoted by an edge. Compared with the attitude dynamics, the trajectory dynamics of each UAV have much larger time constants, which means the attitude controller and trajectory controller can be designed separately. On the formation level, only trajectory control needs to be considered. Therefore, in this brief, the dynamics of each UAV can be described by the following double integrator [18, 21, 22]:

$$\begin{aligned} \dot{x}_i(t) &= v_i(t), \\ \dot{v}_i(t) &= u_i(t), \end{aligned} \tag{1}$$

where $i = 1, 2, \ldots, N$, $x_i(t) \in R^n$ and $v_i(t) \in R^n$ denote the position and velocity vectors of UAV i, respectively, and $u_i(t) \in R^n$ are the control inputs. In the following, for simplicity of description, it is assumed that $n = 1$, if not otherwise specified.

Therefore, UAV swarm systems (1) can be rewritten as

$$\dot{\xi}_i(t) = A\xi_i(t) + Bu_i(t), \tag{2}$$

where $\xi_i(t) = [x_i(t), v_i(t)]^T$, $A = \left[\begin{smallmatrix} 0 & 1 \\ 0 & 0 \end{smallmatrix}\right]$, $B = \left[\begin{smallmatrix} 0 \\ 1 \end{smallmatrix}\right]$.

A formation is specified by a vector $h(t) = [h_1^T(t), h_2^T(t), \ldots, h_N^T(t)]^T \in R^{2N}$ with $h_i(t) = [h_{ix}(t), h_{iv}(t)]^T$ $(i = 1, 2, \ldots, N)$ continuously differentiable and $h_{iv}(t)$ being the derivative of $h_{ix}(t)$. Let $h_x(t) = [h_{x1}^T(t), h_{x2}^T(t), \ldots, h_{xN}^T(t)]^T$ and let $h_v(t) = [h_{v1}^T(t), h_{v2}^T(t), \ldots, h_{vN}^T(t)]^T$; then one has that if $h_v(t)$ are not equal to zeros, the formation is time-varying.

Definition 2 (see [21]). UAV swarm systems (2) are said to achieve formation $h(t)$ if there exists a function $r(t) = [r_x(t), r_v(t)] \in R^2$ with $r_v(t)$ being the derivative of $r_x(t)$ such that

$$\lim_{t\to\infty} (\xi_i(t) - h_i(t) - r(t)) = 0, \quad i = 1, 2, \ldots, N, \tag{3}$$

where $r(t)$ is called a formation center function.

In this paper, the communication topology is molded by a directed graph and we assume that the communication topology is time-varying. Let $\widehat{G} = \{G^1, G^2, \ldots, G^p\}$, $p \geq 1$, be the set of all possible directed topologies. We define the switching signal $\sigma(t)$, where $\sigma(t) : [0, +\infty) \to \mathcal{P} = \{1, 2, \ldots, p\}$. $0 = t_0 < t_1 < t_2 < \cdots$ denote the switching instants of $\sigma(t)$. Let $G^{\sigma(t)} \in \widehat{G}$ be the communication topology at time t. $L^{\sigma(t)}$ stands for the corresponding Laplacian matrix of $G^{\sigma(t)}$.

Assumption 3. Each possible graph $G^{\sigma(t)} \in \widehat{G}$ is fixed and contains a directed spanning tree.

Let $\lambda_i^{\sigma(t)}$ $(i = 1, 2, \ldots, N)$ be the eigenvalues of the Laplacian matrix $L^{\sigma(t)}$. Without loss of generality, it is assumed that $\mathrm{Re}(\lambda_1^{\sigma(t)}) \leq \mathrm{Re}(\lambda_2^{\sigma(t)}) \leq \cdots \leq \mathrm{Re}(\lambda_N^{\sigma(t)})$. Furthermore, from Lemma 1, one can obtain that $\lambda_1^{\sigma(t)} = 0$ and $0 \leq \mathrm{Re}(\lambda_2^{\sigma(t)}) \leq \cdots \leq \mathrm{Re}(\lambda_N^{\sigma(t)})$. Let $\lambda_{\min} = \min\{\lambda_i^m\ (\forall m \in \mathcal{P}; i = 2, 3, \ldots, N)\}$.

Consider the following formation protocol:

$$\begin{aligned} u_i(t) = {} & K_1(\xi_i(t) - h_i(t)) \\ & + K_2 \sum_{j=1}^{N} a_{ij}\left((\xi_j(t) - h_j(t)) - (\xi_i(t) - h_i(t))\right) \\ & + \dot{h}_{iv}(t), \quad i = 1, 2, \ldots, N, \end{aligned} \tag{4}$$

where $i = 1, 2, \ldots, N$, $K_1 \in R^{1\times 2}$ and $K_2 \in R^{1\times 2}$ are constant gain matrices, and a_{ij} is defined as in Section 2.1. Let $\xi(t) = [\xi_1^T(t), \xi_2^T(t), \ldots, \xi_N^T(t)]^T$, let $h_x(t) = [h_{1x}^T(t), h_{2x}^T(t), \ldots, h_{Nx}^T(t)]^T$, and let $h_v(t) = [h_{1v}^T(t), h_{2v}^T(t), \ldots, h_{Nv}^T(t)]^T$. Under protocol (4), the UAV swarm systems (2) can be written in a compact closed-loop form as follows:

$$\begin{aligned} \dot{\xi}(t) = {} & \left(I_N \otimes (A + BK_1) - L^{\sigma(t)} \otimes (BK_2)\right)\xi(t) \\ & - \left(I_N \otimes BK_1 - L^{\sigma(t)} \otimes (BK_2)\right)h(t) \\ & + (I_N \otimes B)\dot{h}_v(t). \end{aligned} \tag{5}$$

This brief mainly investigates how to design the gain matrices in protocol (4) for the UAV swarm systems (5) to achieve the formation $h(t)$.

3. Main Results

Let $z_i(t) = \xi_i(t) - h_i(t)$, $z(t) = [z_1^T(t), z_2^T(t), \ldots, z_N^T(t)]^T$. Then the UAV swarm systems can be rewritten as follows:

$$\dot{z}(t) = \left(I_N \otimes (A + BK_1) - L^{\sigma(t)} \otimes BK_2\right) z(t) + (I_N \otimes A) h(t) - (I_N \otimes I_n) \dot{h}(t) + (I_N \otimes B) \dot{h}_v(t). \quad (6)$$

As for $\dot{h}_{ix}(t) = h_{iv}(t)$, one can obtain that

$$(I_N \otimes A) h(t) - (I_N \otimes I_n) \dot{h}(t) + (I_N \otimes B) \dot{h}_v(t) = 0. \quad (7)$$

Thus (6) can be further rewritten as

$$\dot{z}(t) = \left(I_N \otimes (A + BK_1) - L^{\sigma(t)} \otimes BK_2\right) z(t). \quad (8)$$

It holds directly that UAV swarm systems (2) with directed and switching topologies achieve formation $h(t)$ if and only if system (8) achieves consensus.

Before the consensus analysis of system (8), the following lemmas and definition are introduced.

Lemma 4 (see [23]). *For a Laplacian matrix L of graph G and a full row rank matrix E defined as*

$$E = \begin{bmatrix} 1 & -1 & 0 & \cdots & 0 \\ 0 & 1 & -1 & \cdots & 0 \\ \vdots & \vdots & \ddots & \vdots & \vdots \\ 0 & 0 & 0 & 1 & -1 \end{bmatrix}, \quad (9)$$

there exists a matrix M such that L = ME. Further, if the graph has a directed spanning tree, M is of full column rank and the eigenvalues of EM are equal to the nonzero eigenvalues of L.

Lemma 5 (see [24]). *Suppose that the eigenvalues of* $A \in R^{N\times N}$ *have positive real parts; then there exists a positive definite matrix* $Q > 0$ *such that*

$$A^T Q + QA > 0. \quad (10)$$

Definition 6. For a switching signal $\sigma(t)$ over time interval $[0, t)$, the average dwell time of the switching signal is defined as $\tau_a = t/(N_\sigma(t) + 1)$, where $N_\sigma(t)$ denotes the number of the switches.

Remark 7. In [11, 25], the definition of the average dwell time of a switching signal $\sigma(t)$ over time interval $[0, t)$ can be described as follows. If there exist two positive numbers N_0 and τ_a such that $N_\sigma(t) \le N_0 + t/\tau_a$, where $N_\sigma(t)$ denotes the number of the switches, τ_a is called the average dwell time. It is inaccurate to give the definition by an inequality, but, according to Definition 6, it can be seen that $N_\sigma(t) \le t/\tau_a$.

From Lemma 4, one can obtain that, for each $L^{(i)}$, $i \in \mathscr{P}$, there exists a matrix $M^{(i)}$ such that $L^{(i)} = M^{(i)}E$. Given Assumption 3 and Lemma 1, it can be known that the eigenvalues of each $EM^{(i)}$ $(i \in \mathscr{P})$ have positive real parts. Based on Lemma 5, one can obtain that there exist positive definite matrices $Q^{(i)}$ such that

$$\left(EM^{(i)} - \alpha I\right)^T Q^{(i)} + Q^{(i)} \left(EM^{(i)} - \alpha I\right) > 0, \quad (11)$$

where $\alpha < \lambda_{\min}$. Further, one can obtain that

$$\left(EM^{(i)}\right)^T Q^{(i)} + Q^{(i)} EM^{(i)} > 2\alpha Q^{(i)}, \quad i \in \mathscr{P}. \quad (12)$$

Let $\theta_i(t) = z_{i+1}(t) - z_i(t)$, $i = 1, 2, \ldots, N$, and let $\theta(t) = [\theta_1^T(t), \theta_2^T(t), \ldots, \theta_N^T(t)]^T$. One can obtain that $\theta(t) = (E \otimes I_n) z(t)$, where E is defined as in Lemma 4.

Premultiplying both sides of (8) by $(E \otimes I_n)$ leads to

$$\dot{\theta}(t) = \left(I_{N-1} \otimes (A + BK_1) - EM^{\sigma(t)} \otimes BK_2\right) \theta(t), \quad (13)$$

where $M^{\sigma(t)} = L^{\sigma(t)} E^T (EE^T)^{-1}$.

According to the definition of $\theta(t)$, it is obvious that $z_1(t) = z_2(t) = \cdots = z_N(t)$ if and only if $\theta(t) = 0$. So if system (13) converges to zero, system (8) achieves consensus and UAV swarm systems (2) with directed and switching topologies achieve formation $h(t)$.

Theorem 8. *Suppose that Assumption 3 holds. The formation problem of UAV swarm systems (2) with directed and switching topologies can be solved by controller (4) if there exists a positive definite matrix P such that*

$$(A + BK_1)^T P + P(A + BK_1) - 2\alpha PBB^T P + \beta P \le 0, \quad (14)$$

where $\alpha < \lambda_{\min}$, $\beta > \ln h/\tau_a$, $h = \varphi_1/\varphi_2$, $\varphi_1 = \max_{i\in\mathscr{P}}\{\lambda(Q^{(i)})\}$, $\varphi_2 = \min_{i\in\mathscr{P}}\{\lambda(Q^{(i)})\}$, *and* $Q^{(i)}$ *satisfies (12). The feedback matrix is designed as* $K_2 = B^T P$.

Proof. Consider the following piecewise Lyapunov candidate of system (13):

$$V(t) = \theta(t)^T \left(Q^{\sigma(t)} \otimes P\right) \theta(t), \quad (15)$$

where P is a solution of inequality (14) and $Q^{\sigma(t)}$ are feasible solutions of (12). □

Note that the communication topology is fixed for $t \in [t_i, t_{i+1})$, $i = 0, 1, \ldots$. Then, the derivation of this Lyapunov candidate along the trajectory of system (13) within the interval is

$$\dot{V}(t) = \theta(t)^T \left(I_{N-1} \otimes (A + BK_1) - EM^{\sigma(t)} \otimes BK_2\right)^T \cdot \left(Q^{\sigma(t)} \otimes P\right) \theta(t) + \theta(t)^T \left(Q^{\sigma(t)} \otimes P\right) \cdot \left(I_{N-1} \otimes (A + BK_1) - EM^{\sigma(t)} \otimes BK_2\right) \cdot \theta(t). \quad (16)$$

Substituting $K_2 = B^T P$ into (16) yields

$$\dot{V}(t) = \theta(t)^T \left(Q^{\sigma(t)} \otimes \left((A + BK_1)^T P + P(A + BK_1)\right) - \left(\left(EM^{\sigma(t)}\right)^T Q^{\sigma(t)} + Q^{\sigma(t)} EM^{\sigma(t)}\right) \otimes PBB^T P\right) \cdot \theta(t). \quad (17)$$

It then follows from (12) that

$$\dot{V}(t) \leq \theta(t)^T \left(Q^{\sigma(t)} \otimes \left((A+BK_1)^T P + P(A+BK_1) - 2\alpha PBB^T P\right)\right) \cdot \theta(t). \tag{18}$$

Based on (14), one has

$$\dot{V}(t) < -\beta\theta(t)^T \left(Q^{\sigma(t)} \otimes P\right)\theta(t). \tag{19}$$

Thus, from (15), one can obtain that

$$V(t) < e^{-\beta(t-t_i)} V(t_i). \tag{20}$$

Note that the communication topology switches at $t = t_i$; then one can get

$$V(t_i) \leq hV(t_i^-), \tag{21}$$

where $h = \varphi_1/\varphi_2$, $\varphi_1 = \max_{i\in\mathcal{P}}\{\lambda(Q^{(i)})\}$, and $\varphi_2 = \min_{i\in\mathcal{P}}\{\lambda(Q^{(i)})\}$.

Thus, when $t \in [t_i, t_{i+1})$, from (20) and (21), one has

$$\begin{aligned} V(t) &< e^{-\beta(t-t_i)} hV(t_i^-) < e^{-\beta(t-t_i)} he^{-\beta(t_i-t_{i-1})} V(t_{i-1}) \\ &< e^{-\beta t} h^i V(0). \end{aligned} \tag{22}$$

Since $i \leq N_\sigma(t) \leq t/\tau_a$,

$$V(t) < e^{-(\beta - \ln h/\tau_a)t} V(0). \tag{23}$$

From (15), one can obtain that

$$\begin{aligned} V(0) &\leq \phi_1 \|\theta(0)\|^2, \\ \phi_2 \|\theta(t)\|^2 &\leq V(t), \end{aligned} \tag{24}$$

where $\phi_1 = \varphi_1 \max\{\lambda(P)\}$ and $\phi_2 = \varphi_2 \min\{\lambda(P)\}$.

According to (23) and (24), one has

$$\|\theta(t)\|^2 \leq \frac{\phi_1}{\phi_2} e^{-(\beta - \ln h/\tau_a)t} \|\theta(0)\|^2. \tag{25}$$

Note that $\beta > \ln h/\tau_a$; one has $\theta(t) \to 0$ as $t \to \infty$. This means that the consensus problem of system (8) is solved. Furthermore, formation for UAV swarm systems (2) with directed and switching topologies is achieved.

Corollary 9. *If UAV swarm systems (2) achieve formation $h(t)$, the formation center function $r(t)$ can be determined as follows:*

$$r(t) = e^{(A+BK_1)(t-t_i)} \left(p^{\sigma(t)T} \otimes I\right)\left(x(t_i) - h(t_i)\right), \quad t \in [t_i, t_{i+1}), \tag{26}$$

where $p^{\sigma(t)}$ is the left eigenvector of $L^{\sigma(t)}$ associated with eigenvalue 0 and $(p^{\sigma(t)})^T \mathbf{1}_N = 1$.

Proof. From [7], there exists a left eigenvector $p^{\sigma(t)}$ of $L^{\sigma(t)}$ associated with eigenvalue 0 and $(p^{\sigma(t)})^T \mathbf{1}_N = 1$. For $t \in [t_i, t_{i+1})$, $L^{\sigma(t)}$ is fixed and so is $p^{\sigma(t)}$. □

Premultiplying both sides of (8) by $((p^{\sigma(t)})^T \otimes I)$ results in

$$\begin{aligned} &\left(\left(p^{\sigma(t)}\right)^T \otimes I\right)\dot{z}(t) \\ &= (A+BK_1)\left(\left(p^{\sigma(t)}\right)^T \otimes I\right) z(t). \end{aligned} \tag{27}$$

Based on Definition 2, one can obtain that

$$\lim_{t\to\infty} \left(z(t) - (\mathbf{1}_N \otimes I)\, r(t)\right) = 0. \tag{28}$$

Premultiplying both sides by $((p^{\sigma(t)})^T \otimes I)$, one has

$$\lim_{t\to\infty} \left(\left(\left(p^{\sigma(t)}\right)^T \otimes I\right) z(t) - r(t)\right) = 0. \tag{29}$$

Therefore, a formation center function can be

$$r(t) = \left(\left(p^{\sigma(t)}\right)^T \otimes I\right) z(t). \tag{30}$$

It follows from (27) and (30) that

$$\dot{r}(t) = (A+BK_1)\, r(t). \tag{31}$$

Thus, (26) can be obtained.

Remark 10. As can be seen, the formation center is discontinuous due to the switching of the communication topology. In addition, K_1 can be used to design the motion modes of the formation center function. If $K_1 = 0$, protocol (4) becomes a totally distributed controller. K_2 has no effect on the formation center function.

Remark 11. Compared with [21, 22], the interaction topologies are more common. Formation for UAV swarm systems with directed and switching topologies is solved. Furthermore, the gain matrix was designed by solving an LMI, which is simpler than solving an algebraic Riccati equation in [22]. In fact, undirected topologies are just special cases of directed topologies. So the algorithms presented in this paper are applicable to those cases in [21, 22].

Based on the above results, a design procedure of protocol (4) can be summarized as follows. First, choose K_1 to design the motion modes of the formation center by assigning the eigenvalues of $(A+BK_1)$. Then design K_2 using the conclusion of Theorem 8.

4. Examples

In this section, we provide an example to illustrate the effectiveness of the above theoretical results. UAV swarm

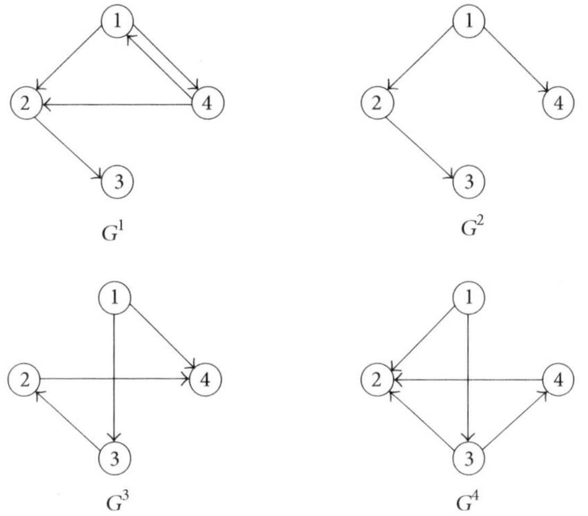

FIGURE 1: Communication topologies.

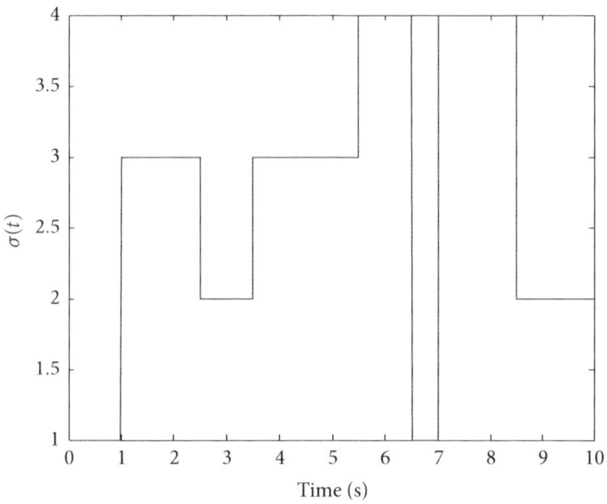

FIGURE 2: Switching signal $\sigma(t)$.

systems consisting of four agents are considered. The system matrices are defined as

$$x_i = \begin{bmatrix} x_{i1} \\ x_{i2} \\ x_{i3} \\ x_{i4} \end{bmatrix},$$

$$A = \begin{bmatrix} 0 & 1 & 0 & 0 \\ 0 & 0 & 0 & 0 \\ 0 & 0 & 0 & 1 \\ 0 & 0 & 0 & 0 \end{bmatrix}, \quad (32)$$

$$B = \begin{bmatrix} 0 & 0 \\ 1 & 0 \\ 0 & 0 \\ 0 & 1 \end{bmatrix},$$

where x_{i1}, x_{i2}, x_{i3}, and x_{i4} stand for east position, east velocity, north position, and north velocity. The directed communication topologies are given in Figure 1. Clearly, each topology contains a directed spanning tree. The switching signal is shown in Figure 2.

Thus, we can obtain $\lambda_{\min} = 1$ and then choose $\alpha = 0.9$. Further, we can get that $\varphi_1 = 3.4175$, $\varphi_2 = 0.2009$, and $h = 17.0075$. From Figure 2, we can get that the average dwell time is 1.25 s and then choose $\beta = 5$.

Assign the eigenvalues of $(A + BK_1)$ at $(\pm i, \pm i)$; we get

$$K_1 = \begin{bmatrix} -1 & 0 & 0 & 0 \\ 0 & 0 & -1 & 0 \end{bmatrix}. \quad (33)$$

Solve LMI (14) with $\alpha = 0.9$ and $\beta = 5$; a feasible solution can be obtained. Accordingly, we can get

$$K_2 = \begin{bmatrix} 0.6565 & 1.3431 & 0 & 0 \\ 0 & 0 & 0.6565 & 1.3431 \end{bmatrix}. \quad (34)$$

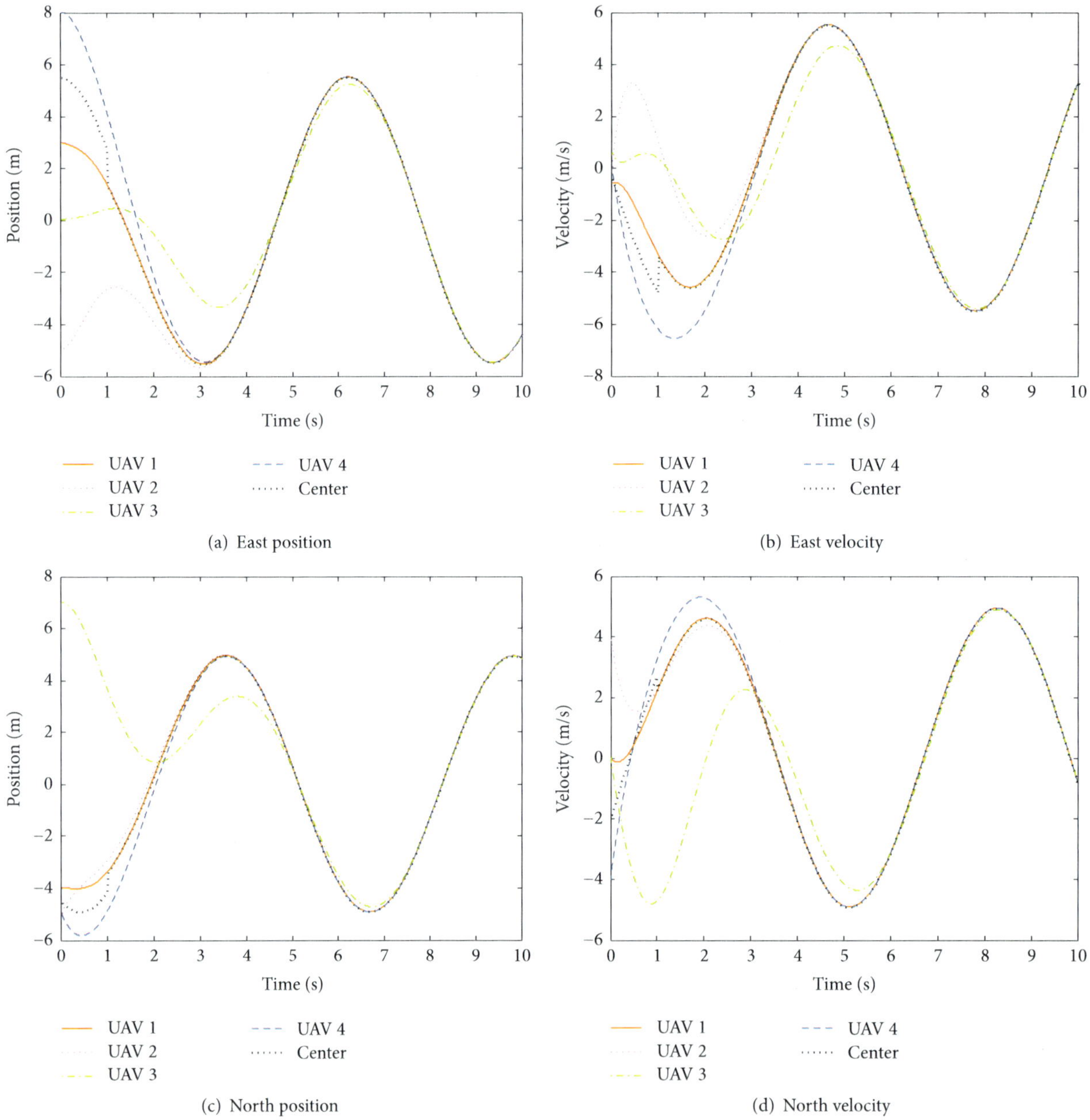

FIGURE 3: Difference of UAV state and time-varying formation.

Choose the following time-varying formation:

$$h_i(t) = \begin{bmatrix} 3\sin\left(0.2t + \dfrac{(i-1)\pi}{2}\right) \\ 0.6\cos\left(0.2t + \dfrac{(i-1)\pi}{2}\right) \\ 8\cos\left(0.5t + \dfrac{(i-1)\pi}{2}\right) \\ -4\sin\left(0.5t + \dfrac{(i-1)\pi}{2}\right) \end{bmatrix} \quad (i = 1, 2, 3, 4). \tag{35}$$

If $h(t)$ is achieved, both the positions and velocities of the four UAVs locate at the vertexes of a rotating parallelogram, respectively. Choose initial states of four UAVs as $x_1(0) = [3\ 0\ 4\ 0]^T$, $x_2(0) = [-2\ 0\ -5\ 0]^T$, $x_3(0) = [0\ 0\ -1\ 0]^T$, and $x_4(0) = [5\ 0\ -5\ 0]^T$.

Figure 3 shows the trajectories of the difference of UAV states and time-varying formation, which are denoted by solid line, dotted line, dash-dotted line, and dashed line. And the bold dotted line denotes the formation center trajectory. It is obvious that the differences achieve consensus after about $t = 7$ s and converge to the formation center. From Definition 2, one can obtain that the time-varying formation problem is solved. Figure 4 shows the snapshots of four UAV

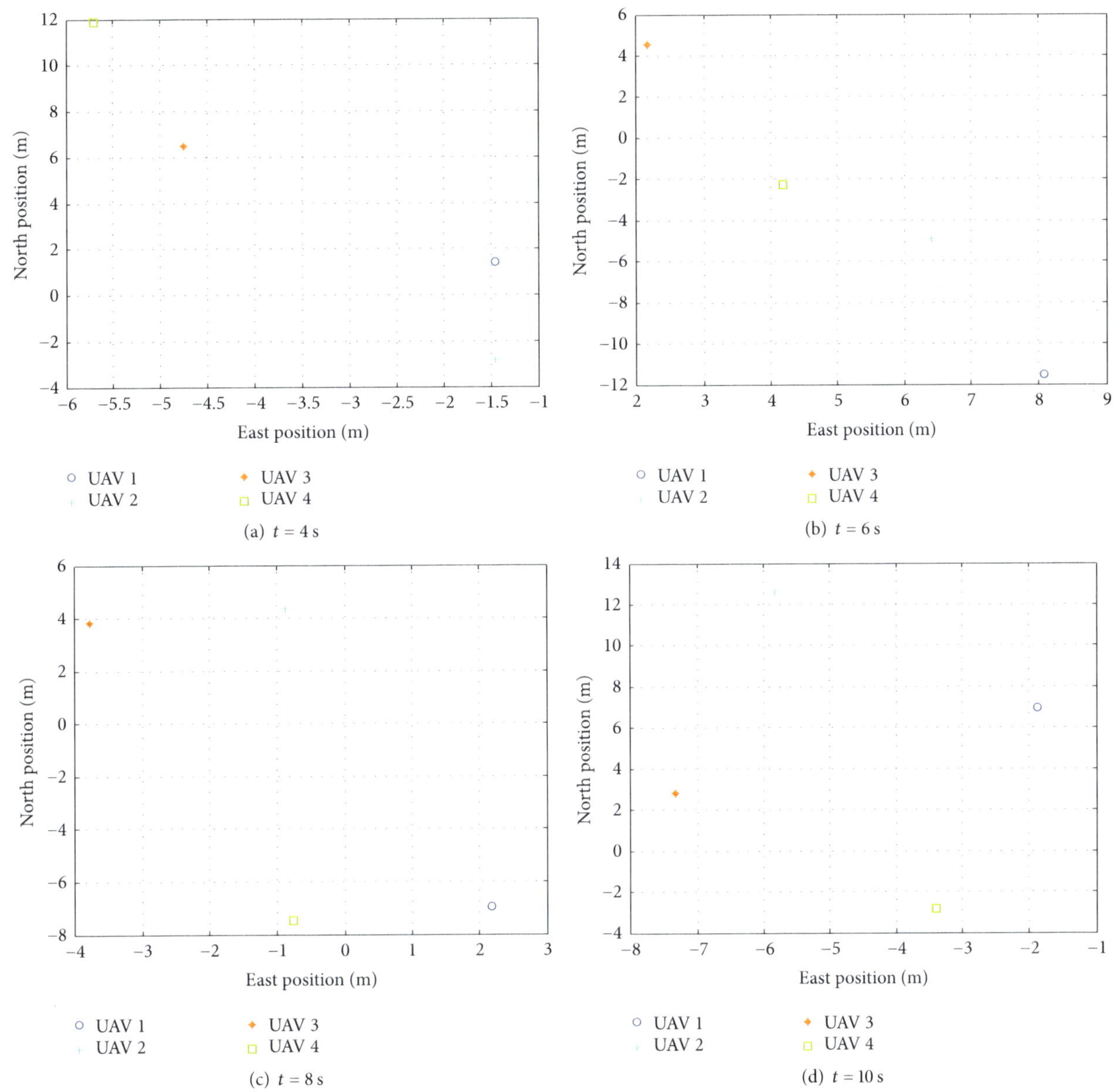

FIGURE 4: Snapshots of UAV positions.

positions at different time. It can be seen that, after $t = 6$ s, the UAV swarm systems achieve a time-varying parallelogram formation. Therefore, the time-varying formation is achieved under the directed and switching topologies.

5. Conclusions

Formation problems for UAV swarm systems with directed and switching topologies are studied. The average dwell time of the switching topologies is introduced, based on which an LMI-based method to design the protocol is proposed. Though the UAV swarm systems can achieve the specified formation with the presented method, there are still problems in real application. As mentioned in Assumption 3, each of the switching topologies is supposed to have a spanning tree, which may not be applicable, so there is still work to do in our future work.

Competing Interests

The authors declare that they have no competing interests.

References

[1] T. Kopfstedt, M. Mukai, M. Fujita, and C. Ament, "Control of formations of UAVs for surveillance and reconnaissance missions," in *Proceedings of the 17th IFAC World Congress*, pp. 6–11, Seoul, South Korea, July 2008.

[2] N. Nigam, S. Bieniawski, I. Kroo, and J. Vian, "Control of multiple UAVs for persistent surveillance: algorithm and flight

test results," *IEEE Transactions on Control Systems Technology*, vol. 20, no. 5, pp. 1236–1251, 2012.

[3] D. Pack, P. DeLima, G. Toussaint, and G. York, "Cooperative control of UAVs for localization of intermittently emitting mobile targets," *IEEE Transactions on Systems, Man, and Cybernetics B*, vol. 39, no. 4, pp. 959–970, 2009.

[4] Z. Lin, W. Ding, G. Yan, C. Yu, and A. Giua, "Leader-follower formation via complex Laplacian," *Automatica*, vol. 49, no. 6, pp. 1900–1906, 2013.

[5] T. Balch and R. C. Arkin, "Behavior-based formation control for multirobot teams," *IEEE Transactions on Robotics and Automation*, vol. 14, no. 6, pp. 926–939, 1998.

[6] M. A. Lewis and K.-H. Tan, "High precision formation control of mobile robots using virtual structures," *Autonomous Robots*, vol. 4, no. 4, pp. 387–403, 1997.

[7] W. Ren and E. Atkins, "Distributed multi-vehicle coordinated control via local information exchange," *International Journal of Robust and Nonlinear Control*, vol. 17, no. 10-11, pp. 1002–1033, 2007.

[8] W. Ren, "On consensus algorithms for double-integrator dynamics," *IEEE Transactions on Automatic Control*, vol. 53, no. 6, pp. 1503–1509, 2008.

[9] Z. Li, Z. Duan, G. Chen, and L. Huang, "Consensus of multiagent systems and synchronization of complex networks: a unified viewpoint," *IEEE Transactions on Circuits and Systems. I. Regular Papers*, vol. 57, no. 1, pp. 213–224, 2010.

[10] S. Su and Z. Lin, "Distributed consensus control of multi-agent systems with higher order agent dynamics and dynamically changing directed interaction topologies," *IEEE Transactions on Automatic Control*, vol. 61, no. 2, pp. 515–519, 2016.

[11] I. Saboori and K. Khorasani, "H_∞ consensus achievement of multi-agent systems with directed and switching topology networks," *IEEE Transactions on Automatic Control*, vol. 59, no. 11, pp. 3104–3109, 2014.

[12] Y. Zhu, C. Chen, X. Guan, and X. Luo, "Finite-time consensus of leader-follower multi-agent systems with fixed and switching topologies," in *Proceedings of the 11th World Congress on Intelligent Control and Automation (WCICA '14)*, pp. 226–231, Shenyang, China, July 2014.

[13] Y. Su and J. Huang, "Stability of a class of linear switching systems with applications to two consensus problems," *IEEE Transactions on Automatic Control*, vol. 57, no. 6, pp. 1420–1430, 2012.

[14] G. Wen, W. Yu, J. Cao, G. Hu, and G. Chen, "Consensus control of switching directed networks with general linear node dynamics," in *Proceedings of the 9th Asian Control Conference (ASCC '13)*, pp. 1–6, IEEE, Istanbul, Turkey, June 2013.

[15] G. Wen and V. Ugrinovskii, "Distributed consensus of linear multi-agent systems with switching directed topologies," in *Proceedings of the Australian Control Conference*, pp. 146–151, IEEE, Canberra, Australia, November 2014.

[16] W. Ren, "Consensus strategies for cooperative control of vehicle formations," *IET Control Theory & Applications*, vol. 1, no. 2, pp. 505–512, 2007.

[17] A. Abdessameud and A. Tayebi, "Formation control of VTOL unmanned aerial vehicles with communication delays," *Automatica*, vol. 47, no. 11, pp. 2383–2394, 2011.

[18] J. Seo, Y. Kim, S. Kim, and A. Tsourdos, "Consensus-based reconfigurable controller design for unmanned aerial vehicle formation flight," *Proceedings of the Institution of Mechanical Engineers, Part G: Journal of Aerospace Engineering*, vol. 226, no. 7, pp. 817–829, 2012.

[19] M. Turpin, N. Michael, and V. Kumar, "Decentralized formation control with variable shapes for aerial robots," in *Proceedings of the IEEE International Conference on Robotics and Automation (ICRA '12)*, pp. 23–30, IEEE, St. Paul, Minn, USA, May 2012.

[20] B. Yu, X. Dong, Z. Shi, and Y. Zhong, "Formation control for quadrotor swarm systems: algorithms and experiments," in *Proceedings of the 32nd Chinese Control Conference (CCC '13)*, pp. 7099–7104, Xi'an, China, July 2013.

[21] Y. Zhou, X. Dong, G. Lu, and Y. Zhong, "Time-varying formation control for unmanned aerial vehicles with switching interaction topologies," in *Proceedings of the International Conference on Unmanned Aircraft Systems (ICUAS '14)*, pp. 1203–1209, Orlando, Fla, USA, May 2014.

[22] X. Dong, B. Yu, Z. Shi, and Y. Zhong, "Time-varying formation control for unmanned aerial vehicles: theories and applications," *IEEE Transactions on Control Systems Technology*, vol. 23, no. 1, pp. 340–348, 2015.

[23] S. Zhou, W. Liu, Q. Wu, and G. Yin, "Leaderless consensus of linear multi-agent systems: matrix decomposition approach," in *Proceedings of the 7th International Conference on Intelligent Human-Machine Systems and Cybernetics (IHMSC '15)*, pp. 327–331, Hangzhou, China, August 2015.

[24] C. T. Cben, *Linear System Theory and Design*, Holt, Rinehart and Winston, Inc, New York, NY, USA, 1970.

[25] D. Liberzon, *Switching in Systems and Control*, Birkhauser, Boston, Mass, USA, 2003.

8

A Fault-Tolerant Multiple Sensor Fusion Approach Applied to UAV Attitude Estimation

Yu Gu,[1] Jason N. Gross,[1] Matthew B. Rhudy,[2] and Kyle Lassak[1]

[1] *Department of Mechanical and Aerospace Engineering (MAE) at West Virginia University (WVU), Morgantown, WV 26506, USA*
[2] *Division of Engineering, Pennsylvania State University, Reading, PA 19610, USA*

Correspondence should be addressed to Yu Gu; yu.gu@mail.wvu.edu

Academic Editor: Enrico C. Lorenzini

A novel sensor fusion design framework is presented with the objective of improving the overall multisensor measurement system performance and achieving graceful degradation following individual sensor failures. The Unscented Information Filter (UIF) is used to provide a useful tool for combining information from multiple sources. A two-step off-line and on-line calibration procedure refines sensor error models and improves the measurement performance. A Fault Detection and Identification (FDI) scheme crosschecks sensor measurements and simultaneously monitors sensor biases. Low-quality or faulty sensor readings are then rejected from the final sensor fusion process. The attitude estimation problem is used as a case study for the multiple sensor fusion algorithm design, with information provided by a set of low-cost rate gyroscopes, accelerometers, magnetometers, and a single-frequency GPS receiver's position and velocity solution. Flight data collected with an Unmanned Aerial Vehicle (UAV) research test bed verifies the sensor fusion, adaptation, and fault-tolerance capabilities of the designed sensor fusion algorithm.

1. Introduction

Sensing through a fusion of diverse but interrelated sensory data could reveal information that is difficult to measure directly. Having complementary and multiperspective viewpoints also allows for real-time evaluation of individual sensors' performance and limitations, further enabling a reconfiguration of the measurement system if necessary. From this perspective, sensor fusion can be viewed as a process of refining internal models of both the measured phenomenon (for improved performance) and the measurement system (for improved reliability) through processing a heterogeneous set of sensory data. Within this context, a three-step sensor fusion design framework is presented in this paper.

Step 1. Combine sensory data from diverse and redundant sources to derive a fused solution that is difficult to measure directly and/or has better quality than the output of each participating sensor.

Step 2. Refine sensor error models with feedback from the sensor fusion algorithm.

Step 3. Reconfigure the measurement system to achieve performance enhancement under nominal conditions and graceful degradation following sensor failures.

Step 1 reflects a "traditional" view of sensor fusion. Many existing approaches are model based, which relies on a set of mathematical models to relate individual measurements to the fused solution. These models represent *a priori* knowledge of the system to be measured and are based either on known relationships or on assumptions and heuristics. For numerical sensory data, a Bayesian filter [1, 2] is often used to derive a fused solution. This solution can either be computed at a centralized location or distributed among sensor nodes [3–5].

Step 2 enables the measurement system to maintain an updated knowledge of its states. At the sensor level, redundant information sources provide references for crosschecking and calibrating individual sensors used within the measurement system. This extends the traditional off-line sensor calibration [6] process to an on-line process. If a certain mathematical structure of the sensor model is assumed, the dynamic sensor calibration reduces to a parameter identification [7, 8] problem. For example, sensor bias or scaling factors are often estimated along with other states within a sensor fusion algorithm [9].

Step 3 integrates results from the first two levels to further improve the performance and robustness of the measurement system in several aspects. First, the sensor error models refined through the dynamic calibration process could improve the overall estimation performance and provide indications of the health condition of each sensor. Second, a comparison of sensory data collected from diverse but interrelated sources provides information for sensor Fault Detection, Identification, and Accommodation (FDIA) [10, 11]. Finally, the robustness of the measurement system could be improved by rejecting [4] low quality or faulty sensor measurements from the measurement update.

The goal of this paper is to demonstrate this three-step sensor fusion approach through a practical application: to achieve reliable and accurate attitude estimation with a low-cost Inertial Measurement Unit (IMU), Global Positioning System (GPS) receiver, and triaxial magnetometers. As an important navigation problem, the 3D attitude for Unmanned Aerial Vehicles (UAVs), mobile robots, and mobile devices were estimated using a variety of information sources. This includes but not limited to dead reckoning with rate gyroscopes, sensing of earth's gravity [12] and magnetic vectors [13], angular position of celestial objects [14], horizon line [15], terrain shape [16], optical flow [17], and known radio sources [18] such as GPS [19] and cellular network [20]. The selection of IMU, GPS, and magnetometers as the primary sensors for this study was mainly due to their widespread availability and popularity in various platforms, as well as the low-computational requirement of implementing these sensors when compared with methods such as computer vision or Lidar based mapping.

The novel contributions of this effort include the following:

(1) A systematic design approach that integrates both sensor calibration and fault tolerance into a multiple sensor fusion system.

(2) A FDI method based on information crosschecking and sensor bias tracking.

The performance of all presented algorithms is evaluated with flight datasets, through processing of real flight data taken from an experimental UAV. Portions of this paper, including some figures and tables, are contained within a chapter of the second author's graduate thesis [21].

The rest of the paper is organized as follows. Section 2 introduces four different sensor fusion formulations for vehicle attitude estimation using Unscented Information Filter (UIF) as the nonlinear estimator. The sensor calibration process for refining sensor error models is described in Section 3. Section 4 presents the FDIA methods. Section 5 discusses the experimental setup used in this study. The main results of this work are presented in Section 6, followed by a conclusion in Section 7.

2. Fusion of GPS, IMU, and Magnetometer Measurements

2.1. Coordinate Systems. Two coordinate systems are used throughout this paper. A local-level Cartesian navigation frame (L) is defined with its origin O^L at an arbitrary point near the vehicle, positive x^L axis pointing toward the geographic north, positive y^L axis pointing east, and positive z^L axis pointing to the center of the earth. A vehicle body-axes coordinate system (B) is defined with its origin at vehicle's Center of Gravity (CG), with positive x^B pointing forward of the vehicle, positive y^B axis toward right, and positive z^B axis toward the bottom of the vehicle. Each sensor on-board of the UAV is assumed are assumed to be aligned with the body axes and to be located relatively close to each other (i.e., a very small lever arm between the GPS and IMU).

The rotation of measurements between the two coordinate systems is calculated through the use of three attitude (Euler) angles: yaw (ψ), pitch (θ), and roll (ϕ). For example, the earth's gravity vector and centrifugal acceleration due to the earth's rotation (i.e., sensed "plumb-bob" acceleration) components in the vehicle body axis are found by assuming that it is parallel to the navigation frame z^L axis and rotating with a Direction Cosine Matrix (DCM):

$$\begin{bmatrix} g_x^B \\ g_y^B \\ g_z^B \end{bmatrix} = \mathrm{DCM}(\phi,\theta,\psi)^T \cdot \begin{bmatrix} 0 \\ 0 \\ g \end{bmatrix} \quad (1)$$

that is defined by the body-axis attitude:

$$\mathrm{DCM}(\phi,\theta,\psi) = \begin{bmatrix} c\psi c\theta & -s\psi c\phi + c\psi s\theta s\phi & s\psi s\phi + c\psi s\theta c\phi \\ s\psi c\theta & c\psi c\phi + s\psi s\theta s\phi & -c\psi s\phi + s\psi s\theta c\phi \\ -s\theta & c\theta s\phi & c\theta c\phi \end{bmatrix}, \quad (2)$$

where "s" and "c" are abbreviated sine and cosine functions, respectively.

2.2. Information Sources. Two types of information sources are generally available for estimating a vehicle's attitude angles: (1) the time-integration of rate gyroscope measurements; (2) the measurement of external vector fields of well-known directions. With readings from a set of 3-axis strapdown rate gyroscopes, the attitude angles are computed with a set of attitude kinematic equations [22]:

$$\begin{aligned} \dot{\phi} &= p + q\sin\phi\tan\theta + r\cos\phi\tan\theta, \\ \dot{\theta} &= q\cos\phi - r\sin\phi, \\ \dot{\psi} &= (q\sin\phi + r\cos\phi)\sec\theta, \end{aligned} \quad (3)$$

where p, q, and r are the roll rate, pitch rate, and yaw rate measured in the body axis, respectively. This attitude solution will diverge over the time due to the accumulation of rate gyroscope biases during the integration process. For typical low-cost microelectromechanical systems- (MEMS-) based gyroscopes, the uncorrected attitude solutions are not directly usable after a short period of time.

A standard approach for controlling the attitude error growth is to regulate it with nondrifting information sources, such as the known direction of an external vector field. Examples of commonly used aiding information include the earth's gravity vector [12], the earth's magnetic vector [13], GPS [19], cellular network [20], celestial map [14], horizon line [15], terrain map [16], computer vision [17], and known radio sources [18]. From this list, earth's gravity and magnetic vector fields are often the easiest to measure and therefore are discussed in detail in this paper. Four sensor fusion formulations are presented in the rest of this section to show that a fusion of any combination of GPS, IMU, and magnetometers is capable of providing nondrifting Euler angle estimates.

2.3. Sensor Fusion Formulation #1: GPS/IMU. The attitude of a stationary vehicle can be directly solved from (1) using measurements from a set of 3-axis accelerometers. For a moving vehicle, the accelerometers measure both the acceleration due to gravity and the vehicle's acceleration in the inertial frame. To isolate the gravity vector from inertial acceleration, a GPS receiver's velocity solution provides an independent observation of the vehicle's inertial acceleration in the local Cartesian coordinates. Since the GPS will not sense the earth's gravity, the relationship between accelerometer measurements $\mathbf{a}^B$ and GPS velocity measurements $\mathbf{V}^L$ can be described by

$$\begin{bmatrix} \dot{V}_x^L \\ \dot{V}_y^L \\ \dot{V}_z^L \end{bmatrix}_{\text{GPS}} = \text{DCM}(\phi, \theta, \psi) \cdot \begin{bmatrix} a_x^B \\ a_y^B \\ a_z^B \end{bmatrix}_{\text{IMU}} + \begin{bmatrix} 0 \\ 0 \\ g \end{bmatrix}. \quad (4)$$

Using the heading information provided by the GPS,

$$\psi = \tan^{-1}\left(\frac{V_y^L}{V_x^L}\right). \quad (5)$$

Equation (4) can be explicitly solved to calculate Euler angles [12]. However, a better approach exists with the use of the stationary gravity vector to regulate the INS integration error with a recursive estimator. Within this formulation, the state, input, and measurement vectors are, respectively, $\mathbf{x} = [\phi \ \theta \ \psi]^T$, $\mathbf{u} = [p \ q \ r]^T$, and $\mathbf{z} = \mathbf{a}^L = [a_x^L \ a_y^L \ a_z^L]^T$.

The nonlinear continuous-time state transition equations $\dot{\mathbf{x}} = \mathbf{f}^c(\mathbf{x}, \mathbf{u}, \mathbf{w})$ are directly based on (3):

$$\begin{bmatrix} \dot{\phi} \\ \dot{\theta} \\ \dot{\psi} \end{bmatrix} = \begin{bmatrix} (p + w_p) + (q + w_q)\sin\phi\tan\theta + (r + w_r)\cos\phi\tan\theta \\ (q + w_q)\cos\phi - (r + w_r)\sin\phi \\ ((q + w_q)\sin\phi + (r + w_r)\cos\phi)\sec\theta \end{bmatrix}, \quad (6)$$

where $w_{p,q,r}$ are the noises associated with the corresponding rate gyroscope measurements, which are assumed to be zero mean white Gaussian: $\mathbf{w}_1 = [w_p \ w_q \ w_r]^T \approx N(0, \mathbf{Q}_1)$. Note that $w_{p,q,r}$ are implemented as nonadditive input noises instead of modeling process noise additive on the state estimates, which is commonly used in Kalman filter formulations.

The continuousobservation equations $\mathbf{z} = \mathbf{h}^c(\mathbf{x}, \mathbf{v})$ are also nonlinear and are modeled by

$$\begin{bmatrix} \dot{V}_x^L + v_{gx} \\ \dot{V}_y^L + v_{gy} \\ \dot{V}_z^L + v_{gz} \end{bmatrix}_{\text{GPS}} = \text{DCM}(\phi, \theta, \psi) \cdot \begin{bmatrix} a_x^B + v_{ax} \\ a_y^B + v_{ay} \\ a_z^B + v_{az} \end{bmatrix}_{\text{IMU}} + \begin{bmatrix} 0 \\ 0 \\ g \end{bmatrix}, \quad (7)$$

where $v_{gx,gy,gz}$ are the noises assumed to be white for the GPS estimated accelerations and $v_{ax,ay,az}$ are noises associated with the accelerometer measurements, with $\mathbf{v}_1 = [v_{gx} \ v_{gy} \ v_{gz} \ v_{ax} \ v_{ay} \ v_{az}]^T \approx N(0, \mathbf{R}_1)$.

2.4. Sensor Fusion Formulation #2: IMU/Magnetometers. The gravity vector used in Formulation #1 can be directly substituted with the Earth's magnetic vector in regulating the growth of the rate gyroscope integration error. For this formulation, the state, input, and measurement vectors are given, respectively, by $\mathbf{x} = [\phi \ \theta \ \psi]^T$, $\mathbf{u} = [p \ q \ r]^T$, and $\mathbf{z} = [M_x^B \ M_y^B \ M_z^B]^T$, where $M_{x,y,z}^B$ are the magnetometer measurements. The state transition equations are the same as in (6), and the nonlinear observation equations are given by

$$\begin{bmatrix} M_x^B + v_{Mx} \\ M_y^B + v_{My} \\ M_z^B + v_{Mz} \end{bmatrix} = \text{DCM}(\phi, \theta, \psi)^T \cdot \begin{bmatrix} Me_x^L \\ Me_y^L \\ Me_z^L \end{bmatrix}, \quad (8)$$

where $Me_{x,y,z}^L$ is the local magnetic vector determined with National Oceanic and Atmospheric Administration's Geomagnetic Online Calculator [23], and $\mathbf{v}_2 = [v_{Mx} \ v_{My} \ v_{Mz}]^T \approx N(0, \mathbf{R}_2)$ are the noises associated with the magnetometer measurements, which are assumed to be zero mean, white, and Gaussian. An appealing feature for the IMU/magnetometers formulation is that it can operate indoors or within other GPS-denied environments. However, the local magnetic field can be distorted by the existence of ferromagnetic materials in the close vicinity, affecting the attitude estimation performance. A solution to this problem is discussed in Section 3.

2.5. Sensor Fusion Formulation #3: GPS/Magnetometers. The body-axis magnetic field measurements, coupled with GPS heading, also provide adequate information for Euler angle estimation at most places on earth that are not near the magnetic poles. In this formulation, the state, input, and measurement vectors are, respectively, $\mathbf{x} = [\phi\ \theta]^T$, $\mathbf{u} = [0\ 0]^T$, and $\mathbf{z} = \left[M_x^B\ M_y^B\ M_z^B\right]^T$. Since the rate gyroscope measurements are not available in this case, the state transition equations are simply defined as follows:

$$\begin{bmatrix} \dot{\phi} \\ \dot{\theta} \end{bmatrix} = \begin{bmatrix} 0 \\ 0 \end{bmatrix} + \mathbf{w}_2, \tag{9}$$

where the unknowns ϕ and θ are assumed to be perturbed with white noises $\mathbf{w}_2 \approx N(0, \mathbf{Q}_2)$. The observation equations are the same as in (8), where ψ is calculated with (5). The observation equations could also be augmented with (7) to incorporate additional gravity vector constraints during the measurement update. The benefit of this formulation is that it provides an independent attitude estimate without the rate gyroscopes; therefore, this approach would not cause stability issues [24] in control systems that rely on rate gyroscopes for inner-loop feedback.

2.6. Sensor Fusion Formulation #4: GPS/IMU/Mag. So far, combinations of any two sensors from the set of GPS, IMU, and magnetometers have been used for attitude estimation. Particularly, the differences between Formulations #1 and #2 are only present in the observation equations. Combining the two sets of observation equations could lead to a tighter regulation of the error growth in the strap-down INS equations. In fact, any measureable external vector field of known direction could be added to the observation equations in a similar fashion. For the GPS/IMU/magnetometers formulation, the state, input, and measurement vectors are, respectively, $\mathbf{x} = [\phi\ \theta\ \psi]^T$, $\mathbf{u} = [p\ q\ r]^T$, and $\mathbf{z} = \left[a_x^L\ a_y^L\ a_z^L\ M_x^B\ M_y^B\ M_z^B\right]^T$. The state transition equations are the same as in (6), and the observation equations are simply a combination of (7) and (8).

2.7. Unscented Information Filter. An information filter approach [5] is used for the fusion of multiple sensor measurements. The main advantage of using the information filter instead of Kalman filter is that the information update can be expressed as a sum:

$$\begin{aligned} \mathbf{I}_{k|k} &= \mathbf{I}_{k|k-1} + \sum_{j=1}^{N} \mathbf{I}_{j,k}, \\ \hat{\mathbf{i}}_{k|k} &= \hat{\mathbf{i}}_{k|k-1} + \sum_{j=1}^{N} \mathbf{i}_{j,k}, \end{aligned} \tag{10}$$

where $\mathbf{I}_{k|k-1}$ is the predicted information matrix, $\hat{\mathbf{i}}_{k|k-1}$ is the predicted information state vector, and $\mathbf{I}_{j,k}$, $\mathbf{i}_{j,k}$ are the measurement information matrix and information vector, respectively, associated with the jth measurement out of a total of N independent measurements. This simple relationship in information update creates a suitable framework for adding/removing sensors and handling unsynchronized measurement updates within a multisensor fusion system.

The state estimation problems presented in Formulations #1–#4 are solved with an Unscented Information Filter (UIF) [3]. The UIF uses the same prediction model as an Unscented Kalman Filter (UKF) [25, 26] to calculate the predicted error covariance matrix, $\mathbf{P}_{k|k-1}$, and state estimation, $\hat{\mathbf{x}}_{k|k-1}$. The predicted information matrix, $\mathbf{I}_{k|k-1}$, and predicted information state vector, $\hat{\mathbf{i}}_{k|k-1}$, are simply defined as follows:

$$\begin{aligned} \mathbf{I}_{k|k-1} &= \mathbf{P}_{k|k-1}^{-1}, \\ \hat{\mathbf{i}}_{k|k-1} &= \mathbf{P}_{k|k-1}^{-1} \hat{\mathbf{x}}_{k|k-1}. \end{aligned} \tag{11}$$

For the measurement update, the nonlinear observation equation for each independent measurement $\mathbf{z}_{j,k} = \mathbf{h}_j(\hat{\mathbf{x}}_{k|k-1}, \mathbf{v}_{j,k})$ is locally linearized with a statistical linear regression method [27]:

$$\mathbf{z}_{j,k} \approx \underbrace{\left[\mathbf{Hx}_{j,k}\ \ \mathbf{Hv}_{j,k}\right]}_{\mathbf{H}_{j,k}} \underbrace{\begin{bmatrix} \hat{\mathbf{x}}_{k|k-1} \\ \mathbf{v}_{j,k} \end{bmatrix}}_{\mathbf{x}^a_{k|k-1}} + \mathbf{b}_{j,k} \tag{12}$$

that minimizes the sum of squared errors $\mathbf{e}_{j,i} = \boldsymbol{\chi}^i_{j,k} - (\mathbf{H}_{j,k}\boldsymbol{\chi}^i_{j,k|k-1} + \mathbf{b}_{j,k})$:

$$\left\{\mathbf{H}_{j,k}, \mathbf{b}_{j,k}\right\} = \arg\min_{\mathbf{H},\mathbf{b}} \sum_{i=0}^{2L} \mathbf{e}_{j,i}^T \mathbf{e}_{j,i}, \tag{13}$$

where $\mathbf{H}_{j,k} = \mathbf{P}_{x_k z_{j,k}}^T {\mathbf{P}^a_{k|k-1}}^{-1}$ and $\mathbf{b}_{j,k} = \hat{\mathbf{z}}_{j,k} - \mathbf{H}_{j,k}\mathbf{x}^a_{k|k-1}$. The mean and covariance of $\mathbf{e}_j$ are given by

$$\begin{aligned} \bar{\mathbf{e}}_j &= 0, \\ \mathbf{P}_{j,ee} &= \mathbf{P}_{z_{j,k} z_{j,k}} - \mathbf{H}_{j,k}\mathbf{P}^a_{k|k-1}\mathbf{H}_{j,k}^T. \end{aligned} \tag{14}$$

The measurement information matrix $\mathbf{I}_{j,k}$ and information vector $\mathbf{i}_{j,k}$ for the jth measurement [3] can then be provided by

$$\begin{aligned} \mathbf{i}_{j,k} &= \mathbf{Hx}_{j,k}^T \overline{\mathbf{R}}_{j,k}^{-1} \left(\mathbf{z}_{j,k} - \mathbf{b}_{j,k}\right), \\ \mathbf{I}_{j,k} &= \mathbf{Hx}_{j,k}^T \overline{\mathbf{R}}_{j,k}^{-1} \mathbf{Hx}_{j,k}, \end{aligned} \tag{15}$$

where $\overline{\mathbf{R}}_{j,k}$ is the covariance matrix for the sum of the linearized actual observation noise $\mathbf{R}_j$ and the linearization noise:

$$\begin{aligned} \overline{\mathbf{R}}_{j,k} &= \mathbf{Hz}_{j,k}\mathbf{R}_j\mathbf{Hz}_{j,k}^T + \mathbf{P}_{j,ee} \\ &= \mathbf{P}_{z_{j,k} z_{j,k}} - \mathbf{Hx}_{j,k}\mathbf{P}_{k|k-1}\mathbf{Hx}_{j,k}^T. \end{aligned} \tag{16}$$

The procedure for deriving (14)–(16) is outlined in [3].

Following the information update equations (10) with all the measurements during each time frame, the estimated posterior covariance matrix and state vector can be recovered using

$$\mathbf{P}_{k|k} = \mathbf{I}_{k|k}^{-1}, \qquad \hat{\mathbf{x}}_{k|k} = \mathbf{P}_{k|k}\hat{\mathbf{i}}_{k|k}. \tag{17}$$

3. Sensor Calibration

The existence of redundant information from different sources provides an opportunity for calibrating each individual sensor within the measurement system. The calibration process may be performed off-line [6], on-line [7, 28], or using a combination of the two. The latter approach is used in this paper where a batch off-line calibration is first performed for the 3-axis magnetometers which then provide the initial condition for the recursive on-line calibration. This procedure has several advantages:

(1) The off-line calibration process is less restricted by the availability of computational resources; therefore, a large set of calibration parameters can be evaluated. Additionally, the off-line calibration can be performed through comparing with temporary sensors of higher quality (to be removed before the operation) and data from deliberately performed maneuvers.

(2) The off-line calibrated sensor parameters provide *a priori* knowledge of the sensor error model. The on-line calibration then starts from off-line calibrated parameters, minimizing the impact of the transient response based on better initial estimates.

(3) The on-line calibration provides the capability for dealing with time varying parameters.

(4) The on-line calibrated parameters can be restricted within a set of prespecified bounds and the calibration can be turned off or revert to the off-line values in the event of unstable conditions.

The concept of this 2-step calibration process is similar to the adaptive augmentation of a baseline controller [29] in control theory. Within this effort, the magnetometers are first calibrated off-line, followed by an on-line estimation of nine sensor biases associated with rate gyroscopes, accelerometers, and magnetometers.

3.1. Magnetometer Error Model.

The magnetometer readings of earth's magnetic field are often distorted by the existence of ferromagnetic materials in the local area, as well as the imperfection in the measurement system itself. The calibration of the magnetometer is a well-studied problem in the literature [27, 28, 30]. However, most calibration research in the past was performed off-line, due to a lack of reference information during the vehicle operation.

Without a loss of generality, the 3-axis magnetometer calibration process is formulated as a nonlinear parameter identification (PID) problem:

$$\begin{bmatrix} \widehat{M}_x^B \\ \widehat{M}_y^B \\ \widehat{M}_z^B \end{bmatrix} = \underbrace{\mathrm{DCM}\left(R_{M\phi}, R_{M\theta}, R_{M\psi}\right)^T}_{\mathbf{R}_M} \cdot \underbrace{\begin{bmatrix} S_{Mx} & 0 & 0 \\ 0 & S_{My} & 0 \\ 0 & 0 & S_{Mz} \end{bmatrix}}_{\mathbf{S}_M} \cdot \begin{bmatrix} M_x^B \\ M_y^B \\ M_z^B \end{bmatrix} - \underbrace{\begin{bmatrix} b_{Mx} \\ b_{My} \\ b_{Mz} \end{bmatrix}}_{\mathbf{b}_M}, \tag{18}$$

where $\widehat{M}_{x,y,z}^B$ are the calibrated magnetometer measurements, $\mathbf{R}_M$ is a rotation matrix parameterized by the three rotation angles: $R_{M\phi}$, $R_{M\theta}$, and $R_{M\psi}$, $\mathbf{S}_M$ is a diagonal scaling matrix, and $\mathbf{b}_M$ is a bias vector. The nine parameters

$$\mathbf{\Theta}_M = \left[R_{M\phi} \; R_{M\theta} \; R_{M\psi} \; S_{Mx} \; S_{My} \; S_{Mz} \; b_{Mx} \; b_{My} \; b_{Mz}\right]^T \tag{19}$$

to be estimated capture all of the soft iron effects, hard iron effects, sensor nonorthogonality, bias, and scaling factor [27].

3.2. Magnetometer Off-Line Calibration.

The off-line calibration process starts with the creation of a set of reference signals to be compared with the actual magnetometer measurements. The reference is created by rotating the known earth's magnetic vector $\mathbf{Me}^L$ from the local Cartesian coordinates to the vehicle's body axis $\mathbf{Me}^B$:

$$\begin{bmatrix} Me_x^B \\ Me_y^B \\ Me_z^B \end{bmatrix} = \mathrm{DCM}\left(\phi, \theta, \psi\right)^T \cdot \begin{bmatrix} Me_x^L \\ Me_y^L \\ Me_z^L \end{bmatrix}, \tag{20}$$

where the Euler angles are provided by a GPS/IMU sensor fusion algorithm discussed earlier. A set of estimated calibration parameters $\widehat{\mathbf{\Theta}}_M$ is then acquired through minimizing a cost function, which spans over an entire set of flight data, using a quasi-Newton method:

$$J = \sum \left[\left(\widehat{M}_x^B - Me_x^B\right)^2 + \left(\widehat{M}_y^B - Me_y^B\right)^2 + \left(\widehat{M}_y^B - Me_y^B\right)^2\right], \qquad \widehat{\mathbf{\Theta}}_M = \underset{\mathbf{\Theta}_M \in R^9}{\arg\min}\,(J). \tag{21}$$

The off-line calibration problem can also be solved with a maximum likelihood method similar to the one discussed in [27].

3.3. Mag., Gyro, and Accelerometer On-Line Calibration. The on-line calibration process is performed by augmenting the sensor fusion Formulation #4 with nine additional bias states, $\mathbf{b} = [b_{Gp}\ b_{Gq}\ b_{Gr}\ b_{Ax}\ b_{Ay}\ b_{Az}\ b_{Mx}\ b_{My}\ b_{Mz}]^T$, one for each magnetometer, rate gyroscope, and accelerometer. In this way, the attitude state estimation and sensor error model parameter identification [8] are performed simultaneously. During the state prediction stage, the state transition equations described in (6) are used for attitude states, with the exception that the estimated rate gyroscope biases are subtracted from the raw IMU measurements:

$$\mathbf{u} = \left[p - \hat{b}_{Gp}\ \ q - \hat{b}_{Gq}\ \ r - \hat{b}_{Gr}\right]^T. \quad (22)$$

The dynamics of the nine bias states are modeled as random walk using:

$$\dot{\mathbf{b}} = 0 + \mathbf{w}_3, \quad (23)$$

where the bias states are only assumed to be perturbed with white noises $\mathbf{w}_3 \approx N(0, \mathbf{Q}_3)$. The initial conditions for the bias states are set to be

$$\mathbf{b}_0 = \left[0\ 0\ 0\ 0\ 0\ 0\ \tilde{b}_{Mx}\ \tilde{b}_{My}\ \tilde{b}_{Mz}\right]^T, \quad (24)$$

where $\tilde{b}_{Mx,My,Mz}$ are off-line calibrated magnetometer bias values.

The nonlinear observation equations are derived from (7) and (8) with added bias terms on the accelerometer measurements as well as the rotated and scaled magnetometer measurements:

$$\begin{bmatrix} \dot{V}_x^L + v_{gx} \\ \dot{V}_y^L + v_{gy} \\ \dot{V}_z^L + v_{gz} \end{bmatrix}_{\text{GPS}} = \text{DCM}(\phi, \theta, \psi) \cdot \begin{bmatrix} a_x^B - \hat{b}_{Ax} + v_{ax} \\ a_y^B - \hat{b}_{Ay} + v_{ay} \\ a_z^B - \hat{b}_{Az} + v_{az} \end{bmatrix}_{\text{IMU}} + \begin{bmatrix} 0 \\ 0 \\ g \end{bmatrix},$$

$$\text{DCM}\left(\tilde{R}_{M\phi}, \tilde{R}_{M\theta}, \tilde{R}_{M\psi}\right)^T \cdot \begin{bmatrix} \tilde{S}_{Mx} & 0 & 0 \\ 0 & \tilde{S}_{My} & 0 \\ 0 & 0 & \tilde{S}_{Mz} \end{bmatrix} \cdot \begin{bmatrix} M_x^B + v_{Mx} \\ M_y^B + v_{My} \\ M_z^B + v_{Mz} \end{bmatrix} - \begin{bmatrix} \hat{b}_{Mx} \\ \hat{b}_{My} \\ \hat{b}_{Mz} \end{bmatrix} = \text{DCM}(\phi, \theta, \psi)^T \cdot \begin{bmatrix} Me_x^L \\ Me_y^L \\ Me_z^L \end{bmatrix}, \quad (25)$$

where $\tilde{\mathbf{R}}_M$ and $\tilde{\mathbf{S}}_M$ are the rotation and scaling matrices acquired from the off-line magnetometer calibration.

4. Fault Detection, Identification, and Accommodation

The sensor FDIA is achieved through two independent approaches: information crosschecking and sensor bias tracking. The first approach detects discrepancies among all information sources to identify outliers, which could be due to temporary low-quality measurements or abrupt sensor failures. The second method allows detection and tracking of "soft" sensor failures that slowly develop over time.

In our information crosschecking approach, it is important to mention that under nominal conditions, we assume that each independent information source can observe valid state estimates and that the optimal state estimate would be the result of fusing all sources. Therefore, because we have multiple redundant information sources, the goal is to identify outlier estimates and exclude their associated information sources from the state estimate altogether. A graphical representation of the information crosschecking approach is shown in Figure 1.

In Figure 1, first, multiple Kalman updated states and error-covariance estimates (or equivalently information matrices and vectors) are derived using the independent sources of information available. In general, $(\mathbf{S}_k, \mathbf{I}_k, \hat{\mathbf{i}}_k)$ are available at each time step k, where $\mathbf{S}_k$ is a set of $N+1$ independent information sources from either prediction or a sensor measurement, and $\mathbf{I}_k$, $\hat{\mathbf{i}}_k$ are sets of the associated Fisher information matrices and vectors. Next, to identify potential outlier information sources, the Mahalanobis distance [31] is used to evaluate the "closeness" between pairwise sets of estimates. For example, to determine if the sth information source is consistent with the rth information source, we evaluate

$$\begin{aligned} \mathbf{P}_{s,k} &= \mathbf{I}_{s,k}^{-1}, \\ \hat{\mathbf{x}}_{s,k} &= \mathbf{P}_{s,k}\hat{\mathbf{i}}_{s,k}, \\ \mathbf{P}_{r,k} &= \mathbf{I}_{r,k}^{-1}, \\ \hat{\mathbf{x}}_{r,k} &= \mathbf{P}_{r,k}\hat{\mathbf{i}}_{r,k} \end{aligned} \quad (26)$$

in order to calculate the square of the Mahalanobis distance between them. Consider

$$F_{s|r,k} = \left(\hat{\mathbf{x}}_{s,k} - \hat{\mathbf{x}}_{r,k}\right)^T \left(\mathbf{P}_{s,k} + \mathbf{P}_{r,k}\right)^{-1} \left(\hat{\mathbf{x}}_{s,k} - \hat{\mathbf{x}}_{r,k}\right). \quad (27)$$

We chose to use Mahalanobis distance because it is an intuitive metric that indicates the statistical agreement between state estimates. Additionally, it takes into consideration estimated error-covariance and is extensible to high dimension state-spaces.

After considering all unique pairs of estimates from individual information sources a total of $N+1$ unique values are available to base our fault detection and identification upon. However, these values represent the "closeness" between estimate pairs, which is not the most convenient for the goal of identifying individual faulty information sources. In particular, it would be more beneficial if we could evaluate how well each individual information source agrees with

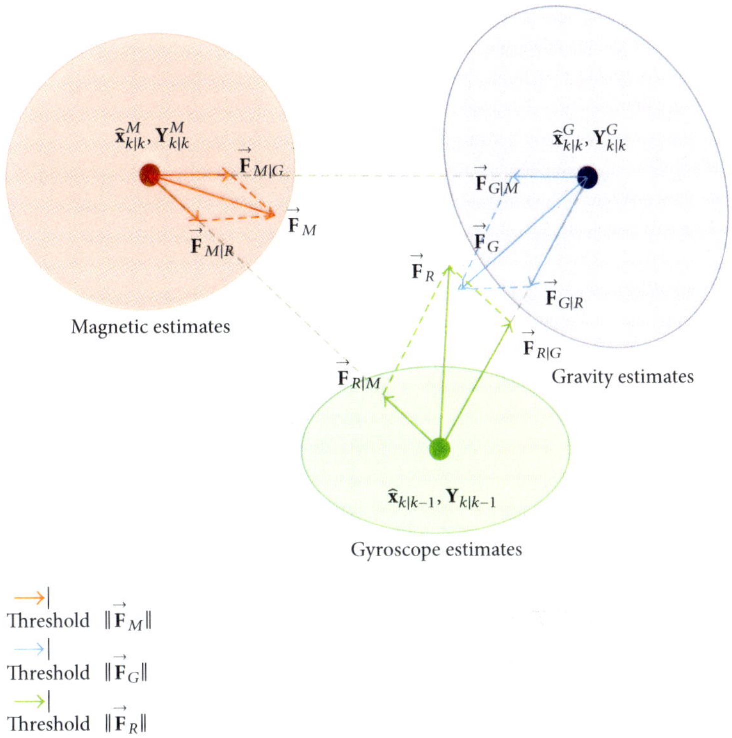

FIGURE 1: Fault Detection, Identification, and Accommodation approach.

all other sources in a single metric. To accomplish this, we employ a simple vector in addition to combining the squared Mahalanobis distance values from each information source with respect to all others. The sum then represents how well a particular information source agrees with all others. As an example, for information source, s, we define all the unit vectors to other information sources:

$$\vec{\mathbf{u}}_{s|i} = \frac{(\mathbf{e}_{sk} - \widetilde{\mathbf{e}}_{ik})}{\left\|\mathbf{e}_{sk} - \widetilde{\mathbf{e}}_{ik}\right\|}, \quad (i = 1 : N, \ i \neq s). \tag{28}$$

Then, using the values calculated in (27), we use their inverse values to scale each unit vector, sum with vector addition, and find the resultant magnitude. The use of the inverse of the squared Mahalanobis distance is analogous to inverse square laws that govern power loss. The resultant $\mathbf{e}_{sk}$ is a sum of a total of $N-1$ vectors associated with $\mathbf{e}_{sk}$ and each component of $\widetilde{\mathbf{E}}_k$:

$$\vec{\mathbf{F}}_s = \sum_{i=1}^{N} \vec{\mathbf{F}}_{s|i}, \quad (i \neq s). \tag{29}$$

Finally, the magnitude of the detection vector, $\|\vec{\mathbf{F}}_s\|$, provides a scalar assessment of the agreement between $\mathbf{e}_{sk}$ and other estimates at the same time. If $\|\vec{\mathbf{F}}_s\|$ is less than a prespecified threshold, Ω_s, that is, set based upon empirical tuning, a failure is declared for the sensor associated with $\mathbf{e}_{sk}$.

For the attitude estimation problem described in this paper, three information sources are available at time step k based on the posterior estimates from the previous step $(\mathbf{I}_{k-1|k-1}, \hat{\mathbf{i}}_{k-1|k-1})$, (1) rate gyroscopes based *a priori* estimates $(\mathbf{I}_{k|k-1}, \hat{\mathbf{i}}_{k|k-1})$; (2) the posterior estimates, $(\mathbf{I}^G_{k|k}, \hat{\mathbf{i}}^G_{k|k})$, updated with the gravity vector information; (3) the posterior estimates, $(\mathbf{I}^M_{k|k}, \hat{\mathbf{i}}^M_{k|k})$, updated with the magnetic vector information. To ensure the independence of the three estimates, the calculation of $(\mathbf{I}^G_{k|k}, \hat{\mathbf{i}}^G_{k|k})$ and $(\mathbf{I}^M_{k|k}, \hat{\mathbf{i}}^M_{k|k})$ are based on a random walk assumption only instead of using the rate gyroscope measurements during the prediction step:

$$\dot{\mathbf{x}} = \mathbf{0} + \mathbf{w}_4, \tag{30}$$

where $\mathbf{w}_4 \approx N(0, \mathbf{Q}_4)$ are white noises.

To complement the information crosschecking approach, the sensor bias tracking method monitors the bias states estimated by the on-line calibration scheme. Under nominal conditions, all sensor biases should be bounded within a prespecified envelope E. The size of E can be determined based on a statistical evaluation of data collected in the past as well as common sensor error specifications, such as the bias instability for the case of accelerometers and rate gyroscopes. During the operation, if a sensor bias grows outside of E, an anomaly warning for this particular sensor is declared and no additional action is taken. If the bias continues to grow beyond $\alpha \cdot$ E, where $\alpha > 1$ is a prespecified threshold value, a sensor failure status is declared. Under this condition, the bias states that are directly correlated or closely coupled with the faulty sensor are capped at or below their current value, and the associated error covariance matrix is scaled at each time step to represent an increasingly uncertain knowledge of the bias states:

$$\mathbf{P}^f_{k|k-1} = \beta \cdot \mathbf{P}^f_{k|k-1}, \tag{31}$$

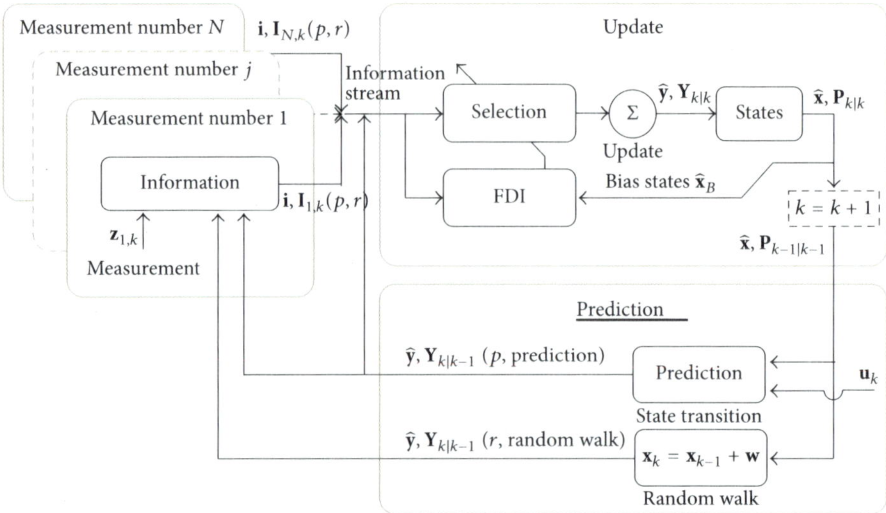

FIGURE 2: Overall sensor fusion algorithm architecture.

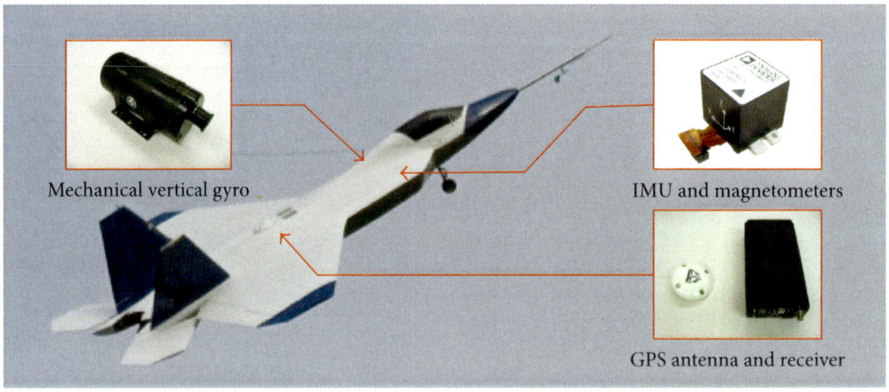

FIGURE 3: WVU YF-22 research aircraft and relevant sensors.

where $\beta > 1$ is empirically selected to be 1.005 in this implementation.

Once a failure is declared with either FDI approach, fault accommodation is based on a simple concept: with the availability of redundant information and with the information update being a sum, the UIF has the freedom of rejecting any sensor measurement it considers a fault or of lower-quality. This approach allows the sensor fusion algorithm to be conservative and only uses the best information for state estimation. A block diagram for the UIF based fault-tolerant multiple sensor fusion algorithm is shown in Figure 2.

5. Experimental Setup

The sensor fusion algorithms outlined in the previous sections are evaluated with the actual flight data from YF-22 unmanned research aircraft [32] developed at West Virginia University. The vehicle, shown in Figure 3, is approximately 2.4 m long with a 2 m wing span and has a take-off weight of approximately 22.5 Kg. The aircraft is powered with a miniature turbine that provides 125 N of static thrust. The cruise speed for the aircraft is approximately 40 m/s.

The aircraft instrumentation [33] includes three sensors directly relevant to this study. An Analog Devices ADIS-16405® tri-axis inertial sensor with magnetometers is used to provide a 14-bit digital output of 3-axis acceleration, angular rate, and magnetic field measurements, with a full scale range of ±18 g, ±150°/s, and ±2.5 Gauss, respectively. The manufacturer reported $1-\sigma$ initial bias errors for the accelerometers, rate gyroscopes, and magnetometers that are ±50 mg, ±3°/s, and ±4 mGauss, respectively. A Novatel OEM4® GPS receiver provides an estimate of the aircraft 3D position and velocity in the Earth-Centered, Earth-Fixed (ECEF) coordinate system independent of the inertial and magnetic sensors, which is transformed into a local Cartesian coordinate system. The manufacturer reported GPS position and velocity accuracies are 1.8 meter Circular Error Probable (CEP) and 0.03 m/s Root Mean Square (RMS), respectively. A Goodrich VG34® mechanical vertical gyroscope is used to provide independent pitch and roll angle measurements and is used as the "truth data" for this sensor fusion study. The VG34 has a ±90° measurement range on the roll axis and a ±60° range on the pitch axis and is sampled with 16-bit resolution. The VG34 has a self-erection system and reported accuracy of within 0.25° of true vertical.

TABLE 1: Statistics of attitude estimation algorithms.

Formulation	$E(\lvert\phi_{err}\rvert)$	$\sigma(\phi_{err})$	$E(\lvert\theta_{err}\rvert)$	$\sigma(\theta_{err})$
GPS/IMU	2.990	1.671	1.590	1.865
GPS/IMU + 6 bias states	1.516	1.524	1.558	1.700
GPS/Mag. (Off-L. Cal.)	1.726	2.008	2.315	2.195
IMU/Mag. (Off-L. Cal.)	15.81	19.59	22.99	27.69
GPS/IMU/Mag. (Raw)	8.061	3.219	5.276	2.279
GPS/IMU/Mag. (Off-L. Cal.)	2.026	1.452	2.415	2.127
GPS/IMU/Mag. (Off-L. + On-L Cal.)	1.598	1.697	1.417	1.679

6. Results

The sensor fusion algorithms are validated using three sets of flight data. The first set is used for magnetometer off-line calibration. A GPS/IMU sensor fusion algorithm (Formulation #1) augmented with six rate gyroscope and accelerometer bias states is used to estimate the attitude angles required by the off-line magnetometer calibration algorithm. The remaining two sets of flight data collected from two different days were used to validate and compare different sensor fusion algorithm performance. Table 1 lists the estimation performance of eight sensor fusion algorithms in terms of mean absolute error and error standard deviation for pitch and roll angle estimates. An average of two flights is used in calculating the value of each entry. The same stochastic noise modeling assumptions of the GPS, rate gyroscope, accelerometer, and magnetometer noises were used for each formulation and no individual tuning was performed.

Table 1 shows that each algorithm is able to provide a pitch and roll estimate. The GPS/IMU and GPS/Mag. formulations both have good performance, but the IMU/Mag. performs poorly with off-line calibrated magnetometers. The performance of IMU/Mag. formulation is found to be highly sensitive to the quality of the calibration.

The combination of all three sensors gives a good attitude estimation performance once a set of off-line calibrated magnetometer parameters are used. The introduction of an on-line calibration scheme to the GPS/IMU/Mag. sensor fusion algorithm provides an additional enhancement of its performance. Figure 4 shows a section of the flight with vertical gyroscope pitch measurements along with estimates from the GPS/IMU/Mag. sensor fusion algorithm with raw magnetometer readings, off-line calibrated parameters, and a combined off-line and on-line sensor calibration.

The fault-tolerant aspect of the sensor fusion algorithm is validated with simulated sensor failures superimposed on the actual flight data. The performance of the sensor fusion algorithm with and without FDI and fault accommodation schemes under nominal condition were first evaluated, along with the estimation performance under a set of simple sensor failure scenarios, where the outputs of GPS, rate gyroscopes, or magnetometers were lost for the second half (50%) of the flight. Table 2 summarizes the results of this test.

TABLE 2: Statistics of UIF-based fault tolerant attitude estimation.

Operating condition	$E(\lvert\phi_{err}\rvert)$	$\sigma(\phi_{err})$	$E(\lvert\theta_{err}\rvert)$	$\sigma(\theta_{err})$
Nominal w/o FDIA	1.598	1.697	1.417	1.679
Nominal w/FDIA	1.585	1.708	1.360	1.610
GPS failure w/FDIA	3.566	5.546	2.205	3.185
Gyro failure w/FDIA	1.814	2.211	1.622	1.821
Mag. failure w/FDIA	1.890	2.111	1.534	1.892

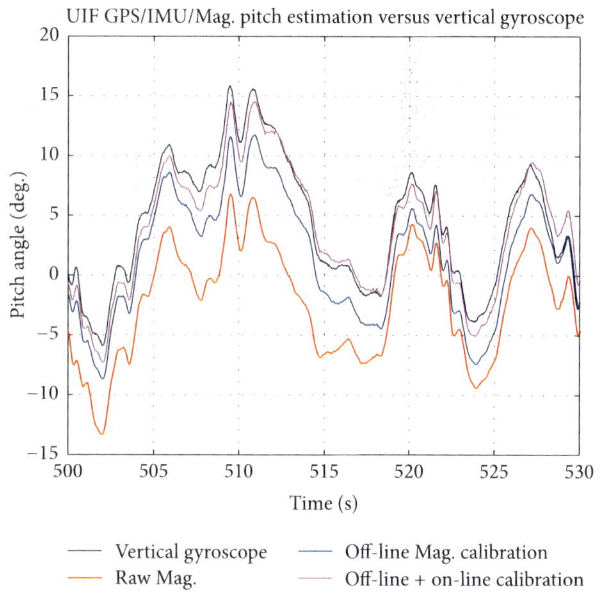

FIGURE 4: GPS/IMU/Mag. sensor fusion with raw, off-line calibration, and a two-stage off-line/on-line calibration.

An interesting observation is that the inclusion of the FDIA scheme slightly improves the overall performance of the sensor fusion algorithm even under nominal conditions without imposed sensor failures. This is due to the fact that the FDIA scheme monitors the quality of each measurement and rejects information of lower quality. This desirable feature is demonstrated in Figure 5, where four GPS data points were rejected by the sensor fusion algorithm during a half-second section of the flight.

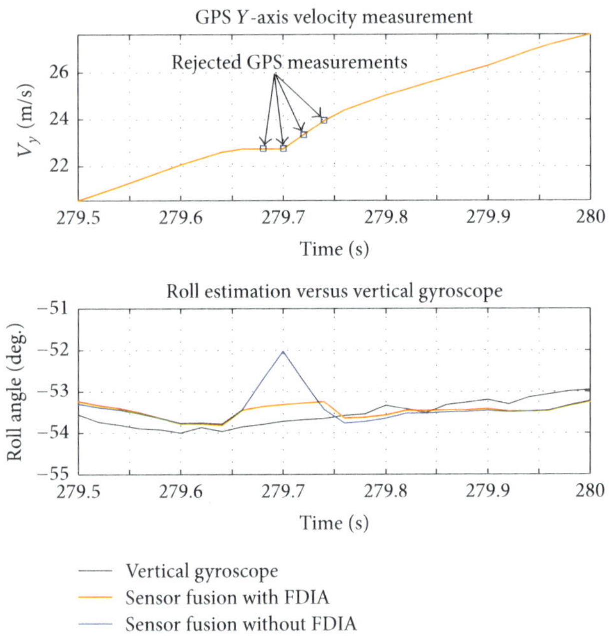

FIGURE 5: Rejection of low-quality GPS measurements by the sensor fusion algorithm.

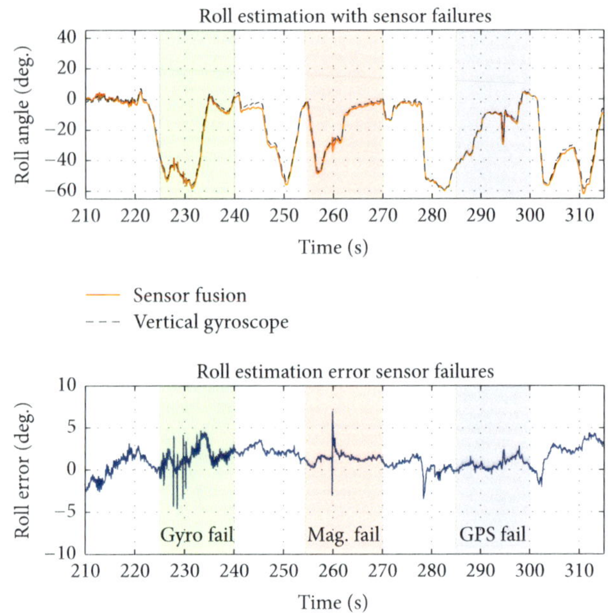

FIGURE 6: Sensor fusion with no failure, GPS failure, rate gyroscope failure, and magnetometer failure.

The ability of the sensor fusion algorithm to detect, accommodate, and recover from different sensor failures is further demonstrated in Figure 6, where a series of sequentially imposed GPS, rate gyroscope, and magnetometer failures are presented. The force field based failure detection signals for each set of sensors are also shown in Figure 7, along with the respective detection threshold indicated as a dotted line.

Figure 6 demonstrates an advantage of this information filter based fault accommodation method for fault accommodation. The transition between nominal and failure conditions in terms of attitude estimation is smooth and seamless. The estimation performance gracefully degrades after a sensor failure and recovers after the failure is removed.

In addition to large and abrupt sensor failures, the ability of the sensor fusion algorithm to handle slowly developing "soft" failures is demonstrated in Figure 8. A random walk bias (with $\sigma = 0.02$ deg/sec) is added to the pitch rate sensor measurements and the pitch rate bias state estimate is used for FDI. The sensor fusion algorithm compensates for the "soft" failure through sensor calibration when the bias is small and through sensor rejection when it becomes large.

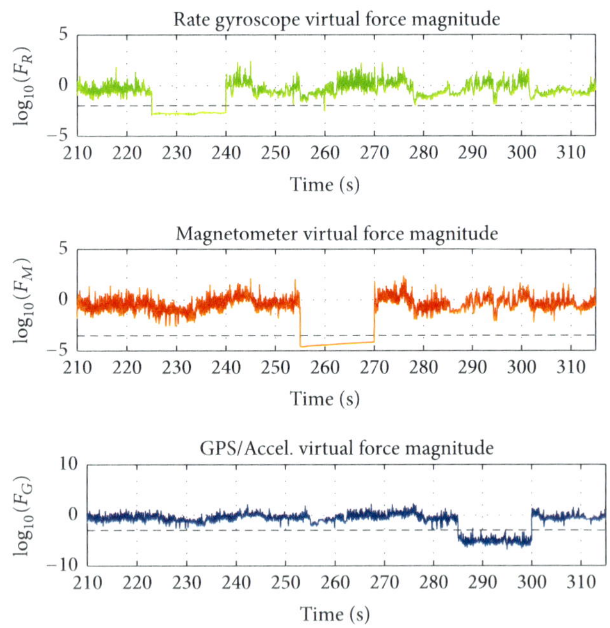

FIGURE 7: "Virtual Force Field" based failure detection signal.

7. Conclusions and Discussion

In this paper, a general 3-step sensor fusion approach is proposed and is applied to a 3D attitude estimation problem. The validation results using sets of UAV flight data show that having multiple redundant information sources allows for on-line calibration of individual sensors within the measurement system, leading to both improved performance and improved understanding of sensor health conditions, especially under "soft" sensor failure conditions. Furthermore, a crosschecking among different information sources allows for identifying "hard" failures or instantaneous faulty measurements. The use of information filter provides a convenient and scalable platform for multiple sensor fusion, information crosschecking, and faulty sensor rejection.

The attitude estimation problem discussed in this paper utilizes three information sources, which is the minimum number required for the presented FDIA approach. With an increased number of sensors, FDI could become more reliable and the overall estimation performance would be

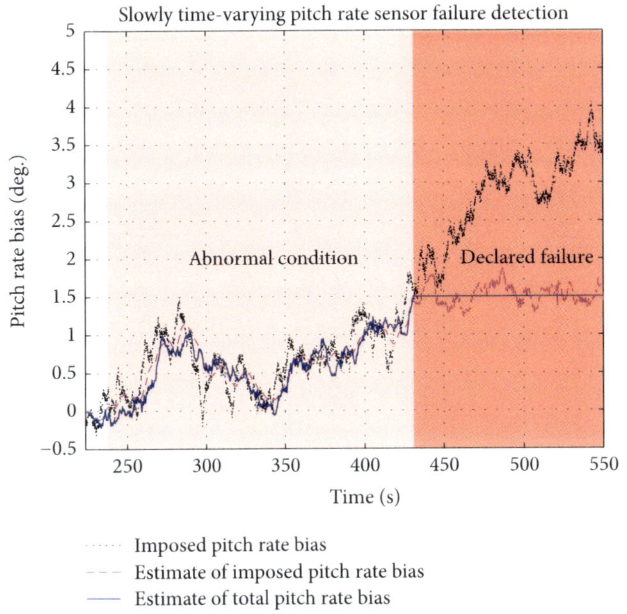

FIGURE 8: Detection and accommodation of slowly building up "soft" failure.

less sensitive to individual sensor failures. The development of a decentralized on-line sensor calibration scheme is the remaining bottleneck before a truly scalable multiple sensor fusion algorithm could be implemented.

Conflict of Interests

The authors declare that there is no conflict of interests regarding the publication of this manuscript.

Acknowledgment

This work was supported in part by NASA under project nos. NNX10AI14G and NNX12AM56A.

References

[1] Z. Chen, "Bayesian filtering: from Kalman filters, to particle filters, and beyond," Tech. Rep., McMaster University, 2003, http://web.maths.unsw.edu.au/~peterdel-moral/chen_bayesian .pdf.

[2] J. L. Junkins and J. L. Crassidis, *Optimal Estimation of Dynamic Systems*, Chapman & Hall/CRC, Washington, DC, USA, 2004.

[3] T. Vercauteren and X. D. Wang, "Decentralized sigma-point information filters for target tracking in collaborative sensor networks," *IEEE Transactions on Signal Processing*, vol. 53, no. 8, pp. 2997–3009, 2005.

[4] D.-J. Lee, "Unscented information filtering for distributed estimation and multiple sensor fusion," in *Proceedings of the AIAA Guidance, Navigation and Control Conference and Exhibit*, AIAA 2008-7426, Honolulu, Hawaii, USA, August 2008.

[5] A. G. O. Mutambara, *Decentralized Estimation and Control for Multisensor Systems*, CRC Press, Washington, DC, USA, 1998.

[6] A. S. Morris, *Measurement & Instrumentation Principles*, Elsevier, Oxford, UK, 2001.

[7] K. Whitehouse and D. Culler, "Calibration as parameter estimation in sensor networks," in *Proceedings of the 1st ACM International Workshop on Wireless Sensor Networks and Applications*, pp. 59–67, New York, NY, USA, September 2002.

[8] V. Klein and E. A. Morelli, *Aircraft System Identification: Theory and Pratice*, AIAA, Reston, Va, USA, 2006.

[9] J. L. Crassidis, "Sigma-point filtering for integrated GPS and inertial navigation," in *Proceednigs of the AIAA Guidance, Navigation and Control Conference and Exhibit*, San Francisco, Calif, USA, August 2005.

[10] P. M. Frank, "Fault diagnosis in dynamic systems using analytical and knowledge-based redundancy: a survey and some new results," *Automatica*, vol. 26, no. 3, pp. 459–474, 1990.

[11] R. Isermann, "Model-based fault detection and diagnosis: status and applications," in *Proceedings of the 16th IFAC Symposium on Automatic Control in Aerospace*, St. Petersburg, Russia, 2004.

[12] D. B. Kingston and R. W. Beard, "Real-time attitude and position estimation for small UAVs using low-cost sensors," in *Proceedings of the AIAA 3rd Unmanned Unlimited Systems Conference and Workshop*, AIAA-2004-6488, Chicago, Ill, USA, September 2004.

[13] U. Kayasal, *Magnetometer Aided Inertial Navigation System: Modeling and Simulation of a Navigation System with an Imu and a Magnetometer*, LAP Lambert Academic, 2009.

[14] Y. H. Li, J. C. Fang, and Z. K. Jia, "Simulation of INS/CNS/GPS integrated navigation," *Journal of Chinese Inertial Technology*, vol. 06, 2002.

[15] S. Winkler, H. W. Schulz, M. Buschmann, T. Kordes, and P. Vorsmann, "Horizon aided low-cost GPS/INS integration for autonomous micro air vehicle navigation," in *Proceedings of the 1st European Micro Air Vehicle Conference and Flight Competition*, Braunschweig, Germany, 2004.

[16] L. D. Hostetler and R. D. Andreas, "Nonlinear kalman filtering techniques for terrain-aided navigation," *IEEE Transactions on Automatic Control*, vol. 28, no. 3, pp. 315–323, 1983.

[17] M. Rhudy, Y. Gu, H. Chao, and J. Gross, "Unmanned aerial vehicle navigation using wide-field optical flow and inertial sensors," *Journal of Robotics*, vol. 2015, Article ID 251379, 12 pages, 2015.

[18] S. Vajda and A. Zorn, "Survey of existing and emerging technologies for strategic submarine navigation," in *Proceedings of the IEEE Position Location and Navigation Symposium*, pp. 309–315, Palm Springs, Calif, USA, April 1998.

[19] M. S. Grewal, L. R. Weill, and A. P. Andrews, *Global Positioning Systems, Inertial Navigation, and Integration*, John Wiley & Sons, Hoboken, NJ, USA, 2nd edition, 2007.

[20] T. S. Rappaport, J. H. Reed, and B. D. Woerner, "Position location using wireless communications on highways of the future," *IEEE Communications Magazine*, vol. 34, no. 10, pp. 33–41, 1996.

[21] J. Gross, *Sensor fusion based fault-tolerant attitude estimation solutions for small unmanned aerial vehicles [Ph.D. thesis]*, West Virginia Univerisity, 2011.

[22] B. L. Stevens and F. L. Lewis, *Aicraft Control and Simulation*, John Wiley & Sons, 2nd edition, 2003.

[23] NOAA's Geophysical Data Center, Geomagnetic Online Calculator, http://www.ngdc.noaa.gov/geomag-web/.

[24] J. C. Doyle and G. Stein, "Robustness with observers," *IEEE Transactions on Automatic Control*, vol. 24, no. 4, pp. 607–611, 1979.

[25] E. Wan and R. van der Merwe, "The unscented Kalman filter for nonlinear estimation," in *Proceedings of the IEEE Adaptive Systems for Signal Processing, Communications, and Control Symposium (AS-SPCC '00)*, Lake Loise, Canada, October 2000.

[26] S. J. Julier and J. K. Uhlmann, "Unscented filtering and nonlinear estimation," *Proceedings of the IEEE*, vol. 92, no. 3, pp. 401–422, 2004.

[27] J. F. Vasconcelos, G. Elkaim, C. Silvestre, P. Oliveira, and B. Cardeira, "A geometric approach to strapdown magnetometer calibration in sensor frame," in *Proceedings of the 2nd IFAC Workshop on Navigation, Guidance, and Control of Underwater Vehicles (NGCUV '08)*, pp. 172–177, Killaloe, Ireland, 2008.

[28] B. J. Anderson, L. J. Zanetti, D. H. Lohr et al., "In-flight calibration of the NEAR magnetometer," *IEEE Transactions on Geoscience and Remote Sensing*, vol. 39, no. 5, pp. 907–917, 2001.

[29] E. Lavretsky, "Robust adaptive inner-loop design for vehicles with uncertain dynamics," in *Proceedings of the American Control Conference (ACC '08)*, pp. 2322–2327, Seattle, Wash, USA, June 2008.

[30] P. F. Guo, H. T. Qiu, Y. C. Yang, and Z. Ren, "The soft iron and hard iron calibration method using extended kalman filter for attitude and heading reference system," in *Proceedings of the IEEE/ION Position, Location and Navigation Symposium*, pp. 1167–1174, Monterey, Calif, USA, May 2008.

[31] K. I. Penny, "Appropriate critical values when testing for a single multivariate outlier by using the Mahalanobis distance," *Journal of the Royal Statistical Society. Series C: Applied Statistics*, vol. 45, no. 1, pp. 73–81, 1996.

[32] Y. Gu, B. Seanor, G. Campa et al., "Design and flight testing evaluation of formation control laws," *IEEE Transactions on Control Systems Technology*, vol. 14, no. 6, pp. 1105–1112, 2006.

[33] Y. Gu, J. Gross, F. Barchesky, H. Chao, and M. Napolitano, "Avionics design for a sub-scale fault-tolerant flight control test-bed," in *Recent Advances in Aircraft Technology*, chapter 21, InTech, Rijeka, Croatia, 2012.

9

Tendon-Sheath Mechanisms in Flexible Membrane Wing Mini-UAVs: Control and Performance

Tegoeh Tjahjowidodo and Shian Lee

School of Mechanical and Aerospace Engineering, Nanyang Technological University, 50 Nanyang Avenue, North Spine (N3), Singapore 639798

Correspondence should be addressed to Tegoeh Tjahjowidodo; ttegoeh@ntu.edu.sg

Academic Editor: Mahmut Reyhanoglu

Flexible membrane wings (FMWs) are known for two inherent advantages, that is, adaptability to gusty airflow as the wings can flex according to the gust load to reduce the effective angle of attack and the ability to be folded for compact storage purposes. However, the maneuverability of UAV with FMWs is rather limited as it is impossible to install conventional ailerons. The maneuver relies only on the rudders. Some applications utilize torque rods to warp the wings, but this approach makes the FMW become unfoldable. In this research, we proposed the application of a tendon-sheath mechanism to manipulate the wing shape of UAV. Tendon-sheath mechanism is relatively flexible; thus, it can also be folded together with the wings. However, its severe nonlinearity in its dynamics makes the wing warping difficult to control. To compensate for the nonlinearity, a dedicated adaptive controller is designed and implemented. The proposed approach is validated experimentally in a wind tunnel facility with imitated gusty condition and subsequently tested in a real flight condition. The results demonstrate a stable and robust wing warping actuation, while the adaptive washout capability is also validated. Accurate wing warping is achieved and the UAV is easily controlled in a real flight test.

1. Introduction

Recently mini-UAVs have been a very important vehicle class in various tasks. Being extremely stealthy and also cheaper and safer to operate than a manned aircraft, a mini-UAV is a suitable aircraft for flying into dangerous zones such as enemy territory for scouting missions. They can be used for border patrol missions, which can be mundane and also costly for human pilots. However, conventional mini-UAVs possess a few critical flaws which hinder the progress of their usage in the aforementioned missions. Due to the size of a mini-UAV being typically 1 m or more in wingspan, a conventional rigid wing mini-UAV is usually disassembled for storage and also transportation to the mission site. Thereafter, the human operator has to reassemble the mini-UAV at the site, while being exposed to the elements. In addition, the time required to assemble the mini-UAV can be a critical disadvantage to a surveillance mission as the target may have left the site. A mini-UAV is also quite difficult to pilot in gusty environments, especially in the climates near the equator with frequent thunderstorms [1]. Due to the very small moment of inertia, a mini-UAV can be a challenge to control in windy conditions. Currently, autopilots with a closed-loop feedback coupled with responsive actuators are being used to stabilize and reject the gusts [2]. This consumes the limited energy on board of a mini-UAV, which is very valuable in order to stay airborne for as long as possible. Actuator saturation may also occur, which lead to a mini-UAV not being able to react quickly enough to reject the gust response.

Flexible membrane wing (FMW) mini-UAVs possess very interesting properties such as adaptive washout [3]; also an FMW can be folded spanwise towards the fuselage for storage and quick deployment purposes. These are huge advantages over the conventional rigid wing. An FMW mini-UAV can be launched immediately by the operator, without any disassembly or assembly process. The operator can also be shielded from the dangerous elements by launching an FMW mini-UAV from inside a vehicle, as illustrated by Prioria

(a) (b)

Figure 1: The FMW in the folded and unfolded states: (a) flexible membrane wing folded around the fuselage of the mini-UAV; (b) flexible membrane wing deployed.

Robotics Maveric UAS. Adaptive washout, which is another desired feature, can dampen the longitudinal gusts response by flexing the airfoil away from the direction of the gusts, thereby decreasing the effects of gusts on an FMW. The ability to flex the wings in the chordwise direction allows an FMW to handle a higher angle of attack (α), therefore delaying the stall response.

However, due to the very thin airfoil, an FMW is unable to house an aileron as the material of the FMW is made of nylon cloth and carbon fiber strips to support the nylon cloth. Previous research has suggested other techniques such as utilizing torque rods and Kevlar lines to warp an FMW [4, 5]. Even though the warping ensues and also the roll rate is satisfactory, the biggest drawback of these systems is that they hinder the downward folding ability of an FMW. A method of warping the wings to provide sufficient roll rate which also does not hinder the two main properties of an FMW (downwards folding and adaptive washout) is required.

In this paper, a novel method that utilizes the tendon-sheath mechanism (TSM) is discussed to fulfill the requirement mentioned above. However, a TSM is not without its own disadvantage, as it experiences severe nonlinear behavior during the actuation. Nevertheless, it has been shown that the TSM nonlinearity behaviors can be modeled and controlled [6]. Similar procedure is carried out in the case of the FMW warping via TSM; the nonlinear hysteresis is identified, modeled, and, subsequently, compensated with a proper controller strategy.

This paper is structured as follows. In Section 2, a brief and concise introduction to the attributes and characteristics of the FMW is provided. The fabrication procedures of the FMW are also illustrated as well. Section 3 includes the design details of the FMW mini-UAV. The characterization of the mini-UAV is also discussed. Section 4 provides some crucial information about the TSM, the nonlinear behaviors, and the procedures carried out to model such behaviors. Section 5 provides details about the control structure implemented in the FMW warping via TSM. Extensive wind tunnel tests using both steady and gusty winds were carried out for the FMW mini-UAV, and the results and discussions are showed in Section 6. In Section 7, the flight test results are presented and discussed. Finally, a conclusion is drawn in Section 8.

Table 1: Specifications of the FMW.

Wingspan	1 m
Chord length	0.25 m
Mean aerodynamic chord	0.22 m
Area	0.19 m^2
Aspect ratio	5.158
Weight	160 g

2. Flexible Membrane Wing

Introduced by [3], the FMW is a piece of thin airfoil wing, similar to the wing of a bat. The ribs of the FMW are made of carbon fiber reinforced plastic (CFRP), while the skin of the FMW is made of nylon cloth. Most of the aerodynamic forces are carried by the nylon cloth, and the cloth is supported and stretched across the CFRP ribs. The detailed specification of the FMW investigated in this paper is given in Table 1. The FMW is able to withstand up to 35 N of lift without buckling when tested in a wind tunnel; therefore, it is considered to be very light for its strength and size.

Many different rib configurations have been tested and an optimized configuration has been identified [3]. Other similar FMWs available in [4, 5] were also created in a similar configuration, with the ribs attached to the main spar at the leading edge of the FMWs. The aerodynamics of the FMW has been studied in detail in [7], and the optimization of the shape of the FMW has been investigated in [8]. The effects of the ratio of ribs and fabric have been investigated thoroughly in [9], and an optimized ratio has been found, which generates the maximum lift over drag performance. The fabricated FMW used in this research, therefore, follows the guideline presented by Hu et al. [9].

2.1. Characteristics and Features. The main feature of the FMW is the downwards folding ability for storage and quick launching purposes (Figure 1 illustrates a typical FMW mini-UAV on storage and unfolding states). Due to its flexible nature in the spanwise direction, coupled with a unique thin airfoil with camber, the FMW is able to only fold downwards easily while resisting aerodynamic load that tries to curl the wing upwards. Experiments in a wind tunnel showed that the

FMW is able to withstand up to 35 N of lift force without buckling upwards. However, only very little force is needed to bend the wing downwards and to fold around the fuselage; hence, the FMW mini-UAV is not suitable in a negative flight condition. More discussions about the wind tunnel experiments will be presented in Section 6.

The chordwise flexibility of the FMW provides the ability to dampen the longitudinal gusts, also known as adaptive washout. This passive phenomenon allows the FMW to adapt to gusty conditions during flight, despite a stable limit cycle oscillation that might appear on the FMW structure [3]. The FMW is able to dampen the gusts, that is, by flexing in the chordwise direction and by changing the effective and airfoil shape. When a gust load attacks from lower side of the wing (e.g., a case of high α), in typical conventional rigid wings, a sudden increase of lift forces will occur. However, for the FMW, the wing will adapt its shape and decrease the lift coefficient (C_L), but the total amount of lift experienced by the FMW will not be perturbed by much. Once the gust is over, the FMW will spring back to its original shape, recovering C_L. Same case applies when there is a perturbation in the headwind. When the airspeed increases or decreases, the FMW will be adapted accordingly to provide a near constant lift force. Tuning is required to have a functional adaptive washout mechanism and the FMW used in this paper has been constructed after numerous times of experimentation to achieve the desired result.

Flexibility in the chordwise direction of the FMW has taken advantage for wing warping. In [5, 11], the FMW was installed with torque rods to induce wing warping. Sufficient roll rates were achieved and the torque rods could induce an almost pure rolling moment without yawing moment. However, the major drawback is that the FMW cannot be folded downwards, as the torque rods are stiff and rigid structures. Kevlar cables have been tried as well [11]; however, the tensioned Kevlar cables produced more spanwise warping (wing curling) than chordwise; therefore, this mechanism is unable to provide sufficient roll rates. In addition, the exposed cables are also prone to breaking during landings and will be tangled easily if the FMW is folded. Wing warping is a very delicate mechanism which relies on an exact amount of flexibility and stiffness. To facilitate wing warping, the structural stiffness of the FMW must be optimized, as the wing has to be flexible enough to allow deformations and yet stiff enough to withstand the aerodynamic loads. Conflicting conditions are presented when choosing the stiffness of the FMW. If a high stiffness is chosen for the FMW, huge amounts of energy will be required to warp the structure. Yet, if a low stiffness is chosen, the wing will not be able to bear the aerodynamic loads [12]. A few iterations have to be carried out to obtain the stiffness value that can compromise the need for warping, folding, and aerodynamic loads. The optimal stiffness is characterized by folding properties and aerodynamics properties that will be discussed in Sections 4 and 6.

2.2. Fabrication Procedures. The fabrication of the carbon fiber membrane wing involves successive steps with different materials, equipment, and handling techniques. First of all, a female mold made of aluminum (Figure 2) is designed and then fabricated. Aluminum is chosen as the material for the mold, as it can withstand the pressure and also the heat of the autoclave curing process.

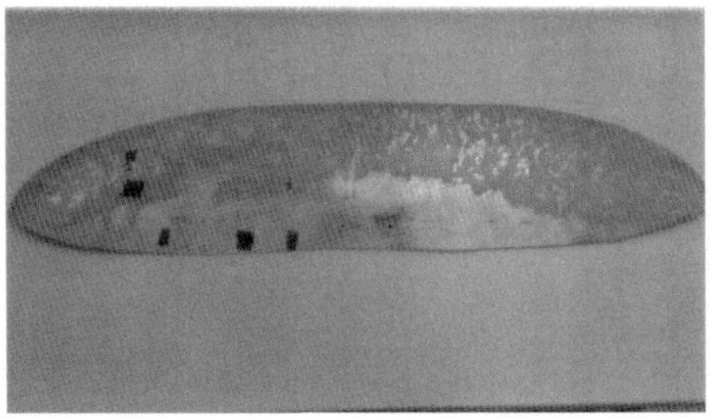

Figure 2: 3D aluminum mold.

A template board is used to guide the cutting of the carbon fiber prepreg. After cutting 3 copies of carbon fiber wing platform, they will be layered on top of each other forming the leading edge and main structure of the wings. Unidirectional carbon fiber strips are cut and then placed on the main carbon fiber wing platform. Six layers of carbon fiber reinforcement strips are used in this fabrication process to provide a suitable stiffness. The nylon cloth is then layered together with the carbon fiber to form the FMW, precured. The whole assembly will be placed on top of the aluminum mold and then taped on to be kept in place. A vacuum bag will be placed around the assembly, and a bag sealant tape is used to secure the vacuum bag with the aluminum mold to ensure an airtight condition. The carbon fiber layup is cured under high temperature and pressure in the autoclave, using the built in program "Autoclave 10743." After the curing cycle is complete, the carbon fiber wing will be separated from the mold and the vacuum bag. Extra bits of the wing are then trimmed off and the FMW fabrication is complete. Although the fabrication process is repeatable, however, the final product depends on the operator's skill; that is, layering the carbon fiber and the nylon cloth, enveloping with a vacuum bag, vacuuming, and demolding (release the parts from the mould) require some experiences.

3. Flexible Membrane Wing Mini-UAV

A mini-UAV was designed and built in order to validate the TSM actuated FMW warping. The design process of the UAV was done by identifying the high-level requirements, for example, payload weight and size, flight endurance, cruising speed, and top speed; and subsequently, the steps described in [13, 14] were followed to guide the aircraft sizing and geometry decisions.

3.1. Design and Build. As the wing is required to fold around the fuselage for storage, the FMW is installed at the top of the fuselage, with a top wing configuration. The payload of the mini-UAV is a visual camera to record the flight. Various electronics, for example, servos, autopilot microcontroller board, and electronic speed controller, are also required, not to mention the source of energy which will be a Lithium-Polymer battery. The location of the electronics and battery

(a)

(b)

FIGURE 3: Design of the FMW mini-UAV and the final prototype: (a) 3D CAD model of the FMW mini-UAV; (b) 3D printed fuselage of the FMW mini-UAV. Carbon fiber rod is used for the tail boom.

are critical as they will affect the center of gravity of the aircraft. The total weight of the aircraft is limited to 1.2 kg in order to fulfill the stall speed, take-off speed, and cruising speed requirements. Careful decisions on the locations of the electronics and battery are made to respect the center of gravity. The 3d CAD model of the mini-UAV is shown in Figure 3(a), and the fabricated FMW mini-UAV is illustrated in Figure 3(b).

The preliminary sizing of the aircraft was done to house all the electronics and payload inside the fuselage. After designing the fuselage, the tail design was done by limiting the horizontal tail span to 0.3 m, for the tube launch capability in the future. The horizontal tail aspect ratio AR_{HT} was recommended to be 2/3 of the wing aspect ratio [15]; therefore,

$$AR_{HT} = \frac{2}{3} AR_W, \qquad S_{HT} = \frac{b_{HT}^2}{AR_{HT}} = 0.0262\,\text{m}^2, \tag{1}$$

where AR_W is the wing aspect ratio, b_{HT} is the horizontal tail span, and S_{HT} is the horizontal tail area. The tail volume ratio C_{HT} was chosen to be 0.5, as normally used in typical mini-UAVs [13–15]. Therefore, the horizontal tail moment arm, l_{HT}, can be found with the tail volume ratio equation:

$$C_{HT} = \frac{l_{HT} S_{HT}}{C'_W S_W}, \qquad l_{HT} = 0.806l, \tag{2}$$

where C'_W is the mean aerodynamic chord and S_W is the wing surface area.

3.2. UAV Characterization via Empirical Methods. A few characteristics of the FMW mini-UAV are difficult to be obtained mathematically; therefore empirical tests are performed; namely, the aerodynamic data, natural frequencies, and moment of inertia of the FMW mini-UAV are obtained experimentally through wind tunnel testing (shown in Section 6). Modal testing was performed through an impact hammer testing to study the natural frequencies of the FMW, and finally, the actual moment of inertia of the entire mini-UAV structure was identified using the bifilar pendulum method.

The fundamental frequency of the FMW is particularly important as the flexibility of the wing will induce fluttering if the gust frequency happens to be the natural frequency of the FMW. For the FMW, the fundamental frequency was determined experimentally via modal testing. In particular, the first mode is the most critical to the FMW, as the deflection will be the greatest. From Figure 4(a), it can be observed that the first mode of the FMW occurs around 7 Hz. The coherence diagram (Figure 4(b)) shows a peak at 7 Hz, signifying high confidence of accuracy on the natural frequency. A slow motion video was captured on the vibration of the FMW, and the vibration frequency of the FMW from the video was shown to be 7 Hz as well.

A bifilar vertical-axis torsional pendulum test was performed to find out the moment of inertia of the mini-UAV in the longitudinal direction. Strings were attached to the center of gravity of the UAV, and then the UAV was hung from the ceiling as a bifilar pendulum. The following equation shows the moment of inertia of the mini-UAV:

$$I = \frac{mgT^2 l_p^{\,2}}{16\pi^2 L} = 0.016975\,\text{kgm}^2, \tag{3}$$

where m is the mass of the mini-UAV, g is the gravitational acceleration, T is the rotational oscillation period of the bifilar pendulum, l_p is the length of the bifilar pendulum, and L is the length of the supporting strings for the bifilar pendulum. The moment of inertia is of interest because the roll rate of the mini-UAV given a rolling moment can be found if the moment of inertia is known.

4. Tendon-Sheath Mechanism

Inspired by the tendon-sheath tissue found inside human hands, engineers have been applying TSM to construct, for example, power transmission systems in robotic hands and fingers, as it can be routed in a very flexible way [16]. The TSM is comprised of a piece of tendon, which slides along inside the sheath covering the tendon. The tendon is made of Teflon coated metal wire, and the sheath is made of a metal coil, much like a spring around the tendon. This helps to reduce the friction between the tendon and the sheath.

4.1. Force Transmission. Only tensile forces can be transmitted via the TSM. When tension is applied at the input side of

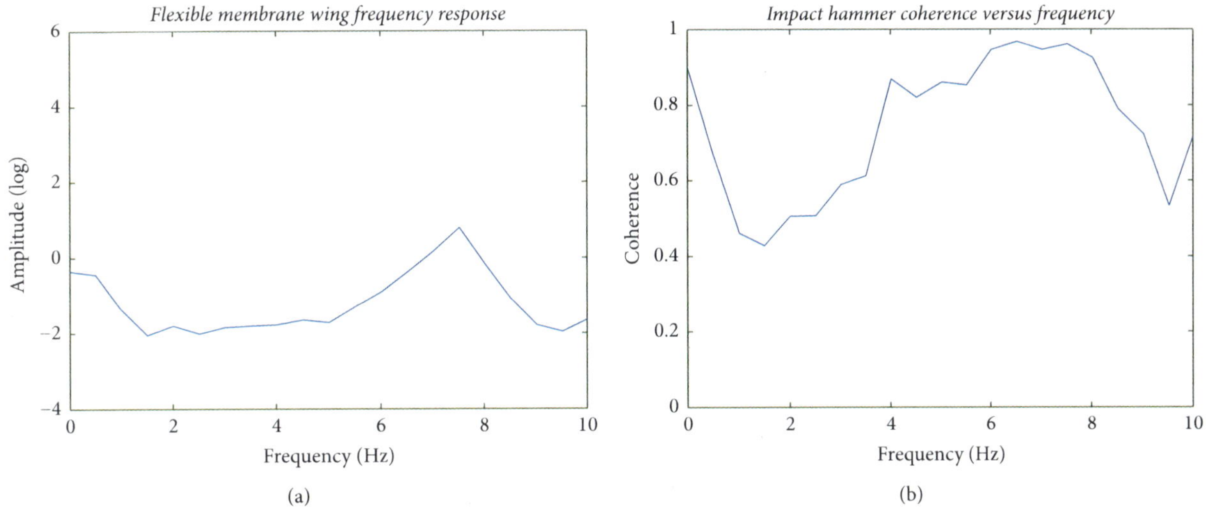

FIGURE 4: Results of the modal testing of the FMW: (a) frequency response of the FMW obtained from modal testing; (b) coherence of the frequency response.

FIGURE 5: The trailing edge of the FMW is fitted with the tendon.

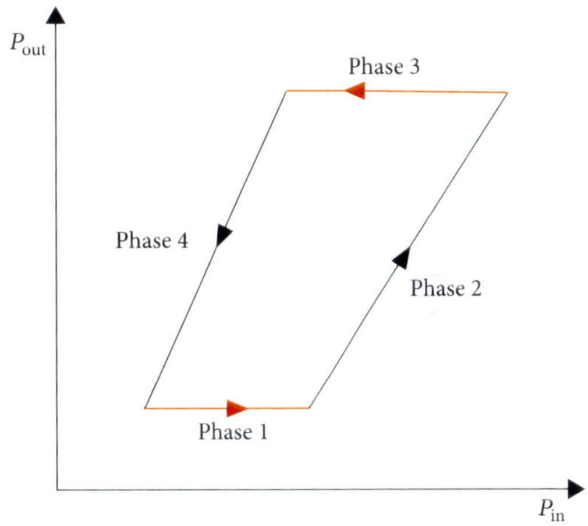

FIGURE 6: Relation between position input and position output [10].

the TSM, the tendon becomes tighter and therefore starts to slide along the sheath, creating a pulling force at the output side of the TSM. The relationship between the tension input and the output is not linear; therefore, a controller is needed when accurate positioning control is required [16]. Typically, a high pretension is given to the tendon, in order to increase the efficiency and to achieve a quicker response. TSM is useful for FMW warping because the tendon-sheath is flexible and is possible to mount it without requiring much space. The trailing edge of the wing is connected to the output side of the TSM (Figure 5), and the servomotor is connected to the input side of the TSM. Therefore, tensile forces can be transmitted from the servomotor placed inside the fuselage to the trailing edge of the FMW in order to control the wing warping. The flexibility of the TSM allows the FMW to fold downwards for storage purposes, which is the primary objective.

4.2. Nonlinear Behaviors. Although the tensile forces can be transmitted via the TSM to induce wing warping, the input-output relationship of the tensile forces is not linear. The nonlinear behavior, which is caused by the friction and also backlash of the TSM, can cause severe inaccuracy in adjusting the wing warping if no appropriate controller is applied. Nonetheless, the nonlinear behaviors can be modeled and controlled, as shown in various works [6].

In [6], the hysteresis behavior of the FMW warping via TSM was modeled using the Generalized Bouc-Wen model. The Generalized Bouc-Wen model from [17] is suitable for the FMW warping via TSM because of the asymmetric properties of the hysteresis curves. Typically, the hysteresis behavior of the TSM is categorized in 4 different phases, as illustrated in Figure 6. Phases 1 and 2 occur when tensile forces are applied from the input side, while Phases 3 and 4 occur when the tensile forces are stopped, and the mechanism is springing back into the original position. For a typical TSM, the hysteresis diagram shows sharp and distinct changes between the phases [10]. However, it was found that the hysteresis behaviors of the FMW warping via TSM were not the same, as the phase change occurs gradually, showing a gradual change in slopes between Phases 1 and 2, as well as in between Phases 3 and 4. This phenomenon has been

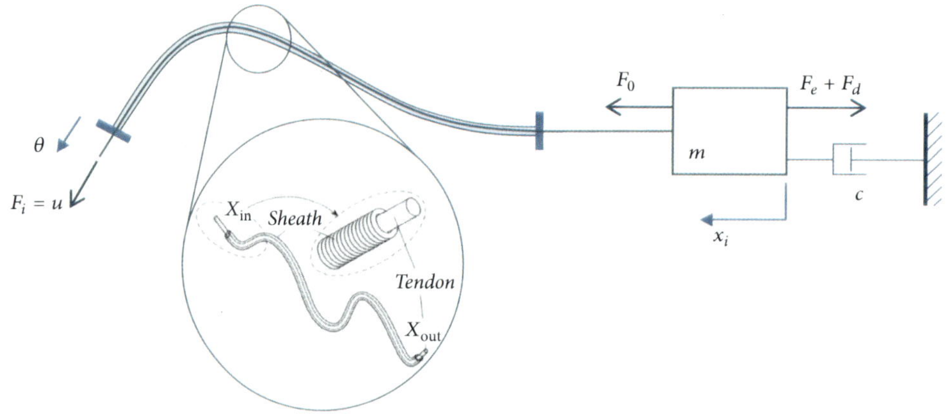

FIGURE 7: Force transmission on a TSM.

investigated in [10] and has been found to be the result of a low pretension in the tendon.

5. Adaptive Backstepping Controller

As the FMW warping via TSM is a highly nonlinear system [6], an appropriate control strategy is necessary to help in controlling the position of the wing warping more accurately. The simplest way to deal with systems with hysteresis nonlinear behavior is by implementing the model in a feedforward loop, combined with a feedback loop to compensate for some disturbances and uncertainties (see, e.g., [10]). In this paper, an adaptive backstepping controller is proposed and the results are benchmarked to the performance of feedforward/feedback controller in [10].

The dynamic friction force, F, for the TSM can be described as follows [18]:

$$F = k_x \phi x_i + k_\zeta \zeta + \upsilon \dot{x}_i + F_\Theta, \tag{4}$$

$$\dot{\zeta} = \rho \left(\dot{x}_i - \sigma \left| \dot{x}_i \right| \left| \zeta \right|^{n-1} \zeta + (\sigma - 1) \dot{x}_i \left| \zeta \right|^n \right), \tag{5}$$

$$\phi = \frac{e^{2\dot{x}_i} + \tanh(x_i) \tanh(\ddot{x}_i)}{e^{2\dot{x}_i} + 1}, \tag{6}$$

where x_i; $\dot{x}_i$; $\ddot{x}_i$ are the position, velocity, and acceleration at the actuator side, respectively; F is the total friction forces at the TSM; k_ζ is a factor that expresses the ratio of the internal state, ζ, to the friction force; k_x is the stiffness factor that controls separate curves of the hysteresis loops; $\rho > 0$; $\sigma > 0.5$; $n \geq 1$ are coefficients that control the shape and size of hysteresis loops for the friction force; $\tanh(\cdot)$ is the hyperbolic tangent function which is defined as $\tanh(\cdot) = (e^{2(\cdot)} - 1)/(e^{2(\cdot)} + 1)$; υ is the viscous coefficient; and F_Θ is a offset point of the friction force; the dot at the top of variables represents the first derivative with respect to time.

Remark 1. It is noted that the variable ζ given by (4)–(6) is uniformly bounded for any piecewise continuous signals x_i; $\dot{x}_i$ (see [19, 20] for more details).

The dynamic model for the TSM-driven side with the mass, m, damping coefficient, c, input torque, F_o, external disturbance torque, F_d, with unknown bound, and environmental torque, F_e, as shown in Figure 7 can be described as [21]

$$m\ddot{\theta} + c\dot{\theta} = F_o - F_e - F_d, \tag{7}$$

where θ is the angular displacement of the output that corresponds to the wing warp.

The relation between the input force $F_i = u$, where u is the control input for the system, and the output force F_o can be expressed as $F_i = F_o + F$ where F is the total friction forces between the tendons and the sheaths. Equation (7) can be rewritten by

$$m\ddot{\theta} + c\dot{\theta} = (u - F) - F_e - F_d. \tag{8}$$

We define the system states as $x = [x_1; x_2]^T = [\theta; \dot{\theta}]^T$; then the dynamics of the joint in the tendon-sheath system can be simplified as

$$\dot{x}_1 = x_2, \tag{9}$$

$$\dot{x}_2 = \frac{1}{m} \left\{ (u - F) - F_e - F_d - c x_2 \right\}, \tag{10}$$

$$\theta = x_1, \tag{11}$$

where the friction force F is described by (4). Then the variable $\dot{x}_2$ given in (10) can be rewritten as

$$\begin{aligned} \dot{x}_2 &= \frac{1}{m} \left\{ \left(u - k_x \phi x_i - \upsilon \dot{x}_i \right) - c x_2 + D \right\} \\ &= \frac{1}{m} (u + D) + \Theta^T \varphi, \end{aligned} \tag{12}$$

where φ; D; Θ are expressed with the following forms:

$$\varphi = [\phi x_i, \dot{x}_i, x_2]^T = \left[x_i \left\{ e^{2\dot{x}_i} + \frac{\tanh(x_i)\tanh(\ddot{x}_i)}{e^{2\dot{x}_i}+1} \right\}, \dot{x}_i, x_2 \right]^T,$$
$$D = -\left(k_\zeta \zeta + F_o\right) - F_e - F_d, \tag{13}$$
$$\Theta = \left[-\frac{k_x}{m}, -\frac{v}{m}, -\frac{c}{m} \right]^T.$$

To design the controller u and suitable adaptive laws, some assumptions are needed: (i) the output and input positions, θ and x_i, are measurable; (ii) the desired trajectories θ_r, $\dot{\theta}_r$, $\ddot{\theta}_r$ are continuous and bounded; (iii) the environmental torque F_e and external disturbance F_d are bounded by unknown bounds. From Remark 1 and assumption (iii), the variable D is bounded by an unknown variable D^* ($|D| \le D^*$) because ζ; F_e, F_d are bounded.

To use the backstepping technique [22, 23], change of coordinate is made to system (9)–(12):

$$\xi_1 = x_1 - \theta_r \tag{14}$$

$$\xi_2 = x_2 - \dot{\theta}_r + \alpha_{v1}\xi_1, \tag{15}$$

where α_{v1} is a positive designed parameter that will be determined later.

From (14) and (15), the new coordinates for the system in (9)–(12) can be expressed by

$$\dot{\xi}_1 = \xi_2 - \alpha_{v1}\xi_1,$$
$$\dot{\xi}_2 = \frac{1}{m}(u + D) + \Theta^T \varphi - \ddot{\theta}_r + \alpha_{v1}\dot{\xi}_1. \tag{16}$$

Let $\widehat{D}^*$, $\widehat{m}$, $\widehat{\Theta}$, and $\widetilde{D}^* = D^* - \widehat{D}^*$, $\widetilde{m} = m - \widehat{m}$, and $\widetilde{\Theta} = \Theta - \widehat{\Theta}$ be the estimates and error estimates of D^*, m, and Θ, respectively. Define a virtual control input, $\overline{u}$, such that $u = \widehat{m}\overline{u} - \widehat{D}^* \tanh(\xi_2/\varepsilon)$, where $\varepsilon > 0$ will be determined later. From (9)–(12), parameter update laws, $\widehat{D}^*$, $\widehat{m}$, $\widehat{\Theta}$, and control input, u, are designed as follows:

$$u = \widehat{m}\overline{u} - \widehat{D}^* \tanh\left(\frac{\xi_2}{\varepsilon}\right), \tag{17}$$

$$\overline{u} = \ddot{\theta}_r - \alpha_{v1}\dot{\xi}_1 - \xi_1 - \alpha_{v2}\xi_2 - \widehat{\Theta}^T \varphi, \tag{18}$$

$$\dot{\widehat{\Theta}} = k_\Theta \xi_2 \varphi - \sigma_1 \widehat{\Theta}, \tag{19}$$

$$\dot{\widehat{m}} = -k_m \xi_2 \overline{u} - \sigma_2 \widehat{m}, \tag{20}$$

$$\dot{\widehat{D}}^* = k_D \xi_2 \tanh\left(\frac{\xi_2}{\varepsilon}\right) - \sigma_3 \widehat{D}^*, \tag{21}$$

where α_{v1}; α_{v2}; k_Θ; k_m; k_D; σ_i ($i = 1;2;3$) are positive parameters that adjust the designed controller u to force the tracking error, $e_r = \theta_r - \theta$, to approach the desired values.

With (17) to (21), the following theorem holds.

Theorem 2. *Consider the nonlinear system (16) with the designed controller, u, from (17) to (18) and adaptive laws (19) to (21); the following statements hold:*

(1) *The tracking error, e_r, and adaptive laws, $\widehat{D}^*$, $\widehat{m}$, $\widehat{\Theta}$, are globally uniformly bounded.*

(2) *In the presence of unknown model parameters and their bounds, the position tracking error converges to a desired compact region $\Omega = \{|e_r| \mid |e_r| \le 2\sqrt{\rho/\Psi}\}$,*

where

$$\Psi = \frac{\sigma_1}{2k_\Theta}\Theta^T\Theta + 0.2785\varepsilon\frac{D^*}{m} + \frac{\sigma_2}{2k_m}m + \frac{\sigma_3}{2mk_D}\left(D^*\right)^2, \tag{22}$$
$$\rho = \max\{2\alpha_{v1}, 2\alpha_{v2}, \sigma_1, \sigma_2, \sigma_3\}.$$

(See Appendix for the proof.)

Remark 3. For validating the proposed controller and adaptive laws given by (17) to (21), a few simulations have been carried out. Some guidelines are recommended to determine the designed control parameters:

(i) The parameters α_{v1} and α_{v2} determine the level of the convergence speed of e_r. Large values of α_{v1} and α_{v2} will lead to faster convergence of the tracking error. However, if these values are too large, it may cause chattering in the controller. High values of parameters k_Θ; k_m; k_D will improve the parameter adaptation speed and tracking performances. In addition, ε is used to guarantee the smoothness and avoid the discontinuity of the controller u given by (17). A small value of ε is desired. Therefore, we suggest to keep α_{v1} and α_{v2} fixed at acceptable large values and ε at a small value and then increase the other parameters such as k_Θ; k_m; k_D until the desired tracking performances are achieved.

(ii) The parameters σ_1; σ_2; σ_3 are used to prevent the estimate values of $\widehat{D}^*$, $\widehat{m}$, $\widehat{\Theta}$ to be too large. However, large values of σ_1; σ_2; σ_3 may suppress the prevention. Hence, small values for σ_1; σ_2; σ_3 are chosen.

To illustrate the proposed controller performance, a dynamic simulation with realistic dynamic friction model parameters are chosen from Do et al. [18], where $k_x = 0.01083$; ρ= 54.658; $n = 2.0458$; $\sigma = 1.58$; $k_\zeta = 0.14368$; $v = 0.02686$; $F_O = 0.0099$. The parameters for the dynamic joint are selected as $m = 0.0349$; $c = 0.0105$. The environment is simulated as an elastic spring $F_e = k_e y$, where $k_e = 0.4185$ and y is the output position. The disturbance is chosen, where $F_d = 0.2\sin(0.2\pi t)$. The designed controller and updated law parameters are selected based on the guidelines given by the Remark 3. With this idea, we choose the control parameters as $\alpha_{v1} = 10$; $\alpha_{v2} = 15$; $k_\Theta = 0.5$; $k_m = 0.5$; $k_D = 1$; $\sigma_i = 0.01$ ($i = 1; 2; 3$); $\varepsilon = 0.05$. The initial condition for updated estimate parameters are chosen such that $\widehat{\Theta}(0) = [0, 0, 0]^T$; $\widehat{m}(0) = 0$; $\widehat{D}^*(0) = 0$.

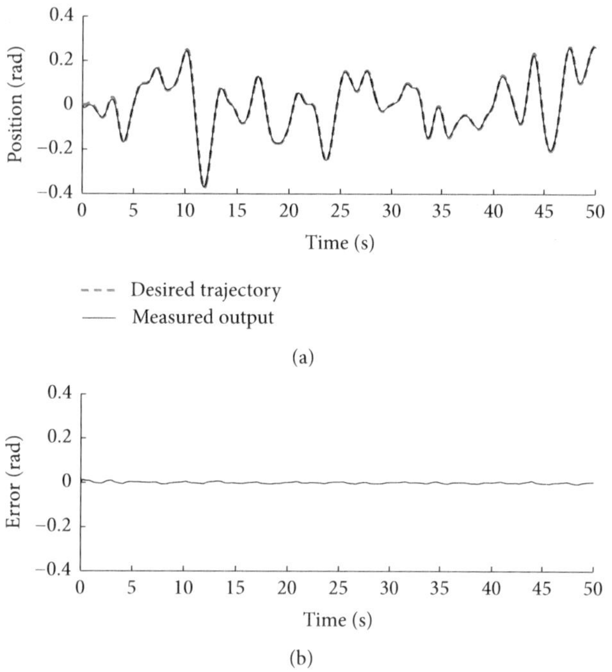

FIGURE 8: Simulation result of the designed adaptive laws for random trajectory case.

Figure 8 shows the result of the control law for a given random trajectory reference. Figure 8(a) illustrates the desired trajectory together with the simulated output for a random signal. The reference trajectory was generated using a uniform random generator filtered at 2 Hz cut-off frequency. Figure 8(a) shows the reference trajectory together with the controlled output, while Figure 8(b) depicts the error of the trajectory. The proposed controller demonstrates a satisfactory performance as reflected by the error depicted in the lower panel of the same figure with MSE = 9.8492×10^{-5}.

6. Wind Tunnel Tests

Three separate wind tunnel tests were performed. The FMW with a backstepping controller was tested in a gusty wind tunnel for robustness, and two different wind tunnel tests were performed to investigate the aerodynamic characteristics of the FMW mini-UAV. Steady uniform airflow was used to identify the aerodynamic properties of the FMW mini-UAV, while controlled gusty airflow was used to study the adaptive washout mechanism of the FMW and to validate the gust response dampening ability of the FMW mini-UAV.

6.1. Adaptive Controller Robustness Test. In order to validate the robustness of the backstepping controller and also to benchmark it with the feedforward/feedback controller in [10], an identical gusty wind tunnel test discussed in the same reference was performed on the FMW. Please refer to the reference for the description of the gust generator used in this experiment. Based on the dynamic analysis as shown in Figure 4, the modal parameters were identified, where the effective mass m = 0.17 kg, c = 0.0026 Ns/rad, and k = 18.9 N/rad.

Figure 9 shows the results of the warping performance in two conditions, that is, no-gust and gusty environment. The left panels of the figure illustrate the performance of the feedforward/feedback controller (FF/PI), while the right ones show the performance of the system with the adaptive backstepping controller.

The backstepping controller is proven to be able to control the warping position when the FMW is subjected to disturbance. In case of minimum disturbance as simulated by the no-gust condition, the output of the system with FF/PI controller is able to track the reference input very well with RMS error of 0.285°. However, the performance slightly drops when the FMW is exposed under gusty environment (RMS = 0.523°). On the other hand, the adaptive backstepping controller performance seems to be more consistent for both cases, where the RMS errors are 0.447° and 0.482° for no-gust and gusty cases, respectively. Please note that, in this study, the parameters of the controllers are not fully optimized as the objective was more to validate the capability of using a controlled TSM in FMW applications rather than investigating the controller itself.

6.2. Aerodynamic Forces Characterization. The wind tunnel used for this experiment is 2.3 m wide, 1.5 m high, and 6 m long. A large test section was required in order to minimize the wall effects of the wind tunnel, which will affect the results of the wind tunnel tests. The FMW mini-UAV was tested at two different wind speeds, which were 12 m/s and 15 m/s. Initially, the FMW mini-UAV was tested at a negative α of -4° and the whole wing fluttered aggressively. A similar incident occurred at $\alpha = 0^{\circ}$, which results in the decision of testing the FMW at an angle of attack, α, of 2° onwards. The fluttering occurred because the FMW is rigid against forces from below the wing, but weak against the forces from above the wing. This property is expected, as it can allow the wing to provide lift while still being able to fold downwards around the fuselage. Even though the FMW was oscillating very aggressively, the TSM stayed intact and was able to still perform wing warping, proving its robustness in very harsh conditions.

The wind tunnel testing results are shown in Figures 10(a)–12(b). All the experimental results were taken three times at the same α and then averaged. The lift coefficient, C_L, drag coefficient, C_D, and pitching moment coefficient, C_M, were measured during the tests. The maximum C_L of the FMW reached close to 1.5 when tested at 12 m/s wind speed. From Figure 11, the highest C_L over C_D occurs when the angle of attack, α, is around 8° when the wind speed is 15 m/s. It can also be observed from the wind tunnel experiment that the rigid wing has a much lower stall angle, α, when compared to the FMW. It is noted that, compared to the conventional rigid wing aircraft, the C_L/C_D ratio of the FMW UAV is considerably small. This is due to the very small area of the wing, as well as the boxy frame of the fuselage structure, contributing to the high C_D of the FMW UAV. Future versions of the FMW should have a more streamlined fuselage, with less wetted area in order to

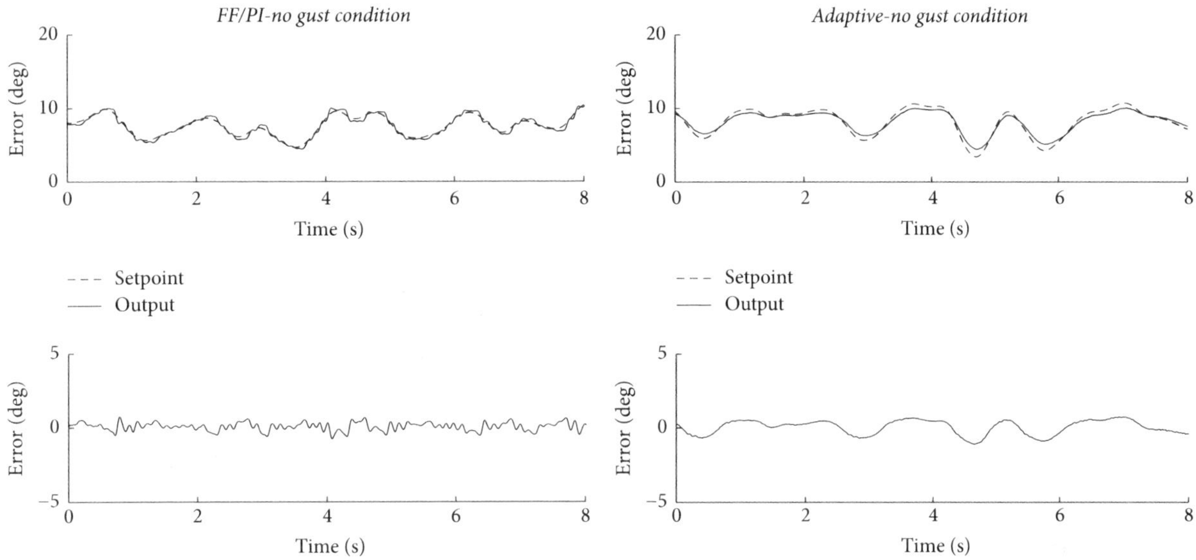

(a) Output and tracking error for no-gust condition

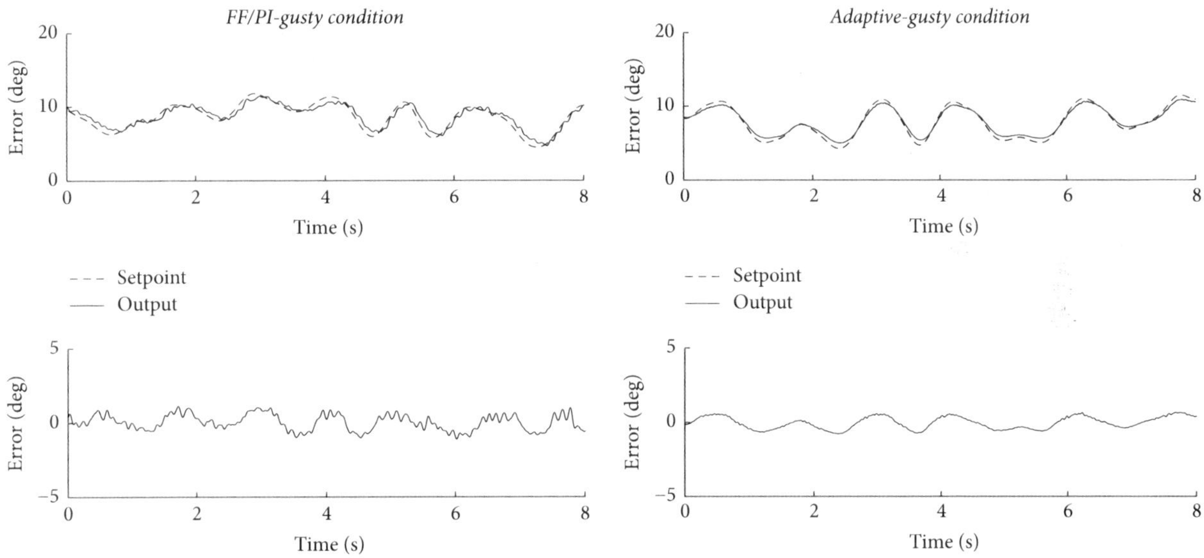

(b) Output and tracking error for gusty condition

FIGURE 9: Controller performances of the wing warping with tendon-sheath actuation under no-gust and gusty environments with wind speed of 7 m/s. The left panels show the case with feedforward/feedback controller; the right panels show the case of adaptive backstepping controller.

minimize the drag coefficient. The increase in wing area is desirable and will be more difficult to implement due to the material properties required of the FMW. Higher wing areas may lead to unwanted or unexpected folding. Further studies will be necessary to investigate the method of increasing the area as well as the aspect ratio of the FMW.

In Figure 12(a), the rigid wing stalls when α reaches 10°, while for the FMW the stall angle is at around 14° when the wind speed is 15 m/s, and the FMW seems to not stall even at 18° when the wind speed is 12 m/s. The rolling moment versus the warping angle is critical for the identification of the FMW roll rate and was obtained in this wind tunnel experiment. Holding the FMW at $\alpha = 2°$, Figure 12(b) shows that the rolling moment is increasing linearly with the warping angle, until warping angle reaches 14°, where the rolling moment starts to decrease. This phenomenon is quite well known as adverse rolling. For conventional wings, adverse rolling will happen when the ailerons are exerting too much force such that the wings itself are twisting into the opposite direction, creating the opposite rolling direction as intended from the pilot. In [24], the relationship between dynamic pressure and roll rate was investigated for rigid and flexible wings, and it was found that as the dynamic pressure increases, after a certain point, the flexible wings will lose the roll rate. For the FMW, the rolling moment starts to decrease when the warping angle is beyond 14° at 15 m/s. Therefore, for wind

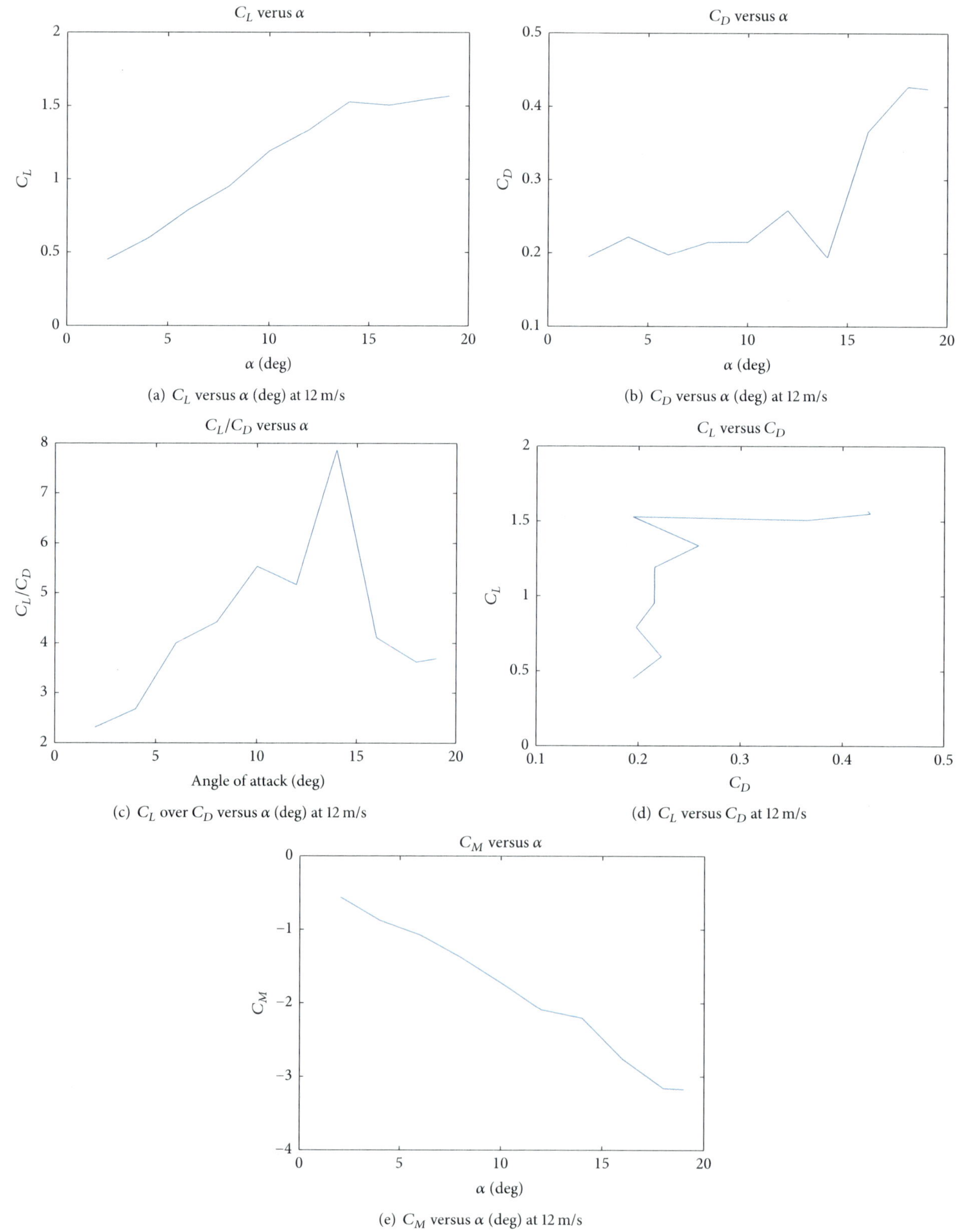

FIGURE 10: Aerodynamic properties of the FMW mini-UAV at 12 m/s wind speed.

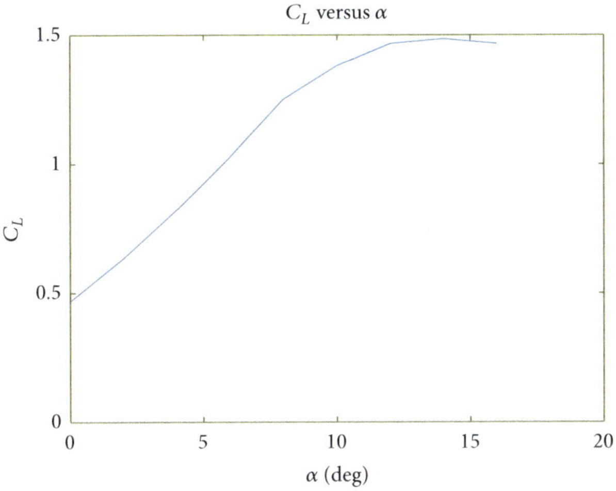

(a) C_L versus α (deg) at 15 m/s

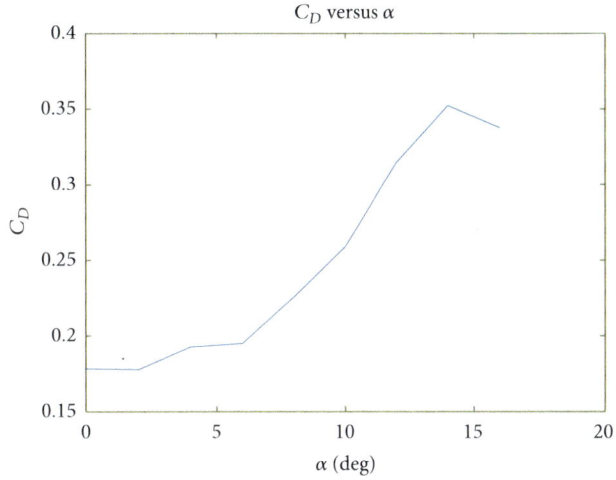

(b) C_D versus α (deg) at 15 m/s

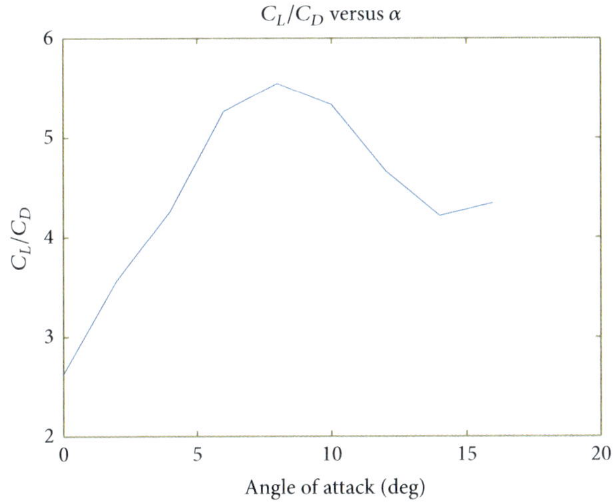

(c) C_L over C_D versus α (deg) at 15 m/s

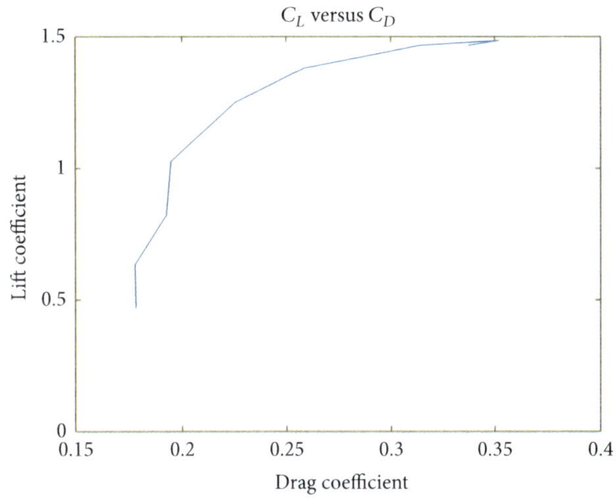

(d) C_L versus C_D at 15 m/s

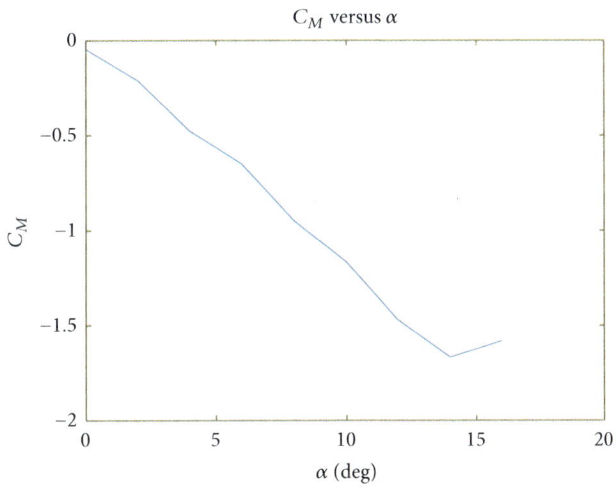

(e) C_M versus α (deg) at 15 m/s

Figure 11: Aerodynamic properties of the FMW mini-UAV in 15 m/s wind speed.

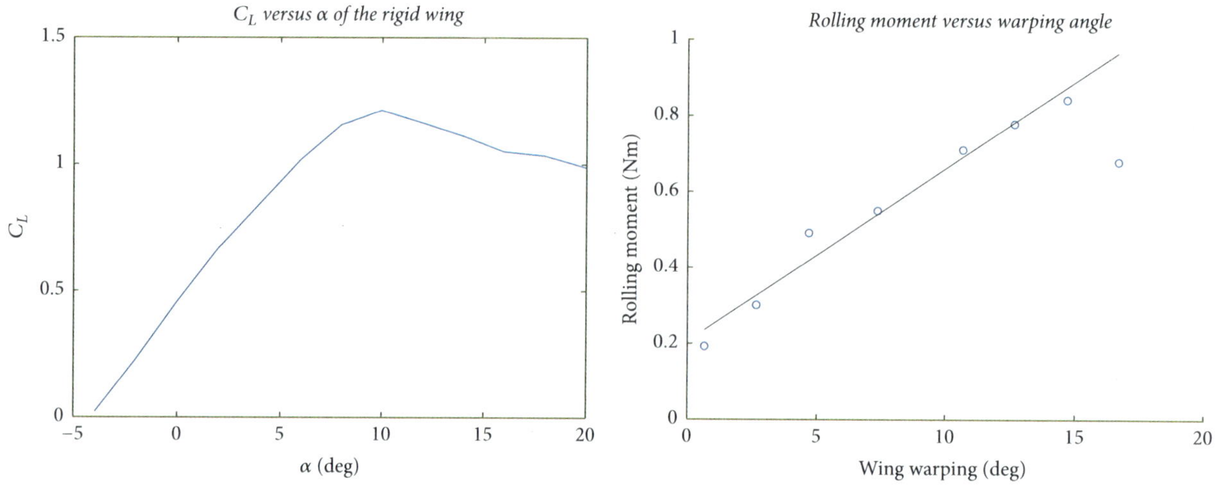

(a) (b)

Figure 12: Wind tunnel test results: (a) aerodynamic property of rigid wing; (b) the rolling moment of FMW.

(a) (b)

Figure 13: Test rigs: (a) the rigid wing and (b) the FMW model, suspended in the middle of the test section of the wind tunnel.

speeds larger than 15 m/s, it is undesirable to warp the FMW more than 14°. Further investigations are required to map the relationship between the roll rate, warping angle, and also the dynamic pressure for the FMW.

6.3. Adaptive Washout Validation. One of the capabilities of the FMW is to dampen the longitudinal gust response via the adaptive washout mechanism. When a longitudinal direction gust hits the FMW from below, it will create an extra lift force, but the lift force will be dampened as the FMW will flex chordwise passively, changing the airfoil shape and also the effective α [3]. Thus, the gust response of the FMW is much lower compared to a conventional rigid wing. When the TSM is installed on the FMW, a validation of the adaptive washout is required. A gusty flow wind tunnel test is performed to obtain data about the gust response of both the FMW mini-UAV and the conventional rigid wing mini-UAV to validate the adaptive washout mechanism of the FMW when installed with the tendon-sheath.

The mini-UAV test model was suspended inside the wind tunnel, with unconstrained pitch and roll degrees-of-freedom, as well as freedom of translational movement along the three axes. Nylon strings were used to suspend the model, and rubber bands were tied to the end of the nylon strings to help increase the deflection and rotational angles of the model. The nylon strings were attached to the nose and the tail of the model, as well as at the aerodynamic center of the model to support the weight of the model. The static margin of the model was set to 5% for a more sensitive response to gusts. The model was fitted with all the payload, with a simulated weight to mimic all electronics and battery suspended in the wind tunnel to obtain a close approximate of the moment of inertias and also to better simulate a real flight situation. Figure 13 illustrates two models, with rigid wings—as a comparison—and FMW, suspended in the test section of the wind tunnel.

To generate longitudinal gusts, a rotating slotted cylinder (RSC) device was used. The RSC is a simple mechanism, which utilizes only one rotating cylinder to generate the simulated gusts. The RSC was designed following [25, 26], where the fundamental frequency of the generated gust is equal to two times the rotational frequency of the cylinder. The chord length of the airfoil, c, in the gust generator is 0.2 m, and the diameter of the RSC is $d = 0.25c$. The thickness of the RSC is $t = 0.0313c$ and the gap between the airfoil trailing edge and the RSC is $e = 0.08c$. NACA 0012 was used as the airfoil for the gust generator. The RSC gust generator is shown in Figure 14. The frequency of the gust is set at 7 Hz as this is

Table 2: Specifications of the FMW mini-UAV.

Wingspan	1 m
CG location	0.25 m behind leading edge
Weight	1060 g
Wing loading	53 g/dm^3
Power	Electric motor + 9″× 6″ propellers
Estimated flight time	8 minutes

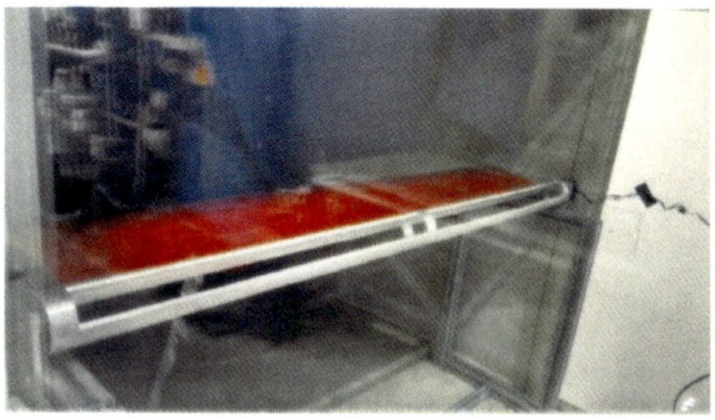

Figure 14: Gust generator with a rotating slotted cylinder.

the fundamental frequency of the FMW. Therefore, the RSC rotates at 3.5 revolutions per second.

The model was fitted with accelerometers and gyroscopes at the center of gravity to detect the movements of the model, and the oscillation data is then streamed and recorded to a data acquisition computer. Figure 15 shows the gust response of the rigid wing and FMW model. From the plots it can be observed that the adaptive washout mechanism of the FMW helped to alleviate the gust response, especially in the z-axis (see Figures 15(e) and 15(f)), as the FMW model has less movements in the gusty airflow compared to the rigid conventional wing model. The standard deviation of the rigid wing movements is generally larger compared to the FMW, with the most significant difference in the z-axis acceleration at 214.82 for the rigid wing and 94.17 for the FMW.

7. Flight Test

To provide autopilot control to the aircraft and to track the flight data, the ArduPilot (APM) 2.5 flight controller is fitted at the CG of the FMW mini-UAV. The main control strategy used for the autopilot is PID feedback for all the flight dynamics. The STM32F4 microcontroller preprogrammed with the control strategy presented in Section 5 is also used to regulate the wing warping and all the sensors and actuators related to the wing warping are connected to this microcontroller. The specifications of the UAV are listed in Table 2.

As the relationship between the rolling moment and the warping position can be modeled as a linear curve from wind tunnel tests, a PID feedback control will be sufficient. From the wind tunnel tests, roll reversal of the FMW has been found at warping angle 14° at 15 m/s wind speed. In this flight test, the wing warping has been saturated at 14° in order to avoid adverse rolling of the mini-UAV. In the future, further investigation will be needed to examine the relationship between rolling moment and dynamic pressure of the FMW.

Figure 16 shows the mini-UAV cruising in the sky. Flying the FMW mini-UAV requires a good piloting skill as the aircraft is very fast and responsive. Therefore, in this following discussion, the flight controllability and maneuverability are assessed qualitatively based on the input from the pilot. According to the pilot, the mini-UAV is very difficult to control without wing morphing strategy (when maneuvering relies only to the rudders), as it is a fast aircraft. There is more control on the rolling movement in a case with the controlled wing warping mechanism, compared to that with only using the rudder input. Rudders are able to provide yaw and banking control only at low speeds, that is, during gliding. During all the flight tests the weather was gusty. Maneuvering the aircraft with an open-loop wing morphing mechanism does not give acceptable performance, either. The aircraft was barely under control. This result is believed to be attributed to the incapability of an open-loop system to manage disturbances. However, the mini-UAV was able to fly steadily under control with the controlled wing warping mechanism.

Figures 17, 18, and 19 illustrate the flight test results to give qualitative measures of the mini-UAV responsiveness under three different control methods, that is, (i) relying only on the rudder input (with no wing warping), (ii) open-loop wing warping, and (iii) backstepping controlled wing warping, to achieve rolling motion for the aircraft. Even though the tests were not conducted exactly at the same instance, those were conducted successively within short duration to minimize the variation of environmental conditions (e.g., gust).

Figures 17(a), 18(a), and 19(a) show the reference input given through the UAV remote controller, while Figures 17(b), 18(b), and 19(b) show the roll rates results. It can be seen easily from the three figures that the wing warping mechanism with the proposed controller demonstrates the best qualitative performance; that is, the output tends to follow the reference input. Rudder input can also induce rolling; however, it is not as effective as the controlled wing warping.

Figure 20 shows the cross correlation between the wing warping input and the roll rate output for the third case (FMW with the controller). It can be seen that the output has a strong correlation to the reference input with an acceptable lag of 0.3 seconds. In gusty condition during the flight test, the mini-UAV was able to achieve a 0.7625 correlation coefficient between the wing warping input and the roll rate. The open-loop wing warping control did not provide a sufficient control of the rolling moment of the mini-UAV, as experienced during flights and also shown in the flight data, with a correlation coefficient of only 0.0945. The rudder inputs did not seem to affect the roll rates of the aircraft either, with the correlation coefficient at 0.1347.

8. Discussion and Conclusion

An FMW was designed to dampen gusts while in flight and also to fold around the fuselage for storage. However, the FMW lacks a suitable actuation method to be able to provide ample rolling moment. A TSM is proven to be a suitable

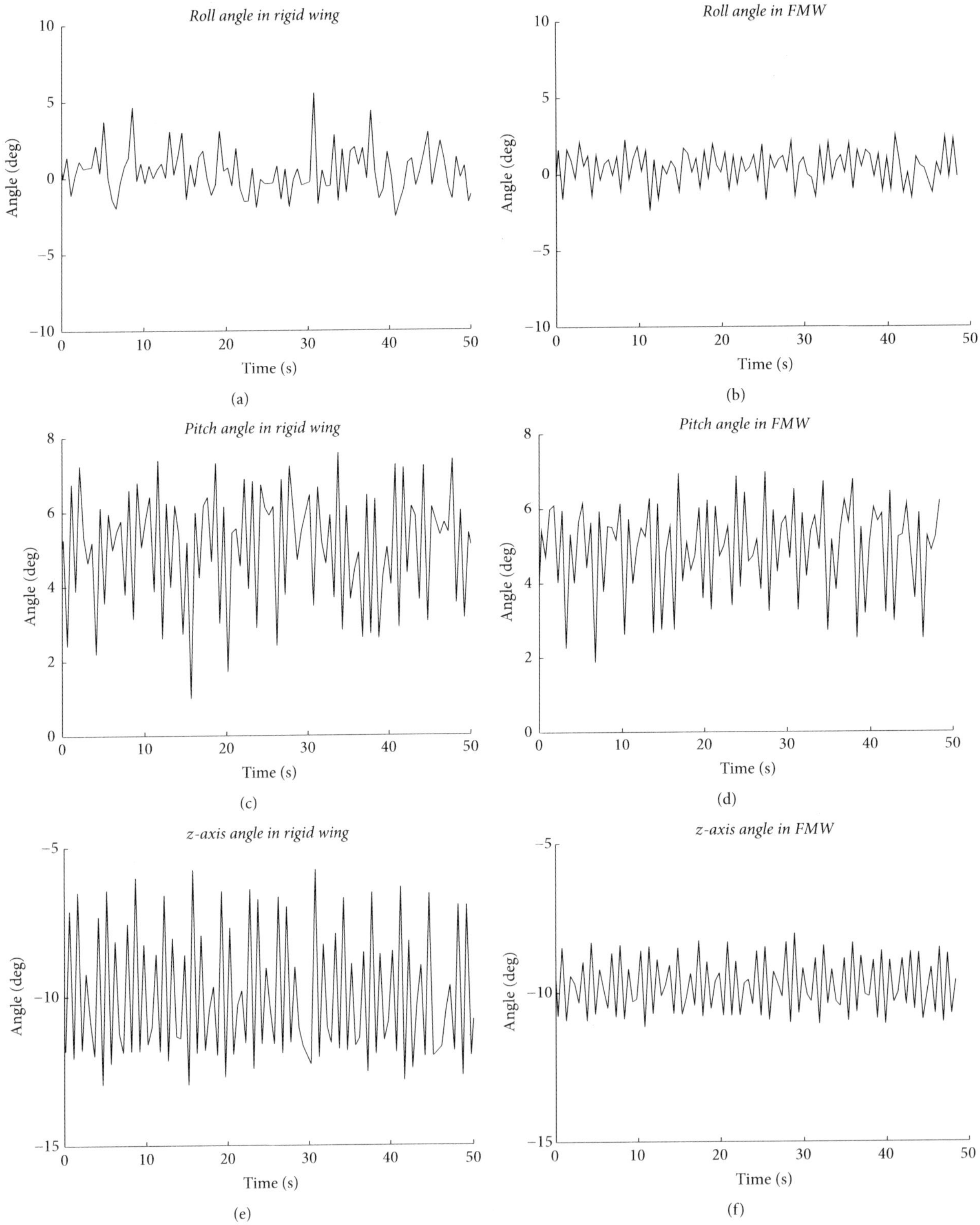

FIGURE 15: The roll and pitch angles and z-axis acceleration of both the rigid wing and FMW model at 7 Hz longitudinal gusts. Rigid wing behavior is shown in (a), (c), and (e), while those from FMW are shown in (b), (d), and (f).

candidate to provide the needed actuation without sacrificing the desired attributes of the FMW.

In this paper, an adaptive backstepping controller was proposed, and it was concluded that the controller is more robust to the disturbance compared to a model-based feedforward controller. Furthermore, a mini-UAV was designed and various tests were performed to identify its characteristics. From the wind tunnel tests, the FMW installed with TSM demonstrated the ability of the adaptive washout, being able to dampen the longitudinal gusts quite well when compared to a rigid, conventional wing. It was also shown that the rudder only input does not induce as much rolling moment as the controlled wing warping. Flight tests were conducted, and the wing warping was proven to be a better method to induce rolling moment when compared to rudder input only. The TSM actuated FMW with the proposed control strategy was proven to work well in a real flight.

One of the findings from the wind tunnel experiments was the roll reversal of the FMW. Due to the flexibility of the wing in the chordwise direction, roll reversal will exist when the dynamic pressure is very high. Careful saturation of the FMW warping is necessary for the safety of the aircraft. Future work could characterize the occurrence of roll reversal in different situations and develop a better control strategy to prevent it.

FIGURE 16: Flight test of the FMW mini-UAV.

Appendix

Proof of Theorem 2

Define a Lyapunov function candidate, V, as follows:

$$V = \frac{\xi_1^2}{2} + \frac{\xi_2^2}{2} + \frac{\widetilde{\Theta}^T\widetilde{\Theta}}{2k_\Theta} + \frac{\widetilde{m}^2}{2mk_m} + \frac{\left(\widetilde{D}^*\right)^2}{2mk_D}. \quad \text{(A.1)}$$

Taking the derivative of V together with (14) and (21), we can obtain

$$\begin{aligned}\dot{V} &= \xi_1\dot{\xi}_1 + \xi_2\dot{\xi}_2 - \frac{\widetilde{\Theta}^T\dot{\widehat{\Theta}}}{k_\Theta} - \frac{\widetilde{m}\dot{\widehat{m}}}{mk_m} - \frac{\widetilde{D}^*\dot{\widehat{D}}^*}{mk_D} \\ &= \xi_1\left(\xi_2 - \alpha_{\upsilon 1}\xi_1\right) \\ &\quad + \xi_2\left(\frac{1}{m}(u + D) + \Theta^T\varphi - \ddot{y}_r + \alpha_{\upsilon 1}\dot{\xi}_1\right) - \frac{\widetilde{m}\dot{\widehat{m}}}{mk_m} \\ &\quad - \frac{\widetilde{\Theta}^T\dot{\widehat{\Theta}}}{k_\Theta} - \frac{\widetilde{D}^*\dot{\widehat{D}}^*}{mk_D}.\end{aligned} \quad \text{(A.2)}$$

Substituting (17) and (18), we can rewrite (A.2) as

$$\begin{aligned}\dot{V} &= -\alpha_{\upsilon 1}\xi_1^2 - \alpha_{\upsilon 2}\xi_2^2 + \left(\widetilde{\Theta}^T\varphi\xi_2 - \frac{\widetilde{\Theta}^T\dot{\widehat{\Theta}}}{k_\Theta}\right) \\ &\quad - \left(\frac{\widetilde{m}\bar{u}\xi_2}{m} + \frac{\widetilde{m}\dot{\widehat{m}}}{mk_m}\right) \\ &\quad + \left(\frac{D\xi_2}{m} - \frac{\widetilde{D}^*\dot{\widehat{D}}^*}{mk_D} - \frac{\widehat{D}^*\xi_2\tanh\left(\xi_2/\varepsilon\right)}{m}\right).\end{aligned} \quad \text{(A.3)}$$

From (19) to (21), the following expression is observed:

$$\begin{aligned}\widetilde{\Theta}^T\varphi\xi_2 - \frac{\widetilde{\Theta}^T\dot{\widehat{\Theta}}}{k_\Theta} &= \widetilde{\Theta}^T\varphi\xi_2 - \frac{\widetilde{\Theta}^T\left(k_\Theta\xi_2\varphi - \sigma_1\widehat{\Theta}\right)}{k_\Theta} \\ &= \frac{\sigma_1\widetilde{\Theta}^T\widehat{\Theta}}{k_\Theta}, \\ \frac{\widetilde{m}\bar{u}\xi_2}{m} + \frac{\widetilde{m}\dot{\widehat{m}}}{mk_m} &= \frac{\widetilde{m}\bar{u}\xi_2}{m} - \frac{\widetilde{m}\left(k_m\xi_2\bar{u} + \sigma_2\widehat{m}\right)}{mk_m} \\ &= -\frac{\sigma_2\widetilde{m}\widehat{m}}{mk_m}, \\ \frac{\widetilde{D}^*\dot{\widehat{D}}^*}{mk_D} &= \frac{\left(D^* - \widehat{D}^*\right)k_D\xi_2\tanh\left(\xi_2/\varepsilon\right)}{mk_D} - \frac{\sigma_3\widetilde{D}^*\widehat{D}^*}{mk_D}.\end{aligned} \quad \text{(A.4)}$$

Subsequently, (A.3) can be rewritten as

$$\begin{aligned}\dot{V} &= -\alpha_{\upsilon 1}\xi_1^2 - \alpha_{\upsilon 2}\xi_2^2 + \frac{\sigma_1\widetilde{\Theta}^T\widehat{\Theta}}{k_\Theta} + \frac{\sigma_2\widetilde{m}\widehat{m}}{mk_m} + \frac{\sigma_3\widetilde{D}^*\widehat{D}^*}{mk_D} \\ &\quad + \frac{D\xi_2}{m} - \frac{D^*\xi_2\tanh\left(\xi_2/\varepsilon\right)}{m}.\end{aligned} \quad \text{(A.5)}$$

Applying Young's inequality for two numbers a and b, that is, $ab \le 0.5(a^2 + b^2)$, the third, fourth, and fifth terms in right hand side of (18) can be reformulated as

$$\begin{aligned}\frac{\sigma_1\widetilde{\Theta}^T\widehat{\Theta}}{k_\Theta} &= \frac{\sigma_1\widetilde{\Theta}^T\left(\Theta - \widetilde{\Theta}\right)}{k_\Theta} \le \frac{\sigma_1\Theta^T\Theta}{2k_\Theta} - \frac{\sigma_1\widetilde{\Theta}^T\widetilde{\Theta}}{2k_\Theta}, \\ \frac{\sigma_2\widetilde{m}\widehat{m}}{mk_m} &= \frac{\sigma_2\widetilde{m}\left(m - \widetilde{m}\right)}{mk_m} \le \frac{\sigma_2 m^2}{2mk_m} - \frac{\sigma_2\widetilde{m}^2}{2mk_m}, \\ \frac{\sigma_3\widetilde{D}^*\widehat{D}^*}{mk_D} &= \frac{\sigma_3\widetilde{D}^*\left(D^* - \widetilde{D}^*\right)}{mk_D} \\ &\le \frac{\sigma_3 D^{*2}}{2mk_D} - \frac{\sigma_3\left(\widetilde{D}^*\right)^2}{2mk_D}.\end{aligned} \quad \text{(A.6)}$$

From (A.6) and using a proven property in [27], $\tanh(\xi_2/\varepsilon)$ obeys $|\xi_2| - \xi_2\tanh(\xi_2/\varepsilon) \le 0.2785\varepsilon$.

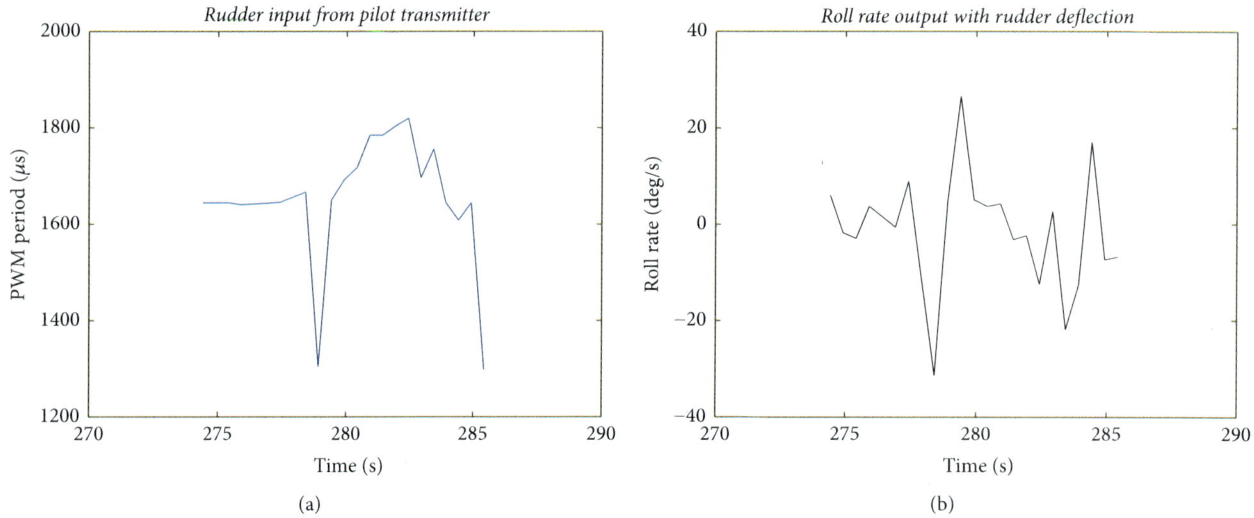

FIGURE 17: Roll rate of the FMW when given rudder doublets: (a) rudder inputs from pilot; (b) roll rates of the FMW mini-UAV.

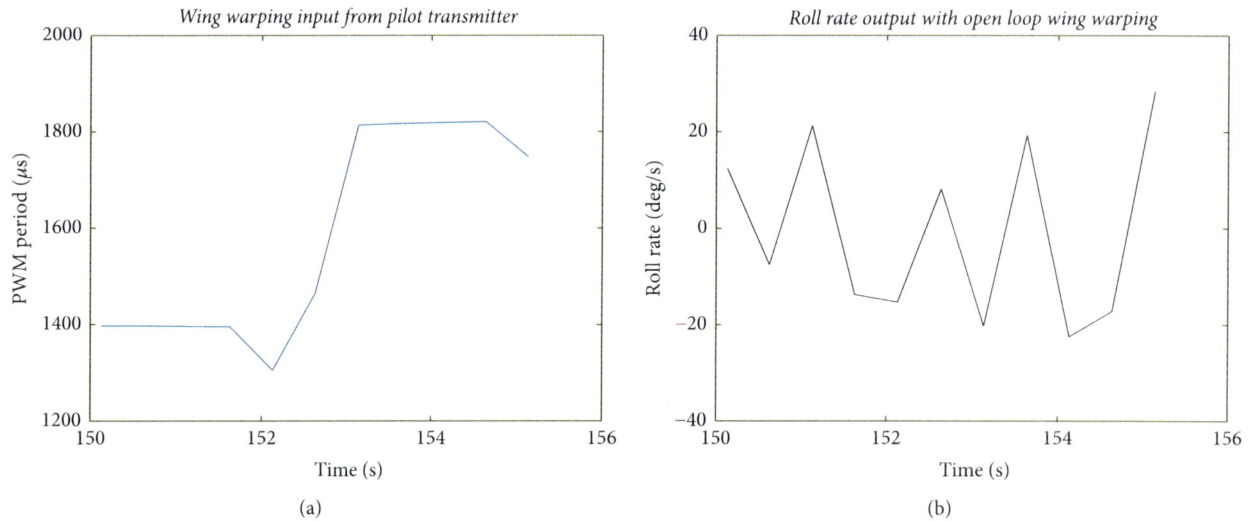

FIGURE 18: Roll rate of the FMW with open-loop wing warping double: (a) rudder inputs from pilot; (b) roll rates of the FMW mini-UAV.

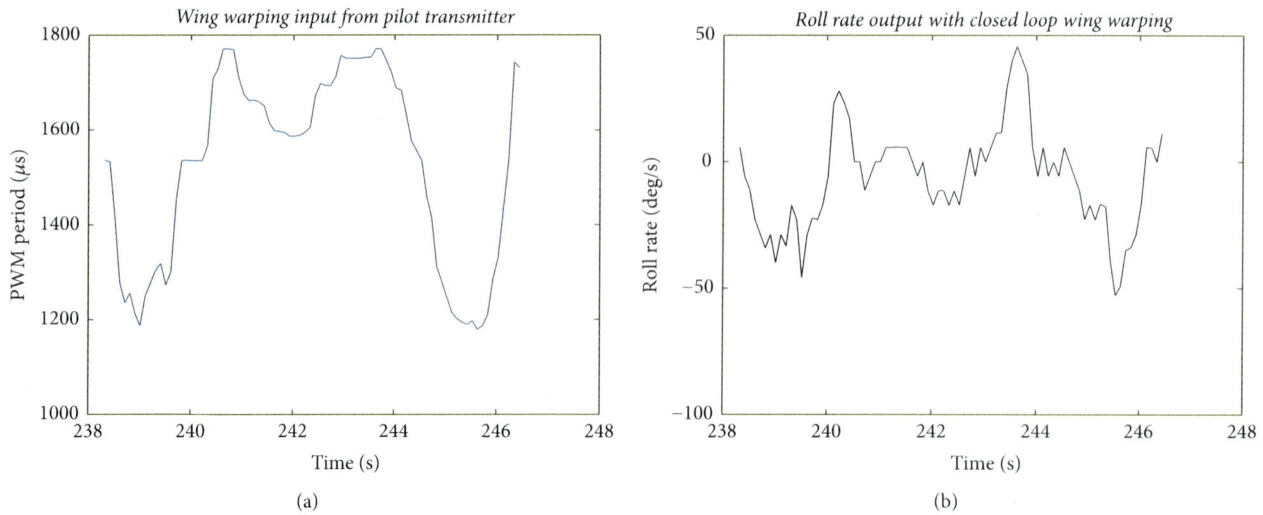

FIGURE 19: Roll rate of the FMW with closed-loop wing warping double: (a) rudder inputs from pilot; (b) roll rates of the FMW mini-UAV.

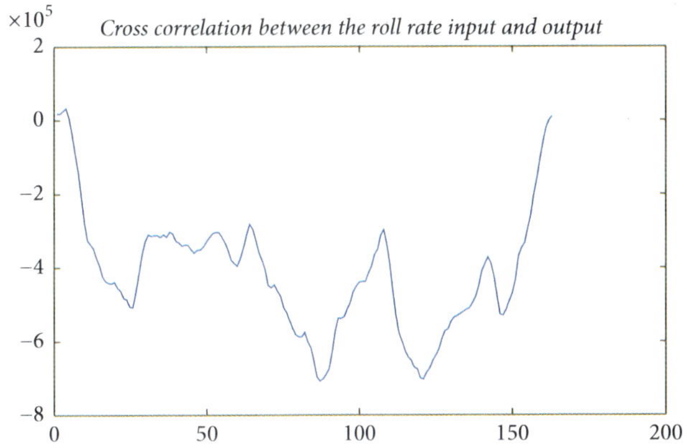

FIGURE 20: Cross correlation between the wing warping input and roll rate output of the FMW mini-UAV with controller.

$\dot{V}$ in (A.5) can be rewritten as

$$\begin{aligned}\dot{V} &\leq -\alpha_{v1}\xi_1^2 - \alpha_{v2}\xi_2^2 - \frac{\sigma_1 \widetilde{\Theta}^T \widetilde{\Theta}}{2k_\Theta} - \frac{\sigma_2 \widetilde{m}^2}{2mk_m} - \frac{\sigma_3 (\widetilde{D}^*)^2}{2mk_D} \\ &+ \frac{\sigma_1 \Theta^T \Theta}{2k_\Theta} + \frac{\sigma_2 m^2}{2mk_m} + \frac{\sigma_3 (D^*)^2}{2mk_D} \\ &+ \frac{D^*}{m}\left(|\xi_2| - \xi_2 \tanh\frac{\xi_2}{\varepsilon}\right) \\ &\leq -\alpha_{v1}\xi_1^2 - \alpha_{v2}\xi_2^2 - \frac{\sigma_1 \widetilde{\Theta}^T \widetilde{\Theta}}{2k_\Theta} - \frac{\sigma_2 \widetilde{m}^2}{2mk_m} - \frac{\sigma_3 (\widetilde{D}^*)^2}{2mk_D} \\ &+ \frac{\sigma_1 \Theta^T \Theta}{2k_\Theta} + \frac{\sigma_2 m^2}{2mk_m} + \frac{\sigma_3 (D^*)^2}{2mk_D} + \frac{0.2785\varepsilon D^*}{m} \\ &= -\rho V + \Psi,\end{aligned} \tag{A.7}$$

where

$$\begin{aligned}\Psi &= \left(\frac{\sigma_1}{2k_\Theta}\right)\Theta^T\Theta + \frac{0.2785\varepsilon D^*}{m} + \left(\frac{\sigma_2}{2k_m}\right)m \\ &+ \left(\frac{\sigma_3}{2mk_D}\right)(D^*)^2,\end{aligned} \tag{A.8}$$

$$\rho = \min\{2\alpha_{v1}, 2\alpha_{v2}, \sigma_1, \sigma_2, \sigma_3\}.$$

Solving V from $\dot{V} \leq -\rho V + \Psi$ and from (A.1), one can obtain

$$0 \leq 0.5\xi_1^2 \leq V \leq \left(V(0) - \frac{\Psi}{\rho}\right)e^{-\rho t} + \frac{\Psi}{\rho}, \tag{A.9}$$

where $V(0) = 0.5(\xi_1(0))^2 + (\widetilde{\Theta}(0))^T\widetilde{\Theta}(0)/2k_\Theta + 0.5(\xi_2(0))^2 + (\widetilde{m}(0))^2/2mk_m + (\widetilde{D}^*(0))^2/2mk_D$.

It can be seen that there exists a time $T > 0$ such that $\forall t > T$, $(V(0) - \Psi/\rho)e^{-\rho t} \rightarrow 0$. Then V is bounded by Ψ/ρ for $\forall t > T$. Hence V is uniformly ultimately bounded (UUB); thus the tracking errors $e_r = \xi_1, \widetilde{D}^*, \widetilde{m}, \widetilde{\Theta}$ are also bounded. This further guarantees the boundedness of $\widehat{D}^*$, $\widehat{m}$, $\widehat{\Theta}$ since D^*, m, Θ are bounded.

From (A.9), the variable ξ_1 can be expressed as

$$|e_r| = |\xi_1| \leq \sqrt{2\left(V(0) - \frac{\Psi}{\rho}\right)e^{-\rho t} + \frac{2\Psi}{\rho}}. \tag{A.10}$$

$\forall t > T$, the tracking error $e_r = \xi_1$ will converge to a compact set $\Omega = \{|e_r| \mid |e_r| \leq 2\sqrt{\rho/\Psi}\}$ since $(V(0) - \Psi/\rho)e^{-\rho t} \rightarrow 0$. Consequently, the control input, u, is also bounded.

Conflicts of Interest

The authors declare that there are no conflicts of interest regarding the publication of this paper.

Acknowledgments

The authors would like to thank Benedict Lai and Van Than Dung for their help and effort during the wind tunnel tests and also Vigneashwara Pandiyan for piloting the FMW mini-UAV during the flight test.

References

[1] E. C. C. Choi, "Extreme wind characteristics over Singapore - An area in the equatorial belt," *Journal of Wind Engineering and Industrial Aerodynamics*, vol. 83, pp. 61–69, 1999.

[2] W. J. Pisano and D. A. Lawrence, "Autonomous gust insensitive aircraft," in *Proceedings of the AIAA Guidance, Navigation and Control Conference and Exhibit*, American Institute of Aeronautics and Astronautics, Honolulu, Hawaii, August 2008.

[3] P. Ifju, D. Jenkins, S. Ettinger, Y. Lian, W. Shyy, and M. Waszak, "Flexible-wing-based micro air vehicles," in *Proceedings of the 40th AIAA Aerospace Sciences Meeting & Exhibit*, American Institute of Aeronautics and Astronautics, Reno, NV, USA, 2002.

[4] M. Abdulrahim, H. Garcia, and R. Lind, "Flight characteristics of shaping the membrane wing of a micro air vehicle," *Journal of Aircraft*, vol. 42, no. 1, pp. 131–137, 2005.

[5] H. Garcia, M. Abdulrahim, and R. Lind, "Roll Control for a Micro Air Vehicle Using Active Wing Morphing," in *Proceedings of the AIAA Guidance, Navigation, and Control Conference and Exhibit*, American Institute of Aeronautics and Astronautics, Austin, Texas, August 2003.

[6] G. Palli and C. Melchiorri, "Model and control of tendon-sheath transmission systems," in *Proceedings of the 2006 IEEE International Conference on Robotics and Automation, ICRA 2006*, pp. 988–993, usa, May 2006.

[7] Y. Lian, W. Shyy, D. Viieru, and B. Zhang, "Membrane wing aerodynamics for micro air vehicles," *Progress in Aerospace Sciences*, vol. 39, no. 6-7, pp. 425–465, 2003.

[8] Y. Lian, W. Shyy, and R. T. Haftka, "Shape Optimization of a Membrane Wing for Micro Air Vehicles," *AIAA Journal*, vol. 42, no. 2, pp. 424–426, 2004.

[9] H. Hu, M. Tamai, and J. T. Murphy, "Flexible-membrane airfoils at low reynolds numbers," *Journal of Aircraft*, vol. 45, no. 5, pp. 1767–1778, 2008.

[10] S. Lee, T. Tjahjowidodo, H. C. Lee, and B. Lai, "Investigation of a robust tendon-sheath mechanism for flexible membrane wing application in mini-UAV," *Mechanical Systems and Signal Processing*, vol. 85, pp. 252–266, 2017.

[11] M. Abdulrahim, H. Garcia, and R. Lind, "Flight Testing a Micro Air Vehicle Using Morphing for Aeroservoelastic Control," in *Proceedings of the 45th AIAA/ASME/ASCE/AHS/ASC Structures, Structural Dynamics & Materials Conference*, American Institute of Aeronautics and Astronautics, Palm Springs, Calif, USA, April 2004.

[12] R. De Breuker, M. M. Abdalla, and Z. Gürdal, "A generic morphing wing analysis and design framework," *Journal of Intelligent Material Systems and Structures*, vol. 22, no. 10, pp. 1025–1039, 2011.

[13] D. P. Raymer, *Aircraft Design: A Conceptual Approach And Rds-Student, Software for Aircraft Design, Sizing, and Performance Set (AIAA Education)*, AIAA, Los Angeles, Calif, USA, 2006.

[14] J. Roskam, *Airplane Design: Determination of stability, control and performance characteristics: far and military requirements*, DARcorporation, Lawrence, Kansas, USA, 2002.

[15] M. H. Sadraey, *Aircraft Design: A Systems Engineering Approach*, John Wiley & Sons, Hoboken, NJ, United States, 2012.

[16] M. Kaneko, T. Yamashita, and K. Tanie, "Basic considerations on transmission characteristics for tendon drive robots," in *Proceedings of the Fifth International Conference on Advanced Robotics, 1991.'Robots in Unstructured Environments', 91 ICAR*, pp. 827–832, IEEE, Pisa, Italy, June 1991.

[17] J. Song and A. Der Kiureghian, "Generalized Bouc-Wen model for highly asymmetric hysteresis," *Journal of Engineering Mechanics*, vol. 132, no. 6, pp. 610–618, 2006.

[18] T. N. Do, T. Tjahjowidodo, M. W. S. Lau, and S. J. Phee, "Dynamic Friction-Based Force Feedback for Tendon- Sheath Mechanism in NOTES System," *International Journal of Computer and Electrical Engineering*, vol. 6, no. 3, pp. 252–258, 2014.

[19] F. Ikhouane, V. Mañosa, and J. Rodellar, "Dynamic properties of the hysteretic Bouc-Wen model," *Systems and Control Letters*, vol. 56, no. 3, pp. 197–205, 2007.

[20] F. Ikhouane, V. c. Mañosa, and J. Rodellar, "Adaptive control of a hysteretic structural system," *Automatica. A Journal of IFAC, the International Federation of Automatic Control*, vol. 41, no. 2, pp. 225–231, 2005.

[21] M. Mahvash and A. Okamura, "Friction compensation for enhancing transparency of a teleoperator with compliant transmission," *IEEE Transactions on Robotics*, vol. 23, no. 6, pp. 1240–1246, 2007.

[22] H. K. Khalil and J. Grizzle, "Nonlinear systems," vol. 3, Prentice hall Upper Saddle River.

[23] M. Krstic, P. V. Kokotovic, and I. Kanellakopoulos, *Nonlinear and adaptive control design*, John Wiley & Sons, Inc., Hoboken, NJ, United States, 1995.

[24] N. S. Khot, F. E. Eastep, and R. M. Kolonay, "Wing twist and camber for the rolling maneuver of a flexible wing without aileron," in *Proceedings of the 38th AIAA/ASME/ASCE/AHS/ASC Structures, Structural Dynamics, and Materials*, pp. 759–768, American Institute of Aeronautics and Astronautics, Reston, Virginia, April 1997.

[25] D. M. Tang, P. G. A. Cizmas, and E. H. Dowell, "Experiments and analysis for a gust generator in a wind tunnel," *Journal of Aircraft*, vol. 33, no. 1, pp. 139–148, 1996.

[26] H. Xu, S. Xing, and Z. Ye, "Numerical study of an airfoil/rotating-slotted-cylinder based flutter exciter," *Journal of Aircraft*, vol. 52, no. 6, pp. 2100–2105, 2015.

[27] Y. Feng, Y.-M. Hu, C. A. Rabbath, and C.-Y. Su, "Robust adaptive control for a class of perturbed strict-feedback nonlinear systems with unknown Prandtl-Ishlinskii hysteresis," *International Journal of Control*, vol. 81, no. 11, pp. 1699–1708, 2008.

10

Synthetic Jet Actuator-Based Aircraft Tracking using a Continuous Robust Nonlinear Control Strategy

N. Ramos-Pedroza,[1] W. MacKunis,[1] and M. Reyhanoglu[2]

[1]*Department of Physical Sciences, Embry-Riddle Aeronautical University, Daytona Beach, FL 32114, USA*
[2]*Department of Engineering, University of North Carolina at Asheville, Asheville, NC 28804, USA*

Correspondence should be addressed to M. Reyhanoglu; reyhanoglu@unca.edu

Academic Editor: Christopher J. Damaren

A robust nonlinear control law that achieves trajectory tracking control for unmanned aerial vehicles (UAVs) equipped with synthetic jet actuators (SJAs) is presented in this paper. A key challenge in the control design is that the dynamic characteristics of SJAs are nonlinear and contain parametric uncertainty. The challenge resulting from the uncertain SJA actuator parameters is mitigated via innovative algebraic manipulation in the tracking error system derivation along with a robust nonlinear control law employing constant SJA parameter estimates. A key contribution of the paper is a rigorous analysis of the range of SJA actuator parameter uncertainty within which asymptotic UAV trajectory tracking can be achieved. A rigorous stability analysis is carried out to prove semiglobal asymptotic trajectory tracking. Detailed simulation results are included to illustrate the effectiveness of the proposed control law in the presence of wind gusts and varying levels of SJA actuator parameter uncertainty.

1. Introduction

The recent surge of interest in applications involving UAVs has motivated the development of low-mass actuators with reduced power requirements. Based on this, the use of SJAs has emerged as a popular tool for UAV control applications. SJAs can be used in a variety of applications, including trajectory tracking control, limit cycle oscillation (LCO) suppression, and boundary-layer flow control. The operation of SJAs is based on an effective combination of electrical, mechanical, and acoustic components [1]. SJAs transfer linear momentum to a flow system through vibration-induced oscillation of fluid flow through a narrow opening (see Figure 1). The oscillations are created by a piezoelectric membrane that operates inside of an air-filled cavity. The oscillating air in the cavity generates fluid vortices (jets) that travel away from the orifice. Since the jets are created using only the air in the surrounding environment, SJAs do not require space for a fuel supply. Moreover, SJAs are capable of transferring momentum to the system through a zero-net mass injection of air across the boundary. These virtues make SJAs an attractive option in UAV applications.

Under the operating conditions characteristic of UAV flight, a laminar separation bubble can form near the boundary layer, and total separation can occur if the angle of attack (AoA) is high enough [2]. This decreases the efficiency (i.e., lift/drag characteristics) of the airfoil. By endowing the airfoil with surface-embedded SJAs, active separation control systems can be developed. Flow separation control can be achieved using SJAs by virtue of their ability to energize the boundary layer by adding or removing momentum to or from the boundary layer [3–5]. SJAs are also capable of decreasing drag by delaying the flow separation point in the airfoil boundary layer [6]. In addition, SJAs can expand the usable range of the AoA, improving aircraft maneuverability [7]. Arrays containing multiple SJAs can be utilized to achieve aircraft tracking control [8, 9]. By using SJAs as replacements for mechanical control surfaces (e.g., elevators and ailerons), radar cross-section can be reduced, and UAV weight, cost, and mechanical complexity can also be reduced.

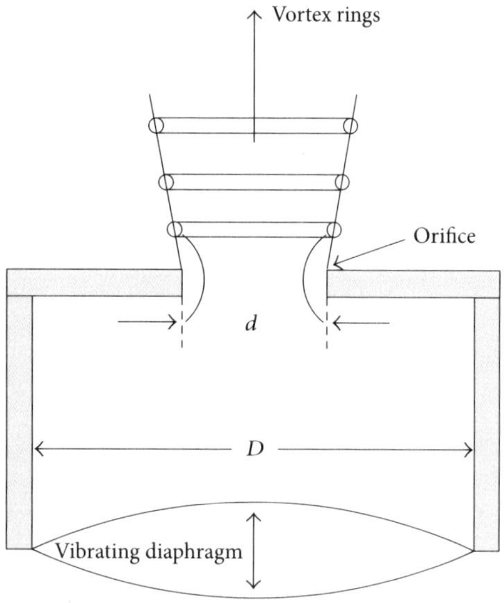

FIGURE 1: Schematic for a synthetic jet actuator.

The challenges in SJA-based control design stem from the fact that the input-output characteristics of SJAs are nonlinear and contain uncertain parameters (see Figure 2). In addition to the challenges involved in control design in the presence of SJA actuator uncertainty, control design for UAV in off-nominal operating conditions (e.g., wind gusts) creates further challenges. Various approaches have recently been developed for aircraft tracking control using SJAs (e.g., see [8, 10–13]), where the SJA actuator uncertainty is compensated using adaptive control methods or neural networks. Other popular approaches for SJA-based control are computational fluid dynamics- (CFD-) based numerical techniques (see [14–28]). Adaptive control, neural network-based control, and numerical CFD methods have been shown to be effective in their respective SJA-based control tasks. However, the focus of this paper is on the design and rigorous performance analysis of a computationally minimal nonlinear SJA-based control method, which can be implemented without adaptive parameter update laws, intelligent control techniques (e.g., neural networks or fuzzy logic rule sets), or heavy computations.

A robust nonlinear control method is presented in this paper that is proven to achieve asymptotic trajectory tracking control for a UAV in the presence of SJA actuator nonlinearity and parametric uncertainty in addition to unmodelled disturbances resulting from wind gusts. The challenge resulting from the uncertain SJA actuator parameters is mitigated through innovative algebraic manipulation in the tracking error system derivation along with a robust nonlinear control law employing constant "best guess" parameter estimates. A key contribution of the proposed control design is a rigorous analysis of the range of SJA actuator parameter uncertainty within which asymptotic UAV trajectory tracking can be achieved. Semiglobal asymptotic trajectory tracking is proven via a Lyapunov-based analysis, and detailed simulation results are provided to illustrate the performance of the proposed control law in the presence of wind gusts and varying levels of SJA actuator parameter uncertainty. A preliminary version of this result was published in the 2013 IEEE Conference on Decision and Control (CDC), but the current result includes the following additions and extensions beyond the CDC result: (1) rigorous stability analysis that now provides a detailed derivation of the operational region within which asymptotic tracking can be proved; (2) a significant extension to the theoretical control law derivation, including the additions of Lemma 1, Property 1, Assumption 3, and Remark 4; (3) a significantly expanded numerical simulation results section, which now includes Monte Carlo-type simulation results of the closed-loop control system under 20 different sets of uncertain SJA parameters that deviate from nominal by up to 35%; (4) the addition of an appendix, which includes a detailed derivation of the control gain conditions required to prove asymptotic stability (i.e., proof of Lemma 2).

2. Dynamic Model and Properties

The dynamic model being considered in this paper incorporates the effects of parametric uncertainty in the aircraft dynamics, along with unmodelled external disturbances, and the inherent SJA actuator nonlinearity and parametric uncertainty. Specifically, the aircraft dynamic model can be expressed as (see, e.g., [5, 8–10, 13, 29–32])

$$\dot{x} = \mathbf{A}x + \mathbf{B}\mathbf{u} + f(x,t), \tag{1}$$

where $\mathbf{A} \in \mathbb{R}^{n\times n}$ and $\mathbf{B} \in \mathbb{R}^{n\times m}$ denote the uncertain state and input matrices, respectively, and $f(x,t) \in \mathbb{R}^n$ represents an unmodelled norm-bounded disturbance. The disturbance term $f(x,t)$ could represent the effects of external disturbances, such as wind gusts, or model inaccuracies resulting from linearization, for example. In (1), the control input $\mathbf{u}(t) \triangleq [\mathbf{u}_1(t)\cdots\mathbf{u}_m(t)]^T \in \mathbb{R}^m$ represents the virtual surface deflections resulting from m arrays of SJA. These virtual surface deflections help create the lift forces on the outer trailing edge of the array [13]. Figure 2 shows the virtual deflection angle versus voltage for four different values of the SJA parameter θ_1^*. A well-accepted empirically determined model of the SJA's dynamics can be expressed as [8, 10, 12, 13]

$$\mathbf{u}_i = \theta_{2i}^* - \frac{\theta_{1i}^*}{v_i}, \quad i = 1, 2, \dots, m, \tag{2}$$

where $v_i(t) = \mathbf{A}_{\mathrm{pp}i}^2(t) \in \mathbb{R}$ denotes the peak-to-peak voltage acting on the ith SJA array and $\theta_{1i}^*, \theta_{2i}^* \in \mathbb{R}$ denote uncertain positive physical parameters. The expression in (2) illuminates the challenges inherent in SJA-based control design: The control inputs $\mathbf{u}_i(t)$ depend nonlinearly on the voltage control signal $v_i(t)$ and include the uncertain parameters θ_{1i}^* and θ_{2i}^*. In the subsequent control development, these challenges will be mitigated through innovative algebraic manipulation in the tracking error system development along with a robust, continuous nonlinear control method.

By substituting (2) into (1), the SJA-based dynamic model can be expressed as

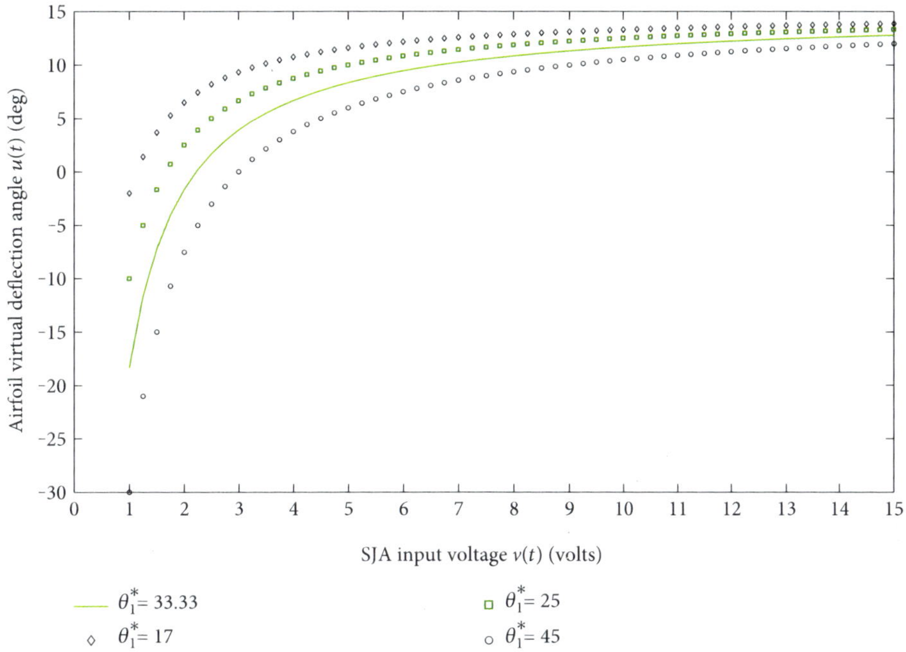

FIGURE 2: The virtual deflection angle versus control voltage for a SJA, showing the high degree of nonlinearity and the effect of parametric uncertainty in the SJA dynamic model. A well-accepted nominal value for the parameter is shown in green, and the three other plots show the SJA dynamic characteristics under off-nominal operating conditions.

$$\dot{x} = \mathbf{A}x + \sum_{i=1}^{m} \mathbf{B}_i \mathbf{u}_i + f(x,t). \tag{3}$$

In (3), $\mathbf{B}_i \triangleq [\mathbf{B}_{1i} \cdots \mathbf{B}_{ni}]^T \in \mathbb{R}^n \ \forall i = 1, \ldots, m$, where $\mathbf{B}_{ij}$ represents the (i,j)th element of the uncertain $\mathbf{B}$ matrix.

Assumption 1. The disturbance $f(x,t)$ is sufficiently smooth in the sense that the first and second time derivatives $\dot{f}(x,t)$ and $\ddot{f}(x,t)$ are bounded, provided that $x(t)$ is bounded.

2.1. Wind Gust Model. This section describes the details of the wind gust model (i.e., the disturbance term $f(x,t)$ introduced in (1)) that is being considered in this paper. The Federal Aviation Regulations (FAR) [33] describe a vertical wind gust as a bounded nonlinearity along the longitudinal axis as

$$f(x,t) = \begin{bmatrix} -11.1 \\ 7.2 \\ 37.4 \\ 0 \end{bmatrix} \frac{1}{V_0} \left\{ \frac{U_{ds}}{2} \left[1 - \cos\left(\frac{\pi s}{H}\right) \right] \right\}. \tag{4}$$

In (4), H denotes the distance (m) along the airplane's flight path for the wind gust to reach its peak velocity, V_0 (m/s) is the forward velocity of the aircraft when it enters the gust, $s \in [0 \quad 2H]$ denotes the distance penetrated into the wind gust (m), and U_{ds} represents the design gust velocity (m/s). The wind gust model used in the subsequent numerical simulation results is based on the mathematical model in (4).

2.2. Robust Nonlinearity Inverse. The main contribution presented here is the mathematical development that demonstrates how a computationally inexpensive, robust nonlinear control method can be designed, which compensates for the parametric uncertainty and nonlinearity present in the SJA actuator dynamics. To achieve this, a robust-inverse control design structure will be utilized for the voltage control signal $v_i(t)$, which contains constant "best guess" estimates of the uncertain SJA parameters θ_{1i}^* and θ_{2i}^*. The robust-inverse control structure is given by [9]

$$v_i(t) = \frac{\widehat{\theta}_{1i}}{\widehat{\theta}_{2i} - \mathbf{u}_{di}(t)}, \quad i = 1, \ldots, m, \tag{5}$$

where $\widehat{\theta}_{1i}, \widehat{\theta}_{2i} \in \mathbb{R}^+$ are constant feedforward estimates of θ_{1i}^* and θ_{2i}^*, respectively, and $\mathbf{u}_{di}(t) \in \mathbb{R} \ \forall i = 1, \ldots, m$ are subsequently defined auxiliary control terms.

Remark 1 (control structure). The robust-inverse control structure is one of the primary contributions of the proposed control design. In contrast to standard adaptive control methods to compensate for parametric actuator uncertainty, it is shown in the current result that this robust nonlinear control method compensates for a significantly higher level of uncertainty in the SJA parameters (see Simulation Results for details).

Remark 2 (avoiding singularities). Based on (5), singularities will occur when $\mathbf{u}_{di}(t) = \widehat{\theta}_{2i}$. To guarantee that the control law in (5) does not encounter these singularities, the auxiliary control terms $\mathbf{u}_{di}(t)$ for $i = 1, 2, \ldots, m$ will incorporate following algorithm [29]:

$$\mathbf{u}_{di}(t) = \begin{cases} \hat{\theta}_{2i} - \epsilon, & \text{if } g(\boldsymbol{\mu}_0(t), \boldsymbol{\mu}_1(t)) \geq \hat{\theta}_{2i} - \varepsilon, \\ g(\boldsymbol{\mu}_0(t), \boldsymbol{\mu}_1(t)), & \text{otherwise}, \end{cases} \tag{6}$$

where $\varepsilon \in \mathbb{R}^+$ is a user-defined design parameter, $g(\cdot)$ is a subsequently defined control input function, and $\boldsymbol{\mu}_0(t)$, $\boldsymbol{\mu}_1(t) \in \mathbb{R}^m$ are subsequently defined auxiliary control signals. Note that the control effectiveness can be ensured over an arbitrarily wide range of voltage input signals $v_i(t)$ through judicious selection of the design parameter ε.

In addition, singularities would be encountered in (2) when the voltage signal $v_i(t) = 0$; however, this singular condition can be avoided by simply selecting $\hat{\theta}_{1i} > 0$ for $i = 1, 2, \ldots, m$.

Remark 3. Numerical simulation results demonstrate that the control term $\mathbf{u}_{di}(t)$ in (5) achieves asymptotic trajectory tracking and disturbance rejection in the presence of a significant deviation between the estimated and actual SJA parameter values $\hat{\theta}_{ji}$ and θ^*_{ji}, $j = 1, 2$.

2.3. Control Development. The objective is to ensure that the actual aircraft state $x(t)$ tracks a model reference (desired) state. Based on the mathematical structure of the dynamic model in (1), the model reference system is designed as

$$\dot{x}_m = \mathbf{A}_m x_m + \mathbf{B}_m \delta, \tag{7}$$

where $x_m(t) \in \mathbb{R}^n$ is the model reference state (i.e., the desired trajectory), $\mathbf{A}_m \in \mathbb{R}^{n\times n}$ denotes the model reference state matrix, $\mathbf{B}_m \in \mathbb{R}^n$ is the model reference input gain matrix, and $\delta(t) \in \mathbb{R}$ is the reference input (e.g., a pilot or autopilot command). The parameters of the reference model in (7) are selected such that the system achieves favorable flight performance characteristics in terms of convergence time and steady-state error, for example.

Assumption 2. The state of the model reference system remains bounded and sufficiently smooth in the sense that $x_m(t), \dot{x}_m(t), \ddot{x}_m(t), \dddot{x}_m(t) \in \mathscr{L}_\infty \ \forall t \geq 0$.

To quantify the control objective, a trajectory tracking error $\mathbf{e}(t) \in \mathbb{R}^n$ is defined as

$$\mathbf{e} = x - x_m. \tag{8}$$

To facilitate the derivation of the error system dynamics, an auxiliary (filtered) error signal $r(t)$ is defined as

$$\mathbf{r} = \dot{\mathbf{e}} + \boldsymbol{\gamma}\mathbf{e}, \tag{9}$$

where $\boldsymbol{\gamma} \in \mathbb{R}^+$ is a constant control gain. By calculating the time derivative of (9) and substituting (1) and (8), the open-loop error system dynamics are obtained as

$$\dot{\mathbf{r}} = \mathbf{A}\dot{\mathbf{e}} + \mathbf{A}\dot{x}_m + \sum_{i=1}^{m} \mathbf{B}_i \left(\frac{\theta^*_{1i}}{\hat{\theta}_{1i}} \dot{\mathbf{u}}_{di}(t) \right) + \dot{f}(x,t) - \ddot{x}_m + \boldsymbol{\gamma}\dot{\mathbf{e}}. \tag{10}$$

Remark 4. Although the constant portion of the SJA actuator model in (2) vanishes upon calculating the time derivative to obtain (10), the complete SJA model is incorporated in implementation by using (2) and (5). Thus, the subsequent simulation results incorporate the full SJA actuator model.

The open-loop error system in (10) can be rewritten in a more compact form as

$$\dot{\mathbf{r}} = \tilde{\mathbf{N}} + \mathbf{N}_d + \boldsymbol{\Omega}\dot{\mathbf{u}}_d(t) - \mathbf{S}\mathbf{e}, \tag{11}$$

where $\boldsymbol{\Omega} \in \mathbb{R}^{n\times m}$ is a constant uncertain matrix, $\mathbf{S} \in \mathbb{R}^{n\times n}$ is a subsequently defined auxiliary matrix, and $\mathbf{u}_d(t) \triangleq [\mathbf{u}_{d1}(t)\cdots\mathbf{u}_{dm}(t)]^T \in \mathbb{R}^m$ is the auxiliary control vector. In (11), the unknown, unmeasurable auxiliary terms $\tilde{\mathbf{N}}(t)$ and $\mathbf{N}_d(t)$ are explicitly defined as

$$\begin{aligned} \tilde{\mathbf{N}} &\triangleq \mathbf{A}\dot{\mathbf{e}} + \boldsymbol{\gamma}\dot{\mathbf{e}} + \mathbf{S}\mathbf{e} + \left(\dot{f}(x,t) - \dot{f}(x_m,t)\right), \\ \mathbf{N}_d &\triangleq \mathbf{A}\dot{x}_m - \ddot{x}_m + \dot{f}(x_m,t). \end{aligned} \tag{12}$$

The motivation for the separation of terms as in (12) is based on the fact that the following bounding inequalities can be developed:

$$\begin{aligned} \|\tilde{\mathbf{N}}\| &\leq \rho(\|\mathbf{z}\|)\|\mathbf{z}\|, \\ \|\mathbf{N}_d\| &\leq \zeta_{\mathbf{N}_d}, \\ \|\dot{\mathbf{N}}_d\| &\leq \zeta_{\dot{\mathbf{N}}_d}, \end{aligned} \tag{13}$$

where $\rho_0(\cdot) \in \mathbb{R}$ is a positive globally invertible nondecreasing function, $\zeta_{\mathbf{N}_d}, \zeta_{\dot{\mathbf{N}}_d} \in \mathbb{R}^+$ are known bounding constants, and $\mathbf{z}(t) \in \mathbb{R}^{2n}$ is an augmented tracking error vector that is defined as

$$\mathbf{z} \triangleq \left[\mathbf{e}^T \, \mathbf{r}^T\right]^T. \tag{14}$$

2.4. Closed-Loop Error System. Based on the open-loop error dynamics in (11) and the subsequent stability analysis, the auxiliary control term $\mathbf{u}_d(t)$ is designed as

$$\mathbf{u}_d(t) = \hat{\boldsymbol{\Omega}}^{\#}(\boldsymbol{\mu}_0 - \boldsymbol{\mu}_1), \tag{15}$$

where $\hat{\boldsymbol{\Omega}} \in \mathbb{R}^{n\times m}$ is a constant estimate of $\boldsymbol{\Omega}$ and $[\cdot]^{\#}$ denotes the matrix pseudoinverse. In (15), $\boldsymbol{\mu}_0(t), \boldsymbol{\mu}_1(t) \in \mathbb{R}^n$ denote feedback control terms defined as the generalized solutions to the differential equations

$$\begin{aligned} \dot{\boldsymbol{\mu}}_0 &= -(\mathbf{k}_s + \mathbf{I}_{n\times n})\mathbf{r}, \\ \dot{\boldsymbol{\mu}}_1 &= -\boldsymbol{\beta}\operatorname{sgn}(\mathbf{e}(t)), \end{aligned} \tag{16}$$

where $\boldsymbol{\beta}, k_s \in \mathbb{R}^{n\times n}$ are constant, positive definite, diagonal control gain matrices.

Remark 5 (control input definitions). The motivation for defining the control terms $\boldsymbol{\mu}_0(t)$ and $\boldsymbol{\mu}_1(t)$ in terms of their time derivatives as in (16) is based on the subsequent Lyapunov-based stability analysis and the desire to design a continuous control law, which can be proven to achieve asymptotic rejection of norm-bounded disturbances. Note that integrating both sides of (16) results in a control expression that is continuous in time.

Remark 6 (integral signum term). Note that the auxiliary control term $\boldsymbol{\mu}_1(t)$ can be shown to be continuous by integrating both sides of the corresponding expression in (16). Mathematically, the integral of the signum of the tracking error $\mathbf{e}(t)$ can be interpreted as a finite-bandwidth signal (i.e., $\boldsymbol{\mu}_1(t)$ is a sawtooth wave with a finite slope). In practical implementation of the proposed control law, the control gain $\boldsymbol{\beta}$ in (16) can be tuned to adjust the slope of the sawtooth wave, thereby compensating for norm-bounded disturbances using high-frequency feedback (finite bandwidth), as opposed to the discontinuous (infinite bandwidth) high-gain feedback that is characteristic of standard sliding mode control methods (i.e., due to direct implementation of the sgn$(\cdot)$ function in standard sliding mode control methods).

After substituting the time derivative of (15) into (11), the error dynamics can be expressed as

$$\dot{\mathbf{r}} = \tilde{\mathbf{N}} + \mathbf{N}_d + \tilde{\boldsymbol{\Omega}}(\dot{\boldsymbol{\mu}}_0 - \dot{\boldsymbol{\mu}}_1) - \mathbf{Se}, \tag{17}$$

where the constant uncertain matrix $\tilde{\Omega} \in \mathbb{R}^{n\times n}$ is defined as

$$\tilde{\boldsymbol{\Omega}} = \boldsymbol{\Omega}\hat{\boldsymbol{\Omega}}^{\#}. \tag{18}$$

Lemma 1 [34]. *Any positive definite matrix $\mathbf{X} \in \mathbb{R}^{n\times n}$ can be decomposed as*

$$\mathbf{X} = \mathbf{ST}, \tag{19}$$

where $S \in \mathbb{R}^{n\times n}$ is a positive definite symmetric matrix and $T \in \mathbb{R}^{n\times n}$ is a unity upper triangular matrix.

Proof. Proof of Lemma 1 can be found in [34] and is omitted here for brevity.

Property 1. Since the matrix S introduced in (19) is positive definite and symmetric, its inverse $\mathbf{S}^{-1}$ is also positive definite and symmetric. This property will be utilized in the subsequent stability analysis.

Assumption 3. Upper and lower bounds on the elements of the uncertain constant matrix $\boldsymbol{\Omega} \in \mathbb{R}^{n\times n}$ are known such that the constant feed forward estimate $\hat{\boldsymbol{\Omega}} \in \mathbb{R}^{n\times n}$ can be chosen to render the product $\tilde{\boldsymbol{\Omega}} = \boldsymbol{\Omega}\hat{\boldsymbol{\Omega}}^{-1}$ positive definite. Further, the estimate $\hat{\boldsymbol{\Omega}}$ is selected such that

$$\tilde{\boldsymbol{\Omega}} = \mathbf{ST}, \tag{20}$$

where the unity upper triangular matrix $\mathbf{T}$ satisfies the diagonal dominance property

$$\varepsilon \le |\mathbf{T}_{ii}| - \sum_{k=i+1}^{n} |\mathbf{T}_{ik}| \le Q, \quad i = 1, \dots, n-1, \tag{21}$$

where $\varepsilon \in (0, 1)$ and $Q \in \mathbb{R}^+$ are known bounding constants and $\mathbf{T}_{ik} \in \mathbb{R}$ denotes the (i, k)th element of the matrix $\mathbf{T}$. In (20), the matrices $\mathbf{S}$ and $\mathbf{T}$ are defined in a manner similar to Lemma 1.

Remark 7. The subsequent numerical simulation results demonstrate that Assumption 3 is satisfied over a significant range of uncertainty between the estimated and actual values of the uncertain input-multiplicative matrix (i.e., deviations between $\hat{\boldsymbol{\Omega}}$ and $\boldsymbol{\Omega}$). Specifically, the results show that asymptotic trajectory tracking is achieved when the constant estimates $\hat{\theta}_{1j}$ and $\hat{\theta}_{2j}\ \forall j = 1, \dots, m$ deviate from the actual values by more than 35%.

After using the decomposition technique in (20), the open-loop error dynamics in (17) can be expressed as

$$\mathbf{S}^{-1}\dot{r} = \tilde{\mathbf{N}}_1 + \mathbf{N}_{d1} + \mathbf{T}(\dot{\boldsymbol{\mu}}_0 - \dot{\boldsymbol{\mu}}_1) - \mathbf{e}, \tag{22}$$

where

$$\tilde{\mathbf{N}}_1 \triangleq \mathbf{S}^{-1}\tilde{\mathbf{N}}, \qquad \mathbf{N}_{d1} \triangleq \mathbf{S}^{-1}\mathbf{N}_d. \tag{23}$$

Since $\mathbf{S}$ is positive definite, $\tilde{\mathbf{N}}_1(t)$ and $\mathbf{N}_{d1}(t)$ satisfy the following inequalities:

$$\begin{aligned} \|\tilde{\mathbf{N}}_1\| &\le \rho_1(\|\mathbf{z}\|)\|\mathbf{z}\|, \\ \|\mathbf{N}_{d1}\| &\le \zeta_{\mathbf{N}_{d1}}, \\ \|\dot{\mathbf{N}}_{d1}\| &\le \zeta_{\dot{\mathbf{N}}_{d1}}, \end{aligned} \tag{24}$$

where $\rho_1(\cdot) \in \mathbb{R}$ is a positive, globally invertible nondecreasing function and $\zeta_{\mathbf{N}_{d1}}, \zeta_{\dot{\mathbf{N}}_{d1}} \in \mathbb{R}^+$ are known bounding constants. By using the fact that the uncertain matrix $\mathbf{T}$ is unity upper triangular, the error dynamics in (22) can be rewritten as

$$\mathbf{S}^{-1}\dot{\mathbf{r}} = \tilde{\mathbf{N}}_1 + \mathbf{N}_{d1} + \dot{\boldsymbol{\mu}}_0 + \bar{\mathbf{T}}\dot{\boldsymbol{\mu}}_0 - \mathbf{T}\dot{\boldsymbol{\mu}}_1 - \mathbf{e}, \tag{25}$$

where $\bar{\mathbf{T}} \triangleq \mathbf{T} - \mathbf{I}_{n\times n}$ is a *strictly upper triangular* matrix, and $\mathbf{I}_{n\times n}$ denotes the $n \times n$ identity matrix. After substituting the control expressions in (16), the closed-loop error system is obtained as

$$\mathbf{S}^{-1}\dot{\mathbf{r}} = \tilde{\mathbf{N}}_1 + \bar{\mathbf{T}}\dot{\boldsymbol{\mu}}_0 + \mathbf{N}_{d1} - (\mathbf{k}_s + \mathbf{I}_{n\times n})\mathbf{r} - \mathbf{T}\dot{\boldsymbol{\mu}}_1 - \mathbf{e}. \tag{26}$$

After utilizing (16), the term $\bar{\mathbf{T}}\dot{\boldsymbol{\mu}}_0$ can be expressed as

$$\bar{\mathbf{T}}\dot{\boldsymbol{\mu}}_0 = \begin{bmatrix} \sum_{j=2}^{n} \bar{\mathbf{T}}_{1j}\dot{\boldsymbol{\mu}}_{0j} \\ \sum_{j=3}^{n} \bar{\mathbf{T}}_{2j}\dot{\boldsymbol{\mu}}_{0j} \\ \vdots \\ \bar{\mathbf{T}}_{(n-1)n}\dot{\boldsymbol{\mu}}_{0n} \\ 0 \end{bmatrix} = \begin{bmatrix} \boldsymbol{\Lambda}_\rho \\ 0 \end{bmatrix}, \tag{27}$$

where the auxiliary signal $\mathbf{\Lambda}_\rho \triangleq [\mathbf{\Lambda}_{\rho 1}, \mathbf{\Lambda}_{\rho 2} \cdots \mathbf{\Lambda}_{\rho(n-1)}]^T \in \mathbb{R}^{n-1}$, with the individual elements defined as

$$\mathbf{\Lambda}_{\rho i} \triangleq -\sum_{j=i+1}^{n} \bar{\mathbf{T}}_{ij}(\mathbf{k}_{sj}+1)\mathbf{r}_j, \tag{28}$$

for $i = 1, \ldots, n-1$ where the subscript j indicates the jth element of the vector. Based on the definitions in (16) and (27), $\mathbf{\Lambda}_\rho$ can be upper bounded as

$$\|\mathbf{\Lambda}_\rho\| \le \rho_{\Lambda 1}\|\mathbf{z}\|, \tag{29}$$

where $\mathbf{z}(t)$ was previously defined in (14) and $\rho_{\Lambda 1} \in \mathbb{R}$ is a known positive bounding constant.

Remark 8. Note that based on (27) and (28), the bounding constant $\rho_{\Lambda 1}$ depends only on elements $i+1$ to n of the control gain matrix $\mathbf{k}_s$ due to the strictly upper triangular nature of $\bar{\mathbf{T}}$. Thus, the element $\dot{\boldsymbol{\mu}}_{01}(t)$ of the control vector $\dot{\boldsymbol{\mu}}_0(t)$ does not appear in the term $\mathbf{\Lambda}_\rho$. This fact will be utilized in the subsequent stability proof [9].

By utilizing (27), the error dynamics in (26) can be expressed as

$$\mathbf{S}^{-1}\dot{\mathbf{r}} = \tilde{\mathbf{N}}_2 + \mathbf{N}_{d1} - (\mathbf{k}_s + \mathbf{I}_{n\times n})\mathbf{r} - \mathbf{T}\dot{\boldsymbol{\mu}}_1 - \mathbf{e}, \tag{30}$$

where

$$\tilde{\mathbf{N}}_2 = \tilde{\mathbf{N}}_1 + \begin{bmatrix} \mathbf{\Lambda}_\rho \\ 0 \end{bmatrix}. \tag{31}$$

Based on (24), (29), and (31), $\tilde{\mathbf{N}}_2(t)$ satisfies the inequality

$$\|\tilde{\mathbf{N}}_2\| \le \rho_2(\|\mathbf{z}\|)\|\mathbf{z}\|, \tag{32}$$

where $\rho_2(\cdot) \in \mathbb{R}$ is a positive, globally invertible nondecreasing function.

To facilitate the subsequent stability analysis, the control gain $\boldsymbol{\beta}$ introduced in (16) is selected to satisfy

$$\boldsymbol{\beta} > \frac{1}{\varepsilon}\left(\zeta_{\mathbf{N}_{d1}} + \frac{1}{\boldsymbol{\gamma}}\zeta_{\dot{\mathbf{N}}_{d1}}\right), \tag{33}$$

where $\zeta_{\mathbf{N}_{d1}}$ and $\zeta_{\dot{\mathbf{N}}_{d1}}$ are introduced in (24) and ε is introduced in (21).

3. Stability Analysis

Let $\mathscr{D} \subset \mathbb{R}^{2n+1}$ be a domain containing $w(t) = 0$, where $w(t) \in \mathbb{R}^{2n+1}$ is defined as

$$w(t) \triangleq \left[\mathbf{z}^T(t)\ \sqrt{P(t)}\right]^T. \tag{34}$$

In (34), the auxiliary function $P(t) \in \mathbb{R}$ is defined as the generalized solution to the differential equation

$$\dot{P}(t) = -L(t), \tag{35}$$

$$P(0) = \boldsymbol{\beta}Q|\mathbf{e}(0)| - \mathbf{e}^T(0)\mathbf{N}_{d1}(0), \tag{36}$$

where the auxiliary function $L(t) \in \mathbb{R}$ is defined as

$$L(t) = \mathbf{r}^T(\mathbf{N}_{d1}(t) - \mathbf{T}\dot{\boldsymbol{\mu}}_1). \tag{37}$$

Lemma 2. *Provided the sufficient condition in (33) is satisfied, the following inequality can be obtained:*

$$\int_0^t L(\tau)d\tau \le \boldsymbol{\beta}Q|\mathbf{e}(0)| - \mathbf{e}^T(0)\mathbf{N}_{d1}(0). \tag{38}$$

Hence, (38) can be used to conclude that $P(t) \ge 0$.

Proof. Proof of Lemma 2 can be found in the appendix.

Theorem 1. *The robust control law given by (5), (15), and (16) achieves asymptotic trajectory tracking in the sense that*

$$\|\mathbf{e}(t)\| \to 0, \quad \text{as} \quad t \to \infty, \tag{39}$$

provided the control gain matrix $\mathbf{k}_s$ introduced in (16) is selected sufficiently large and $\boldsymbol{\beta}$ is selected to satisfy the sufficient condition in (33).

Proof. Let $V(w,t)$: $\mathscr{D} \times [0, \infty) \to \mathbb{R}$ be a continuously differentiable, nonnegative function defined as

$$V = \frac{1}{2}\mathbf{e}^T\mathbf{e} + \frac{1}{2}\mathbf{r}^T\mathbf{S}^{-1}\mathbf{r} + P, \tag{40}$$

which satisfies the inequalities

$$U_1(w) \le V(w,t) \le U_2(w), \tag{41}$$

provided the sufficient condition in (33) is satisfied. In (41), the continuous positive definite functions $U_1(w), U_2(w) \in \mathbb{R}$ are defined as

$$U_1(w) \triangleq \eta_1\|w\|^2, \quad U_2(w) \triangleq \eta_2\|w\|^2, \tag{42}$$

where $\eta_1, \eta_2 \in \mathbb{R}$ are defined as

$$\eta_1 \triangleq \frac{1}{2}\min\{1, \lambda_{\min}(\mathbf{S}^{-1})\}, \quad \eta_2 \triangleq \max\left\{\frac{1}{2}\lambda_{\max}(\mathbf{S}^{-1}), 1\right\}, \tag{43}$$

where $\lambda_{\min}(\cdot), \lambda_{\max}(\cdot)$ denote the minimum and maximum eigenvalues of the arguments, respectively. After taking the time derivative of (40), utilizing (9), (30), (35), and (37), and canceling common terms, $\dot{V}(t)$ can be expressed as

$$\dot{V} = -\boldsymbol{\gamma}\|\mathbf{e}\|^2 - \|\mathbf{r}\|^2 - \mathbf{r}^T(\tilde{\mathbf{N}}_2 - \mathbf{k}_s\mathbf{r}). \tag{44}$$

After using the upper bound for $\tilde{\mathbf{N}}_2(t)$ given in (32) and completing the squares for the parenthetic terms, $\dot{V}$ can be upper bounded as

$$\dot{V} \le -\lambda_0 \|\mathbf{z}\|^2 + \frac{\rho_2(\|\mathbf{z}\|)^2}{4\lambda_{\min}(\mathbf{k}_s)}\|\mathbf{z}\|^2 - \lambda_{\min}(\mathbf{k}_s) \cdot \left(\|\mathbf{r}\|^2 - \rho_2(\|\mathbf{z}\|)\|\mathbf{r}\|\|\mathbf{z}\| + \frac{\rho_2(\|\mathbf{z}\|)^2}{4\lambda_{\min}^2(\mathbf{k}_s)}\|\mathbf{z}\|^2 \right), \quad (45)$$

where $\lambda_0 \triangleq \min\{\gamma, 1\}$ and $\lambda_{\min}(\cdot)$ denotes the minimum eigenvalue of the argument. The upper bound in (45) can be rewritten as

$$\dot{V} \le -\left(\lambda_0 - \frac{\rho_2(\|\mathbf{z}\|)^2}{4\lambda_{\min}(\mathbf{k}_s)}\right)\|\mathbf{z}\|^2. \quad (46)$$

The following expression can be obtained from (46):

$$\dot{V} \le -U(w), \quad (47)$$

where $U(w) = c\|\mathbf{z}\|^2$, for some positive constant $c \in \mathbb{R}$ is a continuous positive semidefinite function that is defined on the domain

$$\mathcal{D} \triangleq \left\{ w(t) \in \mathbb{R}^{2n+1} \middle| \|w\| \le \rho_2^{-1}\left(2\sqrt{\lambda_{\min}(\mathbf{k}_s)\lambda_0}\right) \right\}. \quad (48)$$

The expressions (41) and (46) can be used to prove that $\mathbf{e}(t), \mathbf{r}(t) \in \mathcal{L}_\infty$ in $\mathcal{D}$. Given that $\mathbf{e}(t), \mathbf{r}(t) \in \mathcal{L}_\infty$, (9) can be used to show that $\dot{\mathbf{e}}(t) \in \mathcal{L}_\infty$ in $\mathcal{D}$. Given that $\mathbf{e}(t), \dot{\mathbf{e}}(t) \in \mathcal{L}_\infty$, (8) can be used along with Assumption 2 to prove that $x(t), \dot{x}(t) \in \mathcal{L}_\infty$ in $\mathcal{D}$. Based on the fact that $x(t) \in \mathcal{L}_\infty$, Assumption 1 can be utilized to show that $f(x,t) \in \mathcal{L}_\infty$ in $\mathcal{D}$. Since $x(t), \dot{x}(t), f(x,t) \in \mathcal{L}_\infty$, (1) can be used to show that $\mathbf{u}(t) \in \mathcal{L}_\infty$ in $\mathcal{D}$. Since $\mathbf{e}(t), \mathbf{r}(t) \in \mathcal{L}_\infty$, the expressions in (16) can be used to show that $\dot{\boldsymbol{\mu}}_0(t), \dot{\boldsymbol{\mu}}_1(t) \in \mathcal{L}_\infty$ in $\mathcal{D}$. Given that $\mathbf{e}(t), \mathbf{r}(t), \dot{\boldsymbol{\mu}}_1(t) \in \mathcal{L}_\infty$, (30) can be used along with (32) to show that $\dot{\mathbf{r}}(t) \in \mathcal{L}_\infty$ in $\mathcal{D}$. Since $\dot{\mathbf{e}}(t), \dot{\mathbf{r}}(t) \in \mathcal{L}_\infty$ can be used to show that $\mathbf{e}(t)$ and $\mathbf{r}(t)$ are uniformly continuous in $\mathcal{D}$, thus, $\mathbf{z}(t)$ is uniformly continuous throughout the closed-loop controller operation. Hence, $U(w)$ and $\mathbf{z}(t)$ can be used to prove that $U(w)$ is uniformly continuous in $\mathcal{D}$.

Let $\mathcal{S} \subset \mathcal{D}$ denote a set defined as follows:

$$\mathcal{S} \triangleq \left\{ w(t) \subset \mathcal{D} | U(w(t)) \le \eta_1 \left(\rho_2^{-1}\left(2\sqrt{\lambda_{\min}(\mathbf{k}_s)\lambda_0}\right)\right)^2 \right\}. \quad (49)$$

Theorem 8.4 of [35] can now be invoked to state that

$$c\|\mathbf{z}(t)\|^2 \to 0 \quad \text{as} \quad t \to \infty, \quad \forall w(t_0) \in \mathcal{S}. \quad (50)$$

Based on the definition of $\mathbf{z}(t)$, (50) can be used to show that

$$\|\mathbf{e}(t)\| \to 0 \quad \text{as} \quad t \to \infty, \quad \forall w(t_0) \in \mathcal{S}. \quad (51)$$

Thus, asymptotic regulation of the pitching and plunging displacements can be achieved, provided the initial conditions are within the set $\mathcal{S}$, where $\mathcal{S}$ can be made arbitrarily large by increasing the control gain $\mathbf{k}_s$. Hence, this is a semi-global asymptotic result.

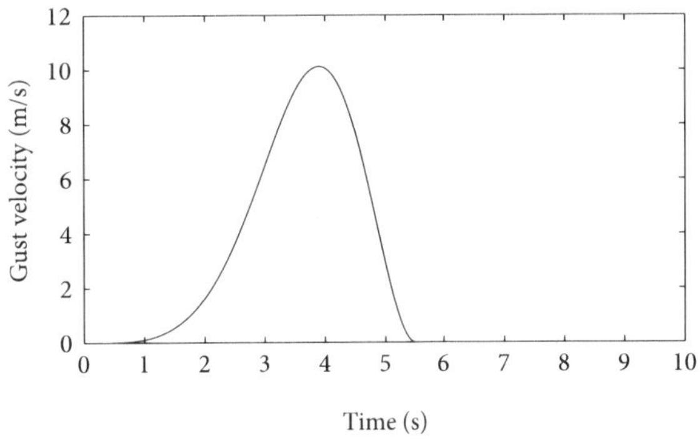

FIGURE 3: Vertical gust velocity.

4. Simulation Results

A numerical simulation was created to test the performance of the control design in (2), (5), (15), and (16). The simulation is based on the dynamic model in (1) and (2), where $n = 3$ and $m = 6$ (i.e., 3-DOF flight control using 6 SJA arrays). The state vector contains the roll, pitch, and yaw rates, and the tracking error vector can be expressed as

$$\mathbf{e}(t) = [\mathbf{e}_1(t), \mathbf{e}_2(t), \mathbf{e}_3(t)]^T. \quad (52)$$

The state and input matrices, **A** and **B**, and reference state and reference input matrices, $\mathbf{A}_m$ and $\mathbf{B}_m$, are defined based on the Barron Associates nonlinear tailless aircraft model (BANTAM) (for further details of the simulation model, see [8]). The 3-DOF linearized model for the BANTAM was obtained analytically during trim conditions, where $M = 0.455$ is the Mach number, $\alpha = 2.7$: (deg) is angle of attack, and $\beta_s = 0$: (deg) denotes the side slip angle. The simulation includes the effects of a wind gust in (4) as described in [33] at a velocity of $U_{ds} = 10.12$: (m/s), $H = 15.24$: (m), and $V_0 = 25$: (m/s).

The reference state and input matrices used in the simulation are explicitly defined as

$$\mathbf{A}_m = \begin{bmatrix} -61.1446 & 0 & -7.5238 \\ 0 & -174.3473 & 0 \\ -7.1579 & 0 & -1.4007 \end{bmatrix}, \quad \mathbf{B}_m = \begin{bmatrix} -1.7517 \\ 0 \\ 0.3096 \end{bmatrix}. \quad (53)$$

The matrices are $\mathbf{A}_m \in \mathbb{R}^{3\times3}$ and $\mathbf{B}_m \in \mathbb{R}^3$. The model reference (desired) state $x_m(t)$ in the simulation represents the desired external body axis motion that is generated in response to a reference command of (see (7))

$$\delta(t) = \sin(t). \quad (54)$$

The matrices **A** and **B** were obtained analytically from the dimensional aerodynamic coefficients of the BANTAM [8]. These matrices are given by

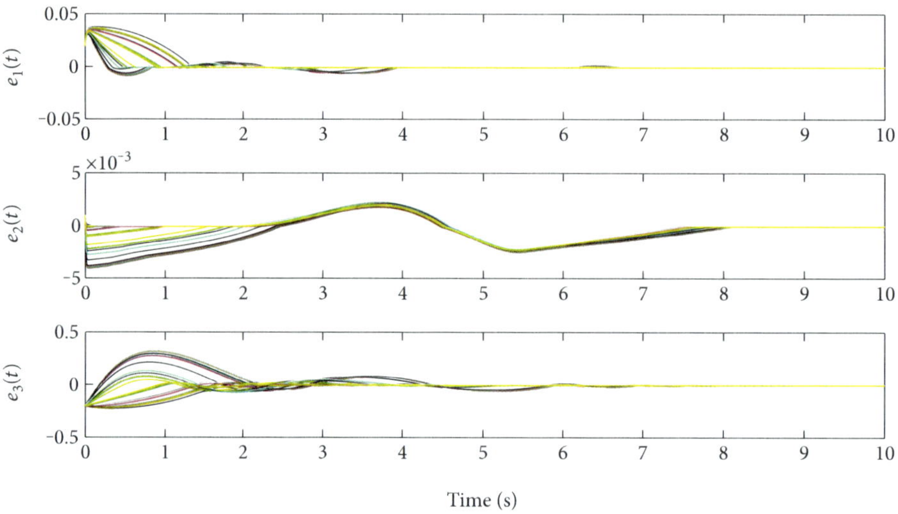

FIGURE 4: Closed-loop regulation of the steady state error.

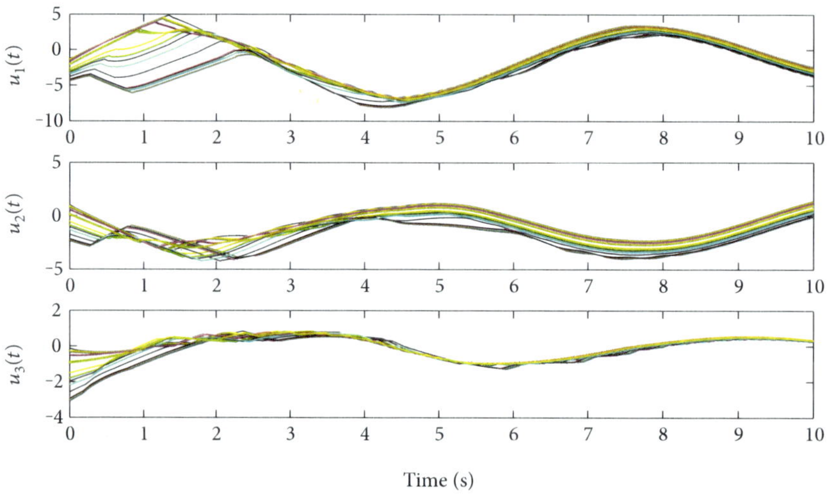

FIGURE 5: Virtual deflection angle control commands for the first three SJA arrays (i.e., $\mathbf{u}_1(t)$, $\mathbf{u}_2(t)$, and $\mathbf{u}_3(t)$) during closed-loop operation.

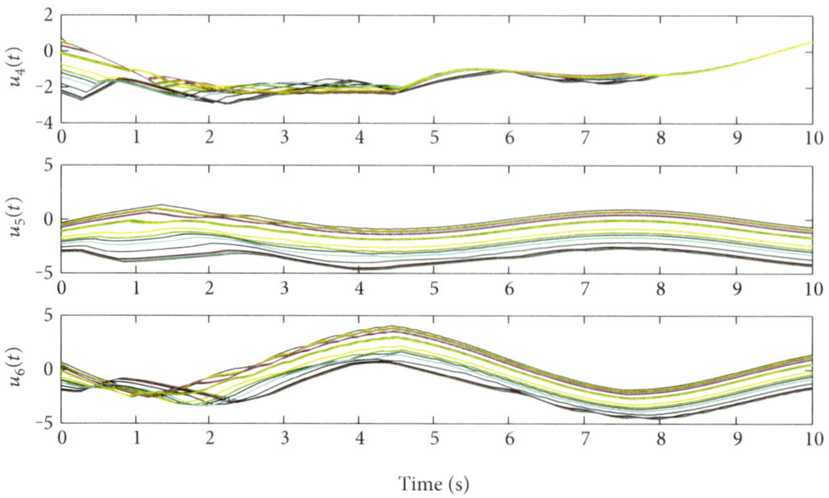

FIGURE 6: Virtual deflection angle control commands for the last three SJA arrays (i.e., $\mathbf{u}_4(t)$, $\mathbf{u}_5(t)$, and $\mathbf{u}_6(t)$) during closed-loop operation.

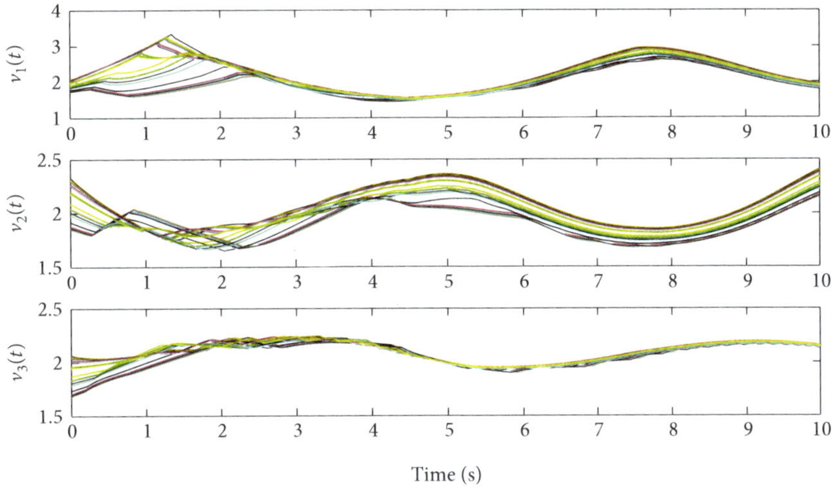

FIGURE 7: Control voltage signals commanded for the first three SJA arrays (i.e., $v_1(t)$, $v_2(t)$, and $v_3(t)$) during closed-loop operation.

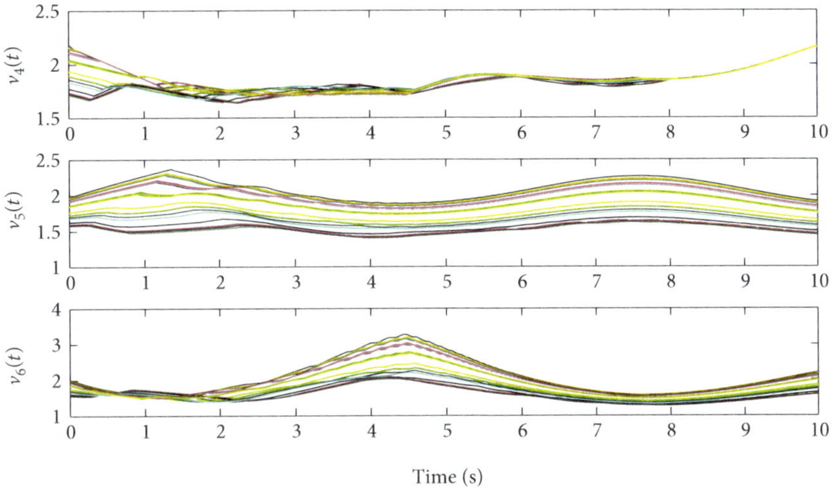

FIGURE 8: Control voltage signals commanded for the last three SJA arrays (i.e., $v_4(t)$, $v_5(t)$, and $v_6(t)$) during closed-loop operation.

$$\mathbf{A} = \begin{bmatrix} -61.1273 & 0 & -7.6409 \\ 0 & -174.3472 & 0 \\ -7.2692 & 0 & -0.4543 \end{bmatrix},$$

$$\mathbf{B} = \begin{bmatrix} -0.2292 & 0.2292 & -0.2292 & 0.2292 & -0.0306 & 0.0306 \\ 0.0599 & 0.0599 & 0.0804 & 0.0804 & -0.0256 & 0.0256 \\ -0.0084 & 0.0084 & -0.535 & 0.0535 & 0.1177 & -0.1177 \end{bmatrix}. \quad (55)$$

The wind gust model used in the simulation is based on the FAR discrete gust model described in ([33]). The simulation model for the wind gust is based on the expression in (4), see Figure 3.

The results of 20 Monte Carlo-type simulations are shown in Figures 4, 5, 6, 7, and 8. The results were obtained using control gains selected as $\mathbf{k}_s = \text{diag}\{0.10, 0.15, 2.3\}$, $\boldsymbol{\beta} = \text{diag}\{3.3, 0.3, 0.8\}$, and $\boldsymbol{\gamma} = 0.3$. Each set of axes shows the control performance for 20 different scenarios, where each plot shows the closed-loop response in the presence of 20 different sets of off-nominal values for the actual (plant) SJA parameters θ_{1i}^* and θ_{2i}^* for $i = 1, \ldots, 6$. The 20 sets of parameter values were generated using a randomization routine, which resulted in deviations of the actual SJA parameter values by up to 35.7% off nominal. The constant estimates (nominal values) used in the simulation are listed in Table 1.

TABLE 1: Constant nominal values of SJA parameters used in simulation.

θ_1^* (deg)	32.9	29.8	26.7	24.0	20.5	17.8
θ_2^* (volt-deg)	14.7	13.8	12.8	11.7	10.0	9.5

Remark 9 (comparison of results). The capability of the proposed robust nonlinear control method to compensate for SJA parameter deviations of more than 35% demonstrates a significant improvement over standard adaptive control approaches (cf. [8, 13]). Specifically, the results using

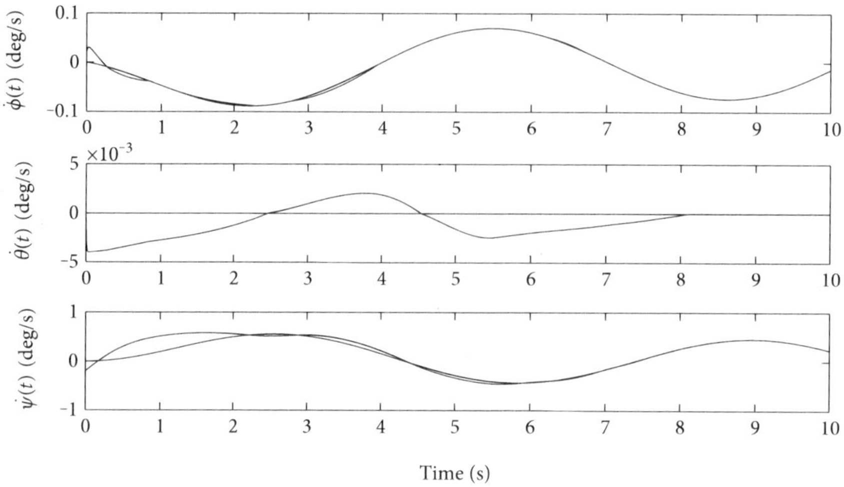

FIGURE 9: Model reference (red) and actual state (blue) during closed-loop controller operation in the presence of 35.75% SJA parameter uncertainty.

standard adaptive control approaches assume that the adaptive parameter estimates are within less than 5% of the actual parameter values.

Figure 4 shows the closed-loop tracking error response and demonstrates rapid convergence of the tracking error to zero in all 20 cases. Figures 5 and 6 show the virtual surface deflection control commands during closed-loop operation, and Figures 7 and 8 show the SJA voltage control inputs commanded during closed-loop operation. The results demonstrate that the closed-loop system remains stable in all 20 cases, and asymptotic tracking is achieved throughout the range of uncertainty tested. Figure 9 shows the convergence of the actual UAV states to the model reference states during closed-loop operation for the first iteration of our Monte Carlo-type simulation. The control commands remain within reasonable limits in all 20 cases.

5. Conclusion

A robust nonlinear control method that achieves asymptotic trajectory tracking for a SJA-based aircraft model is presented. The control method is proven to achieve semi-global asymptotic tracking of a reference trajectory in the presence of SJA actuator parameter uncertainty in addition to external norm-bounded disturbances (i.e., vertical wind gusts). A rigorous stability analysis is carried out to prove that the region of attraction of the closed-loop system can be made arbitrarily large through judicious tuning of a control parameter. The controller is designed to be computationally inexpensive, requiring no function approximators, adaptive laws, or complex computations. By utilizing constant feedforward estimates of the uncertain SJA actuator parameters, a matrix decomposition technique is employed along with a novel error system derivation to compensate for significant SJA parametric uncertainty (i.e., greater than 35% uncertainty in the SJA parameters). Detailed Monte-Carlo-type numerical simulation results are included to illustrate the effectiveness of the proposed control strategy.

Appendix

This appendix provides proof of Lemma 2.

Lemma A.1. *Provided the sufficient gain condition in (33) is satisfied, the following inequality can be obtained:*

$$\int_0^t L(\tau)d\tau \leq \beta Q|\mathbf{e}(0)| - \mathbf{e}^T(0)\mathbf{N}_{d1}(0). \tag{A.1}$$

Hence, (A.1) can be used to conclude that $P(t) \geq 0$, where $P(t)$ is defined in (35) and (36).

To facilitate the following proof, the expression in (37) will be rewritten in a more advantageous form as follows:

$$L(t) = \sum_{i=1}^{m}\left(\mathbf{r}_i(t)\left(\mathbf{N}_{d1i}(t) - \sum_{j=i}^{m}\mathbf{T}_{ij}\boldsymbol{\mu}_{1j}(t)\right)\right) \in \mathbb{R}. \tag{A.2}$$

In (A.2), $\mathbf{r}_i(t)$, $\mathbf{N}_{d1i}(t)$, $\boldsymbol{\mu}_{1i}(t) \in \mathbb{R}$ for $i = 1, \ldots, m$ denote the ith elements of the vectors $r(t)$, $\mathbf{N}_{d1}(t)$, and $\boldsymbol{\mu}_1(t)$; and $\mathbf{T}_{ij} \in \mathbb{R}$ for $i = 1, \ldots, m$ and $j = 1, \ldots, m$ denote the (i, j)th *elements of the matrix* $\mathbf{T}$.

Proof. Integrating both sides of (A.2) yields

$$\int_0^t L(\tau)d\tau = \int_0^t \sum_{i=1}^{m}\mathbf{r}_i(\tau)\left(\mathbf{N}_{d1i}(\tau) - \sum_{j=i}^{m}\mathbf{T}_{ij}\boldsymbol{\mu}_{1j}(\tau)\right)d\tau \in \mathbb{R}. \tag{A.3}$$

Based on the expressions in (8) and (9), the integral in (A.3) can be expressed as

$$\int_0^t L(\tau)d\tau = \int_0^t \sum_{i=1}^m (\dot{\mathbf{e}}_i(\tau)) \left(\mathbf{N}_{d1i}(\tau) - \sum_{j=i}^m \mathbf{T}_{ij}\boldsymbol{\mu}_{1j}(\tau) \right) d\tau + \int_0^t \sum_{i=1}^m (\boldsymbol{\gamma}_i \mathbf{e}_i(\tau) + \boldsymbol{\gamma}_i \eta_i(\tau)) \cdot \left(\mathbf{N}_{d1i}(\tau) - \sum_{j=i}^m \mathbf{T}_{ij}\boldsymbol{\mu}_{1j}(\tau) \right) d\tau, \tag{A.4}$$

where $\boldsymbol{\gamma}_i \in \mathbb{R}$ denotes the ith diagonal element of the control gain matrix $\boldsymbol{\gamma}$. The expression in (A.4) can be rewritten as

$$\int_0^t L(\tau)d\tau = \int_0^t \sum_{i=1}^m \frac{\partial \mathbf{e}_i(\tau)}{\partial \tau} \mathbf{N}_{d1i}(\tau)d\tau - \int_0^t \sum_{i=1}^m \frac{\partial \mathbf{e}_i(\tau)}{\partial \tau} \sum_{j=i}^m \mathbf{T}_{ij}\boldsymbol{\mu}_{1j}(\tau)d\tau + \int_0^t \sum_{i=1}^m \boldsymbol{\gamma}_i \mathbf{e}_i(\tau) \left(\mathbf{N}_{d1i}(\tau) - \sum_{j=i}^m \mathbf{T}_{ij}\boldsymbol{\mu}_{1j}(\tau) \right) d\tau. \tag{A.5}$$

By evaluating the first integral in (A.5) using integration by parts, (A.5) can be expressed as

$$\int_0^t L(\tau)d\tau = \sum_{i=1}^m \mathbf{e}_i(t)\mathbf{N}_{d1i}(t) - \sum_{i=1}^m \mathbf{e}_i(0)\mathbf{N}_{d1i}(0) - \int_0^t \sum_{i=1}^m \mathbf{e}_i(\tau) \frac{\partial \mathbf{N}_{d1i}(\tau)}{\partial \tau} d\tau - \int_0^t \sum_{i=1}^m \frac{\partial \mathbf{e}_i(\tau)}{\partial \tau} \sum_{j=i}^m \mathbf{T}_{ij}\boldsymbol{\mu}_{1j}(\tau)d\tau + \int_0^t \sum_{i=1}^m \boldsymbol{\gamma}_i \mathbf{e}_i(\tau) \left(\mathbf{N}_{d1i}(\tau) - \sum_{j=i}^m \mathbf{T}_{ij}\boldsymbol{\mu}_{1j}(\tau) \right) d\tau. \tag{A.6}$$

After substituting the definition of the auxiliary control term $\mu_1(t)$ given in (16) and rearranging, (A.6) can be expressed as

$$\int_0^t L(\tau)d\tau = \sum_{i=1}^m \left(\mathbf{e}_i(t)\mathbf{N}_{d1i}(t) - \mathbf{e}_i(0)\mathbf{N}_{d1i}(0) - \int_0^t \frac{\partial \mathbf{e}_i(\tau)}{\partial \tau} \sum_{j=i}^m \mathbf{T}_{ij}\boldsymbol{\beta}\operatorname{sgn}(\boldsymbol{\gamma}_j \mathbf{e}_j(\tau))d\tau \right) + \int_0^t \sum_{i=1}^m \boldsymbol{\gamma}_i \mathbf{e}_i(\tau) \left(\mathbf{N}_{d1i}(\tau) - \frac{1}{\boldsymbol{\gamma}} \frac{\partial \mathbf{N}_{d1i}(\tau)}{\partial \tau} - \sum_{j=i}^m \mathbf{T}_{ij}\boldsymbol{\beta}\operatorname{sgn}(\boldsymbol{\gamma}_j \mathbf{e}_j(\tau)) \right) d\tau. \tag{A.7}$$

By using the fact that

$$\sum_{j=i}^m \mathbf{T}_{ij}\boldsymbol{\beta}\operatorname{sgn}(\boldsymbol{\gamma}_j \mathbf{e}_j(\tau)) = \boldsymbol{\beta}\left(\operatorname{sgn}(\boldsymbol{\gamma}_j \mathbf{e}_j(\tau)) + \sum_{j=i+1}^m \bar{\mathbf{T}}_{ij} \operatorname{sgn}(\mathbf{e}_j(\tau)) \right), \tag{A.8}$$

the bounding inequalities in (21) can be used to express (A.7) as

$$\int_0^t L(\tau)d\tau = \sum_{i=1}^m (\mathbf{e}_i(t)\mathbf{N}_{d1i}(t) - \mathbf{e}_i(0)\mathbf{N}_{d1i}(0)) - \int_0^t \sum_{i=1}^m \frac{\partial \mathbf{e}_i(\tau)}{\partial \tau} \boldsymbol{\beta}\delta\operatorname{sgn}(\boldsymbol{\gamma}_i \mathbf{e}_i(\tau))d\tau + \int_0^t \sum_{i=1}^m \boldsymbol{\gamma}_i \mathbf{e}_i(\tau) \cdot \left(\mathbf{N}_{d1i}(\tau) - \frac{1}{\boldsymbol{\gamma}} \frac{\partial \mathbf{N}_{d1i}(\tau)}{\partial \tau} - \delta\boldsymbol{\beta} \operatorname{sgn}(\boldsymbol{\gamma}_i \mathbf{e}_i(\tau)) \right) d\tau, \tag{A.9}$$

where $\delta \in (\varepsilon, Q)$ is a positive constant parameter. By using the property

$$\int_0^t \frac{\partial \mathbf{e}(\tau)}{\partial \tau} \operatorname{sgn}(\boldsymbol{\gamma}\mathbf{e}(\tau))d\tau = |\mathbf{e}(t)| - |\mathbf{e}(0)|, \tag{A.10}$$

the expression in (A.9) can be rewritten as

$$\int_0^t L(\tau)d\tau = -\sum_{i=1}^m \mathbf{e}_i(0)\mathbf{N}_{d1i}(0) + \sum_{i=1}^m \boldsymbol{\beta}\delta|\mathbf{e}_i(0)| + \sum_{i=1}^m \mathbf{e}_i(t)\mathbf{N}_{d1i}(t) - \sum_{i=1}^m \boldsymbol{\beta}\delta|\mathbf{e}_i(t)| + \int_0^t \sum_{i=1}^m \boldsymbol{\gamma}_i \mathbf{e}_i(\tau) \cdot \left(\mathbf{N}_{d1i}(\tau) - \frac{1}{\boldsymbol{\gamma}} \frac{\partial \mathbf{N}_{d1i}(\tau)}{\partial \tau} - \delta\boldsymbol{\beta} \operatorname{sgn}(\boldsymbol{\gamma}_i \mathbf{e}_i(\tau)) \right) d\tau. \tag{A.11}$$

The expression in (A.11) can be upper bounded as

$$\int_0^t L(\tau)d\tau \le -\mathbf{e}^T(0)\mathbf{N}_{d1}(0) + \boldsymbol{\beta}Q|\mathbf{e}(0)| + \sum_{i=1}^m (\zeta_{\mathbf{N}_{d1}} - \varepsilon\boldsymbol{\beta})|\mathbf{e}_i(t)| + \int_0^t \sum_{i=1}^m \boldsymbol{\gamma}_i |\mathbf{e}_i(\tau)| \left(\zeta_{\mathbf{N}_{d1}} - \frac{1}{\boldsymbol{\gamma}} \zeta_{\dot{\mathbf{N}}_{d1}} - \varepsilon\boldsymbol{\beta} \right) d\tau. \tag{A.12}$$

Thus, it is clear from (A.12) that if $\boldsymbol{\beta}$ satisfies the sufficient condition in (33), then

$$\int_0^t L(\tau)d\tau \le \boldsymbol{\beta}Q|\mathbf{e}(0)| - \mathbf{e}(0)^T\mathbf{N}_{d1}(0). \tag{A.13}$$

Hence, $P(t) \ge 0$ from (35), (36), and (A.13).

Conflicts of Interest

The authors declare that they have no conflicts of interest.

Acknowledgments

This research is supported in part by National Science Foundation Award no. 1335405.

References

[1] S. Bhatt, V. V. Golubev, and Y. Tang, "Design, modeling and testing of synthetic jet actuators for MAV flight control," in *52nd Aerospace Sciences Meeting*, pp. 1–17, National Harbor, MD, USA, January 2014.

[2] M. Gul, O. Uzol, and I. S. Akmnador, "Flow control with synthetic jet actuators under adverse pressure gradient laminar boundary layer," in *32nd AIAA Applied Aerodynamics Conference*, pp. 1–17, Atlanta, GA, USA, June 2014.

[3] M. G. De Giorgi, C. G. De Luca, A. Ficarella, and F. Marra, "Comparison between synthetic jets and continuous jets for active flow control: application on a NACA 0015 and a compressor stator cascade," *Aerospace Science and Technology*, vol. 43, pp. 256–280, 2015.

[4] H. Tang, P. Salunkhe, Y. Zheng, J. Du, and Y. Wu, "On the use of synthetic jet actuator arrays for active flow separation control," *Experimental Thermal and Fluid Science*, vol. 57, pp. 1–10, 2014.

[5] V. Golubev, P. Kazarin, W. MacKunis, S. Borener, and D. Hufty, "Analysis of safety implications for SJA-based robust UAS flight control technology," in *2015 IEEE/AIAA 34th Digital Avionics Systems Conference (DASC)*, pp. 1–31, Prague, Czech Republic, 2015.

[6] H. E. Monir, M. Tadjfar, and A. Bakhtian, "Tangential synthetic jets for separation control," *Journal of Fluids and Structures*, vol. 45, pp. 50–65, 2013.

[7] M. Amitay, D. R. Smith, V. Kibens, D. E. Parekh, and A. Glezer, "Aerodynamic flow control over an unconventional airfoil using synthetic jet actuators," *AIAA Journal*, vol. 39, no. 3, pp. 361–370, 2001.

[8] D. Deb, G. Tao, J. O. Burkholder, and D. R. Smith, "Adaptive compensation control of synthetic jet actuator arrays for airfoil virtual shaping," *Journal of Aircraft*, vol. 44, no. 2, pp. 616–626, 2007.

[9] W. MacKunis, S. Subramanian, S. Mehta, J. W. Curtis, M. Reyhanoglu, and C. Ton, "Robust nonlinear aircraft tracking control using synthetic jet actuators," in *52nd IEEE Conference on Decision and Control*, pp. 220–225, Florence, Italy, December 2013.

[10] D. Deb, G. Tao, J. O. Burkholder, and D. R. Smith, "An adaptive inverse control scheme for a synthetic jet actuator model," in *Proceedings of the 2005, American Control Conference, 2005*, pp. 2646–2651, Portland, OR, USA, June 2005.

[11] D. Deb, G. Tao, J. O. Burkholder, and D. R. Smith, "An adaptive inverse control scheme for synthetic jet actuator arrays," in *Infotech@Aerospace*, Arlington, VA, USA, September 2005.

[12] D. Deb, G. Tao, J. O. Burkholder, and D. R. Smith, "Adaptive synthetic jet actuator compensation for a nonlinear tailless aircraft model at low angles of attack," in *2006 American Control Conference*, p. 6, Minneapolis, MN, USA, June 2006.

[13] D. Deb, G. Tao, J. O. Burkholder, and D. R. Smith, "Adaptive synthetic jet actuator compensation for a nonlinear aircraft model at low angles of attack," *IEEE Transactions on Control Systems Technology*, vol. 16, no. 5, pp. 983–995, 2008.

[14] G. Hong, "Numerical investigation to forcing frequency and amplitude of synthetic jet actuators," *AIAA Journal*, vol. 50, no. 4, pp. 788–796, 2012.

[15] M. Kim, S. Kim, W. Kim, C. Kim, and Y. Kim, "Flow control of tiltrotor unmanned-aerial-vehicle airfoils using synthetic jets," *Journal of Aircraft*, vol. 48, no. 3, 2011.

[16] W. Kim, C. Kim, and K. J. Jung, "Separation control characteristics of synthetic jets depending on exit configuration," *AIAA Journal*, vol. 50, no. 3, 2012.

[17] R. B. Kotapati, R. Mittal, O. Marxen, F. Ham, D. You, and L. N. Cattafesta III, "Nonlinear dynamics and synthetic-jet-based control of a canonical separated flow," *Journal of Fluid Mechanics*, vol. 654, pp. 65–97, 2010.

[18] A. Kourta and C. Leclerc, "Characterization of synthetic jet actuation with application to Ahmed body wake," *Sensors and Actuators A: Physical*, vol. 192, pp. 13–26, 2013.

[19] Y. Liu, M. Ciuryla, M. Amitay et al., "Integrated flight control and flow control using synthetic jet arrays," in *AIAA Guidance, Navigation, and Control Conference and Exhibit*, Keystone, CO, USA, 2006.

[20] O. D. L. Mejia, R. D. Moser, D. P. Brzozowski, and A. Glezer, "Effects of trailing-edge synthetic jet actuation on an airfoil," *AIAA Journal*, vol. 49, no. 8, 2011.

[21] Z. B. Luo, Z. X. Xia, and Y. G. Xie, "Jet vectoring control using a novel synthetic jet actuator," *Chinese Journal of Aeronautics*, vol. 20, no. 3, pp. 193–201, 2007.

[22] T. Ozawa, S. Lesbros, and G. Hong, "LES of synthetic jets in boundary layer with laminar separation caused by adverse pressure gradient," *Computers and Fluids*, vol. 39, no. 5, pp. 845–858, 2010.

[23] D. A. Tamburello and M. Amitay, "Active control of a free jet using a synthetic jet," *International Journal of Heat and Fluid Flow*, vol. 29, no. 4, pp. 967–984, 2008.

[24] S. Tardu, "Forcing a low Reynolds number channel flow to generate synthetic turbulent-like structures," *Computers & Fluids*, vol. 55, pp. 101–108, 2012.

[25] Z. Travnicek, L. Nemcova, J. Kordik, V. Tesar, and V. Kopecky, "Axisymmetric impinging jet excited by a synthetic jet system," *International Journal of Heat and Mass Transfer*, vol. 55, no. 4, pp. 1279–1290, 2012.

[26] L.-Y. Tseng, A.-S. Yang, and J.-C. Lin, "Study of a crossflow over a zero-net-mass-flux synthetic jet driven by a vibrating diaphragm," *Journal of Mechanics*, vol. 27, no. 4, pp. 503–510, 2011.

[27] B. Vukasinovic, A. Glezer, S. Gordeyev, E. Jumper, and V. Kibens, "Fluidic control of a turret wake, part I: aerodynamic effects," in *47th AIAA Aerospace Sciences Meeting Including The New Horizons Forum and Aerospace Exposition*, Orlando, FL, USA, 2009.

[28] S. Zhang and S. Zhong, "Experimental investigation of flow separation control using an array of synthetic jets," *AIAA Journal*, vol. 48, no. 3, pp. 611–623, 2010.

[29] S. T. Mondschein, G. Tao, and J. O. Burkholder, "Adaptive actuator nonlinearity compensation and disturbance rejection with an aircraft application," in *Proceedings of the 2011 American Control Conference*, pp. 2951–2956, San Francisco, CA, USA, 2011.

[30] R. Nelson, *Flight Stability and Automatic Control*, McGraw-Hill, 2nd edition, 1998.

[31] A. A. Tchieu, A. T. Kutay, J. A. Muse, A. J. Calise, and A. Leonard, "Validation of a low-order model for closed loop flow control enabled flight," in *4th Flow Control Conference*, Seattle, WA, USA, 2008.

[32] C. Singhal, G. Tao, and J. O. Burkholder, "Neural network-based compensation of synthetic jet actuator nonlinearities for aircraft flight control," in *AIAA Guidance, Navigation, and Control Conference*, Chicago, IL, USA, 2009.

11

Flight-Test Evaluation of Kinematic Precise Point Positioning of Small UAVs

Jason N. Gross, Ryan M. Watson, Stéphane D'Urso, and Yu Gu

Department of Mechanical and Aerospace Engineering, West Virginia University, Morgantown, WV 26506, USA

Correspondence should be addressed to Jason N. Gross; jason.gross@mail.wvu.edu

Academic Editor: Paolo Tortora

An experimental analysis of Global Positioning System (GPS) flight data collected onboard a Small Unmanned Aerial Vehicle (SUAV) is conducted in order to demonstrate that postprocessed kinematic Precise Point Positioning (PPP) solutions with precisions approximately 6 cm 3D Residual Sum of Squares (RSOS) can be obtained on SUAVs that have short duration flights with limited observational periods (i.e., only $\sim\leq 5$ minutes of data). This is a significant result for the UAV flight testing community because an important and relevant benefit of the PPP technique over traditional Differential GPS (DGPS) techniques, such as Real-Time Kinematic (RTK), is that there is no requirement for maintaining a short baseline separation to a differential GNSS reference station. Because SUAVs are an attractive platform for applications such as aerial surveying, precision agriculture, and remote sensing, this paper offers an experimental evaluation of kinematic PPP estimation strategies using SUAV platform data. In particular, an analysis is presented in which the position solutions that are obtained from postprocessing recorded UAV flight data with various PPP software and strategies are compared to solutions that were obtained using traditional double-differenced ambiguity fixed carrier-phase Differential GPS (CP-DGPS). This offers valuable insight to assist designers of SUAV navigation systems whose applications require precise positioning.

1. Introduction

The Precise Point Positioning (PPP) technique was introduced in the late nineties [1, 2] and uses state-space GNSS satellite orbit and clock bias solutions with significantly greater accuracy than their broadcast ephemeris counterparts in order to enable the user-segment to obtain accurate positioning with undifferenced data. The use of undifferenced data means that no GNSS reference station is required to form differential data combinations. To date, PPP technology has matured to the extent that there are now multiple real-time orbit and clock solution products offered by government organizations [3], commercial entities [4, 5], and the International GNSS Service [6]. The PPP technique has revolutionized many geophysical research applications that involve static reference stations and Earth orbiting spacecraft; however, it has not been heavily exploited for applications involving low-cost small UAVs (SUAVs).

Many authors have conducted studies to compare the solution accuracy of the PPP technique with double-difference CP-DGPS solutions or other ground "truths." For example, Colombo et al. show that once a PPP filter has converged, it agrees with double-differenced GPS to within 10 cm [7]. Likewise, for a kinematic vehicular application, Honda et al. demonstrate a few decimeter-level performance with respect to CP-DGPS [8]. In another study, Zhang and Forsberg consider the use of PPP to support missions that require accuracy over very long-ranges (i.e., on the order of many hundreds of kilometers), thereby making double-differences to an individual reference station impractical [9]. In their assessment, Zhang and Forsberg use comparisons of airborne laser altimetry and satellite altimetry products to assess height solution accuracy from PPP and conclude that PPP can produce accuracy at the decimeter-level. In 2009, Bisnath and Gao [10] offered insight on the state-of-the-art of PPP and, in their assessment, demonstrate decimeter-level kinematic PPP of a static reference station after an initial convergence period. Bisnath and Gao conclude that more algorithm development and additional observables are needed to reduce PPP's convergence period before it can be considered as an RTK alternative. Many have shown that PPP can achieve accuracy levels consistent with CP-DGPS but point out its slow convergence.

More recent studies have also considered the impact of multiconstellation GNSS or other navigation systems, such as Inertial Navigation Systems (INS), to yield better accuracy with the PPP approach. For example, Cai et al. evaluated multiconstellation GNSS using the GPS, GLONASS, Galileo, and Beidou and showed improvements in kinematic PPP position solution accuracy and convergence [11]. Likewise, Yigit et al. demonstrated better positioning performance with multiconstellation GNSS, especially whenever there is a relatively short observation duration [12]. In addition, Zhang and Gao include INS within a PPP filter for a kinematic application and show faster solution convergence and accuracy comparable to conventional RTK/INS solutions [13]. More recently, Gross et al. [14] reaffirmed better overall solution accuracy as well as solution robustness to rapid changes in the tropospheric delay induced by abrupt changes in aircraft altitude by comparing both a fused PPP/INS and a PPP-only solution to a postprocessed reference solution that was the NASA Jet Propulsion Laboratory's submission to the US National Geodetic Survey's Kinematic GPS Processing Challenge [15]. Finally, Watson et al. [16] provided a comprehensive evaluation of the benefit of incorporating INS under various common PPP error sources by using a Monte Carlo simulation environment and showed that incorporating INS becomes more important depending on the quality of the troposphere model, multipath environment, and quality of PPP orbit and clock products.

Despite the extensive literature that offers performance comparisons of different GNSS processing techniques, there is a lack of published studies that demonstrate the obtainable PPP positioning accuracy of SUAV flights that have very limited flight durations. Furthermore, while several past studies point to the slow solution convergence as a pitfall of using PPP technique, the need for real-time solutions is often irrelevant for many SUAV scientific applications (e.g., remote sensing, aerial mapping) that can simply wait for postprocessed solutions. However, because some SUAVs have very limited flight durations, which are on the order of fifteen minutes, or often less, some uncertainty remains as to whether PPP's slow solution convergence will impact the accuracy of the postprocessed (i.e., filtered and backwards smoothed) short duration SUAV solutions.

As such, to fill this knowledge gap, the contribution of this paper is to offer an experimental analysis of PPP techniques when compared to CP-DGPS with data collected onboard SUAV that has very short duration flights. This is particularly relevant to the field robotics community as SUAVs are being more regularly used for more ambitious scientific applications that have stringent requirements on platform positioning. For example, recent experimental evaluations have demonstrated the use of Light Detection and Ranging (LIDAR) for fine-scale mapping with UAVs [17–19]. Accurate georegistration requires platform positioning accuracy in the cm-scale, as these LIDAR systems have cm-scale or better resolution. Additionally, researchers at the University of Kansas have recently instrumented SUAV for radar sounding of remotely located Antarctic ice sheets [20]. While the current Kansas ice sounding UAV is operating at radar frequencies with large wavelengths, it is not a stretch to envision a similar SUAV radar system operating higher frequencies (e.g., microwave) and thus needing cm-scale positioning precision, such as a miniaturized version of NASA's L-Band UAVSAR instrument [21]. For these SUAV applications, which are typically cited to offer the benefit of being rapidly deployable and useful for remote regions, eliminating the need for GNSS differential reference station and thus reducing overall cost and complexity of the navigation system by leveraging PPP will further open the use of SUAVs for hosting scientific payloads that have stringent positioning requirements.

The rest of this paper is organized as follows. Section 2 reviews the GNSS observational models for both PPP, which is being evaluated, and double-differenced ambiguity fixed CP-DGPS, which is being used to estimate reference position solutions. In addition, the details of the software implementation employed in this study are discussed. Section 3 discusses the experimental SUAV, GNSS equipment, and flight-test environment used for this study. The results of an experimental comparison study are then detailed in Section 4, and the study's findings are summarized in Section 5.

2. Methodology

2.1. Observational Models. We start by considering the generic observation models for GNSS pseudorange and carrier-phase measurements as shown in (1) and (2), which are found in many textbooks [22, 23]. Consider

$$\rho_i^j = R_i^j + c\delta t_i + T_i^j + I_i^j + c\delta t^j + \epsilon_\rho^j, \quad (1)$$

$$\Phi_i^j = R_i^j + c\delta t_i + T_i^j - I_i^j + c\delta t^j + \lambda_{L_k} N_{L_k}^j + \epsilon_\Phi^j. \quad (2)$$

In (1) and (2), the subscript i is used to denote the user and superscript j denotes the GNSS satellite index. Common to both pseudorange and carrier-phase are several error sources, where t_i is the unknown GNSS receiver clock bias in units of seconds, t^j is the GNSS transmitter's clock bias in units of seconds, c is the speed of light in a vacuum in units of meters per second, T_i^j is the GNSS signal delay due to the refraction in the neutral atmosphere in units of meters, and I_i^j is the phase-advance/pseudorange delay caused by signal refraction through traversing the Earth's ionosphere (note the sign change between (1) and (2)) in units of meters. The remaining unmodeled error sources are denoted by ϵ and are in units of meters. Within the carrier-phase model (2), there is also unknown integer phase tracking ambiguity that is denoted by N_i^j and is taken from units of carrier-phase cycles to meters through a multiplication with the GNSS carrier wavelength λ_{L_k}, where $k = 1, 2$ and $L_1 = 19.0$ cm and $L_2 \approx 24.4$ cm for the case of GPS. Furthermore, in (1) and (2), the geometric range between the user's receiver antenna phase center and the satellite's transmitter antenna phase center is denoted as R_i^j and given by

$$R_i^j = \sqrt{(x^j - x_i)^2 + (y^j - y_i)^2 + (z^j - z_i)^2}, \quad (3)$$

where both the user's position and satellite's position are modeled in the same Cartesian reference frame, typically either an Earth-Centered-Earth-Fixed frame or an Earth-Centered-Inertial frame. For the PPP technique, accurate solutions to each GNSS transmitter's clock bias, t^j, and orbital location (x^j, y^j, z^j) are determined using a global network of tracking reference stations and are used *in lieu* of the GNSS broadcast ephemeris. However, typically, the satellite locations are provided with respect to the center of gravity of each GNSS spacecraft. Therefore, to be consistent with (3), the attitude of each satellite must be modeled such that the lever arm offset between each satellite's antenna phase center and its center of gravity is properly considered. To do this, GNSS satellite type dependent known lever arm offsets and attitude models [24, 25] must be included in the processing strategy.

In this study, we assume access to dual-frequency GNSS data. As such, we use the dispersive nature of the ionosphere's delay and form a linear combination of the dual-frequency signals in order to cancel the effect of the ionosphere signal refraction to the first order. This linear combination is denoted as ionospheric-free (IF) combination [22] and given by (4) and (5), for pseudorange and carrier-phase observations, respectively:

$$\rho_{\text{IF}}^{j} = \rho_{L1}^{j}\left[\frac{f_1^2}{f_1^2 - f_2^2}\right] - \rho_{L2}^{j}\left[\frac{f_2^2}{f_1^2 - f_2^2}\right], \tag{4}$$

$$\phi_{\text{IF}}^{j} = \phi_{L1}^{j}\left[\frac{f_1^2}{f_1^2 - f_2^2}\right] - \phi_{L2}^{j}\left[\frac{f_2^2}{f_1^2 - f_2^2}\right]. \tag{5}$$

Note that the coefficients in (4) and (5) (i.e., 2.546 and −1.546 for GPS) sum to 1.0, and thus the modeled common mode error sources, including clock biases and the troposphere delay, remain unchanged in the IF observations. However, when using the IF combination, the unmodeled random errors denoted by ϵ become amplified. For GNSS, the unmodeled error sources consist of thermal noise and receiver front end noise and multipath. For pseudorange observables ϵ_ρ is on the order of 1 meter and for carrier-phase ϵ_ϕ is on the order of 1 millimeter. As such, when using the IF combination, the measurement noise is ~3 meters for pseudorange and 3 millimeters for carrier-phase (i.e., $3 \approx \sqrt{2.546^2 + 1.546^2}$).

In (1) and (2), the tropospheric delay T_i^j is typically modeled by scaling the zenith direction delay using an elevation angle dependent mapping function to reduce the number of model parameters. Furthermore, this is composed of both wet and dry components of which the dry component is ~90% of the total delay and can be well approximated with a model. As such, the zenith dry delay is typically modeled and the zenith wet delay is estimated as an unknown parameter such that the delay is modeled as shown in

$$T_i^j = T_z\text{Dry}M^{\text{Dry}}\left(\text{el}_i^j\right) + T_z\text{Wet}M^{\text{Wet}}\left(\text{el}_i^j\right), \tag{6}$$

where el_i^j is the elevation angle from the nominal user location to satellite j. Within this study, different mapping functions, $M^{\text{Dry}}, M^{\text{Wet}}$, are utilized depending on the solution strategy and software used as will be detailed in the next section.

The remaining modeled error sources are the carrier-phase ambiguities. Traditionally, with kinematic PPP techniques, despite the fact that these are known to be an integer number, these are modeled as floating point parameters in the PPP filter/smoother. However, PPP ambiguity resolution techniques have been developed [26, 27] and this remains an active research area.

An important unmodeled error source within the ϵ terms is present in (1) and (2) which are the errors induced by multipath reflections. These errors are attributed to reflections of the GNSS signals being present in the received signal in addition to the line of sight signal. Unfortunately, multipath leads to errors that are non-Gaussian and non-White in nature, such that Kalman or least-squares type estimators are not best suited to mitigate them. Instead, a total least-squares processing technique could be employed, such as the one mentioned in Juang [28]. However, the multipath induced errors for carrier-phase are multiple orders of magnitude smaller than those for pseudorange. Further, for many airborne applications, multipath induced errors, due to signal reflections from man-made objects, are less of a potential problem. In this study, to mitigate multipath induced errors, our PPP estimators will heavily rely on carrier-phase data relative to pseudorange data.

2.2. Kinematic PPP with JPL's GIPSY-OASIS. The first PPP approach considered in this study uses Caltech JPL's GNSS-Inferred Positioning System and Orbit Analysis Simulation Software (GIPSY-OASIS) 6.2 [29]. GIPSY has been the primary geodetic and positioning software for NASA's TOPEX/Poseidon [30] and JASON [31] and GRACE [32] low Earth orbiting spacecraft and is operationally used to generate JPL's precise GPS orbits and clock products for the IGS [33]. GIPSY is licensed for free by Caltech to institutions for use in academic research.

When using GIPSY for kinematic PPP, our strategy in this study is to iteratively process the position solution while varying GIPSY configuration parameters in order to converge it to an optimal solution that is free of data outliers. A block diagram of the processing strategy is shown in Figure 1, which requires defining some GIPSY terms.

(i) GNSS Data to Positioning (GD2P). It is GIPSY's main user interface script for PPP.

(ii) Pseudorange Data to Positioning (PR2P). It is GIPSY's script for pseudorange-only point positioning.

(iii) Time Dependent Parameter (TDP). It is GIPSY's output format for positioning solutions and other solved for parameters (e.g., clock biases, troposphere, and phase biases).

(iv) QM File. It is GIPSY's native binary GNSS measurement format.

As indicated in Figure 1, for the first iteration, a position solution is estimated using only pseudorange measurement with GIPSY's PR2P. In addition, the data is translated from the Receiver Independent Exchange Format (RINEX) to a GIPSY

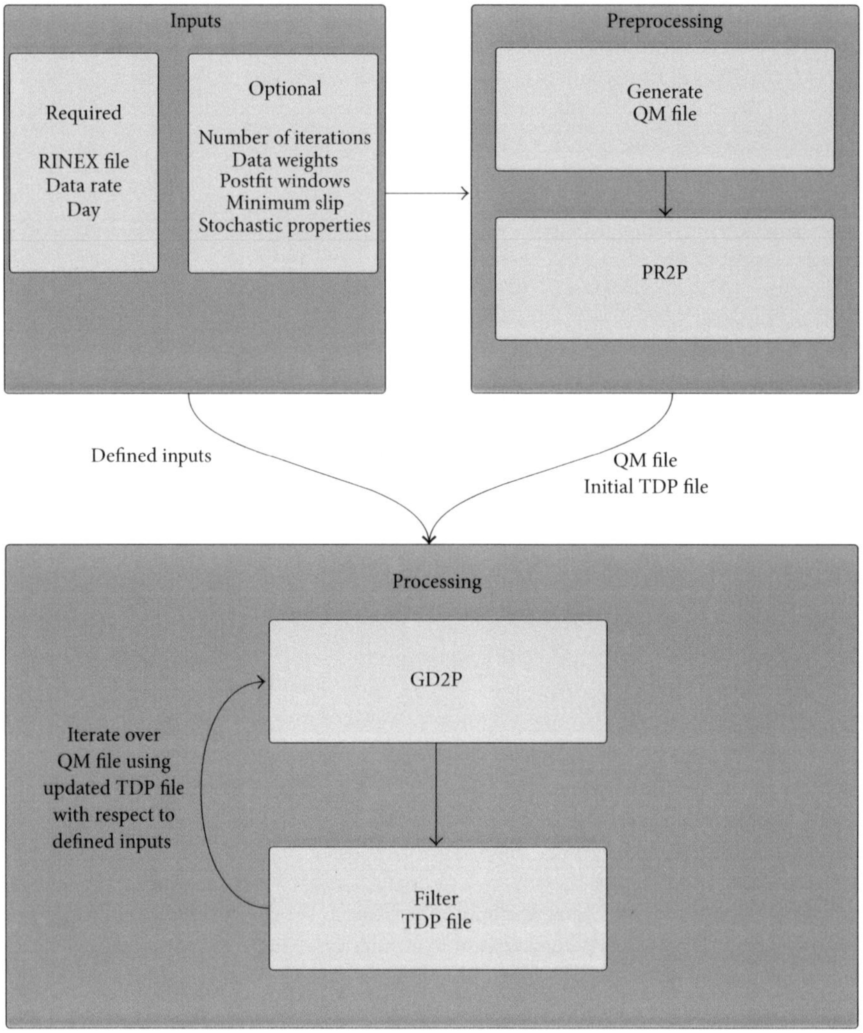

FIGURE 1: Kinematic PPP strategy with GIPSY-OASIS. A wrapper software is used to interactively process the position solution.

binary QM file. During this process, a GNSS data editor is used to flag carrier-phase breaks and remove gross data outliers [34]. For data editing, the GIPSY defaults were used with the exception of the editor requiring a minimum data arc length for a given transmitter. This is due to the short overall observation window of the flights.

For the remaining iterations, a subset of GIPSY processing options are varied while accepting the previous position solution (TDP file) as the *a priori* position solution. Within Figure 1, the configuration options that are varied for each run are as follows.

(i) Data Weights. They are measurement noise ratio between pseudorange and carrier-phase measurements. This starts as one-to-one and varies to one-hundred-to-one, where carrier-phase data are modeled as 100 times more precise than pseudorange data.

(ii) Postfit Residual Window. Within each GIPSY processing run, multiple passes of a Kalman filter and smoother are conducted. Between each pass, postfit data residuals are evaluated and data are marked as outlier based on defined thresholds and excluded from the next pass. At each pass, the residuals of all data, inlier and outlier, are evaluated and either added back in or excluded from the run. This process is repeated until all data meet the postfit window criteria or a maximum number of iterations are exceeded.

(iii) Stochastic Models. The position and wet troposphere delay estimates can be modeled using either white noise about the nominal solution or random walk process noise. Additionally, the *a priori* σ magnitude and rate of process noise updates can be set. In particular, the wet zenith delay was estimated as a random walk process, given that the SUAV evaluated is not flying over great distances. Position is initially estimated using random walk, and after a few iterations it is estimated using white noise about an *a priori* nominal solution.

(iv) Minimum Slip. After each filter iteration, jumps in the postfit phase residuals are used to identify the possibility of

TABLE 1: GIPSY wrapper options of kinematic PPP of SUAV data. Using the process shown in Figure 1.

Iteration #	1	2	3	4	5	6	7	8	9	10
Data weight ratio phase/range	1	1	10	100	100	100	100	100	100	100
Range residual window (m)	1.0e3	1.0e2	10	5.0	5.0	2.5	2.5	2.5	2.5	2.5
Phase residual window (m)	1.0e3	1.0e2	1.0	1.0	0.5	0.25	0.25	0.25	2.5e − 2	2.5e − 2
Per-epoch white noise (m)	1e3	1e3	1e2	1e2	1e2	1e2	50	50	5	5
Postbreak min slip (m)	1e12	1e12	1e12	1e12	1e12	1.0	1.0	1.0	1.0	1.0

a carrier-phase break that was missed by the data editor. New breaks are flagged for the next iteration.

In addition to the configuration parameters listed above that are varied for each iteration, several other GIPSY options were selected and held fixed in this study. In particular, the troposphere mapping function used is the Niell mapping function [35] and the nominal troposphere delays were set using the atmospheric model populated with a nominal height for scaling and pressure and temperature from the Global Pressure and Temperature model [36]. For the several remaining available GD2P options (e.g., elevation cut-off, tide models), the defaults provided by JPL were used.

The specific strategy used for this study with respect to Figure 1 is listed in Table 1, for the 10 iterations conducted.

While the outlier deletion windows are selected *ad hoc* based upon experience, in order to assess the effectiveness of postfit data residual analysis and outlier deletion, the GIPSY log files report an approximate χ^2 statistic of the residuals for each filter/smoother and data editing iteration. The value reported is only approximate, because it is reported as if the entire data set were processed as least-squares batch as opposed to a sequential filter. More formally, the Residual Standard Error (RSE) of the estimator for each outlier editing iteration is reported as follows:

$$\sigma_{\text{RSE}} = \sqrt{\frac{\text{RSOS}}{N - D}}, \tag{7}$$

where RSOS is the Residual Sum of Squares normalized to the data weight of each type of observation, N is the total number of GNSS observations, and D is the total number of parameters estimated. In addition, while it may seem problematic to delete data based upon user selected thresholds, in practice only gross outliers are removed from the filter run. This is because the residual analysis is done in a sequential narrowing window manner, such that such that, first, only gross outliers are deleted and then the residuals are reevaluated, with the possibility of adding data that was flagged as outlier on a previous run as being marked as inlier again (e.g., to handle the case in which one large outlier pollutes all of the residuals for a given epoch). Figure 2 shows an example for the RSE as well as the total percentage of data deleted during a GIPSY run, where the RSE is shown to converge and less than 1.5% of data is shown to have been deleted.

2.3. Kinematic PPP with RTKLIB. The second software package used for PPP in this study is the open-source RTKLIB [37, 38]. When using RTKLIB, several processing options are available for the user. The following list describes the adopted processing strategy of this study.

(i) Kalman Filter Set-Up. The Kalman estimator was configured for both forward filtering and then backwards smoothing.

(ii) Elevation Cut-Off. A 5° elevation angle cut-off was used.

(iii) Troposphere Model. The troposphere was modeled using the Saastamoinen model and a residual wet zenith delay was estimated.

(iv) Ionosphere. The ionosphere free linear combination was used.

(v) Misc. Solid Earth tides and ocean loading tides were modeled. Phase windup was considered.

(vi) Process Noise. The SUAV dynamics were modeled using a double integrator with process noise driving the acceleration states of 10 m/s^2.

These options were selected using RTKLIB's graphical user interface.

2.4. Carrier-Phase Double-Differenced Processing for Reference Solutions. To provide reference position solutions for the kinematic PPP solutions, a carrier-phase double-differenced integer ambiguity fixed processing strategy was used. In this case, the GNSS error sources present in (2), in common with a GNSS base station, which is located at a well-known location within a short baseline separation, are canceled via data differencing [22]. In this scenario, the relative position of the UAV with respect to the base station, $r_{U,B}$, is estimated, and the known location of the base station is added to recover an absolute position estimate of the UAV. For this technique, first, two carrier-phase measurements from the same satellite, j, are differenced between the two user receivers, denoted here as U, for UAV, and B, for base station, to form single-differenced carrier-phase measurements:

$$\Delta\phi^j_{U,B} = \lambda^{-1} r^j_{U,B} + \frac{c}{\lambda}\delta t_{iU,B} + N^j_{U,B} + \epsilon^j_{\phi,U,B}, \tag{8}$$

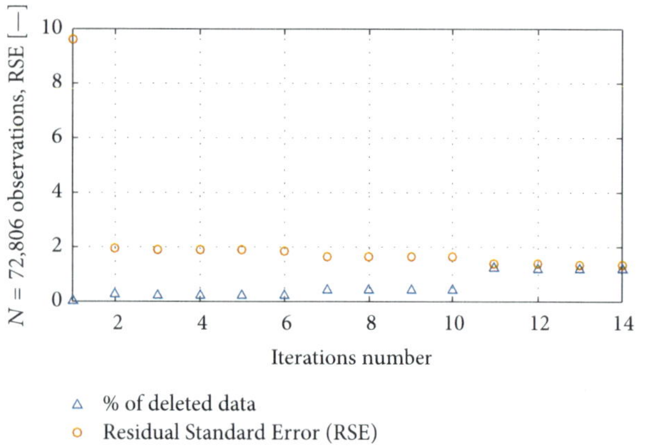

FIGURE 2: Example of GIPSY's iterative outlier deletion strategy.

FIGURE 3: Phastball SUAV research platform. The two GNSS antennas are separated by an 85.3 cm baseline.

where the satellite clock bias errors, the troposphere delay, and ionosphere delay are eliminated, and the ambiguity $N^j_{U,B}$ remains an integer. Next, to further eliminate the error attributed to the combined users' receiver clock biases, $\delta t_{U,B}$, two of the single-differenced carrier-phase measurements from satellites j and k are subtracted to form double-differenced carrier-phase measurements, $\nabla\Delta\phi^{j,k}_{U,B}$. The double-differencing process requires the selection of a reference satellite which is typically selected to be a high-elevation satellite (indicated herein with index k) and subtracting its single-differenced measurement from all other available single-differenced measurements. With double-differenced measurements, the only errors that remain are the multipath errors, which are small for carrier-phase observables and airborne applications, and the ambiguity, which is known to be an integer number of wavelengths of the carrier frequency.

To resolve the integer ambiguity, a Kalman filter first estimates the carrier-phase ambiguities as floating point parameters, and, then, a separate technique is used to to take advantage of the fact that these states are an integer number of wavelengths. In particular, the *de facto* technique that is used in this study is known as the Least-squares AMBiguity Decorrelation and Adjustment (LAMBDA) method [39, 40]. The objective of the LAMBDA method is to find an Integer Least Squared Solution (ILS) with respect to the estimated float ambiguities, $\widehat{N}_\phi$, and a corresponding variance-covariance estimate of the phase ambiguities, $P_{\widehat{N}_\phi,\widehat{N}_\phi}$ [39]. This is done by finding an orthogonal transformation that preservers integer values and decorrelates $P_{\widehat{N}_\phi,\widehat{N}_\phi}$ to a diagonal covariance matrix, such that simple rounding can be employed to estimate the integer states. Following the rounding process, the transformation is reversed to get back into the state-space domain. The cost function that the LAMBDA method optimizes is given by [41]

$$F(N_\phi) = (\widehat{N}_\phi - N_\phi)^T P^{-1}_{\widehat{N}_\phi,\widehat{N}_\phi} (\widehat{N}_\phi - N_\phi) \le \chi^2, \quad N_\phi \in Z^n, \tag{9}$$

where the integer grid search space is defined by χ^2. Once the integer fixed biases have been determined, the relative navigation states are adjusted by assuming that the integer-fixing process is deterministic, such that the nonambiguity states are corrected as

$$x^{\text{fix}}_{\text{position}} = x^{\text{float}}_{\text{position}} - P_{\text{position},\widehat{N}_\phi} P_{\widehat{N}_\phi,\widehat{N}_\phi} (\widehat{N}_\phi - N_\phi), \tag{10}$$

where P refers to the variance-covariance matrix for the floating point ambiguities that were estimated by the Kalman filter, of which particular sections are identified by the subscripts position, which refers to the position states, and $\widehat{N}_\phi$, which refers to phase ambiguity states.

It is important to point out that integer bias fixing will not always be a success in the presence of errors. As such, there must be a validation process. The specific acceptance-test employed in the RTKLIB implementation used in this study was the ratio-test [42]. The ratio-test evaluates how close the float ambiguity estimates are to the best integer ambiguity estimates when compared to the next best integer ambiguity candidate. The best candidate, $N^{1\text{st}}_\phi$, and the second best candidate, $N^{2\text{nd}}_\phi$, are defined as those that minimize the cost function $F(N_\phi)$ of (9):

$$\text{Accept } N^{1\text{st}}_\phi \quad \text{iff } \frac{F(N^{1\text{st}}_\phi)}{F(N^{2\text{nd}}_\phi)} \le \frac{1}{C}, \tag{11}$$

where C is the critical value, which can be derived on the fly to allow a fixed failure rate or set to a constant [42]. For this study, C was set to 3 and held constant. Three is often used in practice, though this is only empirically justified [42]. Throughout this study, only the epochs that were successfully fixed and passed the ratio-test criterion, which ranged from 65 to 85% of all epochs of each flight, were used as the reference solution for conducting the error analysis presented below.

3. Experimental Set-Up

The Phastball SUAV airframe [43] was developed as a modular research platform and has been used for multiple sensor-fusion studies [44–46]. The Phastball Zero SUAV is shown in Figure 3.

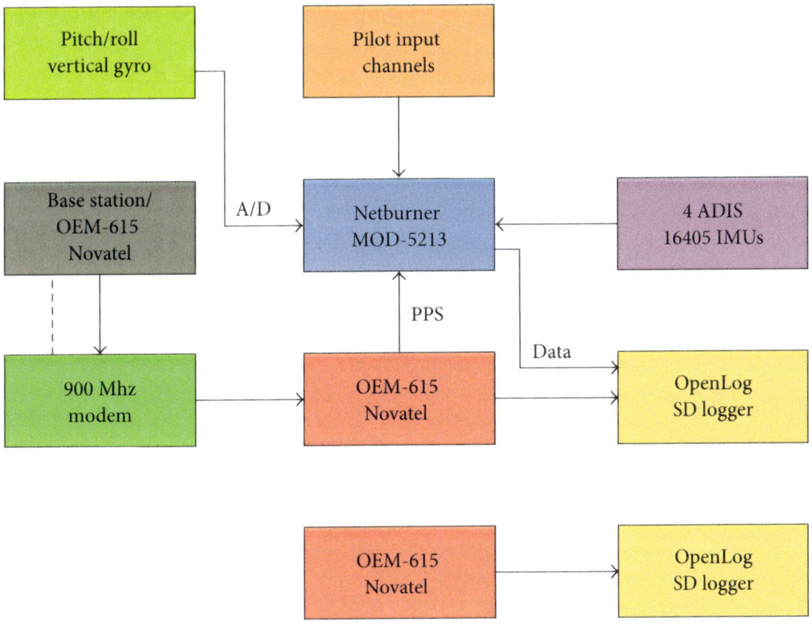

FIGURE 4: Phastball SUAV data acquisition set-up. Two GNSS receivers are logged in addition to IMU, mechanical vertical gyroscope, and ground pilot input data.

For this study, its payload features two Novatel OEM-615 dual-frequency GNSS receivers with antennas separated 85.3 cm along the airframe's longitudinal axis. A block diagram of Phastball Zero's data acquisition structure is shown in Figure 4.

In this set-up, one Novatel Receiver is GPS and GLONASS capable and the other is GPS only. Raw GNSS dual-frequency pseudorange and carrier-phase observables are recorded at a rate of 10 Hz using OpenLog serial microSD data-loggers. The GPS/GLONASS Novatel OEM-615 receiver is wirelessly connected through a 900 MHz modem to a base station that sends up RTK differential correctors. Additionally, a mechanical vertical gyroscope that directly measures the UAV's pitch and roll is included onboard, as well as the pilot input and four Analog Devices ADIS-16405 Microelectromechanical Systems (MEMS) IMUs, which are interfaced through two Netburner MOD-5213 microcontrollers logging data with OpenLog serial stream data-loggers. For postprocessing, the recorded GNSS data and data streams that are interfaced through the Netburner MOD-5213 microcontroller are synchronized by recording the state of the Pulse-Per-Second (PPS) signal from one of the GNSS receivers.

In addition to the Phastball SUAV, the experimental set-up consisted of another Novatel OEM-615 dual-frequency receiver serving as a static reference receiver with the antenna mounted on a tripod. Experimental flight-tests for this study were conducted at WVU's Jackson's Mill airfield. A typical flight pattern is shown in Figure 5.

In total, six data sets of short duration SUAV flights were collected for this study. These consisted of three flight-tests with two data sets per flight. Table 2 lists the total flight duration from take-off to landing of the three flights.

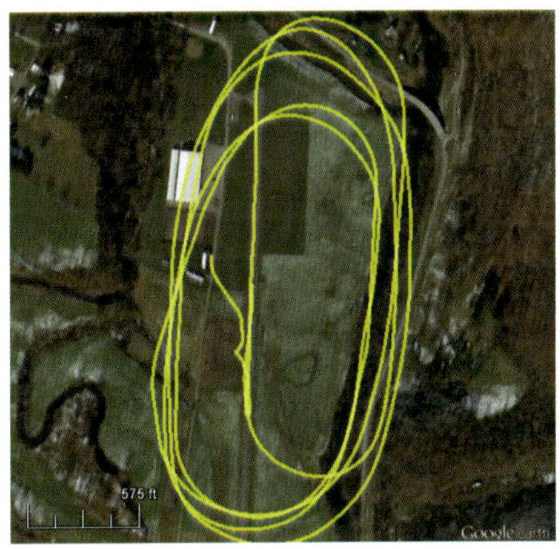

FIGURE 5: Birdseye view of typical Phastball SUAV flight at the WVU Jackson's Mill airfield. Image was generated using Google Earth.

TABLE 2: SUAV flight test durations.

Flight #	Duration (s)
1	156.2
2	342.4
3	264.0

4. Results and Discussion

4.1. Validation of Reference Position Solutions. For comparing the estimation performances of various PPP and single receiver point positioning algorithms, Kalman filter/smoother CP-DGPS solutions were generated for each

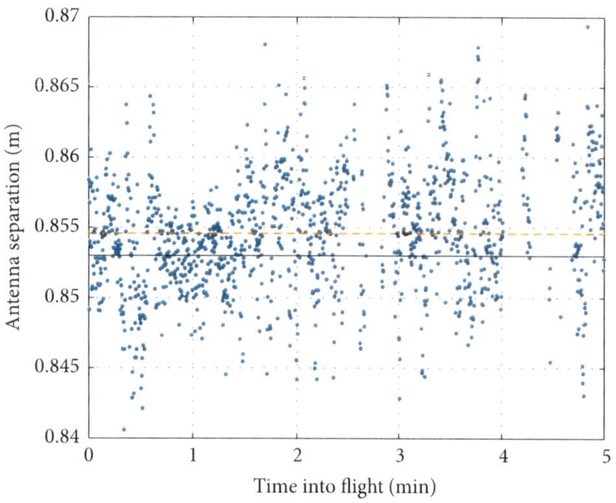

FIGURE 6: Phastball Zero SUAV GNSS antenna baseline separation as estimated by differencing the two RTK GPS solutions that are being used as reference solutions for PPP comparisons and by measuring with a tape measure.

flight-test with respect to a local GPS reference station set-up at the airfield. When processing the CP-DGPS data, in order to obtain the absolute coordinates of the base station's tripod, ~6 hours of static data was processing using GIPSY-OASIS PPP in static mode. The software that was used for generating the CP-DGPS reference solutions was RTKLIB 2.4.2.

While it is well known that double-difference integer fixed CP-DGPS solutions can obtain cm-level accuracy and precision, the Phastball flight-test configuration provides the opportunity to validate this level of accuracy and precision for the reference position solutions. In particular, the Phastball SUAV is outfitted with two GPS antennas separated by a known baseline distance of 83.5 cm. Note that, with this set-up, it is possible to further leverage the known baseline constraint between the multiple antennas to improve estimation performance of position and attitude [47], which can further be extended to improvements with an array of multiple antennas (i.e., more than 2) [48, 49]. While the array-aided PPP is an interesting research direction that the authors intend to pursue with this SUAV platform, for the present study, the goal is to assess the accuracy of single antenna PPP for SUAVs. Therefore, in order to validate the reference solutions used for each separate antenna, the two ambiguity fixed CP-DGPS solutions for each receiver on the SUAV were differenced, and the magnitude of the difference was used to evaluate how well the known antenna separation distance is estimated. Figure 5 shows an example of the result of this analysis for our second SUAV flight-test.

As shown in Figure 6, the separation between the two antennas was estimated with both cm-level accuracy and precision. Table 3 shows the antenna separation estimation error for all three flights, where sub-cm accuracy and cm-level precision were obtained.

TABLE 3: 85.3 cm antenna baseline estimation performance using the RTK reference solutions for the three SUAV flight tests considered.

Flight #	μ (cm)	σ (cm)
1	0.41	1.4
2	0.68	1.09
3	0.22	0.45

4.2. Performance Comparison Study. In order to provide a context, in addition to the PPP solution methods described in Section 2, position solutions obtained by processing the pseudorange data only with the GPS broadcast ephemeris using both the GIPSY software and the RTKLIB software are presented. These solutions represent the expected level of accuracy one would obtain from a GPS receiver position solution without any DGPS or PPP products. Short name identifiers for the processing strategies that were compared are shown in Table 4.

Table 5 shows the mean and standard deviation error statistics for the four processing methods considered over the six data sets, and Table 6 shows the overall average mean error and standard deviation error over the six flights, where all solutions are compared to the ambiguity fixed CP-DGPS reference solutions.

As shown in Tables 5 and 6, when using pseudorange based point positioning, 2-3 meters-level precision and accuracy can be expected. However, when adopting kinematic PPP, decimeter-level accuracy and centimeter-level precision were obtained, despite the fact that these are flight data sets with only a few minutes of observations. Further, to obtain this level of positioning performance, all the user needs to do is collect the raw GNSS measurements for postprocessing.

While the level of precision on short duration SUAV flights is centimeter-level, decimeter-level biases remain in the solution. It is expected that these are likely due to the short duration data set and the known slow convergence of PPP. However, an important insight provided by this study is that when considering short duration postprocessing applications of PPP, the slow convergence yields a constant bias, that is, when considering the error sources that must be modeled or mitigated by PPP in (1) and (2) (e.g., troposphere, orbit, clock, and phase ambiguities). Despite the fact that there is no enough data to converge upon the absolute values of these error sources, each of these error sources remains quite constant over the few minutes' time-scale. Therefore, these decimeter-level biases can easily be mitigated by starting the SUAV at a known location (e.g., by placing a receiver at a take-off location for static PPP) or simply allowing the SUAV to sit static on the runway for an initialization period.

To further substantiate why kinematic PPP with short observational periods is still very precise but biased, an additional five-minute SUAV flight that was simulated using a PPP simulation tool developed for a previous study [16] was analyzed. With this tool, all of the GNSS error sources discussed in Section 2.1 are modeled, but because they are simulated, perfect knowledge about all error sources is available for analysis. As such, the performance with respect to a perfectly known truth was assessed. Figure 7

Table 4: Short names and definitions for the PPP solution strategies evaluated.

Short name	Solution strategy description
PR2P-PP	Dual-frequency pseudorange measurement-only point positioning using broadcast ephemeris
RTKLIB-PP	Single receiver point positioning using RTKLIB with broadcast ephemeris
GD2P-PPP	Kinematic PPP using JPL final GPS orbit and clock products and the processing method described in Section 2.2
RTKLIB-PPP	Kinematic PPP using International GNSS Service combined final GPS orbit and clock products and method described in Section 2.3

Table 5: Mean and standard deviation positioning error statistics by positioning approach for each flight.

Receiver	Flight 1				Flight 2				Flight 3			
	Front		Back		Front		Back		Front		Back	
Algorithm	μ (m)	σ (m)	μ (m)	σ (m)	μ (m)	σ (m)	μ (m)	σ (m)	μ (m)	σ (m)	μ (m)	σ (m)
PR2P-PP	1.98	1.32	3.11	2.80	3.20	3.82	2.57	0.85	1.85	0.84	2.27	0.96
RTKLIB-PP	1.77	1.17	3.68	2.23	3.57	3.75	2.99	1.80	1.69	0.64	2.05	0.83
GD2P-PPP	0.61	0.038	0.91	0.20	0.63	0.03	0.99	0.082	1.391	0.029	1.33	0.026
RTKLIB-PPP	1.99	0.048	2.23	0.18	0.59	0.013	0.65	0.066	0.97	0.018	1.013	0.025

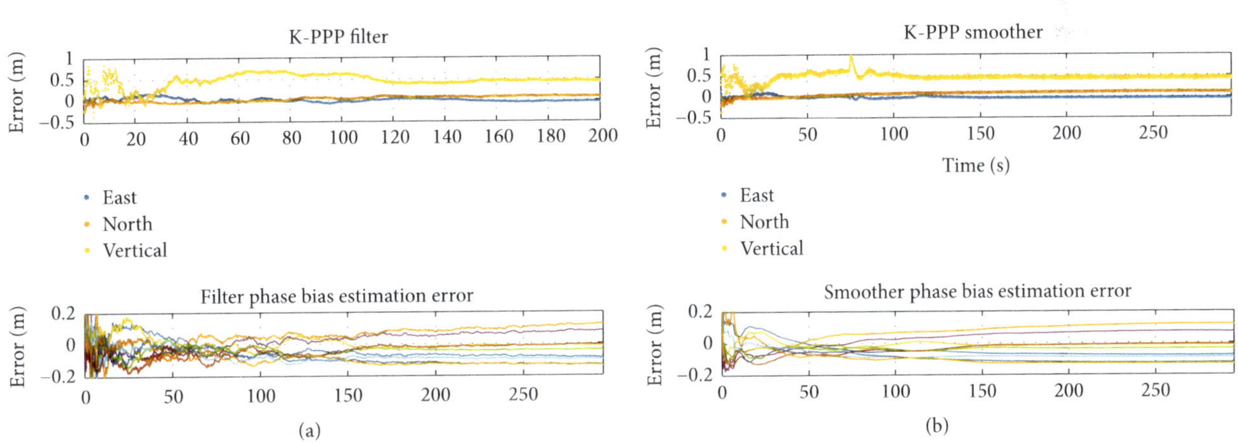

Figure 7: Simulated position and phase bias errors of a 5-minute SUAV PPP filter (a) and the smoothed position and phase bias errors (b).

Table 6: Mean and standard deviation error overall average of the 6 data sets.

Algorithm	Overall average	
	μ (m)	σ (m)
PR2P-PP	2.50	1.76
RTKLIB-PP	2.62	1.74
GD2P-PPP	0.98	0.067
RTKLIB-PPP	1.23	0.058

shows an example kinematic PPP forward filter's position solution errors and phase bias estimation errors, as well as the backwards Rauch-Tung-Striebel [50] Kalman smoothed position and phase bias errors.

As shown in Figure 7, a similar level of positioning performance as presented in the flight data analysis is shown for the simulated flight, that is, decimeter-level accuracy with centimeter-level precision. Further, the phase biases, which are estimated as random constants, do converge but maintain approximately 10 to 15 cm biases after the 5 minutes. Then, during the backwards smoother, the phase bias estimates are nearly constant, leading to considerably smoother positioning errors, as shown, but with decimeter-level biases remaining present. In particular, the mean errors between the forward filter and smoother are only reduced by 1% from 44 cm to 45 cm; however, the standard deviation of the position errors was reduced by more than 25% with a 3D σ reduction from 14.7 cm to 11 cm.

5. Conclusions

An experimental evaluation of kinematic PPP using short duration SUAV data has been presented, as such a study was missing in the literature and the PPP approach has yet to be heavily leveraged in the field robotics and SUAV communities. Through comparison with CP-DGPS reference solutions, it has been shown that approximately 6 cm 3D positioning precision with decimeter-to-meter-level 3D accuracy is obtainable when using PPP even when the flights are only a few minutes in duration. This is of benefit because PPP does not require the user to have access to a differential reference station. This result has been demonstrated with six SUAV flight data sets and two popular GNSS processing software packages. The results of this study are of benefit to many potential SUAV science applications that require precise positioning.

Competing Interests

The authors declare no conflict of interests with the publication of this paper.

Acknowledgments

This research was partially supported by the US National Geospatial Intelligence Agency Academic Research Program (NARP) Grant no. HM0476-15-1-0004. Approved for public release, case no. 16-195.

References

[1] J. F. Zumberge, M. B. Heflin, D. C. Jefferson, M. M. Watkins, and F. H. Webb, "Precise point positioning for the efficient and robust analysis of GPS data from large networks," *Journal of Geophysical Research B: Solid Earth*, vol. 102, no. 3, pp. 5005–5017, 1997.

[2] J. Kouba and P. Héroux, "Precise point positioning using IGS orbit and clock products," *GPS Solutions*, vol. 5, no. 2, pp. 12–28, 2001.

[3] "The global differential gps system," http://www.gdgps.net/.

[4] K. Dixon, "StarFire™: a global SBAS for sub-decimeter precise point positioning," in *Proceedings of the ION GNSS 19th International Technical Meeting of the Satellite Division*, pp. 26–29, Fort Worth, Tex, USA, September 2006.

[5] A. Jokinen, C. Ellum, J. Neumann et al., "Kinematic performance of NovAtel CORRECT with terrastar-D Precise Point Positioning (PPP) service," in *Proceedings of the 27th International Technical Meeting of the Satellite Division of the Institute of Navigation (ION GNSS '14)*, pp. 1020–1034, Tampa, Fla, USA, September 2014.

[6] IGS, "Real-time service," 2016, http://www.igs.org/rts.

[7] O. L. Colombo, A. W. Sutter, and A. G. Evans, "Evaluation of precise, kinematic gps point positioning," in *Proceedings of the ION GNSS-2004 Meeting*, Long Beach, Calif, USA, 2004.

[8] M. Honda, M. Murata, and Y. Mizukura, "GPS precise point positioning methods using IGS products for vehicular navigation application," in *Proceedings of the SICE-ICASE International Joint Conference*, pp. 2212–2217, IEEE, Busan, South Korea, October 2006.

[9] X. Zhang and R. Forsberg, "Assessment of long-range kinematic GPS positioning errors by comparison with airborne laser altimetry and satellite altimetry," *Journal of Geodesy*, vol. 81, no. 3, pp. 201–211, 2007.

[10] S. Bisnath and Y. Gao, "Current state of precise point positioning and future prospects and limitations," in *Observing Our Changing Earth*, pp. 615–623, Springer, 2009.

[11] C. Cai, Y. Gao, L. Pan, and J. Zhu, "Precise point positioning with quad-constellations: GPS, BeiDou, GLONASS and Galileo," *Advances in Space Research*, vol. 56, no. 1, pp. 133–143, 2015.

[12] C. O. Yigit, V. Gikas, S. Alcay, and A. Ceylan, "Performance evaluation of short to long term GPS, GLONASS and GPS/GLONASS post-processed PPP," *Survey Review*, vol. 46, no. 336, pp. 155–166, 2014.

[13] Y. Zhang and Y. Gao, "Integration of INS and un-differenced GPS measurements for precise position and attitude determination," *Journal of Navigation*, vol. 61, no. 1, pp. 87–97, 2008.

[14] J. Gross, R. Watson, V. Sivaneri, Y. Bar-Sever, W. Bertiger, and B. Haines, "Integration of inertial navigation into real-time gipsy-x (rtgx)," in *Proceedings of the 28th International Technical Meeting of The Satellite Division of the Institute of Navigation (ION GNSS+ '15)*, pp. 2560–2569, 2015.

[15] T. M. Damiani, A. Bilich, and G. L. Mader, "Evaluating aircraft positioning methods for airborne gravimetry: results from GRAV-D's Kinematic GPS Processing Challenge," in *Proceedings of the 26th International Technical Meeting of the Satellite Division of the Institute of Navigation (ION GNSS+ '13)*, pp. 3489–3507, Nashville, Tenn, USA, September 2013.

[16] R. M. Watson, V. Sivaneri, and J. N. Gross, "Performance characterization of tightly-coupled GNSS precise point positioning inertial navigation within a simulation environment," in *Proceedings of the AIAA Guidance, Navigation, and Control Conference*, p. 1869, 2016.

[17] Y. Lin, J. Hyyppä, and A. Jaakkola, "Mini-UAV-borne LIDAR for fine-scale mapping," *IEEE Geoscience and Remote Sensing Letters*, vol. 8, no. 3, pp. 426–430, 2011.

[18] L. Wallace, "Assessing the stability of canopy maps produced from UAV-LiDAR data," in *Proceedings of the 33rd IEEE International Geoscience and Remote Sensing Symposium (IGARSS '13)*, pp. 3879–3882, IEEE, Melbourne, Australia, July 2013.

[19] L. Wallace, A. Lucieer, and C. S. Watson, "Evaluating tree detection and segmentation routines on very high resolution UAV LiDAR data," *IEEE Transactions on Geoscience and Remote Sensing*, vol. 52, no. 12, pp. 7619–7628, 2014.

[20] C. Leuschen, J.-B. Yan, A. Mahmood et al., "UAV-based radar sounding of antarctic ice," in *Proceedings of the EGU General Assembly Conference Abstracts*, vol. 16 of *EGU General Assembly Conference Abstracts*, p. 15542, May 2014.

[21] S. Hensley, K. Wheeler, G. Sadowy et al., "The UAVSAR instrument: description and first results," in *Proceedings of the IEEE Radar Conference (RADAR '08)*, pp. 1–6, Rome, Italy, May 2008.

[22] P. Misra and P. Enge, *Global Positioning System: Signals, Measurements and Performance*, Ganga-Jamuna Press, Lincoln, Mass, USA, 2nd edition, 2006.

[23] P. D. Groves, *Principles of GNSS, Inertial, and Multisensor Integrated Navigation Systems*, Artech House, 2013.

[24] Y. E. Bar-Sever, "A new model for GPS yaw attitude," *Journal of Geodesy*, vol. 70, no. 11, pp. 714–723, 1996.

[25] O. Montenbruck, R. Schmid, F. Mercier et al., "GNSS satellite geometry and attitude models," *Advances in Space Research*, vol. 56, no. 6, pp. 1015–1029, 2015.

[26] D. Laurichesse, F. Mercier, J.-P. Berthias, P. Broca, and L. Cerri, "Integer ambiguity resolution on undifferenced GPS phase measurements and its application to PPP and satellite precise orbit determination," *Navigation*, vol. 56, no. 2, pp. 135–149, 2009.

[27] W. Bertiger, S. D. Desai, B. Haines et al., "Single receiver phase ambiguity resolution with GPS data," *Journal of Geodesy*, vol. 84, no. 5, pp. 327–337, 2010.

[28] J.-C. Juang, "On GPS positioning and integrity monitoring," *IEEE Transactions on Aerospace and Electronic Systems*, vol. 36, no. 1, pp. 327–336, 2000.

[29] "Gnss-inferred positioning system and orbit analysis simulation software," https://gipsy-oasis.jpl.nasa.gov/.

[30] T. P. Yunck, W. I. Bertiger, S. C. Wu et al., "First assessment of GPS-based reduced dynamic orbit determination on TOPEX/poseidon," *Geophysical Research Letters*, vol. 21, no. 7, pp. 541–544, 1994.

[31] B. Haines, Y. Bar-Sever, W. Bertiger, S. Desai, and P. Willis, "One-centimeter orbit determination for jason-1: new GPS-based strategies," *Marine Geodesy*, vol. 27, no. 1-2, pp. 299–318, 2004.

[32] R. Kroes, O. Montenbruck, W. Bertiger, and P. Visser, "Precise GRACE baseline determination using GPS," *GPS Solutions*, vol. 9, no. 1, pp. 21–31, 2005.

[33] S. Desai, W. Bertiger, J. Gross et al., "Results from the reanalysis of global gps data in the igs08 reference frame," in *Proceedings of the Fall AGU Meeting*, San Francisco, Calif, USA, 2011.

[34] G. Blewitt, "An automatic editing algorithm for GPS data," *Geophysical Research Letters*, vol. 17, no. 3, pp. 199–202, 1990.

[35] A. E. Niell, "Global mapping functions for the atmosphere delay at radio wavelengths," *Journal of Geophysical Research: Solid Earth*, vol. 101, no. 2, pp. 3227–3246, 1996.

[36] J. Boehm, R. Heinkelmann, and H. Schuh, "Short note: a global model of pressure and temperature for geodetic applications," *Journal of Geodesy*, vol. 81, no. 10, pp. 679–683, 2007.

[37] T. Takasu and A. Yasuda, "Development of the low-cost rtk-gps receiver with an open source program package rtklib," in *Proceedings of the International Symposium on GPS/GNSS*, pp. 4–6, International Convention Centre, Jeju, Korea, 2009.

[38] T. Takasu, *Rtklib: An Open Source Program Package for Gnss Positioning*, 2011.

[39] P. J. G. Teunissen, "The least-squares ambiguity decorrelation adjustment: a method for fast GPS integer ambiguity estimation," *Journal of Geodesy*, vol. 70, no. 1-2, pp. 65–82, 1995.

[40] P. Joosten, *The Lambda-Method: Matlabtm Implementation*, Delft University of Technology, Delft, The Netherlands, 2001.

[41] P. J. G. Teunissen, "An optimality property of the integer least-squares estimator," *Journal of Geodesy*, vol. 73, no. 11, pp. 587–593, 1999.

[42] P. J. G. Teunissen and S. Verhagen, "The gnss ambiguity ratio-test revisited: a better way of using it," *Survey Review*, vol. 41, no. 312, pp. 138–151, 2009.

[43] Y. Gu, F. Barchesky, H. Chao, J. Gross, and M. Napolitano, "Avionics design for a sub-scale fault-tolerant flight control test-bed," in *Recent Advances in Aircraft Technology*, InTech, Rijeka, Croatia, 2012.

[44] M. Rhudy, J. Gross, Y. Gu, and M. R. Napolitano, "Fusion of GPS and redundant IMU data for attitude estimation," in *Proceedings of the AIAA Guidance, Navigation, and Control Conference*, Minneapolis, Minn, USA, August 2012.

[45] H. Chao, Y. Gu, J. Gross, G. Guo, M. L. Fravolini, and M. R. Napolitano, "A comparative study of optical flow and traditional sensors in UAV navigation," in *Proceedings of the 1st American Control Conference (ACC '13)*, pp. 3858–3863, IEEE, Washington, DC, USA, June 2013.

[46] M. B. Rhudy, Y. Gu, H. Chao, and J. N. Gross, "Unmanned aerial vehicle navigation using wide-field optical flow and inertial sensors," *Journal of Robotics*, vol. 2015, Article ID 251379, 12 pages, 2015.

[47] P. J. G. Teunissen, "Integer least-squares theory for the GNSS compass," *Journal of Geodesy*, vol. 84, no. 7, pp. 433–447, 2010.

[48] G. Giorgi, P. J. G. Teunissen, S. Verhagen, and P. J. Buist, "Instantaneous ambiguity resolution in global-navigation-satellite-system-based attitude determination applications: a multivariate constrained approach," *Journal of Guidance, Control, and Dynamics*, vol. 35, no. 1, pp. 51–67, 2012.

[49] G. Giorgi, P. J. G. Teunissen, S. Verhagen, and P. J. Buist, "Testing a new multivariate GNSS carrier phase attitude determination method for remote sensing platforms," *Advances in Space Research*, vol. 46, no. 2, pp. 118–129, 2010.

[50] H. E. Rauch, C. T. Striebel, and F. Tung, "Maximum likelihood estimates of linear dynamic systems," *AIAA Journal*, vol. 3, no. 8, pp. 1445–1450, 1965.

12

Online Adaptive Error Compensation SVM-Based Sliding Mode Control of an Unmanned Aerial Vehicle

Kaijia Xue, Congqing Wang, Zhiyu Li, and Hanxin Chen

College of Automation, Nanjing University of Aeronautics and Astronautics, Nanjing 210016, China

Correspondence should be addressed to Kaijia Xue; 464120586@qq.com

Academic Editor: Kenneth M. Sobel

Unmanned Aerial Vehicle (UAV) is a nonlinear dynamic system with uncertainties and noises. Therefore, an appropriate control system has an obligation to ensure the stabilization and navigation of UAV. This paper mainly discusses the control problem of quad-rotor UAV system, which is influenced by unknown parameters and noises. Besides, a sliding mode control based on online adaptive error compensation support vector machine (SVM) is proposed for stabilizing quad-rotor UAV system. Sliding mode controller is established through analyzing quad-rotor dynamics model in which the unknown parameters are computed by offline SVM. During this process, the online adaptive error compensation SVM method is applied in this paper. As modeling errors and noises both exist in the process of flight, the offline SVM one-time mode cannot predict the uncertainties and noises accurately. The control law is adjusted in real-time by introducing new training sample data to online adaptive SVM in the control process, so that the stability and robustness of flight are ensured. It can be demonstrated through the simulation experiments that the UAV that joined online adaptive SVM can track the changing path faster according to its dynamic model. Consequently, the proposed method that is proved has the better control effect in the UAV system.

1. Introduction

Unmanned Aerial Vehicle (UAV) has drawn a great attention from researchers in the past few years. The characteristics of quad-rotor UAV, such as novel appearance, simple structure, low cost, excellent performance, and a unique way of flight control [1, 2], have enabled it to be the new research hotspot all over the world [3–6].

Quad-rotor UAV consists of four independent propellers attached at each corner of a cross frame. It is relied on the resultant forces and moments of four independent rotors that adjust the flight attitude and position [7]. Compared with the fixed wing aircraft, the quad-rotor UAVs that are equipped with vertical take-off have better landing capabilities even in the narrow space, because this vertical take-off has higher manoeuvrability and hovering capacities [8].

Quad-rotor UAV has all the advantages of unmanned rotor aerial vehicle; thus its application prospect is limitless. UAV has been used more and more in civilian such as rescue missions [9], forest fire surveillance [10], disaster relief support [11], exploration of high buildings or contaminated areas [12], aerial photography [13], area mapping [14], inspection of transmission lines and bridge [15], and assessment of water status [16].

However, the 6-DOF air-frame dynamic of quad-rotor UAV is nonlinear and coupled with each other. In addition, an underactuated system for the aircraft is equipped with four actuators, whose number is smaller than the degrees of freedom. Therefore, it is comparatively difficult to design an excellent controller for aircraft due to the reasons mentioned above.

The conventional sliding mode control (SMC), as a kind of nonlinear system control method, possesses the advantages of wide applicability, fine dynamic characteristics, and insensitivity to interference changes. This research on the sliding model control mainly deals with adaptive sliding model control [17, 18], robust sliding mode control [19], optimal sliding mode control [20], fuzzy slidingmode control

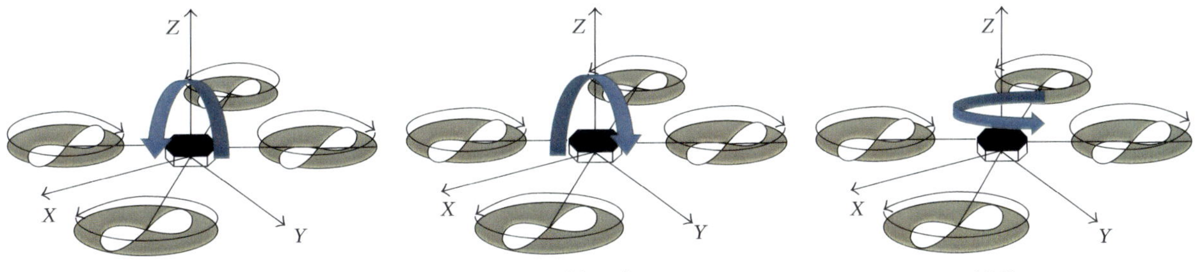

FIGURE 1: Three basic movements of quad-rotor UAV.

[21, 22], neural network sliding mode control [23, 24], and sliding mode control combined with other intelligent controls [25].

To achieve the effective sliding mode control of quad-rotor UAV, the mathematical model of various flight conditions must be designed accurately. Nevertheless, during the process of flight, UAVs are affected by various physical attributes such as aerodynamic force, gravity, gyroscopic effect, and the rotor inertia moment. It is also very vulnerable to the interference of the external environment like airflow. According to these reasons, it is quite difficult to establish a reliable dynamic model. Furthermore, the size and the weight of the rotor, the degree of deformation, and parameters of aerodynamic performance will also directly affect the precision of the model.

In consideration of these problems, how to make the estimator more accurate is very important. Hence the system can reach the sliding surface rapidly in coordination with a sliding mode control law.

Standard offline support vector machine (SVM) is a machine learning based on statistical theory. It can solve small-sample, nonlinear, and high dimension problems by using structural risk minimization. This method can efficiently supply the details without considering the changes of modeling the quad-rotor system. But it only focuses on training sample data within the limited time and its estimated result is predicted one time; hence there are still many dynamic variations that cannot be predicted during the process of flight. The main contribution of this paper is to compensate unknown parameters and to reduce the dynamic error by using online adaptive SVM for the quad-rotor system.

This paper presents a method that standard SVM is combined with online adaptive error compensation SVM to design a sliding mode controller. Through this way, the weights of SVM can be adjusted adaptively to an optimal status. The rest of the paper is divided into five sections. The nonlinear dynamic model of the quad-rotor is described in Section 2. The design procedure of sliding mode controller using the online adaptive error compensation SVM is presented in Section 3. Section 4 analyzes the stability of online adaptive error compensation SVM. Simulation results are shown in Section 5. Section 6 draws the concluding remarks. All in all, this paper proves the effectiveness of this method through the simulation results of the quad-rotor system.

2. Dynamic Model Analysis of Quad-Rotor UAV

The quad-rotor UAV is an underactuated system as it has six degrees of freedom with only four inputs. Pitch movement of quad-rotor UAV is motivated by reducing two motors' speed on either front or back side. And an equivalent increase of the two motors' speed on another side is required simultaneously. According to the symmetrical structure of quad-rotor UAV, the adjustment mode of the roll movement is similar to the pitch movement.

Yaw movement of quad-rotor UAV is realized through reducing one diagonal motors' speed, and equally increasing the two motors' speed of another diagonal at the same time. Based on this theory, the reduced speed on motor torque of the quad-rotor UAV is less than the speed increased; therefore, it makes the aircraft body present in a framework of yaw movement. Three basic movements of quad-rotor are shown in Figure 1.

Two coordinate systems are defined before the dynamic model is established, one for the ground coordinate system and the other for quad-rotor coordinate system. The ground coordinate system is a static coordinate system that relates to a reference ground point. The flight speed and displacement of quad-rotor UAV in different directions are mainly judged by aiming at the ground coordinate system. Pitch angle, roll angle, and yaw angle are mainly considered from the direction of the quad-rotor coordinate system which is based on the geometry center of the aircraft body. The speed of the aircraft under the body coordinate system is described as the following relationship:

$$V = \begin{bmatrix} u \\ v \\ w \end{bmatrix} = \begin{bmatrix} R_1 & R_2 & R_3 \end{bmatrix} \begin{bmatrix} v_{qx} \\ v_{qy} \\ v_{qz} \end{bmatrix} = \begin{bmatrix} R_1 & R_2 & R_3 \end{bmatrix} \begin{bmatrix} \dot{x} \\ \dot{y} \\ \dot{z} \end{bmatrix}. \quad (1)$$

As shown in Figure 2, $V = [u, v, w]^T$ is the velocity vector of the plane with respect to the ground coordinate system and $[v_{qx}, v_{qy}, v_{qz}]^T$ is the velocity vector with respect to the quad-rotor coordinate system. The transformation matrix $R = [R_1, R_2, R_3]^T$ can be denoted as follows:

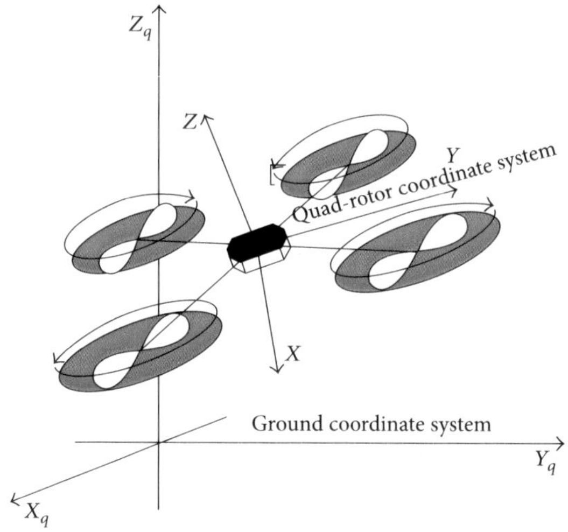

FIGURE 2: Relationship between the quad-rotor coordinate system and the ground coordinate system.

$$R = \begin{bmatrix} \cos\phi\cos\theta & -\sin\phi\cos\varphi + \cos\phi\sin\theta\sin\varphi & \sin\varphi\sin\phi + \cos\phi\sin\theta\cos\varphi \\ \sin\phi\cos\theta & \cos\phi\cos\varphi + \sin\phi\sin\theta\sin\varphi & -\cos\phi\sin\varphi + \sin\phi\sin\theta\cos\varphi \\ -\sin\theta & \cos\theta\sin\varphi & \cos\theta\cos\varphi \end{bmatrix}, \quad (2)$$

where ϕ is the pitch angle, θ is the roll angle, and φ is the yaw angle. According to Newton's second law, the quad-rotor motion equation of ground coordinate system can be written as

$$\frac{d(mV)}{dt} = (F_G + F_T + F_R) = F, \quad (3)$$

where the response of quad-rotor UAV produced by gravity is

$$F_G = mgG_1, \quad (4)$$

where m is the mass of UAV and g is the acceleration of gravity. Each propeller generates vertical lifting force, so the overall lifting force F_T will be

$$F_T = k_b \sum_{i=1}^{4} \Omega_i^2 G_2 = TG_2, \quad (5)$$

where the k_b is lifting force coefficient, $T = k_b(\Omega_1^2 + \Omega_2^2 + \Omega_3^2 + \Omega_4^2)$, and Ω_i means the rotation rate of the ith propeller. G_1 and G_2 are transfer matrices. Air resistance of quad-rotor UAV is expressed as

$$F_R = [-k_{R1}u, -k_{R2}v, -k_{R3}w]^T. \quad (6)$$

Let $G_1 = G_2 = [0, 0, 1]^T$; by taking (3), (4), and (5) into (6), then the motion equations of quad-rotor UAV can be written as

$$\begin{bmatrix} \dot{u} \\ \dot{v} \\ \dot{w} \end{bmatrix} = \begin{bmatrix} -\frac{T}{m}(\sin\phi\sin\varphi + \cos\phi\cos\theta\cos\varphi) - k_{R1}\dot{x} \\ -\frac{T}{m}(-\cos\phi\sin\varphi + \sin\phi\sin\theta\cos\varphi) - k_{R2}\dot{y} \\ g - \frac{T}{m}(\cos\varphi\cos\theta) - k_{R3}\dot{z} \end{bmatrix}. \quad (7)$$

Attitude foundation model can cooperate with the two coordinate systems from (7) by establishing the aircraft angular velocity. On the basis of conversion relations between the aircraft coordinate system and the ground coordinate system, the following quad-rotor attitude kinematics equation can be got:

$$\begin{bmatrix} \dot{\phi} \\ \dot{\theta} \\ \dot{\psi} \end{bmatrix} = \begin{bmatrix} 1 & \sin\phi\tan\theta & \cos\phi\tan\theta \\ 0 & \cos\phi & -\sin\phi \\ 0 & \frac{\sin\phi}{\cos\theta} & \frac{\cos\phi}{\cos\theta} \end{bmatrix} \begin{bmatrix} p \\ q \\ r \end{bmatrix}. \quad (8)$$

According to the moment of momentum theorem, the aircraft torque equation of mass motion can be derived by Newton-Euler equation:

$$\frac{d(J\omega)}{dt} = (M_R + M_G + M_D + M_d) = M, \tag{9}$$

where $\omega = [p, q, r]^T$ is the rotation rate of quad-rotor UAV based on the aircraft coordinate system. $J = \text{diag}[J_x, J_y, J_z]$ means the moment of inertia of quad-rotor UAV. M_R, M_G, M_D, and M_d represent rotor lift, gyroscopic effect generated by the rotor, air resistance torque, and external disturbances torque, respectively. M is overall external torque of quad-rotor UAV.

The torque comes from four rotor wings. The relationship between air dynamic torque and four motors' speeds Ω_1, Ω_2, Ω_3, and Ω_4 is

$$M_R = \begin{bmatrix} \tau_\phi \\ \tau_\theta \\ \tau_\psi \end{bmatrix} = \begin{bmatrix} -\frac{\sqrt{2}}{2}lk_b & \frac{\sqrt{2}}{2}lk_b & \frac{\sqrt{2}}{2}lk_b & -\frac{\sqrt{2}}{2}lk_b \\ \frac{\sqrt{2}}{2}lk_b & \frac{\sqrt{2}}{2}lk_b & -\frac{\sqrt{2}}{2}lk_b & -\frac{\sqrt{2}}{2}lk_b \\ k_d & -k_d & k_d & -k_d \end{bmatrix} \begin{bmatrix} \Omega_1^2 \\ \Omega_2^2 \\ \Omega_3^2 \\ \Omega_4^2 \end{bmatrix}, \tag{10}$$

where τ_ϕ, τ_θ, and τ_φ mean the three-axis torques caused by brush-less motors, l is the distance from the motors to the center of the aircraft, k_b is the lift coefficient, and k_d is the torque coefficient. The torque of the gyroscopic effect M_G is

$$M_G = \begin{bmatrix} \tau_{G\phi} \\ \tau_{G\theta} \\ \tau_{G\psi} \end{bmatrix} = \begin{bmatrix} J_{rx}\Omega' q \\ -J_{ry}\Omega' p \\ 0 \end{bmatrix}, \tag{11}$$

where $\Omega' = \Omega_1 - \Omega_2 + \Omega_3 - \Omega_4$ and J_{rx}, J_{ry} are the rotor moments of inertia that rotate around of the motor shaft. The air resistance torque vector M_D is

$$M_D = \begin{bmatrix} \tau_{D\phi} \\ \tau_{D\theta} \\ \tau_{D\psi} \end{bmatrix} = k_D\omega = \begin{bmatrix} k_{Dx}p \\ k_{Dy}q \\ k_{Dz}r \end{bmatrix}, \tag{12}$$

where $k_D = \text{diag}[k_{Dx}, k_{Dx}, k_{Dx}]$ and k_{Dx}, k_{Dx}, and k_{Dx} are moment coefficients of three-axis air resistance in the aircraft coordinate system. Finally (13) can be obtained by

$$\begin{bmatrix} \dot{p} \\ \dot{q} \\ \dot{r} \end{bmatrix} = \begin{bmatrix} \frac{J_y - J_z}{J_x}qr + \frac{\tau_{G\phi}}{J_x} + \frac{\tau_\phi}{J_x} + \frac{\tau_{D\phi}}{J_x} \\ \frac{J_z - J_x}{J_y}pr + \frac{\tau_{G\theta}}{J_y} + \frac{\tau_\theta}{J_y} + \frac{\tau_{D\theta}}{J_y} \\ \frac{J_x - J_y}{J_z}pq + \frac{\tau_{G\psi}}{J_z} + \frac{\tau_\psi}{J_z} + \frac{\tau_{D\psi}}{J_z} \end{bmatrix}. \tag{13}$$

3. Sliding Mode Controller Design of Unmanned Aerial Vehicle

In this section, the purpose is to carry out the design procedure of basic sliding mode controller. Firstly, sliding surface s is established according to the attitude error that satisfies the necessary sliding condition $s\dot{s} < 0$. Secondly, integrate sliding mode control law with equivalent control part and switch control part will be presented. If the attitude angle variation is small and the speed is slow when the aircraft is hovering, a simplification can be made as $[\dot{\phi}, \dot{\theta}, \dot{\psi}]^T \approx [p, q, r]^T$. The quad-rotor UAV attitude model with sliding mode controller can be written as

$$\begin{aligned} \dot{X} &= X_1, \\ \dot{X}_1 &= \mathbf{f}(X) + \mathbf{g}(X)\mathbf{u} + \Delta\mathbf{f} + \mathbf{d}. \end{aligned} \tag{14}$$

In this equation, $X = [x, y, z, \phi, \theta, \psi]^T$, X_1 is the first derivative of X, and $\mathbf{f}(X)$ is

$$\mathbf{f}(X) = \begin{bmatrix} -k_{R1}\dot{x} \\ -k_{R2}\dot{y} \\ g - k_{R3}\dot{z} \\ \frac{J_y - J_z}{J_x}qr + \frac{J_{rz}}{J_x}\Omega' q + \frac{k_{Dx}p}{J_x} \\ \frac{J_z - J_x}{J_y}pr - \frac{J_{rz}}{J_y}\Omega' p + \frac{k_{Dy}q}{J_x} \\ \frac{J_x - J_y}{J_z}pq + \frac{k_{Dz}r}{J_x} \end{bmatrix}. \tag{15}$$

$\mathbf{g}(X)$ is the input matrix, $\Delta\mathbf{f} + \mathbf{d} = [d_x, d_y, d_z, d_\phi, d_\theta, d_\psi]^T$, $\Delta\mathbf{f} \in R^6$ is an uncertainty vector, and $\mathbf{d} \in R^6$ is a noise vector. U_1, U_2, U_3, and U_4 as the inputs of the quad-rotor are adjusted by four motor rotor speeds. Control input is selected as

$$\mathbf{u} = \begin{bmatrix} -\frac{1}{m}U_1U_x \\ -\frac{1}{m}U_1U_y \\ \frac{1}{m}(\cos\phi\cos\theta U_1) \\ \frac{l}{J_x}U_2 \\ \frac{l}{J_y}U_3 \\ \frac{1}{J_z}U_4 \end{bmatrix}. \tag{16}$$

3.1. Sliding Mode Controller of Quad-Rotor UAV. Let the desired state of quad-rotor be $X_d = [x_d, y_d, z_d, \phi_d, \theta_d, \psi_d]^T$. Then the state tracking error of the system $\mathbf{e}_X^{(\alpha)}$ can be described as follows:

$$\mathbf{e}_X^{(\alpha)} = X_d^{(\alpha)} - X^{(\alpha)}, \tag{17}$$

where $\mathbf{e}_X^{(\alpha)} \in R^6$ and α is the factorial number. Sliding surface equation is given by

$$\mathbf{s}_X = \mathbf{e}_X^{(\alpha)} + \sum_{i=1}^{\alpha-1} \lambda_{X(\alpha-i)} \frac{d^i \mathbf{e}_X}{dt^i} + \lambda_{X\alpha} \mathbf{e}_X$$
$$\text{s.t.} \quad \mathbf{s}_X \dot{\mathbf{s}}_X < -\beta_X \left|\mathbf{s}_X\right| < 0. \tag{18}$$

Because $\lambda_{X(\alpha-i)}$ conforms to Hurwitz stable polynomial, $\beta_X > 0$, the sliding mode motion is asymptotically stable. The input includes two parts can be written as

$$U_{\mathrm{eg}} = \mathbf{g}(X)^{-1} \left[-\mathbf{f}(X) - \Delta \mathbf{f}(X) + \frac{d^{(\alpha+1)} X_d}{dt^{(\alpha+1)}} - \sum_{i=1}^{\alpha-1} \lambda_{X(\alpha-i)} \frac{d^{i+1} \mathbf{e}_X}{dt^{i+1}} - \lambda_{X\alpha} \frac{d\mathbf{e}_X}{dt} \right],$$
$$U_{\mathrm{sw}} = -\mathbf{g}(X)^{-1} \mathbf{k} \operatorname{sign}(s_X), \tag{19}$$

where U_{eg} is the equivalent control part and U_{sw} is the switch part for the purpose of suppressing systematic disturbances. Let $\mathbf{k} = \mathbf{d} + \boldsymbol{\eta}$, where $\boldsymbol{\eta}$ is greater than zero.

It can be written as $\mathbf{u} = U_{\mathrm{eg}} + U_{\mathrm{sw}}$. Similarly, U_2, U_3, and U_4 can be obtained by

$$U_2 = \frac{J_x}{l} \left[-\frac{J_y - J_z}{J_x} qr - \frac{J_{rz}}{J_x} \Omega' q - \frac{k_{Dx} p}{J_x} - d_\phi + \lambda_{\phi 1} \frac{d^2 \phi_d}{dt^2} - \lambda_{\phi 2} \frac{de_\phi}{dt} \right] - \frac{J_x}{l} k_\phi \operatorname{sign}(s_\phi),$$
$$U_3 = \frac{J_y}{l} \left[-\frac{J_z - J_x}{J_y} pr + \frac{J_{rz}}{J_y} \Omega' p - \frac{k_{Dy} q}{J_x} - d_\theta + \lambda_{\theta 1} \frac{d^2 \theta_d}{dt^2} - \lambda_{\theta 2} \frac{de_\theta}{dt} \right] - \frac{J_y}{l} k_\theta \operatorname{sign}(s_\theta) \tag{20}$$
$$U_4 = J_z \left[-\frac{J_x - J_y}{J_z} pq - \frac{k_{Dz} r}{J_x} - d_\psi + \lambda_{\psi 1} \frac{d^2 \psi_d}{dt^2} - \lambda_{\psi 2} \frac{de_\psi}{dt} \right] - J_z k_\psi \operatorname{sign}(s_\psi),$$

where $k_\phi > 0$, $k_\theta > 0$, and $k_\psi > 0$. This system will tend toward origin point along the sliding surface s by the action of the sliding mode control inputs.

The sliding mode controller receives the difference between desired position (x_d, y_d, z_d) and the real-time position (x_t, y_t, z_t) of quad-rotor UAV. After calculating the control input U_1, the two virtual control inputs U_x, U_y can be obtained. Virtual controllers are joined into the attitude algorithm module and then the desired attitude angles ϕ_d, θ_d are calculated. Further, the desired attitude angles ϕ_d, θ_d, ψ_d together with the returned attitude angles ϕ, θ, ψ are sent to the attitude control system, where they can be promoted to control the quad-rotor UAV flight attitude. Similar to the sliding mode control design for attitude system, the position errors are defined as follows:

$$e_x = x_d - x_t,$$
$$e_y = y_d - y_t, \tag{21}$$
$$e_z = z_d - z_t.$$

Derivation process is similar to (18) and (19) and the sliding mode surfaces (s_x, s_y, s_z) of (x, y, z) can be got by using the position error and their derivatives. Finally, control inputs U_1, U_x, and U_y are designed, respectively, as follows:

$$U_1 = \frac{m}{(\cos\phi \cos\theta)} \left[g - k_{R3} \dot{z} + \lambda_{z1} \frac{d^2 z_d}{dt^2} - \lambda_{z2} \frac{de_z}{dt} - d_z \right] - \frac{m}{(\cos\phi \cos\theta)} k_z \operatorname{sign}(s_z),$$
$$U_x = \frac{m}{U_1} \left[-k_{R1} \dot{x} + \lambda_{x1} \frac{d^2 x_d}{dt^2} - \lambda_{x2} \frac{de_x}{dt} - d_x \right] - \frac{m}{U_1} \cdot k_x \operatorname{sign}(s_x), \tag{22}$$
$$U_y = \frac{m}{U_1} \left[-k_{R2} \dot{y} + \lambda_{y1} \frac{d^2 y_d}{dt^2} - \lambda_{y2} \frac{de_y}{dt} - d_y \right] - \frac{m}{U_1} \cdot k_y \operatorname{sign}(s_y).$$

According to (22), through the virtual control inputs U_x, U_y to get $\ddot{x}$, $\ddot{y}$, thus the desired pitch angle and roll angle are calculated. The inverse equations are given by

$$\phi_d = \arcsin \left[-\frac{(\sin\psi m\ddot{x} - \cos\psi m\ddot{y})}{T} \right],$$
$$\theta_d = \arcsin \left[-\frac{(\cos\psi m\ddot{x} + \sin\psi m\ddot{y})}{T\cos\phi} \right]. \tag{23}$$

3.2. Offline SVM Modeling of Quad-Rotor UAV with Unknown Parameters. Some nonlinear terms and parameters of a quad-rotor UAV are usually unknown, which may disturb the precision of the model. Aiming at the disadvantages of physical model, this paper adopts the offline SVM to establish the quad-rotor UAV model. For a given training sample data set $\{(\mathbf{x}_i, \mathbf{f}_i) \mid \mathbf{x}_i \in R^n,\ y_i \in R,\ i = 1, 2, \ldots, L\}$, L is the number of sample data and SVM prediction function equation can be written as

$$\mathbf{f}(\mathbf{x}) = w^T K(\mathbf{x}_i, \mathbf{x}) + b, \tag{24}$$

where $K(\mathbf{x}_i, \mathbf{x}) = \exp(-\|\mathbf{x}_i - \mathbf{x}\|^2 / 2\sigma^2)$ is a RBF kernel function, b is the intercept number, and σ is the width parameter of the RBF kernel function. According to the structural risk minimization principle

$$\begin{aligned} \min \quad & J(w) = \frac{1}{2}\langle w, w\rangle + C\sum_{ti=1}^{L}(\xi_i + \xi_i^*), \\ \text{s.t.} \quad & f(\mathbf{x}_i) - w^T K(\mathbf{x}_i, \mathbf{x}) - b \le \varepsilon + \xi_i, \\ & w^T K(\mathbf{x}_i, \mathbf{x}) + b - f(\mathbf{x}_i) \le \varepsilon + \xi_i^*, \\ & i = 1, 2, \dots, L, \end{aligned} \quad (25)$$

where C is the penalty parameter, ε is the insensitive loss function, and ξ_i, ξ_i^* are the relaxation factors. Lagrange function is established as follows to solve the above optimization problem:

$$\begin{aligned} & L(w, b, \varepsilon, \xi_i, \xi_i^*, \lambda_i, \lambda_i^*) \\ & = J(w) - \sum_{i=1}^{L}\lambda_i\left[f(\mathbf{x}_i) - w^T K(\mathbf{x}_i, \mathbf{x}) - b - \varepsilon - \xi_i\right] \\ & \quad - \sum_{i=1}^{L}\lambda_i^*\left[w^T K(\mathbf{x}_i, \mathbf{x}) + b - \varepsilon - \xi_i^* - f(\mathbf{x}_i)\right], \end{aligned} \quad (26)$$

where λ_i, λ_i^* are Lagrange multipliers. According to the KKT (Karush-Kuhn-Tucker) condition, eliminating ε, ξ_i, ξ_i^*, and w, the following SVM nonlinear function can be obtained by

$$\mathbf{f}(\mathbf{x}) = \sum_{i=1}^{m}(\lambda_i - \lambda_i^*) K(\mathbf{x}_i, \mathbf{x}) + b. \quad (27)$$

3.3. Online Adaptive SVM Error Compensations of Uncertainties and Noises. It is relatively hard to obtain a completely precise estimation of model utilizing the offline SVM once regression. Other uncertain factors such as modeling error and uncertainties in the flight process will also influence the stability of the UAV during the flight. Therefore, this paper combines the online adaptive SVM with offline SVM for compensation of modeling errors which are caused by the uncertainties and noises during the flight process. Define $D(\mathbf{x}) = \mathbf{f}(X) - \widehat{\mathbf{f}}(\mathbf{x} \mid w') + \Delta\mathbf{f} + \mathbf{d}$, where $\widehat{\mathbf{f}}(\mathbf{x} \mid w')$ represents the estimated value of $\mathbf{f}(X)$ and $w' = \lambda_i - \lambda_i^*$. According to the characteristic value $\mathbf{x}_i$ in the sample data, the predicted value $\widehat{D}_i$ with respect to the D_i is given by

$$\begin{aligned} \widehat{D}_i(\mathbf{x} \mid w_i'') &= \sum_{i=1}^{L}(\lambda_{Di} - \lambda_{Di}^*) K(\mathbf{x}_i, \mathbf{x}) + b' \\ &= \sum_{i=1}^{L} w_i'' K(\mathbf{x}_i, \mathbf{x}) + b', \end{aligned} \quad (28)$$

where λ_{Di}, λ_{Di}^* are the computed Lagrange multipliers of $\widehat{D}_i(\mathbf{x} \mid w_i'')$ and $w_i'' = \lambda_{Di} - \lambda_{Di}^*$. Then the boundary function of the ith sample can be defined as

$$\begin{aligned} h(\mathbf{x}_i \mid w_j'') &= D(\mathbf{x} \mid w_j'') - \widehat{D}(\mathbf{x} \mid w_j'') \\ &= \sum_{j=1}^{m} w_j'' K(\mathbf{x}_i, \mathbf{x}_j) + b - \widehat{D}(\mathbf{x} \mid w_j''). \end{aligned} \quad (29)$$

According to the KKT condition, a congruent relationship between $h(\mathbf{x}_i \mid w_j'')$ and w_j'' can be obtained as follows [26]:

$$\begin{aligned} & h(\mathbf{x}_i \mid w_j'') \ge +\varepsilon \quad w_j'' = -C, \\ & h(\mathbf{x}_i \mid w_j'') = +\varepsilon \quad w_j'' \in [-C, 0], \\ & h(\mathbf{x}_i \mid w_j'') \in [-\varepsilon, +\varepsilon] \quad w_j'' = -C, \\ & h(\mathbf{x}_i \mid w_j'') = -\varepsilon \quad w_j'' \in [0, +C], \\ & h(\mathbf{x}_i \mid w_j'') \le -\varepsilon \quad w_j'' = +C. \end{aligned} \quad (30)$$

According to the size of $D_i(\mathbf{x})$, it can be divided into the following three kinds of circumstances: support vector error set $E = \{i \mid |w_j''| = C\}$, support vector set $S = \{i \mid 0 < |w_j''| < C\}$, and retain sample set $R = \{i \mid w_j'' = 0\}$.

If a new sample $\mathbf{x}_l$ is added here, parameters $\Delta w''$, $\Delta b'$ can be obtained by computing. Because each sample satisfies KKT condition, the change of available boundary function value is given by

$$\begin{aligned} \Delta h(\mathbf{x}_i \mid w_j'') &= K(\mathbf{x}_i, \mathbf{x}_l)\Delta w_l'' \\ &\quad + \sum_{j=1}^{L} K(\mathbf{x}_i, \mathbf{x}_j)\Delta w_j'' + \Delta b'. \end{aligned} \quad (31)$$

And because the sum of $\Delta w''$ is always equal to zero,

$$\sum_{j\in S} w_j'' + w_l'' = 0. \quad (32)$$

According to (31) and (32), it can be found in the support vector set S as follows:

$$\sum_{j\in S} K(\mathbf{x}_i, \mathbf{x}_j)\Delta w_j'' + \Delta b = -K(\mathbf{x}_i, \mathbf{x}_l)\Delta w_l'', \quad (33)$$

$$\sum_{j\in S}\Delta w_j'' = -\Delta w_l''. \quad (34)$$

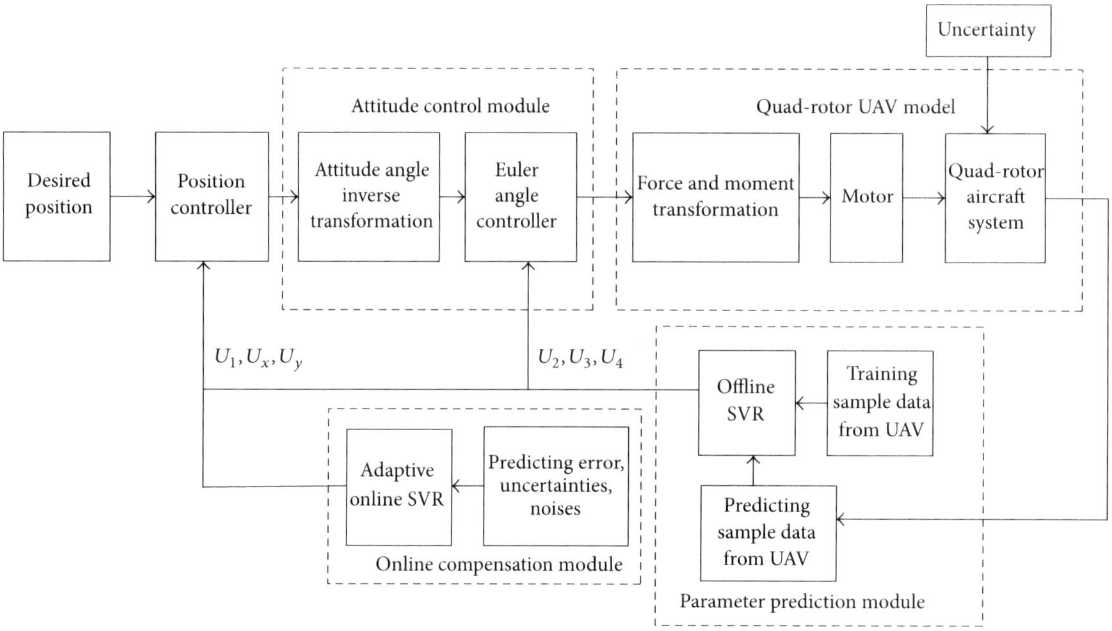

FIGURE 3: Position control system structure for quad-rotor UAV.

The support vector set is defined as $S = \{s_1, s_2, \ldots, s_L\}$. According to (33), the following matrix can be given by

$$\begin{pmatrix} \Delta b \\ \Delta w''_{s_1} \\ \vdots \\ \Delta w''_{s_m} \end{pmatrix} = -\begin{pmatrix} 0 & 1 & \cdots & 1 \\ 1 & K(\mathbf{x}_{s_1}, \mathbf{x}_{s_1}) & \cdots & K(\mathbf{x}_{s_1}, \mathbf{x}_{s_m}) \\ \vdots & \vdots & \ddots & \vdots \\ 1 & K(\mathbf{x}_{s_m}, \mathbf{x}_{s_1}) & \cdots & K(\mathbf{x}_{s_m}, \mathbf{x}_{s_m}) \end{pmatrix}^{-1} \cdot \begin{pmatrix} 1 \\ K(\mathbf{x}_{s_1}, \mathbf{x}_{s_l}) \\ \vdots \\ K(\mathbf{x}_{s_m}, \mathbf{x}_{s_l}) \end{pmatrix} \Delta w''_l = \eta \Delta w''_l, \tag{35}$$

where

$$\eta = -\begin{pmatrix} 0 & 1 & \cdots & 1 \\ 1 & K(\mathbf{x}_{s_1}, \mathbf{x}_{s_1}) & \cdots & K(\mathbf{x}_{s_1}, \mathbf{x}_{s_m}) \\ \vdots & \vdots & \ddots & \vdots \\ 1 & K(\mathbf{x}_{s_m}, \mathbf{x}_{s_1}) & \cdots & K(\mathbf{x}_{s_m}, \mathbf{x}_{s_m}) \end{pmatrix}^{-1} \cdot \begin{pmatrix} 1 \\ K(\mathbf{x}_{s_1}, \mathbf{x}_{s_l}) \\ \vdots \\ K(\mathbf{x}_{s_m}, \mathbf{x}_{s_l}) \end{pmatrix}. \tag{36}$$

It can change w''_j but cannot change the h of the support vectors set. However, it will alter h for the support vector error set and retain sample set. Assume that $N = E \cup R = \{n_1, n_2, \ldots, n_N\}$; the changes of h can be got from (31), (35), and (36)

$$\begin{pmatrix} \Delta h(\mathbf{x}_{n_1} \mid w''_{s_l}) \\ \Delta h(\mathbf{x}_{n_2} \mid w''_{s_l}) \\ \vdots \\ \Delta h(\mathbf{x}_{n_N} \mid w''_{s_l}) \end{pmatrix} = \begin{pmatrix} \Delta K(\mathbf{x}_{n_1}, \mathbf{x}_{s_l}) \\ \Delta K(\mathbf{x}_{n_2}, \mathbf{x}_{s_l}) \\ \vdots \\ \Delta K(\mathbf{x}_{n_N}, \mathbf{x}_{s_l}) \end{pmatrix} \Delta w''_l + \begin{pmatrix} 0 & K(\mathbf{x}_{n_1}, \mathbf{x}_{s_1}) & \cdots & K(\mathbf{x}_{n_1}, \mathbf{x}_{s_m}) \\ 1 & K(\mathbf{x}_{n_2}, \mathbf{x}_{s_1}) & \cdots & K(\mathbf{x}_{n_2}, \mathbf{x}_{s_m}) \\ \vdots & \vdots & \ddots & \vdots \\ 1 & K(\mathbf{x}_{n_N}, \mathbf{x}_{s_1}) & \cdots & K(\mathbf{x}_{n_N}, \mathbf{x}_{s_m}) \end{pmatrix} \eta \Delta w''_l. \tag{37}$$

So every set can be renewed and adjusted to achieve the objective that online SVM learns the noises and the uncertainties of the model through the above equations. Position control system structure diagram of the corresponding sliding mode controller is shown in Figure 3.

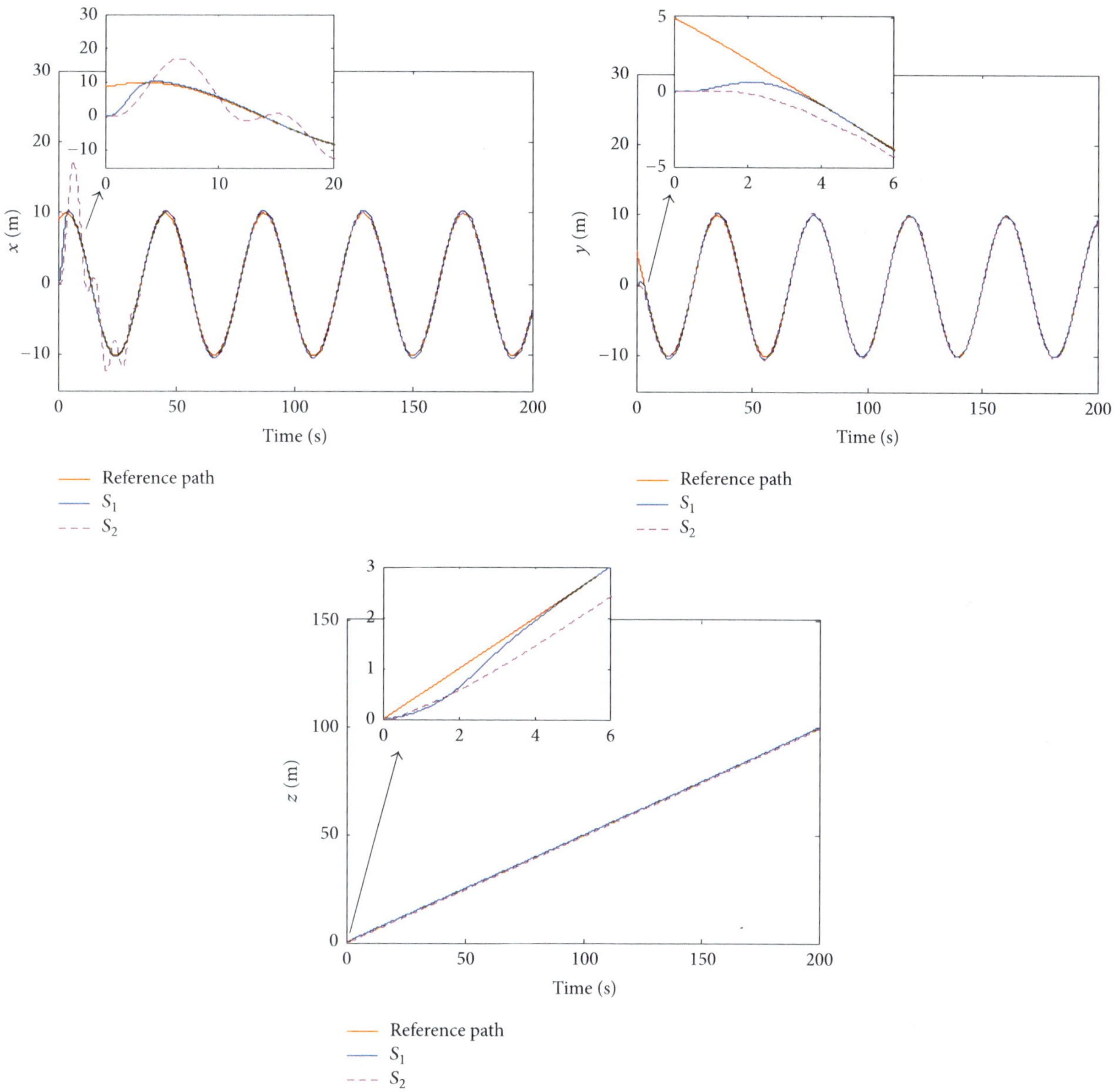

FIGURE 4: UAV flight path tracking results of the positions (x, y, z).

4. Stability Analysis of Online Adaptive Error Compensation SVM

According to (19), (27), and (28), the sliding mode controller is redesigned as

$$\mathbf{u} = \mathbf{g}(X)^{-1}\left[-\hat{\mathbf{f}}\left(X \mid w'\right) - \widehat{D}\left(X \mid w''\right) + \frac{d^{(\alpha+1)}X_d}{dt^{(\alpha+1)}} - \sum_{i=1}^{\alpha-1}\lambda_{X(\alpha-i)}\frac{d^{i+1}\mathbf{e}_X}{dt^{i+1}} - \lambda_{X\alpha}\frac{d\mathbf{e}_X}{dt} - \mathbf{k}\,\mathrm{sign}\left(s_X\right)\right]. \tag{38}$$

Then

$$\hat{\mathbf{f}}\left(X \mid w'\right) + \widehat{D}\left(X \mid w''\right) = \sum_{i=1}^{sv} w' K\left(\mathbf{x}_i, \mathbf{x}_j\right) + \sum_{i=1}^{sv'} w'' K\left(\mathbf{x}_i, \mathbf{x}_j\right), \tag{39}$$

where w' and w'' are support vector weights of offline SVM and online SVM, respectively.

Assumption 1. Outputs $\hat{\mathbf{f}}(X \mid w')$ and $\widehat{D}(X \mid w'')$ of offline SVM and online SVM are continuous and bounded.

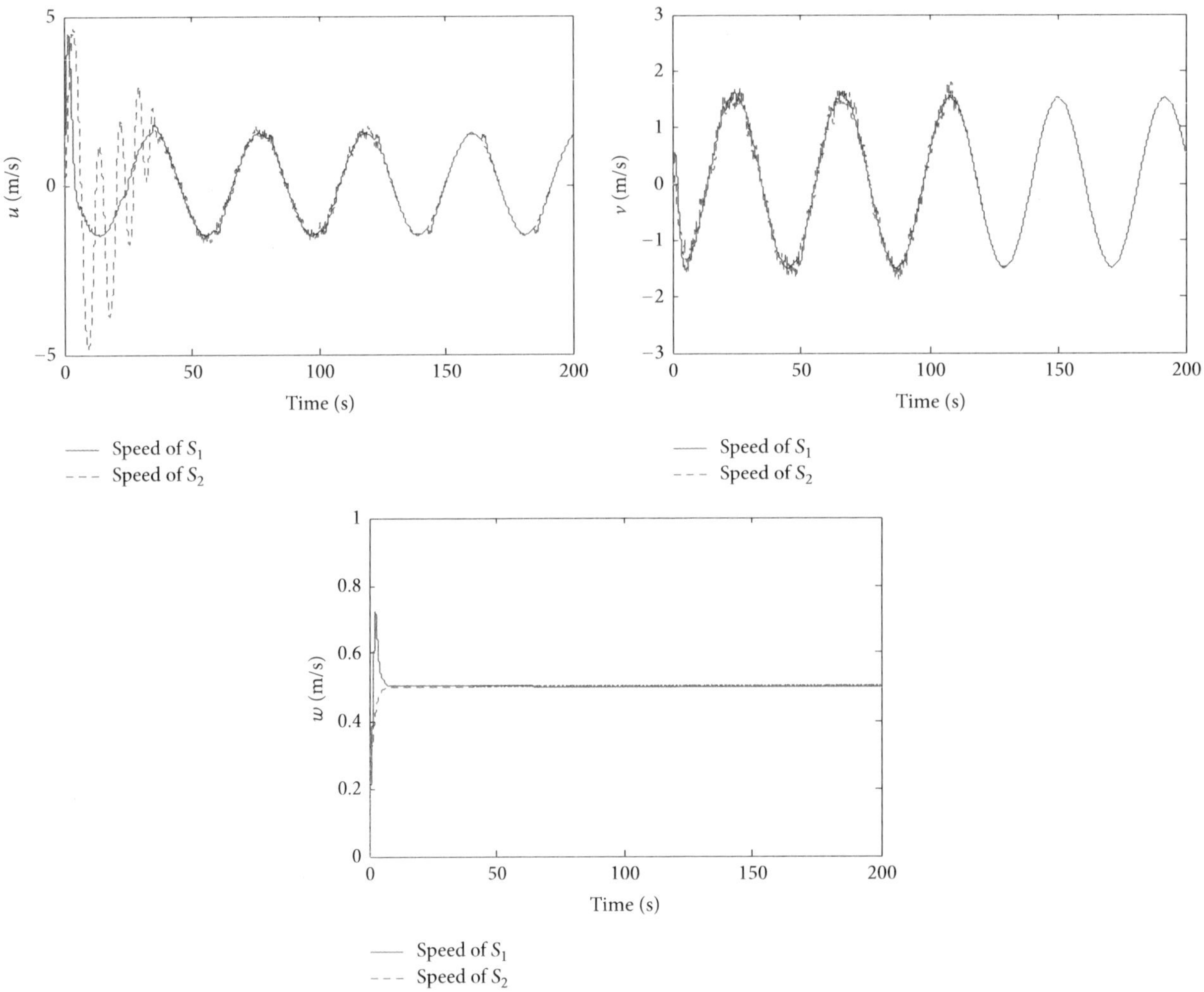

FIGURE 5: UAV flight path tracking results of the speeds (u, v, w).

Assumption 2. Outputs $\widehat{\mathbf{f}}(X \mid w')$ and $\widehat{D}(X \mid w'')$ can approximate $F(X) = \mathbf{f}(X) + \Delta\mathbf{f} + \mathbf{d}$, if there is a very small positive number ε_0 that satisfies $\max\|F - \widehat{\mathbf{f}}(X \mid w') - \widehat{D}(X \mid w'')\| \le \varepsilon_0$.

If $\mathbf{f}^*(X \mid w^*) = \max\|\widehat{\mathbf{f}}(X \mid w') + \widehat{D}(X \mid w'')\| = \sum_{i=1}^{sv} w^* K(\mathbf{x}_i, \mathbf{x}_j)$, where $w^* = \arg\min[\sup|F(X) - \widehat{\mathbf{f}}(X \mid w') - \widehat{D}(X \mid w'')|]$ is the optimal weight vector, then $F(X) = \mathbf{f}^*(X \mid w^*) + \zeta$ that satisfies $\zeta \le \varepsilon_0$; let $\widetilde{w} = w^* - w' - w''$.

Theorem 3. *For the nonlinear system (14), using the modified sliding mode control law (38), the sliding mode control system is asymptotically stable if the adaptive adjusting law is* $\dot{\widetilde{w}} = -\gamma K(\mathbf{x}_i, \mathbf{x}_j)s^T$, *where* $\gamma > 0$ *is the gain parameter of adaptive rate.*

Proof. The following Lyapunov function is considered:

$$V = \frac{1}{2}{s_X}^T s_X + \frac{1}{2\gamma}\widetilde{w}^T\widetilde{w}. \tag{40}$$

The derivative of (18) is given by

$$\begin{aligned} \dot{s}_X &= \mathbf{e}_X^{(\alpha+1)} + \sum_{i=1}^{\alpha-1}\lambda_{X(\alpha-i)}\frac{d^{i+1}\mathbf{e}_X}{dt^{i+1}} + \lambda_{X\alpha}\frac{d\mathbf{e}_X}{dt} \\ &= \frac{d^{(\alpha+1)}X}{dt^{(\alpha+1)}} - \frac{d^{(\alpha+1)}X_d}{dt^{(\alpha+1)}} \\ &\quad + \sum_{i=1}^{\alpha-1}\lambda_{X(\alpha-i)}\frac{d^{i+1}\mathbf{e}_X}{dt^{i+1}} + \lambda_{X\alpha}\frac{d\mathbf{e}_X}{dt}. \end{aligned} \tag{41}$$

Using (14) and letting $d = 0$, then

$$\begin{aligned} \dot{s}_X &= F(X) + g(X)\,\mathbf{u} - \frac{d^{(\alpha+1)}X_d}{dt^{(\alpha+1)}} \\ &\quad + \sum_{i=1}^{\alpha-1}\lambda_{X(\alpha-i)}\frac{d^{i+1}\mathbf{e}_X}{dt^{i+1}} + \lambda_{X\alpha}\frac{d\mathbf{e}_X}{dt}. \end{aligned} \tag{42}$$

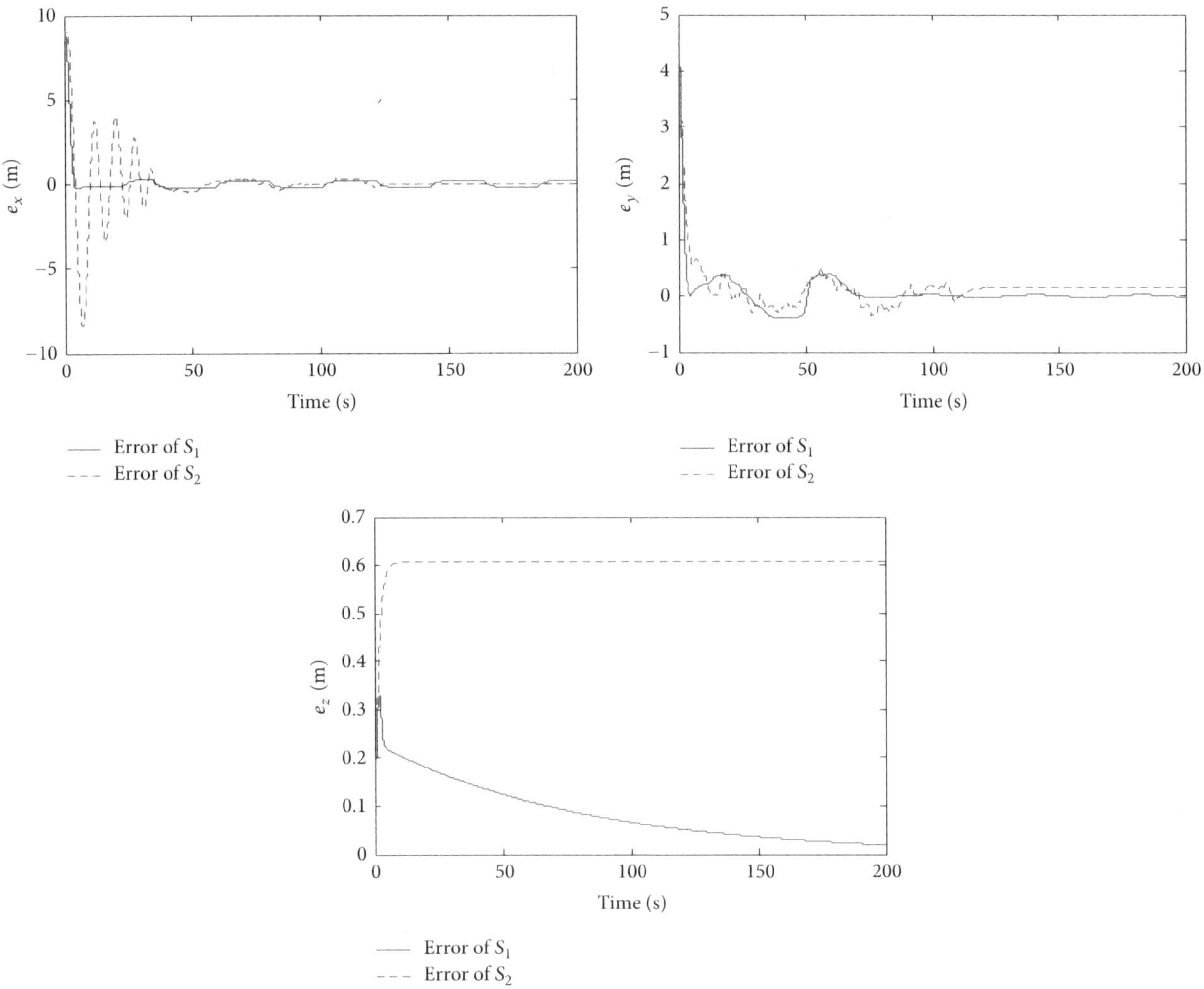

FIGURE 6: The tracking errors of the positions (x, y, z).

By substituting (38) into (42), (42) becomes

$$\begin{aligned}\dot{s}_X &= F(X) - \hat{\mathbf{f}}(X \mid w') - \hat{D}(X \mid w'') - \mathbf{k}\,\mathrm{sign}(s_X) \\ &= \mathbf{f}^*(X \mid w^*) - \hat{\mathbf{f}}(X \mid w') - \hat{D}(X \mid w'') \\ &\quad - \mathbf{k}\,\mathrm{sign}(s_X) + \zeta \\ &= \widetilde{w}K(\mathbf{x}_i, \mathbf{x}_j) - \mathbf{k}\,\mathrm{sign}(s_X) + \zeta.\end{aligned} \tag{43}$$

Using (43) and taking derivative of (40) yield

$$\begin{aligned}\dot{V} &= s_X{}^T\dot{s}_X + \frac{1}{\gamma}\widetilde{w}^T\dot{\widetilde{w}} \\ &= s_X{}^T(\zeta - \mathbf{k}\,\mathrm{sign}(s_X)) \\ &\quad + \frac{1}{\gamma}\mathrm{tr}\left(\dot{\widetilde{w}}\widetilde{w}^T + \gamma K(\mathbf{x}_i, \mathbf{x}_j)\, s_X{}^T\widetilde{w}^T\right).\end{aligned} \tag{44}$$

Because $\dot{\widetilde{w}}_\phi = \dot{w}_\phi^* - \dot{w}_\phi' - \dot{w}_\phi''$ and $\dot{w}_\phi^* = \dot{w}_\phi = 0$, the adaptive adjusting law is selected as

$$\dot{\widetilde{w}} = -\dot{w}' = -\gamma K(\mathbf{x}_i, \mathbf{x}_j)\, s_X{}^T. \tag{45}$$

Then (44) can be written as

$$\dot{V} = s_X{}^T(\zeta - \mathbf{k}\,\mathrm{sign}(s_X)). \tag{46}$$

The boundary value of ζ is ζ_0. It can be known that $\dot{V} \le 0$ if $k \ge \zeta_0$. Not only the asymptotic stability of the system and the boundedness of $\widetilde{w}$ are ensured, but also the system outputs tend to the desired target X_d. □

5. Simulation Results

To verify the effectiveness of the control method, Matlab/Simulink simulation platform is introduced. This platform is responsible for simulating the position tracking control of the quad-rotor UAV system with multiple flight aim points under the condition of high frequency noises

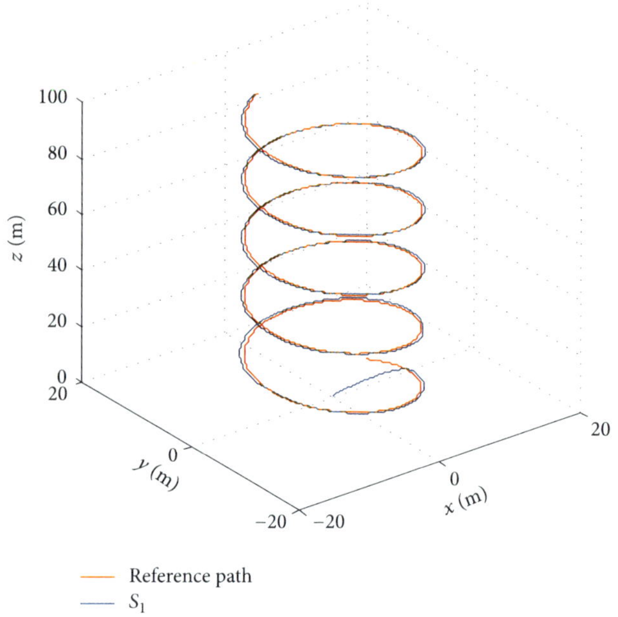

FIGURE 7: The 3D path tracking of S_1.

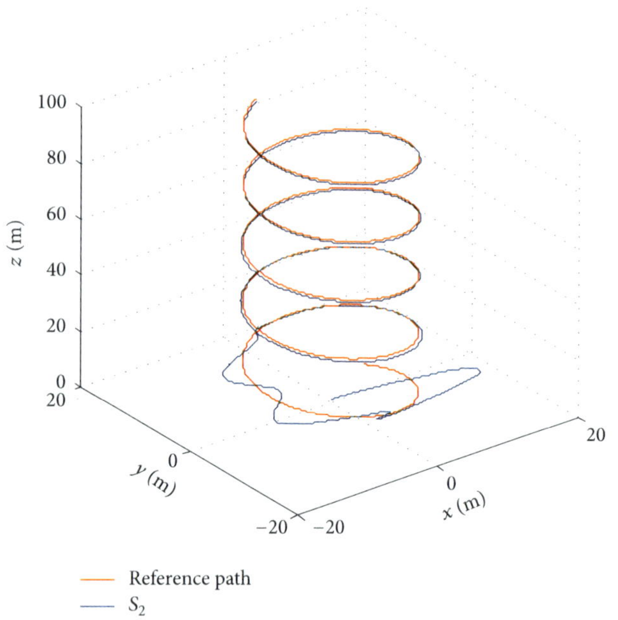

FIGURE 8: The 3D path tracking of S_2.

and uncertainties. At the same time, the attitude controller output is recorded. Control goal is to make the position track reference instructions.

First, part parameters of quad-rotor Unmanned Aerial Vehicle are given in Table 1. The sliding mode controller (19) is established by using the offline SVM to predict the model when there are some unknown parameters in the UAV system. Training sample data include the speed of each motor and the aircraft attitude. Sampling period is 0.01 s and 2000 points are taken as training sample data.

TABLE 1: Quad-rotor parameters in simulation.

Parameter	Value
m	0.468 kg
g	9.81 m/s^2
l	0.225 m
k_b	2.98×10^{-6} N·s/rad^2
k_d	1.14×10^{-7} N·m·s/rad^2
J_x	4.856×10^{-3} kg·m^2
J_y	4.856×10^{-3} kg·m^2
J_z	8.801×10^{-3} kg·m^2
J_{rx}	6.15×10^{-5} kg·m^2
J_{ry}	6.15×10^{-5} kg·m^2

At the same time, the sliding mode controller is reconstructed by joining online adaptive SVM compensation. There are 30 online SVM training samples and replaced by the new data real-timely. Adaptive parameter γ of online SVM is taken as 1.5.

The start point of flight is (0, 0, 0) m and UAV tracks the desired paths $x_d = 10\sin(0.15t + \pi/3)$ m, $y_d = 10\sin(0.15t + \pi * 5/6)$ m, and $z_d = 0.5t$ m. The simulation time is 200 s. Both controllers are joined to the same UAV system under the conditions of existing noise which is equal to $0.4 * \sin(50 * X)$ and 20% uncertainties including mass, inertia, and aerodynamic coefficients. UAV flight path with online adaptive error compensation SVM-based sliding mode control was labeled as S_1, and the flight path with only offline SVM-based sliding mode control was labeled as S_2. The flight path tracking results of two conditions are shown in Figure 4.

And the speeds u, v, and w of the quad-rotor are shown in Figure 5.

It can be seen that the UAV with the sliding mode control based on online adaptive SVM compensation can track the path more rapidly and accurately even if there are some noises and uncertainties. It also can be seen that the UAV which joins online adaptive SVM compensation has the good performance from the tracking error. Changes of the position errors e_x, e_y, and e_z are shown in Figure 6.

The 3D path tracking with two different controllers is shown in Figures 7 and 8.

In order to test the step signal response of UAV with online adaptive error compensation SVM-based sliding mode control, let UAV fly from the initial position (0, 0, 0) m through points (3, 3, 8) m, (12, 7, 8) m, and (13, 14, 8) m in turn. The simulation time is 60 s and position simulation results are shown as in Figure 9.

Figure 9 shows that the UAV with online adaptive SVM compensation reaches the three target points at 9 s, 28 s, and 47 s. And the path without online adaptive SVM compensation is more slow. Three-dimensional path tracking diagrams are shown in Figure 10, Figure 10(a) is the expected route, and Figures 10(b), 10(c), and 10(d) are the actual paths of UAV in the different time points.

As shown in Figure 9, although there are uncertainties and noises, it still can achieve the satisfactory tracking. Good results have two reasons. First, offline SVM and online

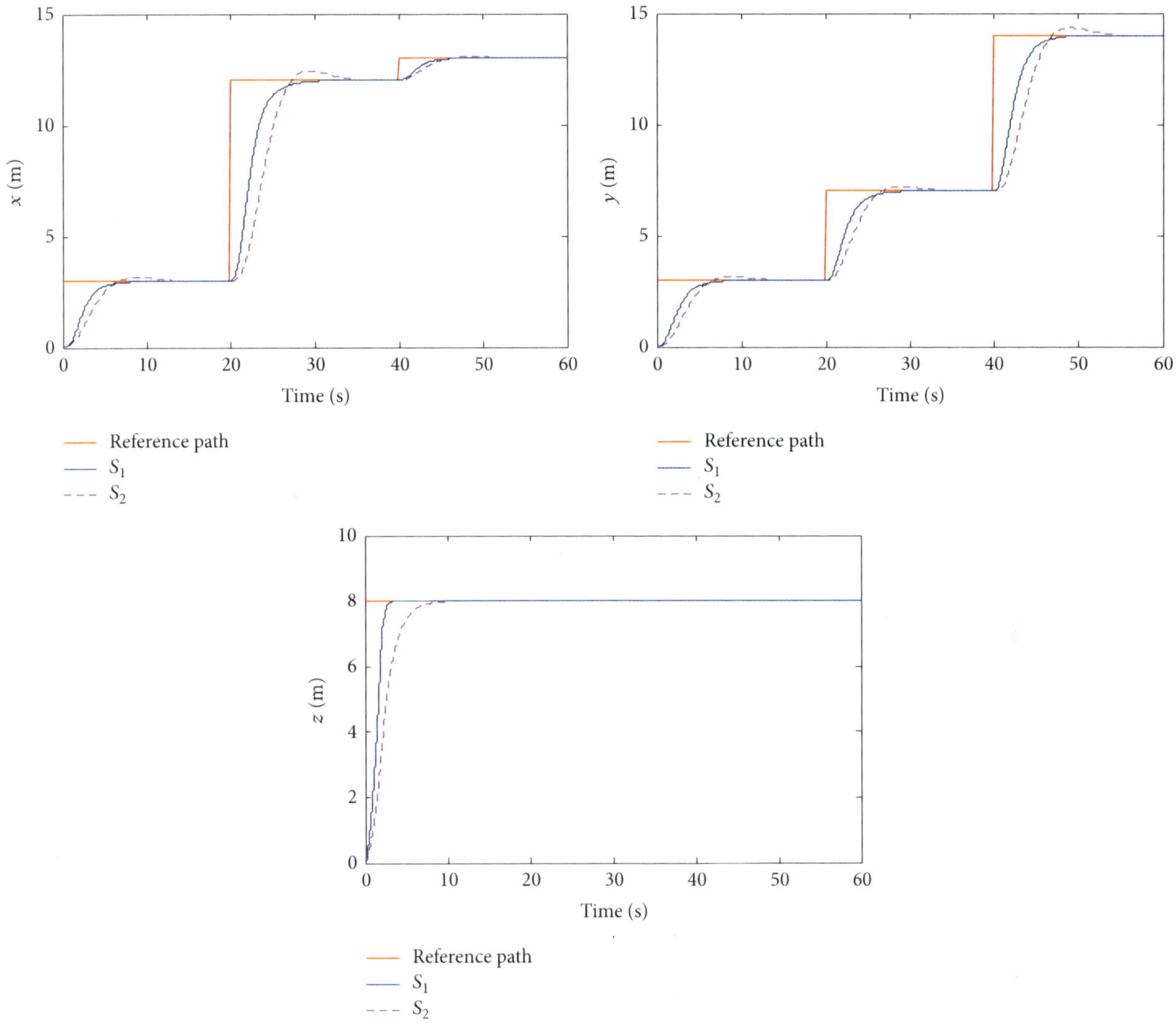

FIGURE 9: UAV path tracking of the positions (x, y, z).

adaptive SVM used together can establish the model rapidly and the errors will be compensated real-timely even if there are noises and uncertainties. Second, as the adaptive online SVM real-time adjustment method, it will not lead to a decline in tracking performance even if the predictions have deviation or drift.

It can be seen from the results of simulation, whatever, modeling errors, parameter uncertainties or noises do not affect the tracking accuracy of the controller. And the robustness of the controller can be proved.

6. Conclusion

In this paper, the control problem of quad-rotor UAV system is discussed during tracking the procedure by using the sliding mode controller. To ensure the stabilization and the navigation of UAV with unknown parameters, a sliding mode controller with online adaptive error compensation SVM is presented. Sliding mode controller is established through analyzing quad-rotor dynamics model in which the unknown parameters are estimated by offline SVM. But offline SVM cannot guarantee the precision of model in the process of flight because modeling errors and external disturbances always exist. So the online adaptive SVM method is added for the controller to compensate noises and uncertainties. Control law is adjusted in real-time by introducing new training sample data to online adaptive SVM in the control process and thereby it guarantees the stability and the robustness of flight.

Two experiments about above control scheme are performed. In the first experiment, it can be shown that the control performance with only offline SVM modeling is worse than the control performance with both offline SVM and online adaptive SVM. And the tracking speed is much slower. In the second experiment, the step signal response of UAV is tested with online adaptive error compensation SVM-based sliding mode control. The simulation results also verify its superiority and the better results are obtained in this case.

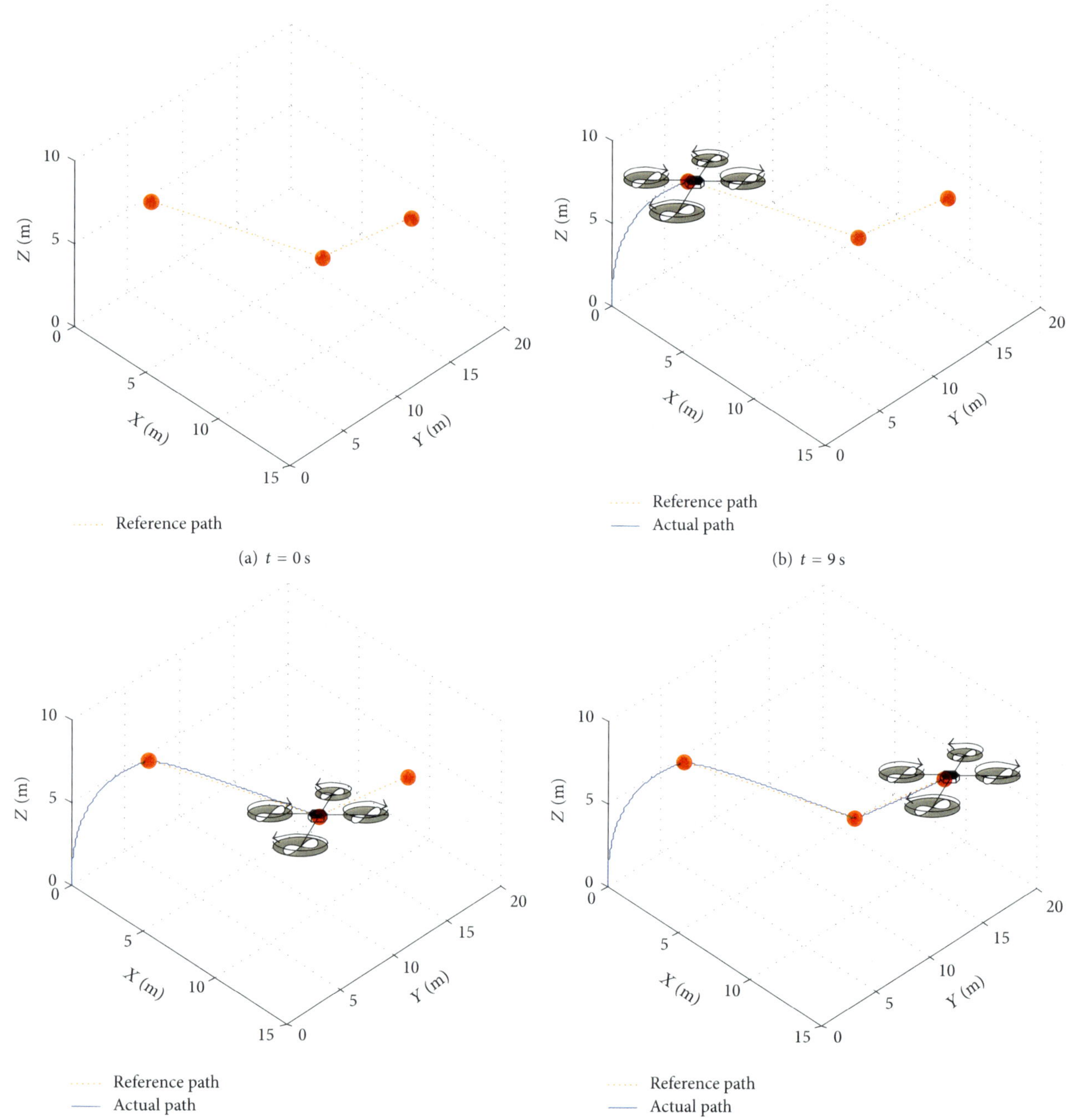

(a) $t = 0$ s
(b) $t = 9$ s
(c) $t = 28$ s
(d) $t = 47$ s

FIGURE 10: Three-dimensional UAV path tracking.

The simulation results demonstrate that the proposed method effectively improves the path tracking performance. They also prove the good performance of online adaptive error compensation SVM-based sliding mode control in a quad-rotor UAV system with noises and uncertainties.

Conflict of Interests

The authors declare that there is no conflict of interests regarding the publication of this paper.

Acknowledgment

This work was supported by JiangSu Scientific Support Program of China (Grant no. BE2014712).

References

[1] C. Ha, Z. Zuo, F. B. Choi, and D. Lee, "Passivity-based adaptive backstepping control of quadrotor-type UAVs," *Robotics and Autonomous Systems*, vol. 62, no. 9, pp. 1305–1315, 2014.

[2] A. Modirrousta and M. Khodabandeh, "A novel nonlinear hybrid controller design for an uncertain quadrotor with disturbances," *Aerospace Science and Technology*, vol. 45, pp. 294–308, 2015.

[3] A. Nagaty, S. Saeedi, C. Thibault, M. Seto, and H. Li, "Control and navigation framework for quadrotor helicopters," *Journal of Intelligent & Robotic Systems*, vol. 70, no. 1–4, pp. 1–12, 2013.

[4] G. Nikolakopoulos and K. Alexis, "Switching networked attitude control of an unmanned quadrotor," *International Journal of Control, Automation and Systems*, vol. 11, no. 2, pp. 389–397, 2013.

[5] K. Alexis, G. Nikolakopoulos, and A. Tzes, "Experimental constrained optimal attitude control of a quadrotor subject to wind disturbances," *International Journal of Control, Automation and Systems*, vol. 12, no. 6, pp. 1289–1302, 2014.

[6] V. Roldão, R. Cunha, D. Cabecinhas, C. Silvestre, and P. Oliveira, "A leader-following trajectory generator with application to quadrotor formation flight," *Robotics and Autonomous Systems*, vol. 62, no. 10, pp. 1597–1609, 2014.

[7] S. A. Belokon', Y. N. Zolotukhin, A. S. Mal'tsev, A. A. Nesterov, M. N. Filippov, and A. P. Yan, "Control of flight parameters of a quadrotor vehicle moving over a given trajectory," *Optoelectronics, Instrumentation and Data Processing*, vol. 48, no. 5, pp. 454–461, 2012.

[8] Y. Yali, S. Feng, and W. Yuanxi, "Controller design of quadrotor aerial robot," *Physics Procedia*, vol. 33, pp. 1254–1260, 2012.

[9] H. A. F. Almurib, P. T. Nathan, and T. N. Kumar, "Control and path planning of quadrotor aerial vehicles for search and rescue," in *Proceedings of the SICE Annual Conference (SICE '11)*, Tokyo, Japan, September 2011.

[10] K. Alexis, G. Nikolakopoulos, A. Tzes, and L. Dritsas, "Coordination of helicopter UAVs for aerial forest-fire surveillance," in *Applications of Intelligent Control to Engineering Systems*, K. P. Valavanis, Ed., vol. 39 of *Intelligent Systems, Control, and Automation: Science and Engineering*, pp. 169–193, 2009.

[11] C.-C. Li, G.-S. Zhang, T.-J. Lei, and A.-D. Gong, "Quick image-processing method of UAV without control points data in earthquake disaster area," *Transactions of Nonferrous Metals Society of China*, vol. 21, no. 3, pp. s523–s528, 2011.

[12] F. Fraundorfer, L. Heng, D. Honegger et al., "Vision-based autonomous mapping and exploration using a quadrotor MAV," in *Proceedings of the 25th IEEE/RSJ International Conference on Robotics and Intelligent Systems (IROS '12)*, pp. 4557–4564, IEEE, Vilamoura, Portugal, October 2012.

[13] P. J. Zarco-Tejada, R. Diaz-Varela, V. Angileri, and P. Loudjani, "Tree height quantification using very high resolution imagery acquired from an unmanned aerial vehicle (UAV) and automatic 3D photo-reconstruction methods," *European Journal of Agronomy*, vol. 55, pp. 89–99, 2014.

[14] F. Nex and F. Remondino, "UAV for 3D mapping applications: a review," *Applied Geomatics*, vol. 6, no. 1, pp. 1–15, 2014.

[15] N. Metni and T. Hamel, "A UAV for bridge inspection: visual servoing control law with orientation limits," *Automation in Construction*, vol. 17, no. 1, pp. 3–10, 2007.

[16] J. Baluja, M. P. Diago, P. Balda et al., "Assessment of vineyard water status variability by thermal and multispectral imagery using an unmanned aerial vehicle (UAV)," *Irrigation Science*, vol. 30, no. 6, pp. 511–522, 2012.

[17] F. Chen, B. Jiang, and G. Tao, "An intelligent self-repairing control for nonlinear MIMO systems via adaptive sliding mode control technology," *Journal of the Franklin Institute*, vol. 351, no. 1, pp. 399–411, 2014.

[18] R. Ashok and Y. Shtessel, "Control of fuel cell-based electric power system using adaptive sliding mode control and observation techniques," *Journal of the Franklin Institute*, vol. 352, no. 11, pp. 4911–4934, 2015.

[19] Y. Singh and M. Santhakumar, "Inverse dynamics and robust sliding mode control of a planar parallel (2-PRP and 1-PPR) robot augmented with a nonlinear disturbance observer," *Mechanism and Machine Theory*, vol. 92, pp. 29–50, 2015.

[20] V. Azhmyakov, "On the set-valued approach to optimal control of sliding mode processes," *Journal of the Franklin Institute*, vol. 349, no. 4, pp. 1323–1336, 2012.

[21] A. Al-Khazraji, N. Essounbouli, A. Hamzaoui, F. Nollet, and J. Zaytoon, "Type-2 fuzzy sliding mode control without reaching phase for nonlinear system," *Engineering Applications of Artificial Intelligence*, vol. 24, no. 1, pp. 23–38, 2011.

[22] T.-C. Lin, M.-C. Chen, and M. Roopaei, "Synchronization of uncertain chaotic systems based on adaptive type-2 fuzzy sliding mode control," *Engineering Applications of Artificial Intelligence*, vol. 24, no. 1, pp. 39–49, 2011.

[23] C.-F. Hsu, "Adaptive neural complementary sliding-mode control via functional-linked wavelet neural network," *Engineering Applications of Artificial Intelligence*, vol. 26, no. 4, pp. 1221–1229, 2013.

[24] F. G. Rossomando, C. Soria, and R. Carelli, "Adaptive neural sliding mode compensator for a class of nonlinear systems with unmodeled uncertainties," *Engineering Applications of Artificial Intelligence*, vol. 26, no. 10, pp. 2251–2259, 2013.

[25] Y. G. Kao, C. H. Wang, H. R. Karimi, and W. Li, "H_∞ sliding mode control for uncertain neutral-type stochastic systems with Markovian jumping parameters," *Information Sciences*, vol. 314, pp. 200–211, 2015.

[26] M. Castro-Neto, Y.-S. Jeong, M.-K. Jeong, and L. D. Han, "Online-SVR for short-term traffic flow prediction under typical and atypical traffic conditions," *Expert Systems with Applications*, vol. 36, no. 3, pp. 6164–6173, 2009.

13

Design and Performance Evaluation of Propeller for Solar-Powered High-Altitude Long-Endurance Unmanned Aerial Vehicle

Donghun Park,[1] **Yunggyo Lee**,[2] **Taehwan Cho**,[2] **and Cheolwan Kim**[2]

[1] *Department of Aerospace Engineering, Pusan National University, Busan 46241, Republic of Korea*
[2] *Aerodynamics Research Team, Korea Aerospace Research Institute, Daejeon 34133, Republic of Korea*

Correspondence should be addressed to Donghun Park; parkdh@pusan.ac.kr

Academic Editor: Mahmut Reyhanoglu

Design, wind tunnel test, computational fluid dynamics (CFD) analysis, and flight test data analysis are conducted for the propeller of EAV-3, which is a solar-powered high-altitude long-endurance unmanned aerial vehicle developed by Korea Aerospace Research Institute. The blade element momentum theory, in conjunction with minimum induced loss, is used as a basic design method. Airfoil data are obtained from CFD analysis, which takes into account the low Reynolds number effect. The response surface is evaluated for design variables by using design of experiment and kriging metamodel. The optimization is based on desirability function. A wind tunnel test is conducted on the designed propeller. Numerical analyses are performed by using a commercial CFD code, and results are compared with those obtained from the design code and wind tunnel test data. Flight test data are analyzed based on several approximations and assumptions. The propeller performance is in good agreement with the numerical and measurement data in terms of tendency and behavior. The comparison of data confirms that the design method, wind tunnel test, and CFD analysis used in this study are practically useful and valid for the development of a high-altitude propeller.

1. Introduction

High-altitude long-endurance (HALE) unmanned aerial vehicles (UAVs) have gained considerable interest because their use could replace several of the functions of a satellite, such as earth observation, telecommunication, reconnaissance, and surveillance. HALE UAV has significant advantages over satellites in terms of low production cost, effective operation, and easy maintenance through landing/takeoff. In addition, as interest in green technology increases, a number of countries are actively researching and developing electric-powered or solar-powered HALE UAVs [1–4]. Among the well-known examples is Helios [5, 6], developed by NASA and the Aeronautical Environment Corporation (UAS). In 2003, it ascended to an altitude of over 29 km, the highest climb record in the world achieved by a propeller-driven UAV at that time. Another successful and widely known example is Zephyr [7, 8], which was designed and built by the British company, QinetiQ. It holds the official endurance world record. It is currently being developed as a part of the Airbus high-altitude pseudosatellite (HAPS) program. In recent years, large IT companies have given considerable attention to HALE UAVs for the purpose of internet distribution. For example, Google acquired Titan Aerospace and is developing Solara 50 and 60, whereas Facebook is developing Aquila. Other examples include the Solar Eagle, which was developed by Boeing (DARPA Vulture II Program) [9].

Recently, the Korea Aerospace Research Institute (KARI) built a series of solar-powered HALE UAVs through a research program. In this program, KARI is aimed at exploring, developing, and demonstrating the underlying technologies needed for the development of an electric-powered HALE UAV. The scale-down version, EAV-2H+, was developed and flight tested during which it ascended up to an altitude of 10 km [10, 11]. Thereafter, KARI developed the full-scale version, EAV-3, which successfully soared above

18 km during its test flight [12–14]. The flight of the EAV-3 is third in the world record for an electric-powered UAV to ascend at an altitude of over 18 km.

In developing HALE UAV, state-of-art technologies in all fields, including aerodynamics and structural design, composite materials, solar cells and rechargeable battery (energy system), electric-motors, propellers, flight stability and control, and system integration, must be incorporated. From the perspective of aerodynamics, its performance could be significantly degraded due to low Reynolds number flows formed by the low air density at altitudes of over 10 km. To improve performances such as the lift-to-drag ratio and drag characteristics, the sizing and aerodynamic configuration design must be carefully conducted, taking into consideration the low Reynolds number environment. Additionally, since an electric-powered HALE UAV uses propellers for propulsion, the high efficiency of these propellers is essential for the endurance and climb capability of HALE. However, at high altitudes, propellers are vulnerable to efficiency reduction due to low Reynolds number conditions. Moreover, a design solely focused on maximizing the cruising efficiency at high altitudes may result in the considerable degradation of the performance at low altitudes or during climb operations.

Therefore, the proper design of propeller, considering various operating conditions and system constraints, is extremely important to achieve both climbing performance and endurance. Moreover, the accurate prediction and analysis of the propeller performance enables the establishment of a reliable and efficient mission profile. Thus, the design and performance prediction of the propeller are among the key technologies required for HALE UAV development.

Concepts on design, analysis, and testing of propellers for conventional aircrafts and UAVs have been well established through enormous research efforts expended for approximately a century. Although several solar-powered HALE UAVs have been designed and built, only a few literature pertain to design and performance evaluation of their propellers [15, 16]. As mentioned by Monk [15], the history of propellers is still sparse in the area of high-altitude applications. He reported and summarized findings from efforts in conjunction with the NASA ERAST program. The efforts include investigations into the existing propellers designed for height altitude flight condition, the analysis capability to performance estimation for a given design, and the acquisition of the performance data through experimental testing. Several examples of propeller-driven high-altitude aircrafts, design approaches and limitations, analysis approaches, and wind tunnel testing for 2D airfoil and 3D propellers are introduced. More recently, Monk [16] evaluated a propeller design based on minimum induced loss concept and analysis for performance prediction based on momentum theory for a HALE fight vehicle. A conceptual propeller was designed, and analyses were carried out for altitudes up to 15 km. The results from a wind tunnel testing for a scaled model showed reasonable agreement with the prediction.

In particular, literature that contain detailed information and data on propellers mounted on an actual HALE UAV are rather rare. With the EAV program, KARI acquired and demonstrated the fundamental technologies related to propellers for solar-powered HALE UAVs and evaluated their reliability and feasibility. In this study, the design and performance evaluation of propellers, conducted during the development of EAV-3, are presented. The performance analysis includes CFD analysis, wind tunnel test, and flight test data analysis.

In Section 2, the design methodology and procedure for the EAV-3 propeller are introduced. The wind tunnel test and CFD analysis results are presented in Sections 3 and 4, respectively. A simple analysis of the flight test data to evaluate the propeller performance is given in Section 5. Finally, conclusion follows in Section 6.

2. Design of Propeller

Propellers for HALE UAVs operate under various operating conditions, ranging from the sea level to stratosphere altitudes. Apparently, it is appropriate to adopt a variable pitch system to provide the optimal propulsive efficiency under the aforementioned conditions. However, its adoption imposes additional weight and complexity due to the addition of actuators and pitch links. Additionally, these pitch links and actuators will practically be exposed to external flows at low temperatures from −70 to −80°C at stratospheric altitudes. The extreme environment and mechanical complexity may lead to an increased possibility of malfunctions and uncertainty. Consequently, the demand for reliability and being ultra-lightweight, which are top-level constraints of HALE UAVs, makes it difficult to adopt the variable pitch system. Therefore, fixed-pitch propellers are generally used. When the fixed-pitch propellers are optimized for aerodynamic performance at high-altitude operation, the required torque, approximately at sea level, becomes considerably large and exceeds the specification for electric motors. This can lead to low climbing performances or, sometimes, the inability to climb. On the other hand, as altitude increases, the rotational speed of the propeller gradually increments, which consequently results in an increase of the required power. Thus, the maximum required power occurs under high-altitude climbing conditions. In this respect, the design of HALE UAV propellers must not only take into account the two conflicting constraints but also simultaneously maximize efficiency under the desired operating condition.

2.1. Design Requirement and Constraints. The ultra-lightweight aircraft, EAV-3, has a wingspan, total length, and design total weight of approximately 19.5 m, 8.9 m, and 46.5 kg, respectively. It uses twin propellers mounted on each wing. The specifications, configuration, geometry, and mission profile of EAV-3 can be found in literature [14]. The operating conditions and required thrust for each propeller are summarized in Table 1. The specifications of the direct current (DC) motor, which is developed and customized for EAV-3, are listed in Table 2. The maximum available torque should correspond to the climb condition at sea level, requiring the highest thrust. The maximum power condition should correspond to the climb operation at 18 km, where the highest rotational speed is required. Considering the motor diameter, the design spinner diameter was fixed at 0.16 m

Table 1: Design operating condition and required thrust for each propeller.

Altitude	Speed (m/s)	Thrust (N)
Sea level	5.8	38.8
10 km	10.0	26.6
15 km	14.6	21.3
18 km	18.5	18.8

Table 2: Specification of DC motor.

Power (W)	RPM	Torque (Nm)	Note
314	1200	2.5	Rated condition
351	880	3.8	Maximum torque
369	2200	1.6	Maximum power

as a geometry constraint. In conformance with the EAV-3 mission profile, which is mainly aimed at climbing to high altitudes, the climb condition of 15 km was set as the propeller design point.

2.2. Basic Design Methodology. The blade element momentum theory (BEMT) [17] is a useful basic methodology for propeller design and analysis. It is the most common engineering model and still widely used due to its simplicity and easy applicability with various corrections and modifications [18–21]. There are several free software that use BEMT for propeller design and analysis. For example, JBLADE [22] or JavaProp [23] are easy to get online. However, these cannot be modified by the user, airfoil selection is limited to a given list, and airfoil aerodynamic data in the low Reynolds number range are less reliable. Consequently, a BEMT-based design and analysis code was independently developed and utilized for EAV-3 propellers. This code allowed the easy modification of details and use of any airfoil by supplying customized aerodynamic data. The code was based on the theory of Adkins and Liebeck using BEMT and minimum energy loss condition [24]. To take into account the tip loss effect, the tip loss corrections of Glauert [17] were implemented. A graphical user interface was constructed so as to enhance the user convenience in the early design stage and off-design analysis. The concept and input/output of the code are shown in Figure 1. Further details on the design code can be found in [25].

Operating conditions such as forward flight velocity, rotational speed (rpm), and air properties (density) at design point are essential input parameters. The input parameters for the geometry are the diameter, spinner diameter, and section airfoil. The distribution of the section lift coefficient along the blade span also needs to be prescribed. The design can be conducted when the foregoing input parameters and target performance are given. The target performance can be the desired thrust, torque, or power. The results of the design yield blade geometry information, specifically the distribution of the sectional chord and twist angle.

2.3. Blade Airfoil and Airfoil Data. Airfoil aerodynamic data are essential for BEMT calculations and are key factors in determining the accuracy of design and performance prediction. Since blade elements operate at various Reynolds numbers and angle of attacks, it is efficient to construct the set of aerodynamic data in advance. The required lift and drag coefficient at a specific condition of interest can be simply and quickly obtained through table look-up or interpolation of the data.

In general, fast analysis tools such as XFOIL [26] and XFLR5 [27] are reasonable options for obtaining airfoil data due to their low computational costs. XFOIL is the most widely used tool in previous studies [19–21]. However, in the case of EAV-3 propellers, blade elements will be placed under conditions where the Reynolds number is less than 10^5. The authors of the present study attempted to evaluate the reliability of airfoil data under these low Re conditions prior to the use of fast analysis tools.

Figure 2 shows result examples of the airfoil analysis by using XFOIL and CFD. The airfoil is DAE51, and the Reynolds number and angle of attack are 10^5 and 0°, respectively (chord length 0.2 m). A two-dimensional, C–H-type structured grid with quadrilateral cells is used. The total number of cells is 165600, and 300 grid points are used along each upper and lower surfaces. The height of the first cell from the surface is set as $y+$ which remains below 1 over most of the airfoil surface. The computational domain extends to 30 chord to upstream, top, and bottom directions and 60 chord to downstream direction. CFD analyses were carried out by using Fluent 17.0 [28] and for the two cases of k-ω SST and γ-Re_θ [29] based (four-equation) transition-SST turbulence models. The inlet turbulence intensity is set to 0.1%. The distributions of pressure and skin friction coefficients on both upper and lower surfaces are compared in the figure.

From previous studies [30–33], it is well known that the characteristics of flow around an airfoil at low Reynolds numbers are considerably different from those at moderate and high Reynolds numbers. The separation of the laminar boundary layer and transition in the separated flow region followed by a reattachment form a separation bubble on the airfoil surface. A typical feature of low Reynolds number flow is the formation of a large separation bubble at low angle of attacks. Moreover, the streamwise extent and position of the bubble dynamically change with the angle of attack.

As can be observed in Figure 2, a typical characteristic at low Reynolds number flow is represented by a practically flat region in the pressure coefficient distribution followed by an abrupt break, which corresponds to the existence of the separation bubble. However, it can be seen from the figure that the result of the k-ω SST model does not show this feature, which indicates that this model cannot capture the flow feature and bubble properly. Additionally, the friction coefficient values are large, which correspond to a fully turbulent boundary layer and results in a high viscous drag. Since the k-ω SST model assumes a fully turbulent flow, it cannot accurately reflect the actual phenomena occurring under low Reynolds number conditions. On the other hand, results from the transition SST model and XFOIL show practically similar distributions of surface pressure and skin friction. It

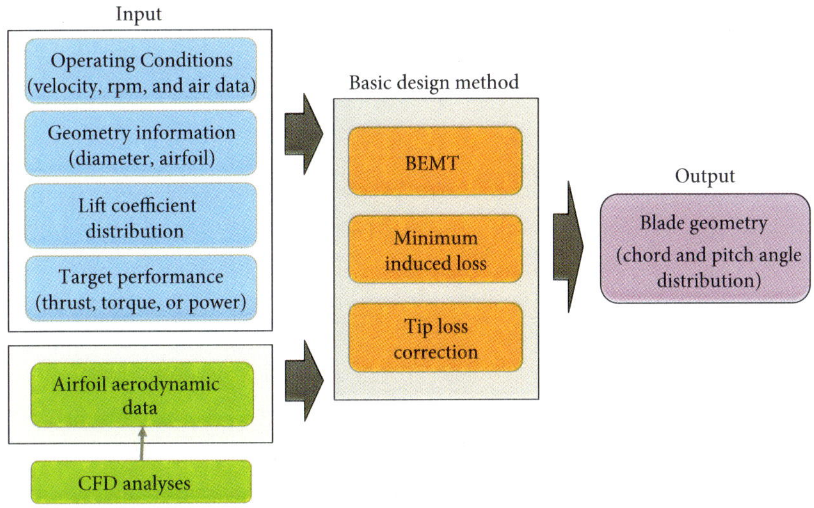

FIGURE 1: Schematic of the basic propeller design code.

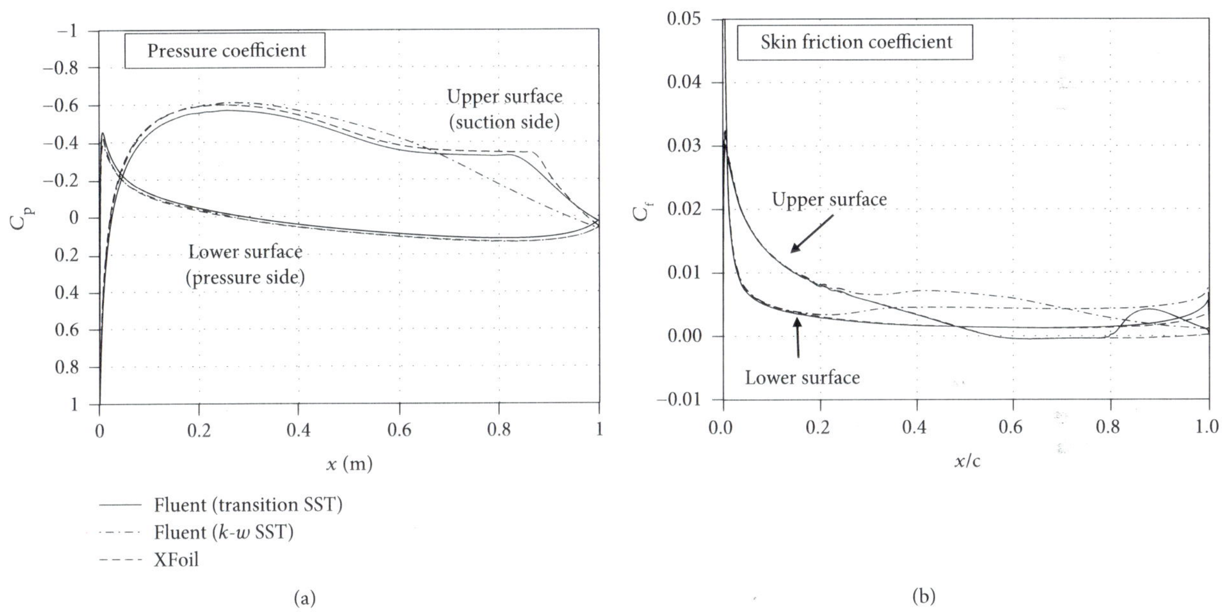

FIGURE 2: Comparison of XFOIL and CFD results for an airfoil at low Reynolds number.

can be deduced that essential elements, such as the laminar boundary layer separation, transition, and reattachment, are properly captured in those two calculations. However, XFOIL results exhibit notable differences in the predicted friction at the turbulent boundary layer region after the reattachment.

The differences of aerodynamic coefficients are shown in Figure 3, which compare the drag polar calculated by three methods. The results of the k-ω SST model corresponds to a typical parabolic drag polar at high Reynolds numbers and are not suitable for the calculation of airfoil data at low Reynolds number regions. Although two results of the transition SST model and the XFOIL show the typical drag polar at low Reynolds numbers, their difference in drag increases as the angle of attack increases. The same calculations were conducted for several different Reynolds numbers. From the results, low Reynolds number characteristics began to distinctly appear as the Reynolds number becomes lower as 2×10^5. It was confirmed that the aerodynamic coefficient and drag polar considerably change with respect to the Reynolds number in the region around 10^5 and below. Moreover, the difference between the results of the transition SST model and XFOIL was found to increase gradually as the Reynolds number decreased. From the comparison with available wind tunnel testing data [30, 31] and previous numerical study [34], it was deduced that the CFD result with the transition SST model was more reliable than that of XFOIL for the friction in the turbulent boundary layer after reattachment. Accordingly, it was decided to use CFD to calculate and construct airfoil data for the EAV-3 propeller design.

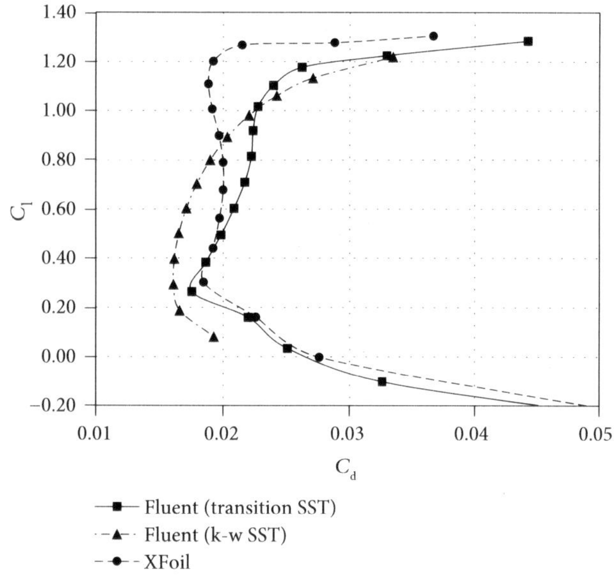

FIGURE 3: Comparison of drag polar calculated by three methods.

Analyses of several candidate airfoils used in low Reynolds numbers were performed, and aerodynamic characteristics were compared within the operating range of the angle of attack near the design point. At Reynolds numbers of 3.0×10^5 or less, SG6043 and DAE51 were considered to be superior airfoils in terms of the lift-to-drag ratio. However, DAE51 and E387 exhibited rather higher performance at Reynolds numbers below 1.5×10^5. From preliminary studies, we expect that the most of section Reynolds number is lower than 1.0×10^5 at design point. As an example, Figure 4 shows CFD results for several airfoils at Reynolds number of 1.0×10^5. All calculations were conducted by using the transition SST model. We readily see that DAE51 shows a good lift-to-drag ratio over a wide range of angle of attack for the linear region. Although E387 also exhibits a good lift-to-drag ratio, it is limited to only an angle attack range for lift coefficients less than 0.8 and has a rather lower lift slope and the maximum lift coefficient. Similar comparison and examination were made for the other Reynolds numbers. By taking into account the desired lift coefficient which is between 0.6 and 1.0 for low Reynolds number operation, we concluded that DAE51 is a reasonable choice for the propeller blade airfoil.

Airfoil data of DAE51 as an input for a propeller design code was obtained by using CFD with the transition SST model. The incompressible and unsteady (pressure-based and transient in Fluent, resp.) calculations were conducted using a second-order scheme for both time and spatial discretization. The calculations for Reynolds numbers ranging from 10^4 to 10^5 were carried out at intervals of 10^4. Calculations for Reynolds numbers at 1.5, 2.0, 3.0, 4.0, and 5.0×10^5 were also conducted. For each Reynolds number, angle of attacks, from −10 to 20°, at intervals of 1°, were considered.

Figure 5 shows an example of DAE51 airfoil data calculated for several Reynolds number conditions. Aerodynamic data beyond the range of computation were extended following the method given in [35]. The correction for compressibility is applied using formulae described in [36]. However, it has little influence in our design because the Mach number is not sufficiently high to have a compressibility effect over the entire operating condition range.

2.4. Design Variables and Effect of Lift Coefficients. Along with the design code and airfoil data described above, design variables were identified and defined for the operating condition at design point. The diameter (d) and rpm of the propeller are essential design variables. The remaining information that needs to be specified is the lift coefficient distribution along the blade span.

In traditional aircraft propeller design, a lift coefficient of 1.0 or greater is generally used to procure sufficient lift-to-drag ratio, which guarantees good aerodynamic efficiency of blade elements. However, as previously discussed, the lift coefficient and lift-to-drag ratio considerably deteriorate at high altitudes, where the Reynolds number is 10^5 or less (see Figure 5). When high lift coefficients are imposed even under this condition, there is a possibility that the lift coefficient cannot reach the given value as a design result. Otherwise, it can yield the design of lower efficiency originated from the low lift-to-drag ratio of the blade elements due to derivation of short chord length.

To avoid the foregoing possibility, the lift coefficient must be specified at lower values of certain degree than those in the traditional approach. Thus, for design, we assumed a lift coefficient distribution as a parabolic profile with its minimum value at a certain location in the spanwise direction along the blade. Recall that most of conventional design approaches use a constant lift coefficient along the span. The lift coefficient values at three different spanwise locations define a particular parabolic profile. In the present study, the lift coefficient at the location of the spinner was fixed at 1.0, taking into consideration the geometrical and structural constraints. The remaining two locations were chosen as 60% of radius ($r/R = 0.6$) and at the blade tip ($r/R = 1.0$). We note here that all parabolic profiles can be defined by specifying proper lift coefficient values at any three spanwise locations. The choice of $r/R = 0.6$ seems arbitrary; however, variation of the lift coefficient at this location can consider all the possible parabolic profiles for given values at the spinner and tip. A parabolic profile with a certain lift coefficient value at $r/R = 0.6$ would be identical to the profile with a different lift coefficient at some other spanwise location. Therefore, changing the lift coefficients at $r/R = 0.6$ and 1.0 is sufficient to explore all parabolic profiles of a fixed value at the spinner. Consequently, the lift coefficients at $r/R = 0.6$ and at the blade tip ($r/R = 1.0$) were chosen as additional design variables, clm and clt, respectively. The four design variables used in this study are listed in Table 3.

Example cases with several parabolic lift coefficient distributions were studied to evaluate the influence of the lift coefficient distribution on design results. The designs were conducted under the condition of design point with a fixed diameter and rpm and required thrust. Table 4 summarizes the values of clm and clt for five cases, and Figure 6(a)

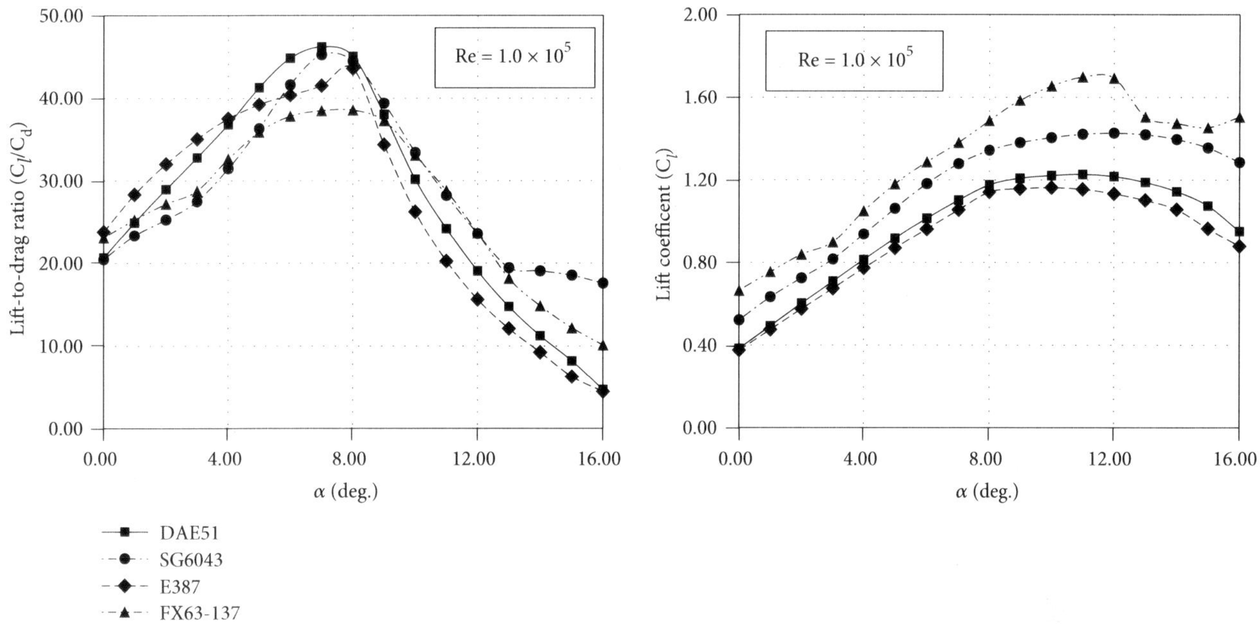

FIGURE 4: Example of CFD results for several airfoils at Re = 10^5.

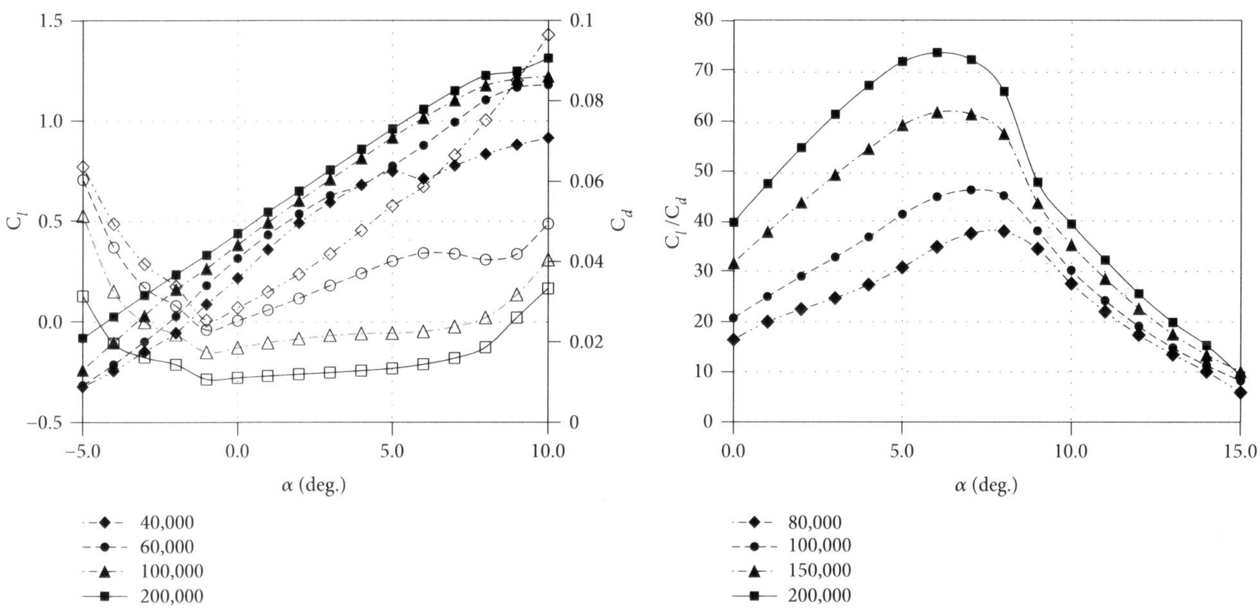

FIGURE 5: Example of aerodynamic data of DAE51 obtained by CFD.

TABLE 3: Design variables.

Symbol	Description
d	Propeller diameter
rpm	Rotational speed at design condition (rev./min)
clm	Lift coefficient at $r/R = 0.6$
clt	Lift coefficient at the blade tip

TABLE 4: Values of design variables for example cases.

Case	clm	clt
A	0.50	0.55
B	0.55	0.60
C	0.60	0.65
D	0.70	0.75
E	0.80	0.85

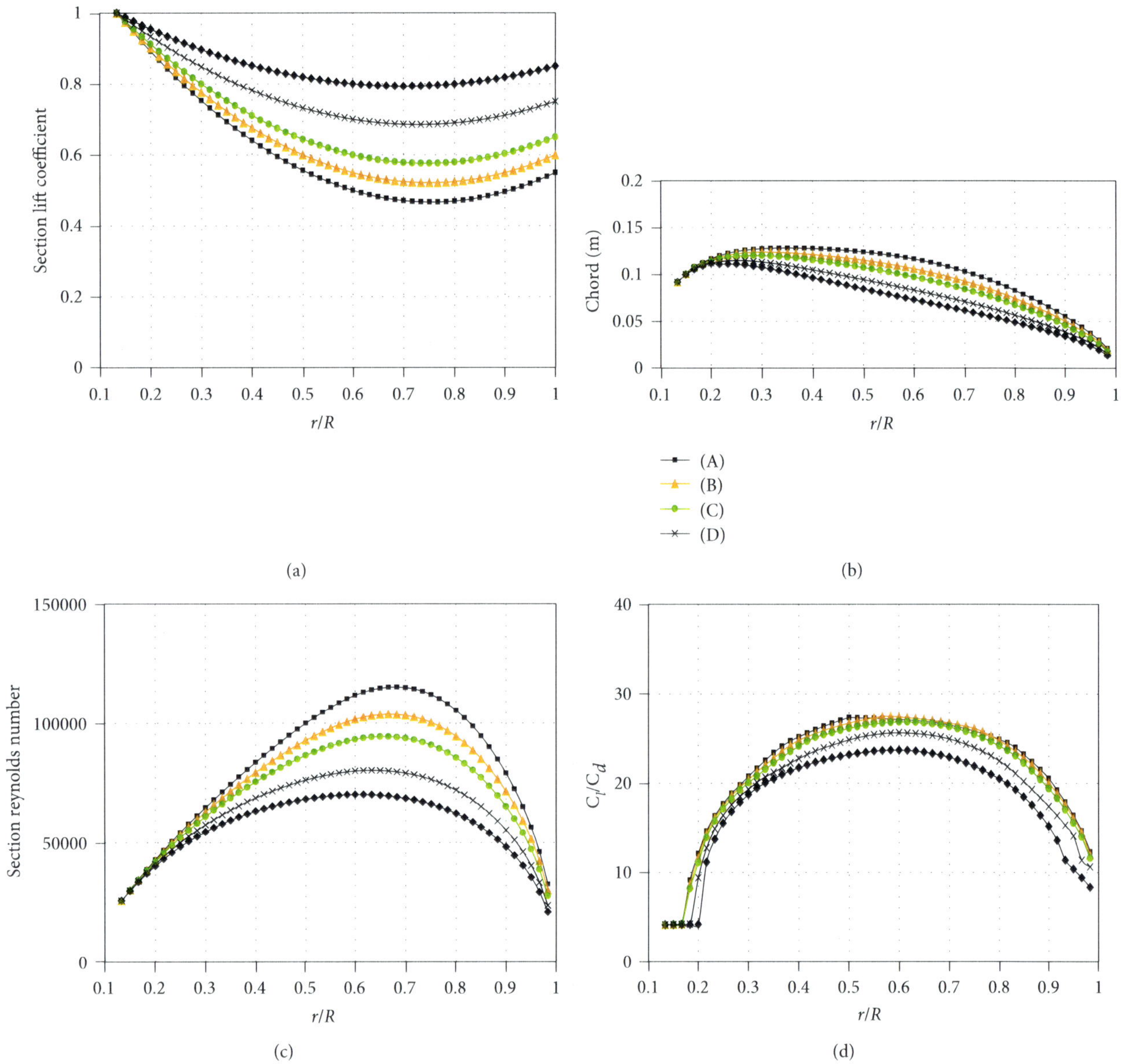

FIGURE 6: Effect of lift coefficient distribution on design.

shows the corresponding lift coefficient distributions. It is observed from Figure 6(b) that the design using the lower lift coefficient distribution yields a large chord length to produce the same thrust, which leads to a blade geometry with a relatively large planform area. It can be confirmed from Figures 6(c) and 6(d) that the higher chord Reynolds number yields a higher lift-to-drag ratio, which directly results to a higher efficiency of the propeller. It can be easily deduced that the chord length below a certain level can deteriorate the performance at high altitudes due to low chord Reynolds numbers. This means that a chord length greater than that at a certain level should be secured in order to improve efficiency. It is consistent with the observation that previous HALE UAVs have propeller blades with large chord length (large planform area) than conventional aircraft propellers. The same observations and discussions were made by Monk [16].

It was confirmed that it is possible to improve the efficiency of the propeller by imposing low values of the lift coefficient distribution. However, it should be that the benefits derived from the use of lower lift coefficients discussed above are limited up to a certain level of chord the Reynolds number only. The degree of improvement in the lift-to-drag ratio is gradually decreased as the Reynolds number increases (see Figure 6(d)). An excessively low lift coefficient will yield a large chord length that requires a high torque exceeding the limitation of the electric motor at low-altitude conditions. Therefore, it is necessary to carefully evaluate whether the flow is within the sufficiently low Reynolds number range at the design point, and whether imposing low lift coefficients will definitely improve performance.

2.5. Response Surface Based on Kriging Metamodel. For a given combination of design variables, the propeller design

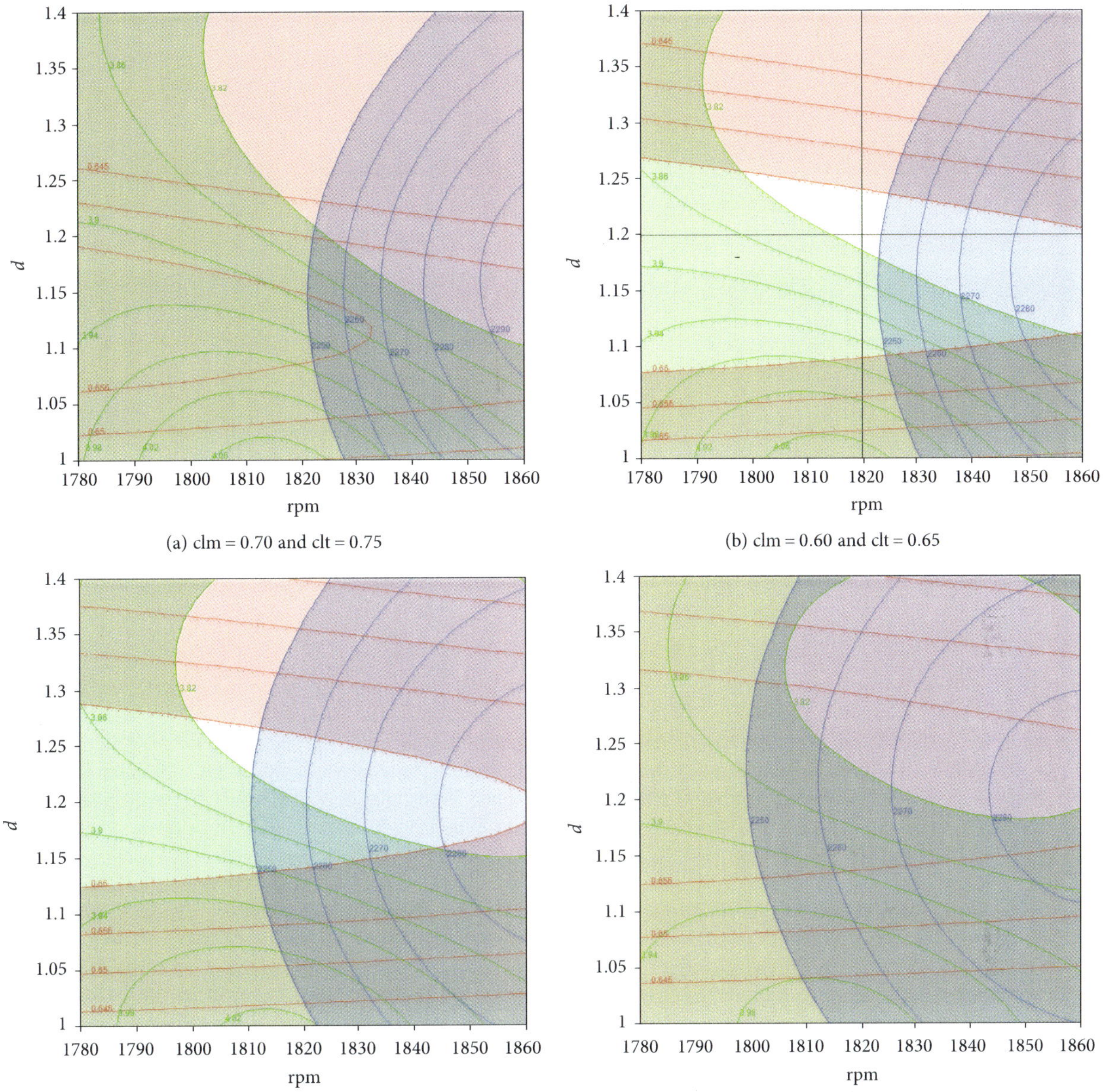

(a) clm = 0.70 and clt = 0.75

(b) clm = 0.60 and clt = 0.65

(c) clm = 0.50 and clt = 0.55

(d) clm = 0.45 and clt = 0.50

FIGURE 7: Contours of response surface from kriging metamodel (rpm-*d* plane).

is uniquely determined by the methods described above. The efficiency of the propeller at the design point, required torque at sea level, and required rpm at altitude of 18 km were set as responses for each design. All three responses are evaluated for the climb condition at rate of climb of 1 m/s.

The parameter space of design variables was examined through a response surface methodology (RSM) [37]. Test points were generated through the design of experiment (DOE) [37], and the Latin hypercube sampling [38] is used as DOE method. The propeller design is conducted, and responses are calculated by off-design analysis for each test point. Using the responses, the kriging metamodel [39] was generated to construct the response surface for each response. The DOE and Gaussian process for the metamodel generation were carried out using the commercial program JMP 9.0 [40].

Thirty-two design variable combinations (test points) were extracted by using DOE of Latin hypercube option implemented in JMP. Using the propeller design and analysis code, a propeller for each combination was designed and efficiency at the design point was obtained. For each propeller, the required torque at sea level and rpm at 18.5 km are computed from off-design analysis. Then, the data of three responses for all propellers are mounted to JMP for generation of the kriging metamodel. Since design and off-design analyses are needed only for 32 propeller cases, the interaction of JMP 9.0 with the other tools was not automated. The design, off-design analysis, data recording, synthesis, and transfer were carried out manually.

Figure 7 shows the response surface contours of three responses on an rpm-*d* plane for several cases of lift coefficient distributions. In each figure, the abscissa is rpm,

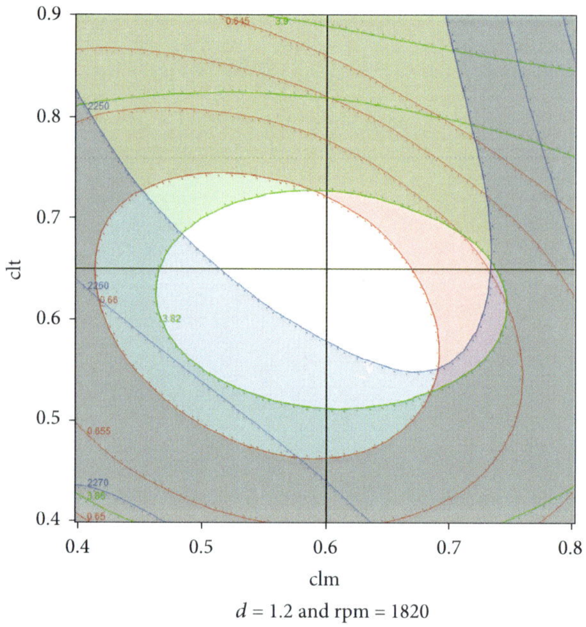

FIGURE 8: Example contour of response surface from the kriging metamodel (clm-clt plane).

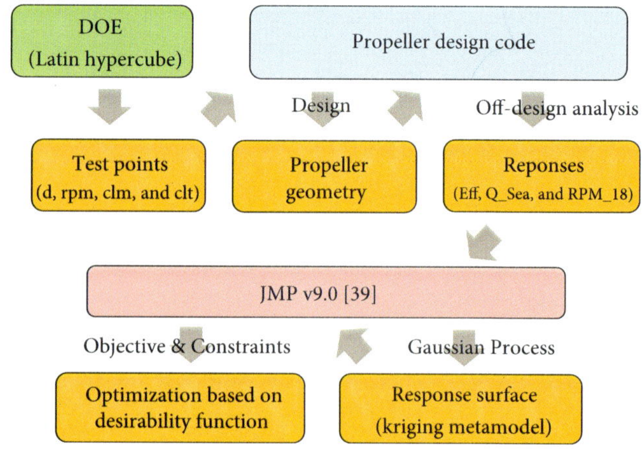

FIGURE 9: Schematic of optimization process.

centered at 1820 with an interval of 10, whereas the ordinate is d, centered at 1.2 m with an interval of 0.05 m. Each shaded area represents the combination of design variables, corresponding to an unallowable design due to a constraint on any one of the three responses. The area shaded in red corresponds to designs in which the efficiency is less than 66% at the design point. The green shaded area represents designs in which the required torque exceeds 3.82 N·m at sea level. Similarly, the blue shaded area shows the designs in which rpm at an altitude of 18 km exceeds 2250. The available designs which satisfy all the constraints with efficiency greater than 66% are designs corresponding to the unshaded area only. It is observed from Figure 7 that only a limited range of design variables yields an allowable design. There is no possible design for the lift coefficient distribution greater than 0.7 (Figure 7(a)). It was confirmed that only clm and clt values, approximately between 0.5 and 0.7, allow possible designs with limited combination of diameter and rpm (see Figures 7(b) and 7(c)). Moreover, there are no available designs for values less than 0.5 (see Figure 7(d)). Changing the diameter and rpm in the direction of the maximum efficiency at the design point is bounded by constraints of the required motor performance at sea level and altitude of 18.5 km.

Figure 8 shows another response surface contour on the clm-clt plane for $d = 1.2$ m and rpm = 1820. We see again that only limited ranges of clm and clt give allowable design for a given diameter and rpm. Similarly, we examined response surfaces for several diameters and rpms. The diameter and rpm with an interval of 0.01 m and 20, respectively, were considered. It was confirmed that available designs exist only for diameters within 1.17~1.29 m. For an rpm lower than 1820, although a higher efficiency design can be possible due to increased chord length, the required torque at sea level exceeds the maximum torque of the motor. On the other hand, for an rpm higher than 1820, although the maximum torque constraint is satisfied, the efficiency is decreased due to the shortened chord length, and the rpm at an altitude of 18 km exceeded the motor specification.

2.6. Optimization Based on Desirability Function. The optimization was conducted based on the concept of the desirability function by using JMP. Figure 9 depicts the basic procedure for the optimization. The objective functions, constraint functions, and response surfaces for responses must be defined prior to exploring an optimal solution based on desirability. The objective function was set to maximize the propeller efficiency at the design point. Two constraint functions were defined for the required torque at the sea level and required rpm at an altitude of 18 km. Considering motor specifications, the torque and rpm constraints were set so as not to exceed 3.82 Nm and 2250, respectively. The kriging metamodel described in the previous subsection is used as response surfaces.

The desirability functions must be set to achieve the objective and constraints for each response surface. The desirability was set in the direction of maximizing the efficiency and minimizing the torque and rpm. In addition, the weight for each desirability function needs to be imposed according to importance. The importance was set to 1.0, 0.8, and 0.6 for efficiency, torque, and rpm, respectively. For the desirability function and weights provided, we explored and determined the optimal solution. Figure 10 shows the determination of optimal solution conducted by using JMP. The optimal solution was found as 1.22, 1817.9, 0.482, and 0.648 for d, rpm, clm, and clt, respectively. It was observed that the optimal solution can be slightly changed by the imposed values on importance. The solutions corresponding to various imposed values on importance for the torque and rpm constraints were obtained and examined. Since the maximizing efficiency at the design point is the main objective of the present design, importance for efficiency was fixed to 1.0. The results indicated that the influence of changing importance between 0.4 and 1.0 does not noticeable in the solution within 3 digits.

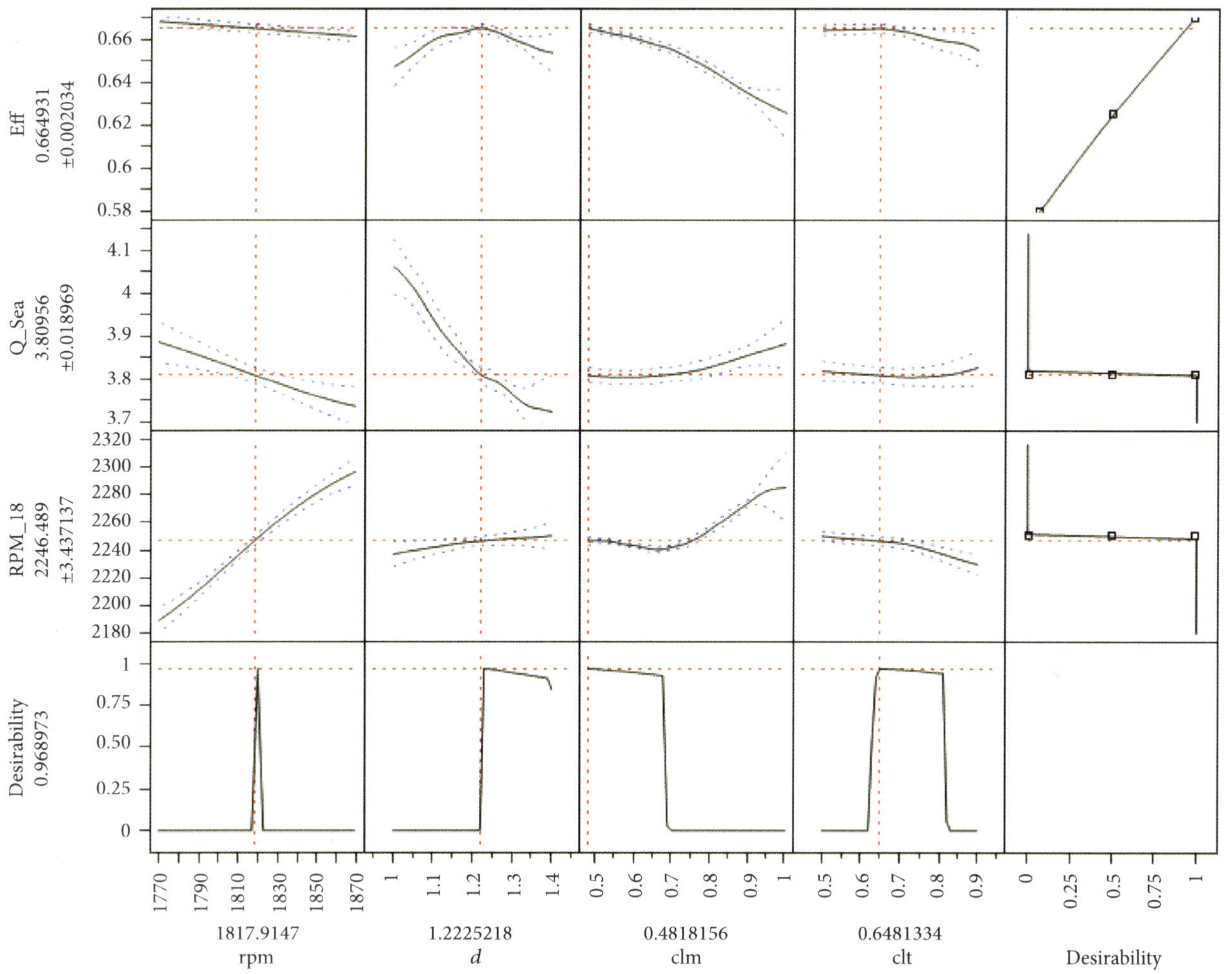

FIGURE 10: Optimization based on desirability function in JMP.

TABLE 5: Comparison of optimization and final adjusted design of the EAV-3 propeller.

Case	d	rpm	clm	clt	Efficiency
Optimization	1.22	1817.9	0.482	0.648	0.6633
Final design (adjust)	1.2	1820	0.60	0.65	0.6604

2.7. Final Design and Manufacturing. The result of the optimization described in the previous subsection is summarized in Table 5. The optimized result was examined, and the design was slightly adjusted by considering manufacturing, system integration, and requirements from other disciplines and so on. The diameter and rpm were truncated to 1.2 m and 1820, respectively, which are the nearest value from the optimization. We reviewed the response surface for $d = 1.2$ m and rotation speed at 1820 rpm, which is already shown in Figure 8. The optimized value of clm (0.482) is located around the left boundary of the region for allowable design, and approximately lies at the boundary defined by rpm constraint at 18 km. Since the development of a customized motor was under progress at that time, it was decided to impose a margin on the propeller design considering the uncertainty of motor specification. The clm value was adjusted to a higher value of 0.60 to relieve the motor requirement. The value is approximately at the center of the region of available design, which is marked as a vertical solid line (see Figure 8). The clt value is truncated to 0.65, which is marked as a horizontal solid line in the figure. These adjustments are compromised by the efficiency reduction of 0.29%p at the design point, as given in Table 5.

Figure 11 illustrates the comparison of chord (planform geometry) and twist angle distributions between the optimization and final adjusted design. It can be seen from Figure 11 that adjusted design has a slightly short chord length at the midspan region. This is primarily due to the adjustment of clm to a higher value. This behavior is discussed in Section 2.4. In the adjusted design, the twist angle at the midspan region is slightly increased to compensate for the decrease in chord length (see Figure 11(b)). Furthermore, in order to achieve sufficient structural robustness, the structural design, analysis, and static load test were independently conducted. To avoid structural interference by the motor and achieve short length of rotational axis, the design twist angle was adjusted by limiting it to 50° in the transition region at spinner location ($r = 0.08$–0.12 m,

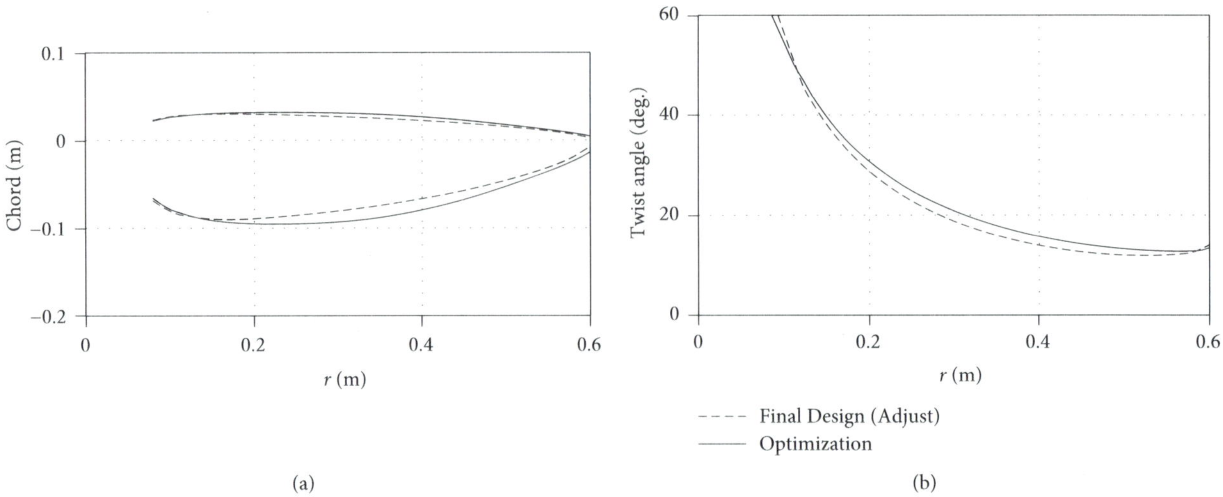

FIGURE 11: Comparison of geometry for optimization and final design.

FIGURE 12: Manufactured and mounted propeller for EAV-3.

$r/R = 0.133–0.2$). Figure 12 shows the final modeling for manufacturing and actual propeller fabricated with composite materials. For the accuracy of manufactured geometry of the propeller, an outer mold is made of aluminum with precision machining. Carbon fiber-reinforced plastic is carefully laminated and fabricated on the outer mold. A total eight pieces of propeller were produced, and quality of product is examined individually.

For an overview, a flow chart which briefly illustrates the design procedure used in the present study is shown in Figure 13.

3. Wind Tunnel Test

3.1. Test Facility and Measurement Device. The wind tunnel test is conducted to evaluate the performance of the EAV-3 propeller [41, 42]. The test was carried out at KARI low speed wind tunnel with a closed-circuit test section, whose width and height are 4 m and 3 m, respectively. The maximum speed is 120 m/s, and the turbulence intensity is 0.07% at 88 m/s.

A measurement device capable of measuring the performance of propellers having diameters of 1–2 m was designed and installed [41]. Figure 14 shows schematic and modeling of the device. The validity of the measuring concept implemented on the device was confirmed in our previous experimental study conducted for a propeller with a diameter that is less than 1 m by using the same concept [43]. A torque sensor and motor properly arranged behind the propeller constitute the driver module. This module is placed on a pair of cylinder guides, which enable it to freely move back and forth, and connected to a base via a load cell. The force pulling the module forward is directly measured by the load cell as thrust. The torque exerted on the driving shaft is directly measured by the torque sensor between the motor and propeller.

The motor, torque sensor, and load cell were selected based on the prediction of the thrust, torque, and rpm range. To generate the required torque and power output and be capable of precise rpm control, an industrial alternating current motor was used. The motor selected was Mitsubishi's HG-JR 153, with a maximum rpm of 6000, a maximum

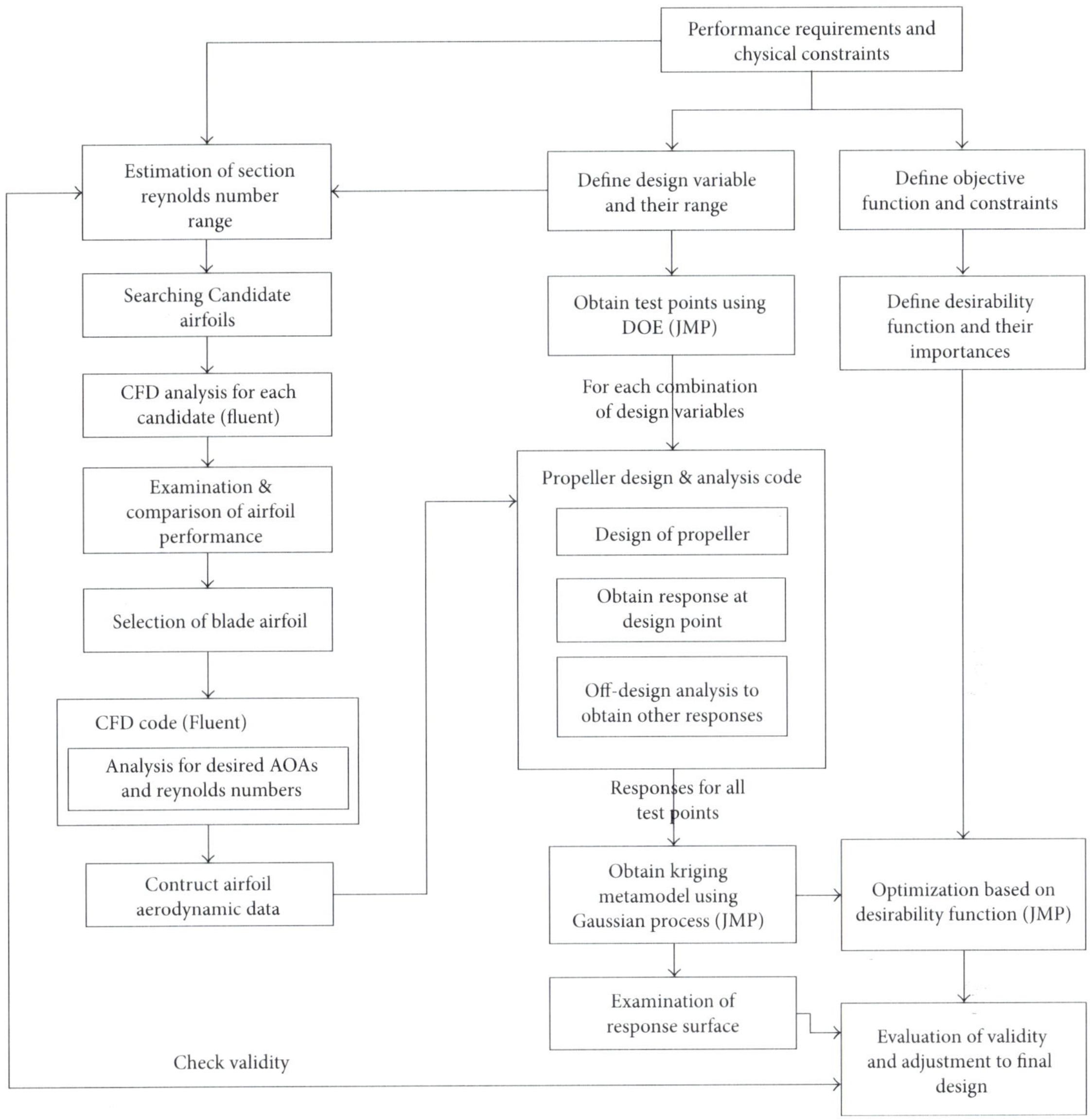

FIGURE 13: Flow chart of design of propeller for EAV-3.

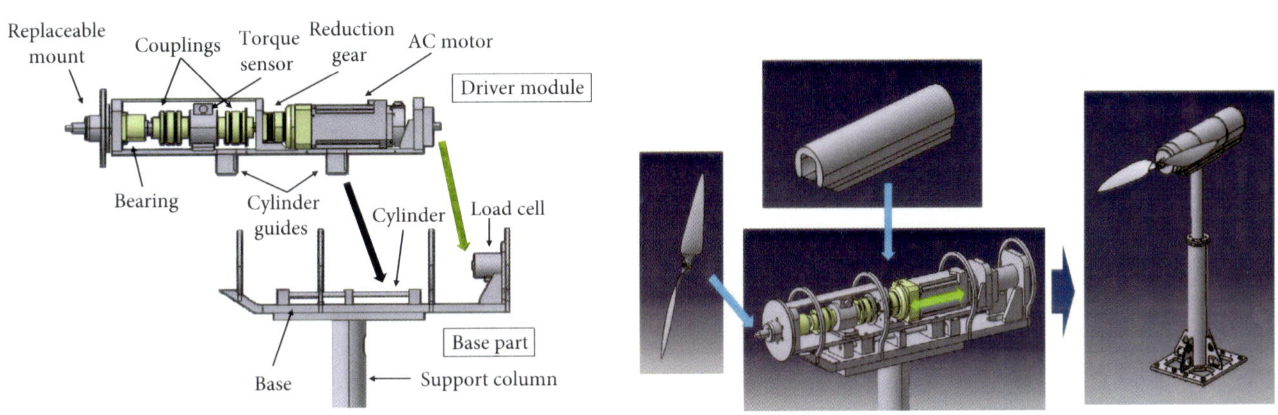

FIGURE 14: Schematic and modeling of measurement device.

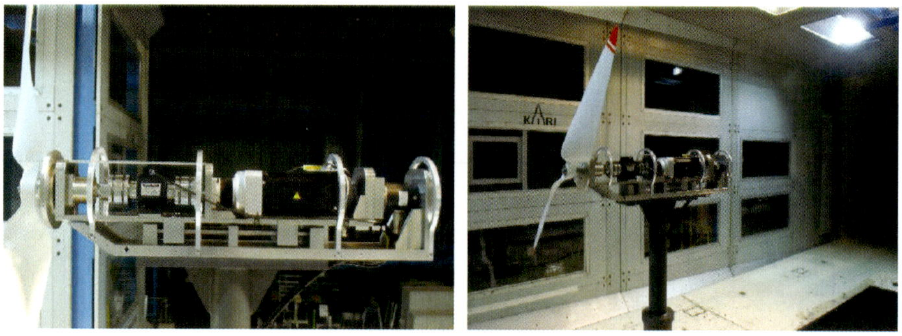

Figure 15: Installation of measurement device.

torque of 14.3 N·m, and a rated output of 1.5 kW. A reduction gear with a ratio of 3 : 1 was attached to the motor to allow the operation under low rpm and high torque conditions. The load cell and torque sensor chosen were Dacell UMI (20 kgf) and HBM T-22 (2 kgf·cm), respectively. The rpm is measured using Autonics BF4RP optical sensor with a rotating plate attached to the coupling.

The main structure of the driver module and base is designed as a single part with precision machining to minimize tolerances in the assembly and alignment. The steel use stainless material was used to ensure sufficient strength for minimizing deformation. Starting from the design stage, the assembly method and components carefully considered the alignment accuracy of the rotational axis. The cable connection ports of the torque sensor and motor were modified/remodeled to minimize the outer diameter of the device. The control rack and panel were separately designed and manufactured for the accurate speed control of the motor. Figure 15 shows the manufactured/installed measurement device. The length of the main body is 791 mm, and the outer diameter is 196 mm.

3.2. Calibration and Data Acquisition. The sensor calibration was performed using a weight. The load in the direction of the thrust was applied to the load cell in a stepwise manner and thereafter removed. The slope of the obtained signal was then taken as the conversion factor. The load cell signal showed a hysteresis of approximately 0.5~1.0 N, mainly due to the friction force in the cylinder guide. Since the applied torque and signal from the torque sensor are in good agreement with each other, no conversion factor was applied. The first and second zeroes of the sensor were recorded for each run and used as zero offset correction.

The measurement uncertainty of the device was estimated from calibration tests. The errors from both the calibration factor and zero offset correction were estimated to be within 0.1% for both the thrust and torque. Moreover, the same level of errors was confirmed for repeatability and nonlinearity. However, due to the friction in the sliding guide, the hysteresis error for the thrust measurement was relatively high and confirmed as the dominant factor of uncertainty. The total uncertainty in the measured thrust was estimated to be approximately 1.7% at a maximum. As for torque, additional error factors were not identified, and the overall uncertainty was estimated to be within 0.1%. Because the load cell and torque sensor are independent, and the manufacturing is precisely performed considering the axial alignment, it is considered that the occurrence of the interference error between the sensors and other axes is negligible.

Different from conventional aerodynamic tests, the dynamic pressure is considerably low at 20–50 Pa, because the test of the EAV-3 propeller is usually performed at low wind speeds of less than 10 m/s. In this case, the accurate measurement of the dynamic pressure of freestream is required to ensure reliability and repeatability of results, since the calculated wind speed of the test section will substantially vary according to a small difference in dynamic pressure. The dynamic pressure was determined by measuring the static pressure difference between the stagnation chamber and inlet of the test section. The differential pressure was measured using an MKS Baratron sensor and a Mensor model 2106 transducer. The accuracy is found to be within 1 Pa. The zero values of the differential pressure were recorded in several preliminary tests at various days and times. The data indicated that these zero values change from −3 to 3 Pa, depending on the day and time, even under the wind tunnel stop state. The differential pressures were recorded at the start and end of each run, and the dynamic pressure in the test section was corrected using this value.

The flow temperature was measured with a 100 Ω resistance temperature detector installed through the side wall of the inlet of the test section. The difference in temperature readings between this detector and another temperature sensor installed in the device was found to be less than approximately 0.2°C. The air density was calculated according to the measured temperature and humidity, and wind speed was calculated from the density and the measured dynamic pressure.

When all sensor signals become stable, data acquired for 5 seconds with sampling rate of 10 Hz were averaged. Thereafter, the thrust and torque of the propeller were determined by applying the corrections.

3.3. Test Results. The test was conducted at a fixed rpm and varying wind speed to measure performances at various advance ratios (J). Figure 16(a) shows the measured thrust and torque at 700, 867, 950, and 1050 rpm. In the figure, the filled and nonfilled symbols represent the thrust and torque, respectively. Performance coefficients calculated

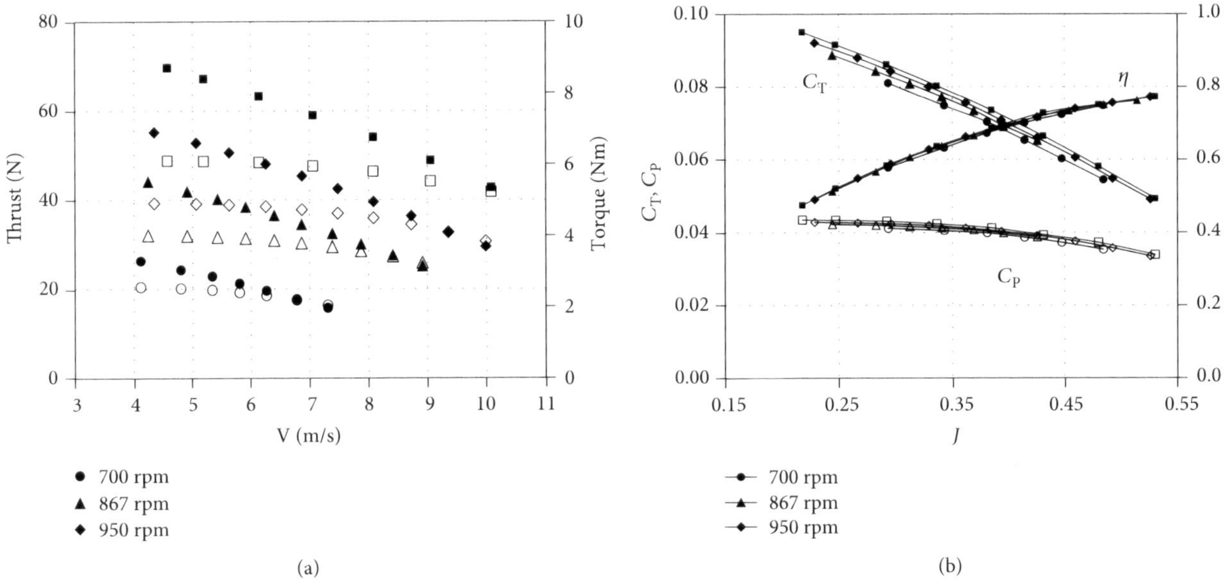

FIGURE 16: (a) Measured thrust and torque and (b) performance coefficients.

from measured data are shown in Figure 16(b). The thrust coefficient (C_T), power coefficient (C_P), and efficiency (η) are as defined in (1).

$$J = \frac{V}{nD},$$
$$C_T = \frac{T}{\rho n^2 D^4},$$
$$C_P = \frac{Q\omega}{\rho n^3 D^5}, \quad (1)$$
$$\eta = J\frac{C_T}{C_P}.$$

The difference among the performance coefficients is observed according to the rpm; the higher the rpm is, the higher the thrust and power coefficients are. The efficiency was practically the same at more than 867 rpm, but slightly lower by approximately 1%p at 700 rpm. In our previous study, it was confirmed that efficiency degradation may occur if the blade chord Reynolds number decreases below 10^5 [43]. Similar observations were made in other studies that measured propeller performance in the low Reynolds number range [44, 45]. Considering the rotational speed and chord length of the blade, the condition of 700 rpm ($Re_c = 10^5 - 2 \times 10^5$) is expected to effect a slight decrease in efficiency. However, it is estimated that a considerable decrease in efficiency is further expected at lower rpms.

3.4. Correction of Wall Blockage Effect. Even if the propeller generates thrust in the closed test section, the continuity equation at any cross-sectional area of the test section must be satisfied. This yields the same effect as the propeller was under a lower wind speed when it operates in the absence of the wall. As a result, the propeller attains the thrust increasing effect. Glauert correction [46] is commonly used as a method for correcting the blockage effect [45, 46]. As shown in (2), this method predicts the equivalent freestream velocity (V') for the case without the wall, which yields the same thrust generated in the wind tunnel.

$$V' = V\left(1 - \frac{\tau_4 \alpha_1}{2\sqrt{1 + 2\tau_4}}\right),$$
$$\tau_4 = \frac{T}{\rho A_{disk} V^2}, \quad (2)$$
$$\alpha_1 = \frac{A_{disk}}{A_{tunnel}}.$$

In addition, various correction techniques have been studied and proposed. However, in order to apply recent methods, the static pressure distribution on the test section wall in the downstream direction must be measured separately [47]. Glauert correction is also known to provide a similar accuracy to the methods that use static pressure measurements in the general advance ration range ($J = 0.2$–0.6) [47].

Figure 17 shows the results of the case of 867 rpm with Glauert correction. For comparison, data without correction are shown as symbols with no fill. As mentioned above, the correction yields a lower freestream speed than the measured wind speed in the test. It can be observed that the performance coefficient curve slightly shifts to the left, and the change in the advance ratio is approximately 0.005–0.01. The results of the two 867 rpm test cases are plotted together in Figure 17. It can be confirmed that the measurement system has good repeatability. It is noted here that the possibility of a change in the performance curve by the wall effect needs to be taken into consideration when the test is conducted in the closed test section. If the influence is not negligible, the test result can be improved by applying the blockage effect correction.

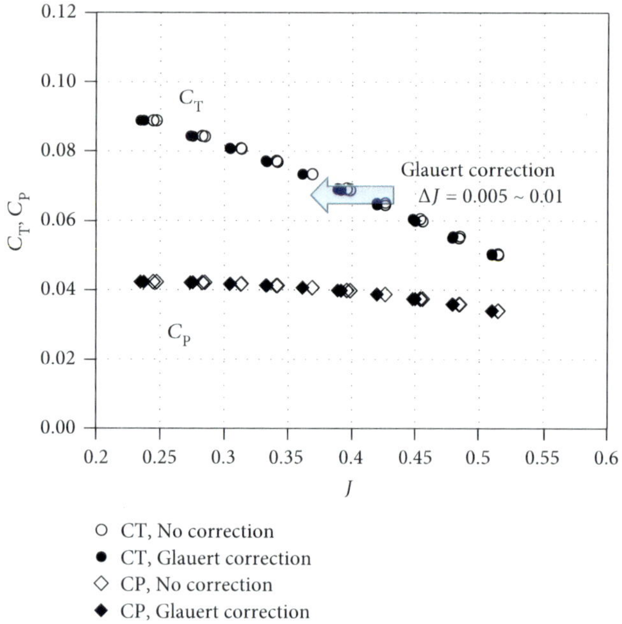

FIGURE 17: Change of performance coefficient due to Glauert correction.

4. CFD Analysis

To evaluate the performance of the propeller, numerical analyses were performed by using a commercial CFD code. It was intended to compare the calculation results with the test data to mutually confirm the reliability and validity of the methods of the wind tunnel test and numerical analysis. The basic approach and method of analysis are the same as that in our previous study on a 0.7 m diameter propeller [43].

4.1. Computational Grid. The computational grids are generated by using GAMBIT and T-GRID [46]. For the surface mesh, the blade is composed of structured-type quadrilateral cells. The hub, blade tips, and connecting regions consist of unstructured triangular cells. The total number of surface mesh is approximately 3×10^5. To properly capture the viscous boundary layer on the surface, 18 layers of boundary layer cells are generated with an initial height of 4×10^{-5} m. The height of the first cell from the surface corresponds to a $y+$ value of approximately 1.

Figure 18 shows the size of the computational domain and types of boundary conditions. The far boundary is cylindrical in shape. The diameter and height were set to 10 and 15 times the diameter of the propeller, respectively. The inlet and outlet boundaries are located at distances of 5D upstream and 10D downstream, respectively. The velocity inlet boundary condition is imposed at the inlet and side boundaries, whereas the pressure outlet boundary condition is imposed at the outlet boundary.

The region around the propeller is designated as a cylindrical-shaped region, and the boundary between the inside and outside regions is set as an interface. This mesh configuration enables the use of both the multiple reference frame (MRF), which adds a source term to simulate the rotation, and the sliding mesh model (SMM), which directly rotates the inside region. In the vicinity of the propeller, an unstructured grid of tetrahedral cells was generated at a growth rate of 1.2. The structured-type grid with hexahedron cells are generated for the outer region by dividing it into several blocks. The total number of volume cells is approximately 14.2 million.

4.2. Method of Analysis. The commercial CFD code FLUENT [48] was used for the analysis. Since the expected blade tip Mach number which is estimated based on the velocity and rpm did not exceed 0.2, the pressure-based solver was used assuming an incompressible flow. The semi-implicit method for pressure-linked equation (SIMPLE) algorithm was selected for the pressure-velocity coupling. For spatial discretization, the standard method was chosen for pressure, and second-order upwind schemes were used for the other variables. A four-equation model, transition-SST, which is based on γ-Re_θ, is used as turbulence model. The turbulence intensity at the far boundary and intermittency were set to 0.1% and 0.2, respectively.

In order to reduce the computational cost, the rotation of the propeller is simulated by steady-state calculation of the MRF method. On the other hand, unsteady analysis using the SMM method was individually performed to evaluate the validity of the MRF method. The thrust and torque from the two methods only showed a difference of less than 0.1%. The iteration was continued until the residuals of the continuity equation and other equations were reduced below 10^{-3} and 10^{-5}, respectively, while the thrust and torque were kept constant.

4.3. Analysis Results. The test calculation is carried out for the condition of the design point (15 km, 1820 rpm, $V = 14.6$ m/s). The results are summarized in Table 6 for comparison with prediction from the design code. The design code result is the same with the "final design" result given in Table 5. We note here that the design code result considered only blades without adjustment of the twist angle near the spinner (Section 2.7). On the other hand, the geometry for CFD analyses included the hub, connecting region, and the adjustment of twist angle in the transition region near the spinner. Despite the difference in the geometry, it is evident that the two results are in good agreement. It must be noted that at the design point, the stall is minimized over the entire area of the blade, and the flow at each cross section in CFD becomes practically identical to that of BEMT prediction. Additionally, since the aerodynamic coefficients for BEMT were obtained by the same CFD analysis, we can definitely expect that the CFD results and design code predictions approximate each other.

For an altitude of 15 km and at sea level, analyses at various rpms with a fixed flight speed are conducted. The performance coefficient results with respect to the advance ratio are shown in Figure 19. As previously observed in Table 6, Figure 19(a) shows that, at an altitude of 15 km (design point), the design code predictions and the CFD results agree well, both qualitatively and quantitatively. However, under sea level condition, which corresponds to the off-design point, the design code predictions showed that,

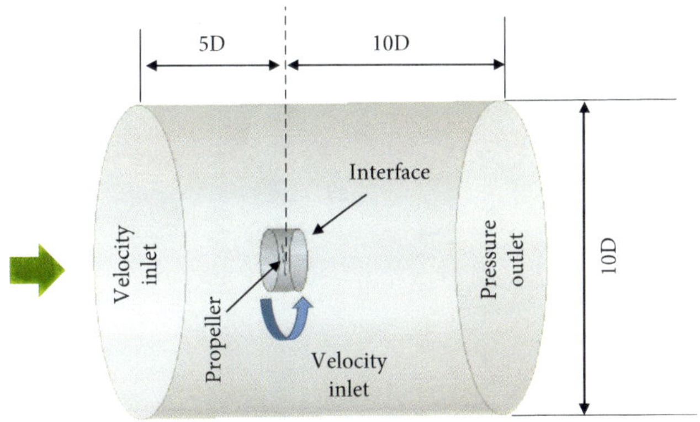

FIGURE 18: Computational domain and boundary conditions for the case of the propeller alone.

TABLE 6: Comparison of performance at design point.

	Thrust (N)	Torque (Nm)	Efficiency (%)
Design code	21.31	2.47	66.0
CFD	21.17	2.48	65.4

to a moderate extent, the thrust coefficient has higher values than those of the CFD, although their tendency agrees well. It is concluded that the design code yields a more optimistic prediction in thrust due to the limitation of BEMT, which does not properly reflect the effect of wake and its contraction at high load conditions.

Figure 19(b) shows the results for cases under various forward speed conditions at 867 rpm. For comparison, the wind tunnel test results of the previous section are shown together with them. The results agree well with test data qualitatively, but the thrust and power coefficients are found to be slightly lower than that of test data. In order to analyze the cause of this difference, the level of uncertainty that can inherently arise from the CFD was evaluated. From grid tests and analyses using other CFD coeds, such as OpenFOAM [49] and Star-CCM+ [50], the uncertainties that can arise from CFD are found to be significantly lower than the difference observed in Figure 19(b) [51].

In the wind tunnel test, the measurement devices and support column located downstream of the propeller can cause a blockage effect. Thus, their influence can be considered as one of the causes in the difference between the analysis results and those of the test. Since the results in Figure 19 correspond to cases of propeller-alone analysis, the blockage effect and its level of influence on the wind tunnel test were evaluated.

For convenience and to reduce the number of grids, the geometry of the device and support column are simplified, as shown in Figure 20. The connection between the propeller and device is ignored from the analysis. Approximately 112000 surface meshes and 16 layers of boundary layer cells were generated over the device and column geometry. As depicted in Figure 20(b), the interface surface of cylindrical shape surrounding the propeller was set to separate the inside and outside regions. Different from the case of the propeller alone analysis, the flow symmetry with respect to the rotational axis is not guaranteed in the presence of the device. Therefore, the unsteady analysis with the SMM technique was performed to directly rotate the propeller in the calculation.

Analyses for several forward speeds with a fixed rpm of 867 were performed. For each case, a total of 12 rotations were calculated with a time step size corresponding to 2° of rotation. The steady solution with the MRF method was obtained first and thereafter used as the initial condition for the unsteady calculation. Figure 21 illustrates the time history of the thrust and torque for several cases. It is observed that the thrust and torque fluctuate with period, which coincides with that of rotation. For the purpose of comparison, the analysis result of the propeller alone case at 5.8 m/s is plotted together with the dashed line. We readily observe that the thrust and torque slightly increase due to the presence of the device and support column.

The thrust and torque were determined as time-averaged values during the last two rotations of the calculation. The summarized results in Tables 7 and 8 are compared with that of the propeller alone analysis. The thrust increased by approximately 1.8–8.9%, and the torque increased by approximately 1.8–7.5%. The level of increase becomes higher as the speed increases (higher advance ratio). The results, including the blockage effect, are shown in Figure 22 together with those in Figure 19. It can be confirmed that the thrust and torque, including the blockage effect, are considerably close to the wind tunnel test data (Figure 22(a)). It is confirmed that the blockage effect has no significant influence on the efficiency (Figure 22(b)) since it increases both thrust and torque. The analysis results show the efficiency to be approximately 1–1.5%p higher than that from the wind tunnel data and that their tendency agrees well.

The results imply that the level of the blockage effect from the device and support column should be evaluated properly when constructing a propeller test system. By taking into account the blockage effect in analyzing and processing measured data, accuracy can be improved. In the previous studies on propeller testing [44], the blockage effect was corrected by using the method proposed by Barlow et al. [52]. However, the measured thrust and torque can vary considerably depending on the shape and position of the device. Therefore, it is necessary to evaluate the accuracy of the correction method for each system.

Figure 23 shows the efficiency curves obtained on the basis of the propeller alone analysis under sea level, 15 km,

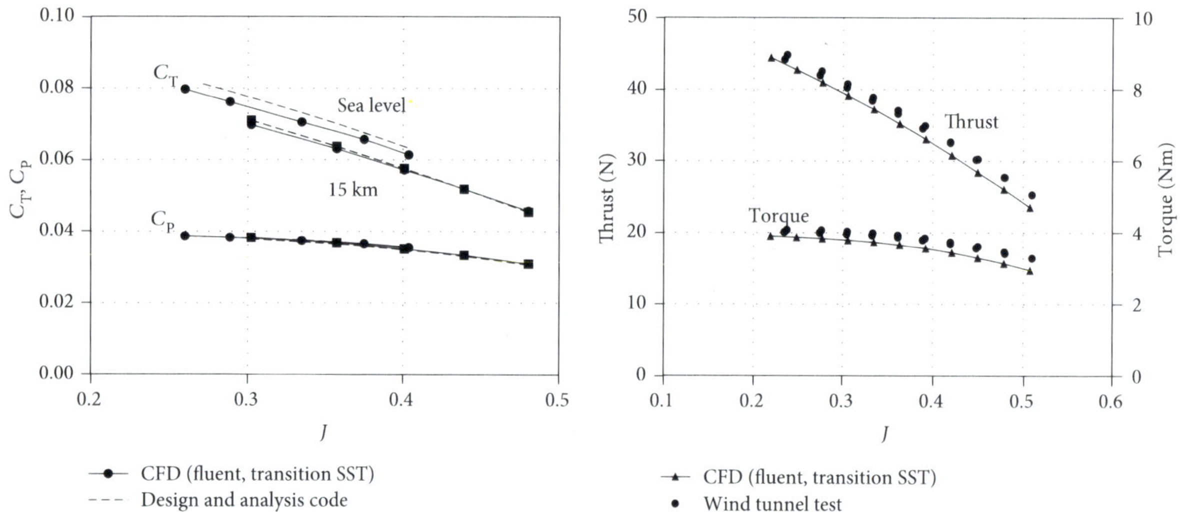

(a) (b)

FIGURE 19: Comparison of CFD results with (a) design code and (b) wind tunnel test.

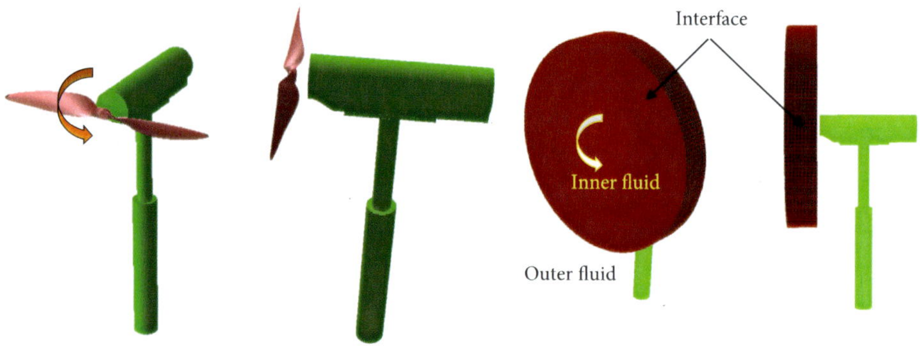

(a) (b)

FIGURE 20: Comparison of CFD results with (a) design code and (b) wind tunnel test.

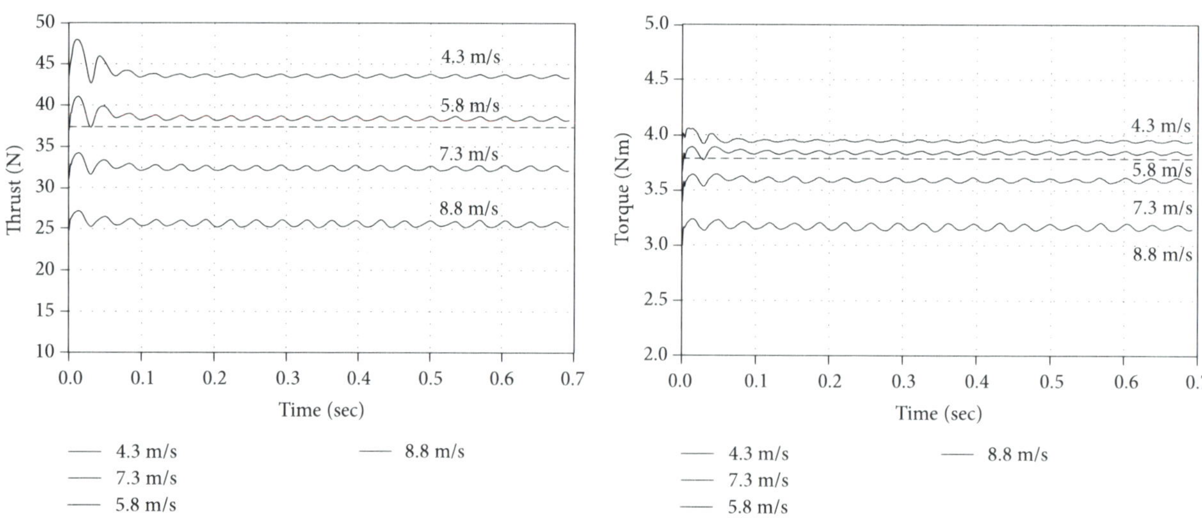

(a) (b)

FIGURE 21: Comparison of CFD results with (a) design code and (b) wind tunnel test.

TABLE 7: Comparison of thrust.

V (m/s)	Propeller only	With device/column	Difference
4.3	42.72	43.52	+0.80
5.8	37.22	38.38	+1.16
7.3	30.78	32.38	+1.60
8.8	23.54	25.57	+2.03

TABLE 8: Comparison of torque.

V (m/s)	Propeller only	With device/column	Difference
4.3	3.87	3.94	+0.07
5.8	3.73	3.84	+0.11
7.3	3.44	3.59	+0.15
8.8	2.94	3.16	+0.22

and 18 km altitude conditions. Numerical results clearly show that under high-altitude conditions, the efficiency decreases over the entire range of the advance ratio due to the decrease in the Reynolds number. The reduction in the thrust coefficient and efficiency in the low Reynolds number range have been well known from previous studies on small propellers [44, 45, 53]. However, the design operating advance ratio of EAV-3 increases as the altitude increases. The advance ratios for the sea level, 15 km, and 18 km flights are estimated as $J = 0.334$, 0.401, and 0.411, respectively. In spite of the reduction in efficiency, it can be observed from Figure 23 that the efficiency of the designed propeller remains at more than 65% even at high altitudes.

5. Flight Test Data

In 2015 and 2016, KARI conducted a flight test of EAV-3. The takeoff and flyover of EAV-3 at a low altitude are shown in Figure 24. During the flight, data from various sensors were recorded at 10 Hz in an onboard storage. After landing, these data were collected for analysis. This section intends to briefly discuss the attempt to evaluate the performance of the propeller based on available data and several assumptions.

In the wind tunnel test, propeller performances were directly measured through the sensors. In case of the CFD, the performance was directly obtained through the analysis result. However, in the flight test, the aerodynamic performance of both the propeller and aircraft cannot be measured directly. In order to evaluate the propeller performance accurately, reliable aerodynamic data of the aircraft during its flight are needed and vice versa.

Additionally, during the actual flight, there are practically no steady-state conditions in which parameters, such as speed and attitude, are kept constant. The limitation on the flyable airspace imposed by the government (10 km radius) makes the acquisition of steady-state data considerably difficult since the aircraft must continuously change direction to remain in the airspace. These factors make flight data analysis difficult. Because the aerodynamic performance can be predicted based only on indirect data acquired from sensors, the uncertainty and prediction error can significantly increase due to the underlying assumptions.

5.1. Method of Analysis. The atmospheric density, flight speed, rpm, and thrust and torque values are required to obtain the performance coefficient with respect to the advance ratio. However, all quantities except rpm, which is measured from the rpm sensor, are not directly measured in the flight test but should be estimated from available data. In this study, we attempted to estimate these values according to simple assumptions and approximations.

There are two kinds of altitudes obtainable from the flight data. These are the pressure altitude, which is calculated from the atmospheric pressure measured by the air data system (ADS), and the recorded GPS altitude. The difference between these two altitudes was confirmed to be less than 15 m for the entire flight. The density is assumed to be the standard atmospheric value according to the GPS altitude data. The flight speed is computed using the dynamic pressure measured from ADS and the estimated density. The rpm of each motor was determined by averaging sensor data at 1-second intervals at a particular time of interest. The rpm for data analysis was obtained as average rpm value of the two motors.

The power consumption of the propeller was calculated by using the recorded voltage and current data supplied to the motor electronic speed controller (ESC). The actual power delivered to the propeller was estimated by assuming an ESC efficiency as 96%. The torque is calculated using the rpm and the estimated power. The climb speed (rate of climb in m/s) was determined by average climb speed for 2 s centered at a particular time of interest using the GPS altitude.

To estimate the drag of EAV-3, the drag polar of the main wing was obtained by CFD analysis under several altitude conditions. The basic principle and method of analysis are the same as those described in Section 4. Although the main wing is a flexible structure, it is assumed as rigid body in the analysis. The results are curve fitted to obtain the polynomial equation for the drag coefficient with respect to the angle of attack. The increment in the drag with altitude, which is different for each angle of attack, is assumed to be a quadratic variation, from the sea level to the 18 km analysis results. Based on further analyses, the additional drag by the tail wing and fuselage was approximated to 130 counts at sea level. The degree of the additional drag increase with altitude is assumed to be the same as that of the main wing. The summation of contributions from the main wing and the other parts results in the estimation of the total drag coefficient of EAV-3. The total drag was calculated by multiplying the drag coefficient, dynamic pressure, and wing area.

The power required during the climb flight is calculated by using the drag (D), flight speed (V), weight (W), and climb speed (V_c), as given in (3). The propeller thrust is simply predicted from the relationship $P = TV$ by assuming that the propeller generates the required power.

$$P = DV + WV_c. \tag{3}$$

As mentioned above, it is difficult to maintain a steady state with a fixed attitude due to the continuous maneuvering

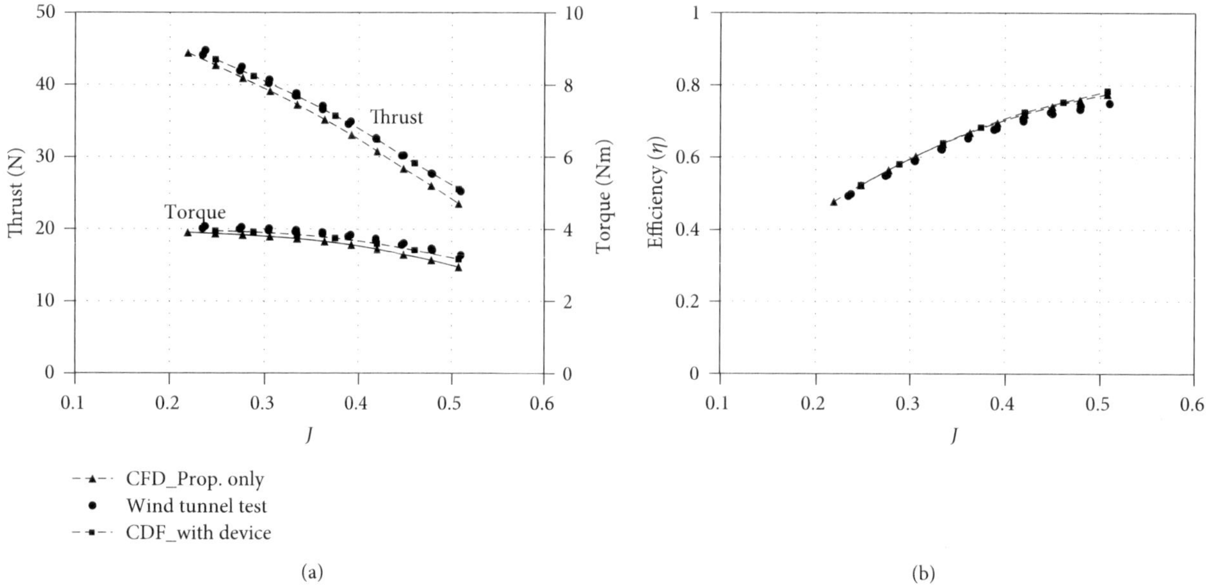

FIGURE 22: Comparison of CFD results with (a) design code and (b) wind tunnel test.

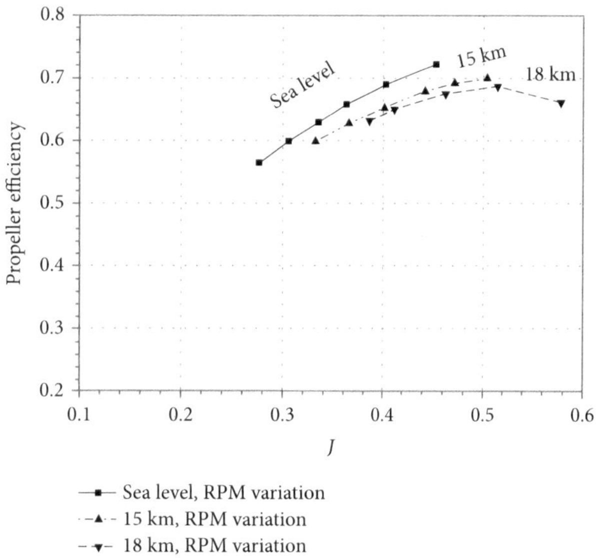

FIGURE 23: Efficiency curves at several altitude conditions.

to remain in the allowable airspace, as well as maintain a desired flight under disturbances of the atmosphere. As an example, Figure 25 depicts several flight data for a duration of 2 min during a climb from an altitude of approximately 3400–3900 m. During this period, the equivalent airspeed and angle of attack fluctuate continuously. Moreover, it is observed that the aircraft performed two major turn maneuvers during the time shown in the figure.

To take into account the approximately steady and stable state to the extent possible, data during the turn maneuver were excluded from the analysis of the propeller performance. For this purpose, only data corresponding to the moments of the roll angle from −0.015 to 0.015 rad, roll rate from −0.006 to 0.006 rad/s, and yaw rate from −0.01 to 0.01 rad/s were considered. For altitudes below 17 km, additional restrictions are imposed on deflection angles of control surfaces. Data are filtered based on the deflection angles of −2 to 2° and −3 to 3° for the aileron and rudder, respectively. Since EAV-3 adapts rpm differential control, the data on the rpm difference of less than 20 are filtered and chosen for the analysis.

5.2. Results of Analysis. Figure 26(a) shows the time history of two altitudes obtained during the climb of up to 18.5 km. As previously mentioned, the two altitudes are practically identical. It is observed that it took approximately 7.5 hours after takeoff to reach the 18 km altitude. The climb speed with respect to altitude is plotted in Figure 26(b) based on the filtered data. In general, during the flight, the climb speed varied from 0.5 to 1.0 m/s. For convenience, the time region was divided into four sections, designated as A–D in the figure.

The predicted thrust coefficient with respect to the advance ratio at each moment of filtered data is shown in Figure 27. The results corresponding to the four sections are illustrated by symbols with different colors. For the purpose of comparison, CFD results of the propeller alone analysis at several altitudes and wind tunnel test data are plotted together. It can be observed that the estimated values from the flight data are scattered over a wide range around the wind tunnel and CFD data. It is difficult to make a strict quantitative comparison because the degree of dispersion is large.

Nevertheless, several meaningful qualitative tendencies and characteristics can be observed from the comparison. It is evident that the values at lower altitudes (section A) are distributed around the wind tunnel test data and exhibit similar tendencies. It can be conjectured that the flight data contain the blockage effect of the main wing similar to that of the support column in the wind tunnel test. This can be regarded as a possible explanation on the observation that flight data values are slightly higher than CFD results of the propeller alone analysis at sea level.

Figure 24: Flight test of EAV-3.

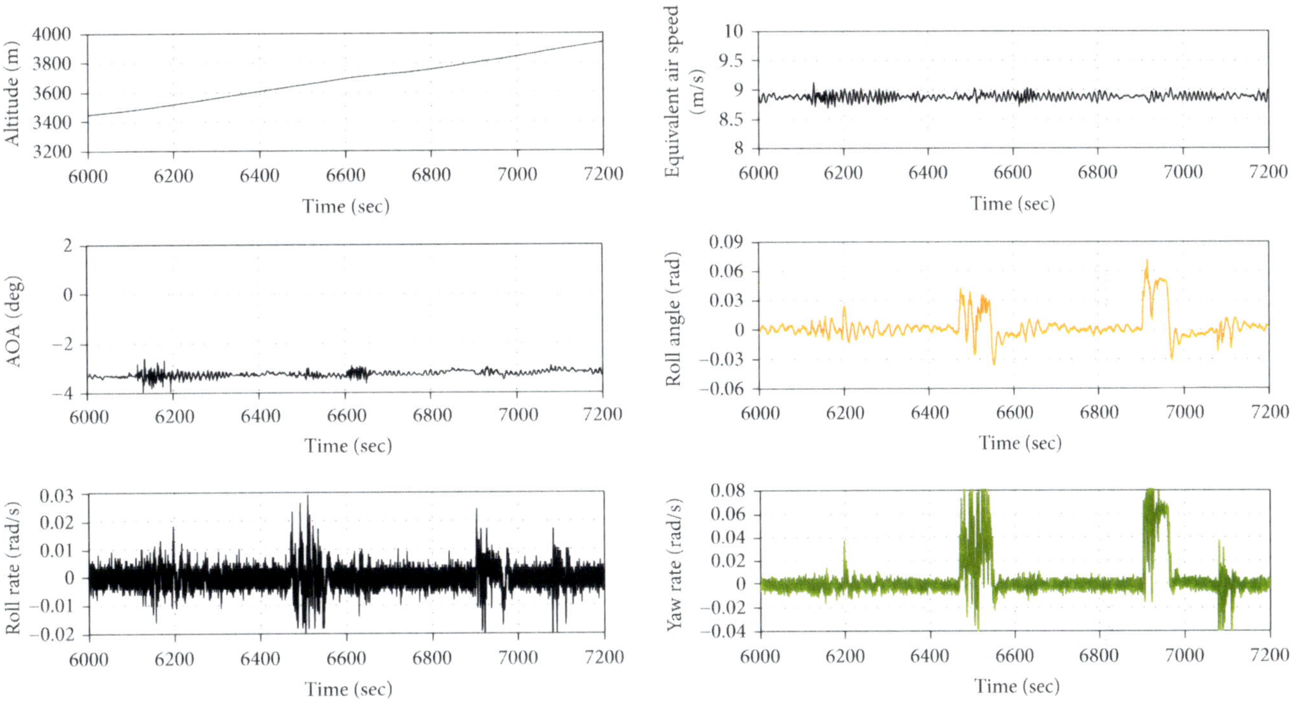

Figure 25: Example of flight data (altitude, equivalent air speed, attitude angle, and rate).

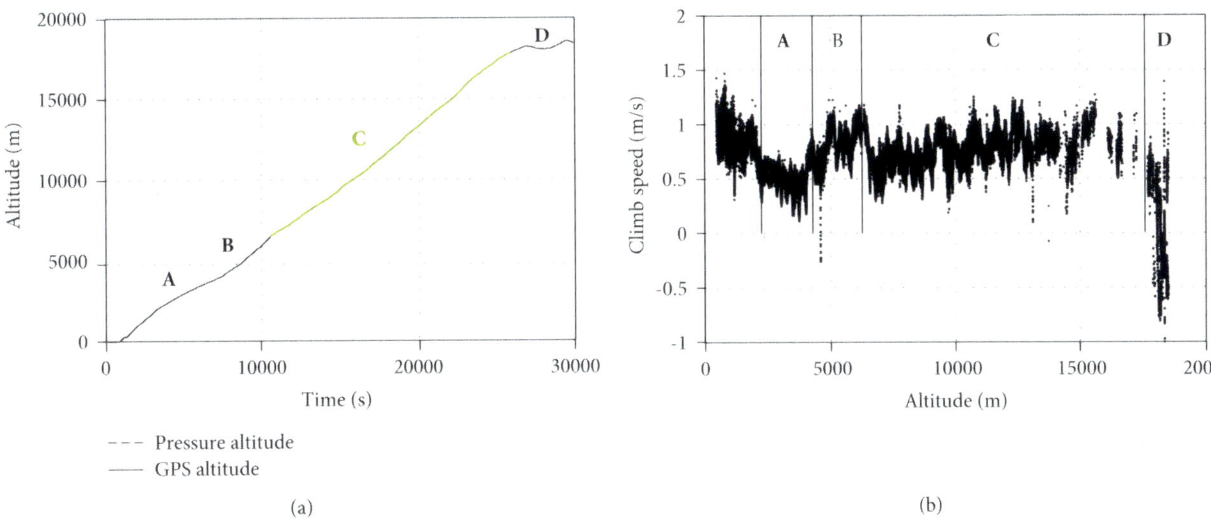

Figure 26: (a) Time history of altitude and (b) climb speed.

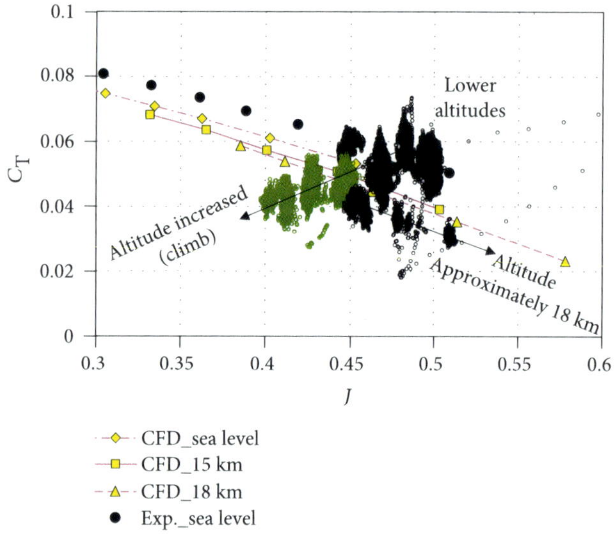

Figure 27: Comparison of thrust coefficient results.

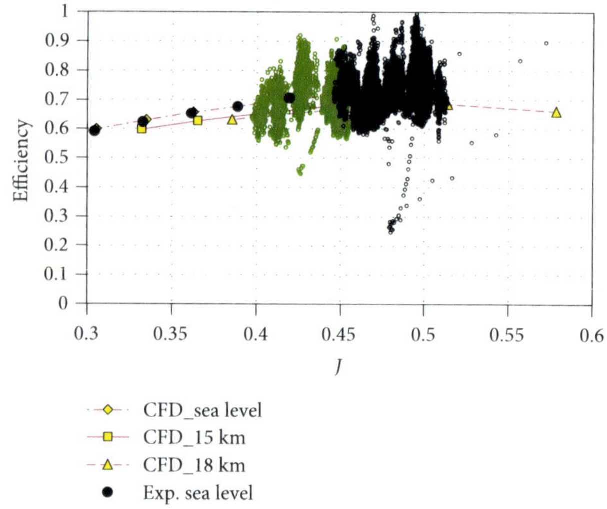

Figure 28: Comparison of thrust coefficient results.

The operating advance ratios during climb are in the range of approximately 0.4 to 0.5 and gradually decrease as the altitude increases. This means that the rpm increase was greater than the increase in the flight speed as the altitude increases. It is a different behavior compared with what was expected in the propeller design stage. We note here that the weight of EAV-3 was increased by approximately 10% from its original design after the finalization of its fabrication. Moreover, the final performance of the motor was also enhanced by over 10% in torque compared to that given as constraint of the propeller design. Therefore, it was expected that the propeller was operated under an off-design condition at some extent in the flight test. The flight control algorithm, which is optimized not only for the propeller design philosophy, is also regarded as a reason for the unexpected behavior.

It is evident from the figure that the thrust coefficient considerably decreases as the altitude increases (as changing from A to B, and B to C). This tendency is consistent with CFD prediction and is interpreted as the low Reynolds number effect, which was also confirmed in wind tunnel tests. In section D, which is approximately 18 km in altitude, the propeller operated under various flight conditions, including climb and cruising, resulting in a wide range of operating rpms. Thus, the thrust coefficient is distributed over a wide range of advance ratios for significantly limited altitudes, as shown in the figure. It can be seen that the distribution follows well the tendency of the curve of CFD results from the propeller alone analysis at 18 km. It must be emphasized that the thrust coefficient further decreases to even lower values than that from CFD results. This implies that the degree of decrement that occurred in the actual flight is considerably greater than that in the prediction of CFD using the transition SST model. It must be mentioned here that the evaluation of the thrust coefficient is directly affected by the drag estimation and it can be shifted up and down in the figure depending on the assumptions and approximations.

Figure 28 shows the comparison of results on efficiency. Although the prediction from flight data exhibits a high level of scattering, it is distributed around the wind tunnel test and CFD results and follows well their tendencies along a wide range of advance ratios. However, it is noted that the flight data seem slightly shifted to higher values. It is suspected that this optimistic prediction is primarily due to the inaccuracy arises from the estimation of power consumption. Since the estimation is simply based on voltage and current data, it is anticipated that a relatively large uncertainty would be introduced. Apparently, a more reliable strategy to obtain the power consumption would be necessary to improve accuracy. On the other hand, it is observed that the propeller operated in the advance ratio ranges, where the maximum efficiency is expected from the wind tunnel test and CFD data. This implies that the fundamental objective of the propeller design is successfully achieved.

6. Conclusion

In this paper, the design and performance evaluation of the propeller of KARI's HALE UAV were introduced.

For the design, a basic methodology was established based on BEMT and minimum induced loss. The aerodynamic data of the airfoil were constructed by using CFD to ensure the reliability under low Reynolds number conditions. The blade airfoil was chosen based on the evaluation of the lift-to-drag ratio for expected Reynolds numbers under the design condition. The effect of the design lift coefficient distribution on high-altitude conditions was investigated, and the characteristics of low Reynolds number behavior were analyzed. Design variables and responses for the design were selected, and the response surface was examined based on DOE and kriging metamodel. After defining the objective function and constraints, the optimization was conducted based on the desirability function. The final design was determined from the examination of the optimization result.

The wind tunnel test and numerical analysis were conducted to evaluate the performance of the propeller and validity of the design. The measurement device was designed, constructed, and calibrated. The method of data acquisition and correction was established. The performance of the designed propeller was measured under several conditions. The changes in performance as the low Reynolds number decreases were confirmed from the test data. The numerical analysis was conducted by using a commercial CFD code, and the results showed that the tendency was in good agreement with the test data. However, the thrust and torque are slightly lower than those in the test. The analyses, including the device and support column in the wind tunnel test, were further carried out to confirm that the thrust and torque were increased. It was found that the results approximated the test data, and the experimental and numerical results showed good agreement. It can be inferred from the comparison that the blockage effect must be appropriately considered to either interpret or correct measured data.

Flight test data are analyzed to evaluate the performance of the propeller in an actual flight. Due to inherent limitations in the flight test data, several assumptions and approximations were made to obtain the variables required to calculate performance. Approximately 8 h of data during the climb of up to 18 km altitude were analyzed. In spite of dispersion of analyzed data, the results for the thrust coefficient and efficiency generally exhibited good agreement in their tendencies, and satisfactory agreement in their quantitative comparison. The degradation of the thrust coefficient as the altitude increased was clearly observed in the flight data. Moreover, the level of decrease with respect to the altitude was found to be greater than the prediction obtained from CFD analysis. It was confirmed that the propeller operated over the advance ratio ranges, where the maximum efficiency is expected from both wind tunnel test and CFD analysis. This implies that the design methodology used in this study is practically useful and valid for the development of high-altitude propellers.

Conflicts of Interest

The authors declare that there is no conflict of interest regarding the publication of this paper.

Acknowledgments

This work was supported by the National Research Foundation of Korea (NRF) grant funded by the Korea government (MSIP) (no. 2017R1C1B1010109) and Pusan National University Research Grant, 2016.

References

[1] G. Romeo, G. Frulla, E. Cestino, and G. Corsino, "HELIPLAT: design, aerodynamic, structural analysis of long- endurance solar-powered stratospheric platform," *Journal of Aircraft*, vol. 41, no. 6, pp. 1505–1520, 2004.

[2] C. L. Nickol, M. D. Guynn, L. L. Kohout, and T. A. Ozoroski, "High altitude long endurance air vehicle analysis of alternatives and technology requirements development," in *45th AIAA Aerospace Sciences Meeting and Exhibit*, pp. 1–17, Reno, NV, USA, 2007, AIAA 2007-1050.

[3] B. S. de Mattos, N. R. Secco, and E. F. Salles, "Optimal design of a high-altitude solar-powered unmanned airplane," *Journal of Aerospace Technology and Management*, vol. 5, no. 3, pp. 349–361, 2013.

[4] L. Sai, Z. Wei, and W. Xueren, "The development status and key technologies of solar powered unmanned air vehicle," *IOP Conference Series: Materials Science and Engineering*, vol. 187, article 012011, 2017.

[5] L. J. Ehernberger, C. Donohue, and E. H. Teets, "A review of solar-powered aircraft flight activity at the Pacific missile range test facility, Kauai, Hawaii," in *11th AMS Conference on Aviation, Range, and Aerospace Meteorology*, pp. 1–7, Hyannis, MA, USA, 2004.

[6] https://en.wikipedia.org/wiki/Helios_Prototype.

[7] https://www.airbus.com/defence/uav/zephyr.html.

[8] A. Rapinett, *Zephyr: a high altitude long endurance unmanned air vehicle, [M.S. thesis]*, University of Surrey, 2009.

[9] https://en.wikipedia.org/wiki/Boeing_SolarEagle.

[10] Y. Lee, S. Hwang, S. Kim, C. Kim, S. Ahn, and D. Lee, "Development of an electrically powered HALE UAV," in *Proceeding of the 2013 KSAS Spring Conference*, vol. 1, pp. 1097–1100, Jeongseon, Korea, 2013.

[11] Y. Lee, S. Hwang, S. Kim et al., "Development and flight tests of a scaled electrically powered HALE UAV," in *Proceeding of the 2014 KSAS Fall Conference*, pp. 797–800, Jeju, Korea, 2014.

[12] Y. Lee, S. Hwang, S. Kim, and S. Kim, "Development of a full-scale solar powered HALE UAV," in *Proceeding of the 2015 KSAS Fall Conference*, pp. 181–184, Jeju, Korea, 2015.

[13] S. Hwang, S. Kim, Y. Lee, and C. Kim, "Design of the solar powered high altitude long endurance (HALE) unmanned aerial vehicle," in *Proceeding of the 2015 KSAS Fall Conference*, pp. 950–954, Jeju, Korea, 2015.

[14] S. J. Hwang, S. G. Kim, C. W. Kim, and Y. G. Lee, "Aerodynamic design of the solar-powered high altitude long endurance (HALE) unmanned aerial vehicle (UAV)," *International Journal of Aeronautical and Space Sciences*, vol. 17, no. 1, pp. 132–138, 2016.

[15] A. Colozza, *High Altitude Propeller Design and Analysis Overview*, Federal Data Systems, Cleveland, OH, USA, 1998.

[16] J. S. Monk, *A Propeller Design and Analysis Capability Evaluation for High Altitude Application, [M.S. Thesis]*, University of the Witwatersrand, 2010.

[17] H. Glauert, "Airplane propellers," in *Aerodynamic Theory*, W. F. Durand, Ed., pp. 169–360, Springer, Berlin, Heidelberg, 1935.

[18] http://web.mit.edu/drela/Public/web/qprop/.

[19] M. A. R. Silvestre, J. Morgado, and J. C. Pascoa, "JBLADE: a propeller design and analysis code," in *2013 International Powered Lift Conference*, pp. 1–12, Los Angeles, CA, USA, 2013, AIAA 2013-4220.

[20] M. H. McCrink and J. W. Gregory, "Blade element momentum modeling of low-*Re* small UAS electric propulsion systems," in *33rd AIAA Applied Aerodynamics Conference*, pp. 1–23, Dallas, TX, USA, 2015, AIAA 2015-3296.

[21] M. K. Rwigema, *The Evaluation of a Low-Order Propeller Performance and Wake Prediction Capability for the Calculation of Power Effects, [M.S. thesis]*, University of the Witwatersrand, 2014.

[22] https://sites.google.com/site/joaomorgado23/downloads.

[23] http://www.mh-aerotools.de/airfoils/javaprop.htm.

[24] C. N. Adkins and R. H. Liebeck, "Design of optimum propellers," *Journal of Propulsion and Power*, vol. 10, no. 5, pp. 676–682, 1994.

[25] D. Park and C. Kim, "Development of HALE propeller design and analysis program," in *Proceedings of KSAS 2014 Fall Conference*, pp. 1039–1042, Jeju, Korea, 2014.

[26] M. Drela, "XFOIL: an analysis and design system for low Reynolds number airfoils," in *Low Reynolds Number Aerodynamics*, T. J. Mueller, Ed., pp. 1–12, Springer, Berlin, Heidelberg, 1989.

[27] http://www.xflr5.com/xflr5.htm.

[28] https://www.ansys.com/Products/Fluids/ANSYS-Fluent.

[29] F. R. Menter, R. B. Langtry, S. R. Likki, Y. B. Suzen, P. G. Huang, and S. Völker, "A correlation-based transition model using local variables: part I — model formulation," in *ASME Turbo Expo 2004: Power for Land, Sea, and Air*, pp. 57–67, Vienna, Austria, June 2004, ASME GT2004-53452.

[30] R. J. McGhee, B. S. Walker, and B. F. Millard, "Experimental results for the Eppler 387 airfoil at low Reynolds numbers in the Langley low-turbulence pressure tunnel," Tech. Rep. NASA-TM-4062, NASA, 1988.

[31] M. S. Selig and B. D. McGranahan, "Wind tunnel aerodynamic tests of six airfoils for use on small wind turbines," in *42nd AIAA Aerospace Sciences Meeting and Exhibit*, pp. 1–19, Reno, NV, USA, 2004, AIAA 2004-1188.

[32] H. Hu and Z. Yang, "An experimental study of the laminar flow separation on a low-Reynolds-number airfoil," *Journal of Fluids Engineering*, vol. 130, no. 5, article 051101, 2008.

[33] L. W. Traub and E. Cooper, "Experimental investigation of pressure measurement and airfoil characteristics at low Reynolds numbers," *Journal of Aircraft*, vol. 45, no. 4, pp. 1322–1333, 2008.

[34] M. Sahin, J. Hall, K. Mohseni, and K. Hillewaert, "Direct numerical simulation of separated low-Reynolds number flows around an Eppler 387 airfoil," in *46th AIAA Aerospace Sciences Meeting and Exhibit*, pp. 1–21, Reno, NV, USA, 2008, AIAA 2008-422.

[35] J. Morgado, M. A. R. Silvestre, and J. C. Páscoa, "Full range airfoil Polars for propeller blade element momentum analysis," in *2013 International Powered Lift Conference*, pp. 1–10, Los Angeles, CA, USA, 2013, AIAA 2013-4379.

[36] J. Dorfling and K. Rokhsaz, "Integration of airfoil stall and compressibility models into a propeller blade element model," *Journal of Aerospace Engineering*, vol. 29, no. 4, article 04016014, 2016.

[37] D. C. Montgomery, *Design and Analysis of Experiments*, John Wiley & Sons, Inc., 6th edition, 2005.

[38] G. G. Wang, "Adaptive response surface method using inherited Latin hypercube design points," *Journal of Mechanical Design*, vol. 125, no. 2, pp. 210–220, 2003.

[39] J. P. C. Kleijnen, "Kriging metamodeling in simulation: a review," *European Journal of Operational Research*, vol. 192, no. 3, pp. 707–716, 2009.

[40] https://www.jmp.com/en_us/home.html.

[41] T. Cho, D. Park, S. Lee, Y. Kim, and Y. Lee, "Development of wind tunnel test system for solar-powered HALE UAV Propeller," in *Proceeding of the 2015 KSAS Spring Conference*, Jeongseon, Korea, 2015.

[42] C. Teahwan, K. Yangwon, and P. Donghun, "Wind tunnel test for the propeller performance of the high altitude UAV," *Journal of the Korean Society for Aeronautical & Space Sciences*, vol. 46, no. 3, pp. 189–196, 2018.

[43] D. Park, T. Cho, C. Kim, Y. Kim, and Y. Lee, "Performance evaluation of propeller for high altitude by using experiment and computational analysis," *Journal of the Korean Society for Aeronautical & Space Sciences*, vol. 43, no. 12, pp. 1035–1047, 2015.

[44] J. B. Brandt and M. S. Selig, "Propeller performance data at low Reynolds numbers," in *49th AIAA Aerospace Sciences Meeting including the New Horizons Forum and Aerospace Exposition*, pp. 1–18, Orlando, FL, USA, 2011, AIAA 2011-1255.

[45] M. A. R. Silvestre, J. Morgado, P. Alves, P. Santos, P. Gamboa, and J. C. Páscoa, "Propeller performance measurement at low Reynolds numbers," *International Journal of Mechanics*, vol. 9, pp. 154–166, 2015.

[46] H. Glauert, *The Element of Aerofoil and Airscrew Theory*, Cambridge University Press, Second edition, 1948.

[47] R. E. Fitzgerald, *Wind Tunnel Blockage Corrections for Propellers, [M.S. Thesis]*, University of Maryland, 2007.

[48] Fluent, https://www.ansys.com.

[49] OpenFOAM, https://www.openfoam.org.

[50] STAR-CCM+, https://mdx.plm.automation.siemens.com/star-ccm-plus.

[51] D. H. Park, C. W. Kim, and Y. G. Lee, "Comparison of commercial and open source CFD codes for aerodynamic analysis of flight vehicles at low speeds," *Journal of Computational Fluids Engineering*, vol. 21, no. 2, pp. 70–80, 2016.

[52] J. B. Barlow, W. H. Rae, and A. Pope, *Low-Speed Wind Tunnel Testing*, John Wiley and Sons, New York, NY, USA, Third edition, 1999.

[53] M. P. Merchant, *Propeller Performance Measurement for Low Reynolds Number Unmanned Aerial Vehicle Applications, [M.S. Thesis]*, Wichita State University, 2004.

14

Hankel Matrix Correlation Function-Based Subspace Identification Method for UAV Servo System

Minghong She[1,2] **and Pengju Zhao**[2]

[1]*School of Automation, Chongqing University, Chongqing, China*
[2]*Computer College, Chongqing College of Electronic Engineering, Chongqing, China*

Correspondence should be addressed to Minghong She; mh_she@126.com

Academic Editor: Hikmat Asadov

For the identification problem of closed-loop subspace model, we propose a zero space projection method based on the estimation of correlation function to fill the block Hankel matrix of identification model by combining the linear algebra with geometry. By using the same projection of related data in time offset set and LQ decomposition, the multiplication operation of projection is achieved and dynamics estimation of the unknown equipment system model is obtained. Consequently, we have solved the problem of biased estimation caused when the open-loop subspace identification algorithm is applied to the closed-loop identification. A simulation example is given to show the effectiveness of the proposed approach. In final, the practicability of the identification algorithm is verified by hardware test of UAV servo system in real environment.

1. Introduction

In the past twenty years, the subspace model identification (SMI) has received great attention, not only because of its excellent convergence and simple numerical calculation, but also the suitability for the application in the estimation, prediction, and control algorithm. In the early references, most subspace identification methods have open-loop recognition characteristics. Considering stability, safety, and control-oriented identification problems, researchers have been trying to apply these subspace methods to closed-loop identification [1–4].

The main difficulty in closed-loop system identification is that the correlation between device input and interference leads to a bias in the parameter estimation of the system model [5]. So far, it has developed many subspace identification methods, such as the literature [6–11], which can obtain consistent estimates and closed-loop data. It is noticed that most subspace system identification methods are based on the input and output data in time domain, and the frequency response methods of some linear time invariant systems are often based on the signal correlation function [12, 13].

Subspace method is extended to frequency response function estimation, and continuous and discrete time models are determined by auxiliary variables [14]. Two frequency statistical properties and subspace convergence analysis methods are proposed in the literature [15]. For a linear closed-loop system, where the external input is independent of the observed noise, the cross correlation function of the output and the external input signals is equal to the cross correlation function of the input and external signals of the dynamic system [16]. By using correlation function sequence as the interface function, the important information has been carried by the correlation function in data compression which has been hidden in the sequence, and by extracting the parameter information of interface function, it provides the basis for parameter identification [17].

The above algorithms can solve the problem of correlation between input and interference of the equipment to some extent. The unbiased parameter estimation can be obtained under arbitrary noise characteristics [18]. But the subspace identification algorithm can solve the relationship between the input and interference at the same time, because the relation between the two related function sequences cannot be completely determined, which leads to the higher dimension of the input matrix of the subspace computation matrix [19].

In order to simplify the calculation and improve the calculation accuracy, we design the input deletion strategy based on the zero space projection on the basis of block Hankel matrix and related data estimation separately [20]. This paper also uses LQ framework to solve the identification process and simplifies the calculation of the algorithm [21]. In this paper, a new subspace identification algorithm based on the estimation of correlation function is proposed, and the unbiased parameter estimation of closed-loop dynamics under linear closed-loop conditions is obtained.

For the EIV (errors-in-variables) model structure, namely, the input and output, both are affected by noise pollution. Chou and Verhaegen put forward a new method of subspace identification [12]. The method eliminates noise effects by regarding the past input/output data as auxiliary variables. Gustafsson [22] changed the steps of the traditional subspace identification method and proposed a new subspace auxiliary variable method (subspace-based identification using instrumental variables (SIVs)). The algorithm presented in the literature [12] was included, and the identification precision was improved after the algorithm was modified. In terms of the algorithm itself, the input being independent of noise assumption is not involved in the two methods; thus, it seems that they can be applied for identification in a closed-loop system. However, this is not the case based on simulation examples; these two types of algorithms used in closed-loop identification do not obtain consensus estimates in some cases.

On the other hand, for closed-loop systems subject to the input and output measurement noises, [23] developed a new closed-loop subspace identification algorithm (SOPIM) by adopting the EIV model structure of SIMPCA and proposed an orthogonal projection approach to avoid identifying the parity space of the feedback controller. In the literature [24], it proposed the use of parity space and principal component analysis (SIMPCA) for EIV identification with colored input excitation can also be applied to closed-loop identification. Instead of preestimating the Markov parameters or eliminating them via noncausal projections, SIMPCA reformulates the SIM problem in parity space. In the process of algorithm implementation, SVD should be applied twice to solve the orthogonal complement space in the two methods. It is difficult to determine the dimensions of the orthogonal complement space when the system order is unknown. The contribution of this paper is to use a zero space projection method based on the estimation of correlation function to fill the block Hankel matrix of identification model by combining the linear algebra with geometry. By using the same projection of related data in time offset set and LQ decomposition, the multiplication operation of projection is achieved and dynamics estimation of the unknown equipment system model is obtained. Simulation shows that this algorithm has higher accuracy, which is another contribution of this paper.

The outline of this paper is as follows: we start in Section 2 with the statement of the problem and notation, and then a state space model based on estimation of correlation function is obtained. In Section 3, we present a block Hankel matrix, related data estimation equations, and deletion of input items based on null space projection for subspace model identification process. In Section 4, both a simulation example and a real hardware verification experiment are presented. We end this paper with our conclusions in Section 5.

2. Closed-Loop Subspace Identification

2.1. Statement of the Problem and Notation. In the UAV servo closed-loop control system, there is an unknown device, as shown in Figure 1. The device model contains deterministic part P, and random parts are obtained by filtering white noise sequence η_y with the noise filter H_p. Therefore, the device model can be represented as

$$\mathbf{y}(k) = \mathbf{P}(z)\mathbf{u}(k) + H_p(z)\boldsymbol{\eta}_y(k). \quad (1)$$

In the formula, $\mathbf{u} \in \mathscr{R}^{n_u}$ and $\mathbf{y} \in \mathscr{R}^{n_y}$ are input and output vectors, respectively, $\boldsymbol{\eta}_y \in \mathscr{R}^{n_y}$ is the zero mean and covariance matrix, and $\Delta_{\eta_y} > 0$ is the white noise.

The system operates under the controller C and shows closed-loop characteristics, so that the closed-loop system can be stably controlled in the whole control trajectory. The resulting control output signals are as follows:

$$u(t) - r_2(t) = C(z)[r_1(t) - y(t)] + H_c(z)\eta_u(t). \quad (2)$$

In the formula, the variance matrix is the zero mean of $\Delta_{\boldsymbol{\eta}_u} > 0$. The white noise sequence $\eta_u \in \mathscr{R}^{n_u}$ is filtered by H_c. Exogenous inputs of $r_1 \in \mathscr{R}^{n_y}$ and $r_2 \in \mathscr{R}^{n_u}$ meet persistent excitation conditions and are independent of white noise η_u and η_y. The internal signals ω and v can be represented by η_u and η_y, which are statistically independent quantities of arbitrary colors and known external sequences r_1 and r_2. In Figure 1, the system scheme based on the following loop states recognition problem of this study: given the exogenous input of $r_1(k)$ and $r_2(k)$, and the input and output sequences of $u(k)$ and $y(k)$, the goal is to determine the unknown devices in the closed-loop system without the state space model of partial parameter description, as shown in Figure 1.

The signal $s(t) \in \mathscr{R}^{n_s}$ and $t \in \mathscr{Z}$ are quasi steady processes satisfying the following two conditions:

$$E(s(t)) = m_s(t), \quad \|m_s(t)\|_2 \le C, \forall t \in \mathscr{Z},$$
$$R_{ss}(\tau) = \lim_{N\to\infty} \frac{1}{N} \sum_{t=0}^{N-\tau-1} E\left(s(t+\tau)s(t)^T\right). \quad (3)$$

In the formula, $\|R_{ss}(\tau)\|_2 \le C, \ \forall\tau \in \mathscr{Z}$. E indicates expected operation. The function $R_{ss}(\tau)$: $\mathscr{Z} \to \mathscr{R}^{n_s \times n_s}$ is called the autocorrelation function of $s(t)$. Similarly, if $\omega(t) \in \mathscr{R}^{n_\omega}$ is a quasi stationary signal, then the cross-correlation function $R_{sw}(\tau)$: $\mathscr{Z} \to \mathscr{R}^{n_s \times n_w}$ of $s(t)$ and $\omega(t)$ is computed as

$$R_{sw}(\tau) = \lim_{N\to\infty} \frac{1}{N} \sum_{t=0}^{N-\tau-1} E\left(s(t+\tau)\omega(t)^T\right) \quad (4)$$

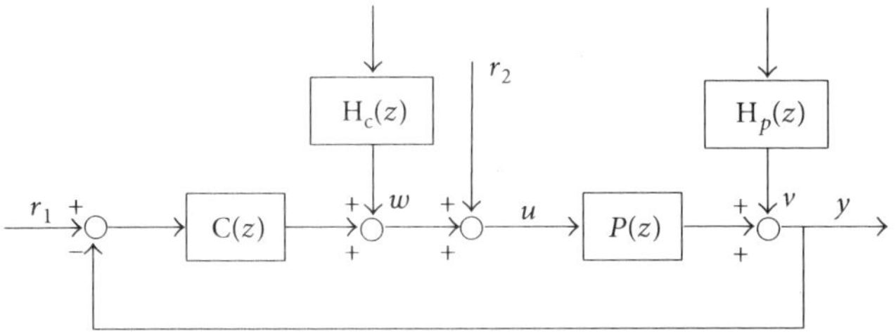

FIGURE 1: Schematic diagram of closed-loop system.

If only N data is available, the estimates of the autocorrelation function and cross-correlation function can be calculated as

$$\widehat{R}_{ss}(\tau) = \frac{1}{N}\sum_{t=0}^{N-\tau-1} s(t+\tau)s(t)^T,$$
$$\widehat{R}_{sw}(\tau) = \frac{1}{N}\sum_{t=0}^{N-\tau-1} s(t+\tau)\omega(t)^T. \quad (5)$$

At the same time, it is assumed that in $N \to \infty$ case, $R_{ss}(\tau)$ and $R_{sw}(\tau)$ are convergents, respectively.

2.2. State Space Model Based on Estimation of Correlation Function. The subspace identification process adopts the method of calculating the state space matrix to identify the system parameters. Therefore, the first step of the algorithm is to represent the system model into the state space model. Here, the unknown device model P shown in Figure 1 is rewritten as the following state space model form:

$$\mathbf{x}_p(k+1) = \mathbf{A}_p\mathbf{x}_p(k) + \mathbf{B}_p\mathbf{u}(k),$$
$$\mathbf{y}(k) = \mathbf{C}_p\mathbf{x}_p(k) + \mathbf{D}_p\mathbf{u}(k) + \mathbf{v}(k). \quad (6)$$

In the formula, $\mathbf{x}_p \in \mathscr{R}^{n_p}$ is the state vector of the device, and the system matrices are $\mathbf{A}_p \in \mathscr{R}^{n_p\times n_p}$, $\mathbf{B}_p \in \mathscr{R}^{n_p\times n_u}$, $\mathbf{C}_p \in \mathscr{R}^{n_y\times n_p}$, and $\mathbf{D}_p \in \mathscr{R}^{n_y\times n_u}$. If the exogenous signal $r = r_1$ or $r = r_2$ is related to the input u, the output y, and the noise v, the cross-correlation functions $R_{yr}(\tau) \in \mathscr{R}^{n_y\times n_y}$, $R_{ur}(\tau) \in \mathscr{R}^{n_u\times n_y}$, and $R_{vr}(\tau) \in \mathscr{R}^{n_y\times n_y}$ exist. The covariance $R_{x_pr}(\tau) \in \mathscr{R}^{n_p\times n_y}$ of the states x_p and $r(t)$ is defined, and then the correlation function of the device can be represented as in the state space matrix.

$$R_{x_pr}(\tau+1) = A_pR_{x_pr}(\tau) + B_pR_{ur}(\tau),$$
$$R_{yr}(\tau) = C_pR_{x_pr}(\tau) + D_pR_{ur}(\tau) + R_{vr}(\tau). \quad (7)$$

It is assumed that the correlation functions $R_{yr}(\tau)$ and $R_{ur}(\tau)$ are known within a certain interval $\tau \in [\tau_0, \tau_1]$, so that $\widehat{R}_{yr}(\tau)$ is the estimate of the correlation function $R_{yr}(\tau)$, and the definitions of $\widehat{R}_{ur}(\tau)$ and $\widehat{R}_{vr}(\tau)$ are similar. Then, the correlation function shown in (7) can be rewritten as

$$R_{x_pr}(\tau+1) = A_pR_{x_pr}(\tau) + B_p\widehat{R}_{ur}(\tau),$$
$$\widehat{R}_{yr}(\tau) = C_pR_{x_pr}(\tau) + D_p\widehat{R}_{ur}(\tau) + \widehat{R}_{vr}(\tau). \quad (8)$$

Since the external input instrument signals r are uncorrelated with noise of η_y and v, all elements of $\widehat{R}_{vr}(\tau)$ along with $N \to \infty$ will gradually tend to 0. Then, the correlation function (8) will converge to

$$R_{x_pr}(\tau+1) = A_pR_{x_pr}(\tau) + B_p\widehat{R}_{ur}(\tau),$$
$$\widehat{R}_{yr}(\tau) = C_pR_{x_pr}(\tau) + D_p\widehat{R}_{ur}(\tau). \quad (9)$$

Based on the consistency of correlation function estimation, the direct method of correlation function estimation can be obtained, instead of the original input and output data consistent with the system estimate.

3. The Correlation Function Recognition Based on Subspace Estimation

Through the correlation function estimation and zero space followed by projection, block Hankel matrix of identification framework for filling, we can obtain the scope of the extended observability matrix, which is the basic step of subspace identification methods. Then, an estimate of the dynamics of an unknown device system model can be obtained based on the same projection on the time offset set of the correlation data. Finally, the operation of projection multiplication can be realized by numerical calculation of LQ decomposition.

3.1. Block Hankel Matrix and Related Data Estimation Equations. Construct the correlation function to estimate the partitioned Hankel matrices of $\widehat{\mathbf{R}}^{yr}_{\tau_0|\tau_{i-1}}$ and $\widehat{\mathbf{R}}^{ur}_{\tau_0|\tau_{i-1}}$, including i rows and j columns, as defined below:

$$\widehat{\mathbf{R}}^{yr}_{\tau_0|\tau_{i-1}} = \begin{bmatrix} \widehat{\mathbf{R}}_{yr}(\tau_0) & \widehat{\mathbf{R}}_{yr}(\tau_1) & \cdots & \widehat{\mathbf{R}}_{yr}(\tau_{j-1}) \\ \widehat{\mathbf{R}}_{yr}(\tau_1) & \widehat{\mathbf{R}}_{yr}(\tau_2) & \cdots & \widehat{\mathbf{R}}_{yr}(\tau_j) \\ \vdots & \vdots & \ddots & \vdots \\ \widehat{\mathbf{R}}_{yr}(\tau_{i-1}) & \widehat{\mathbf{R}}_{yr}(\tau_i) & \cdots & \widehat{\mathbf{R}}_{yr}(\tau_{j+i-2}) \end{bmatrix},$$
$$\widehat{\mathbf{R}}^{ur}_{\tau_0|\tau_{i-1}} = \begin{bmatrix} \widehat{\mathbf{R}}_{ur}(\tau_0) & \widehat{\mathbf{R}}_{ur}(\tau_1) & \cdots & \widehat{\mathbf{R}}_{ur}(\tau_{j-1}) \\ \widehat{\mathbf{R}}_{ur}(\tau_1) & \widehat{\mathbf{R}}_{ur}(\tau_2) & \cdots & \widehat{\mathbf{R}}_{ur}(\tau_j) \\ \vdots & \vdots & \ddots & \vdots \\ \widehat{\mathbf{R}}_{ur}(\tau_{i-1}) & \widehat{\mathbf{R}}_{ur}(\tau_i) & \cdots & \widehat{\mathbf{R}}_{ur}(\tau_{j+i-2}) \end{bmatrix}. \quad (10)$$

In the formula, $\widehat{\mathbf{R}}^{yr}_{\tau_0|\tau_{i-1}} \in \mathscr{R}^{n_y i\times n_r j}$, $\widehat{\mathbf{R}}^{ur}_{\tau_0|\tau_{i-1}} \in \mathscr{R}^{n_u i\times n_r j}$, each element of $\widehat{\mathbf{R}}^{yr}_{\tau_0|\tau_{i-1}}$ and $\widehat{\mathbf{R}}^{ur}_{\tau_0|\tau_{i-1}}$ is the correlation function data, and i and j are user-defined indexes. According to (9), the above matrix satisfies the relation.

$$\widehat{\mathbf{R}}^{yr}_{\tau_0|\tau_{i-1}} = \mathbf{\Gamma}\mathbf{R}^{xr}_{\tau_0} + \mathbf{T}_{0|i-1}\widehat{\mathbf{R}}^{ur}_{\tau_0|\tau_{i-1}}. \quad (11)$$

In the formula, vector $\mathbf{R}^{xr}_{r0}$ consists of data from the state cross-correlation function.

$$\mathbf{R}^{xr}_{r0} = \left[\mathbf{R}_{xr}(\tau_0)\mathbf{R}_{xr}(\tau_1)\cdots\mathbf{R}_{xr}(\tau_{j-1})\right]. \quad (12)$$

The extended observation matrix and the lower triangular block Toeplitz matrix are defined as follows:

$$\boldsymbol{\Gamma} = \begin{bmatrix} \mathbf{C}_p & \mathbf{C}_p\mathbf{A}_p & \cdots & \mathbf{C}_p\mathbf{A}_p^{i-1} \end{bmatrix}^T, \quad (13)$$

$$\mathbf{T}_{0|i-1} = \begin{bmatrix} \mathbf{D}_p & 0 & \cdots & 0 \\ \mathbf{C}_p\mathbf{A}_p & \mathbf{D}_p & \cdots & 0 \\ \vdots & \vdots & \ddots & \vdots \\ \mathbf{C}_p\mathbf{A}_p^{i-2}\mathbf{B}_p & \mathbf{C}_p\mathbf{A}_p^{i-3}\mathbf{B}_p & \cdots & \mathbf{D}_p \end{bmatrix}. \quad (14)$$

By applying the one-step displacement process, the displacement equation corresponding to (11) can be defined as

$$\widehat{\mathbf{R}}^{yr}_{\tau_1|\tau_i} = \boldsymbol{\Gamma}\mathbf{A}_p\mathbf{R}^{xr}_{\tau_0} + \mathbf{T}_{0|i}\widehat{\mathbf{R}}^{ur}_{\tau_0|\tau_i}. \quad (15)$$

In the formula, the matrix $\mathbf{T}_{0|i}$ can be added to the left of the $\mathbf{T}_{0|i-1}$ with a column of zeros, as follows:

$$\mathbf{T}_{0|i} = \begin{bmatrix} 0 & \mathbf{D}_p & 0 & \cdots & 0 \\ 0 & \mathbf{C}_p\mathbf{A}_p & \mathbf{D}_p & \cdots & 0 \\ \vdots & \vdots & \vdots & \ddots & \vdots \\ 0 & \mathbf{C}_p\mathbf{A}_p^{i-2}\mathbf{B}_p & \mathbf{C}_p\mathbf{A}_p^{i-3}\mathbf{B}_p & \cdots & \mathbf{D}_p \end{bmatrix}. \quad (16)$$

Similarly, a zero line extension is added to the bottom of the $\widehat{\mathbf{R}}^{ur}_{\tau_0|\tau_{i-1}}$ to obtain the representation of $\widehat{\mathbf{R}}^{ur}_{\tau_0|\tau_i}$.

3.2. Deletion of Input Items Based on Null Space Projection. Although the range of the extended observable matrix $\boldsymbol{\Gamma}$ and the Toeplitz matrix $\mathbf{T}_{0|i-1}$ is contained in the $\widehat{\mathbf{R}}^{yr}_{\tau_0|\tau_{i-1}}$ of (11), the range of $\mathbf{T}_{0|i-1}$ can be deleted by zero space projection. So the range of the extended observability matrix $\boldsymbol{\Gamma}$ can be obtained. The main reason is that the null space projection; before feature extraction, the feature of zero space projection extracts the feature projection on nonzero vertex position, and the characteristics of complex single vertex extraction problem were simplified into any nonzero vertex extraction, which is a simplified algorithm of computing process. Since the orthogonal projection matrix of the row space of the partitioned Hankel matrix $\widehat{\mathbf{R}}^{ur}_{\tau_0|\tau_{i-1}}$ can be computed, the projection matrix is first defined as follows:

$$\prod_{\tau_0|\tau_{i-1}}^{ur} = \mathbf{I}_{n_r j} - \left(\widehat{\mathbf{R}}^{ur}_{\tau_0|\tau_{i-1}}\right)^T \left(\widehat{\mathbf{R}}^{ur}_{\tau_0|\tau_{t-1}} \left(\widehat{\mathbf{R}}^{ur}_{\tau_0|\tau_{i-1}}\right)^T\right)^{\dagger} \widehat{\mathbf{R}}^{ur}_{\tau_0|\tau_{i-1}}. \quad (17)$$

Theorem 1. *$\widehat{\mathbf{R}}^{yr}_{\tau_0|\tau_{i-1}}$, $\widehat{\mathbf{R}}^{ur}_{\tau_0|\tau_{i-1}}$, and $\mathbf{T}_{0|i-1}$ are defined by the above formula, and $\prod^{ur}_{\tau_0|\tau_k}$ is defined according to (17) as follows:*

$$\prod_{\tau_0|\tau_k}^{ur} = \mathbf{I}_{n_r j} - \left(\widehat{\mathbf{R}}^{ur}_{\tau_0|\tau_k}\right)^T \left(\widehat{\mathbf{R}}^{ur}_{\tau_0|\tau_k} \left(\widehat{\mathbf{R}}^{ur}_{\tau_0|\tau_k}\right)^T\right)^{\dagger} \widehat{\mathbf{R}}^{ur}_{\tau_0|\tau_k}. \quad (18)$$

Then,

$$\mathbf{T}_{0|i-1}\widehat{\mathbf{R}}^{ur}_{\tau_0|\tau_{i-1}} \prod_{\tau_0|\tau_k}^{ur} = \mathbf{0}. \quad (19)$$

In the formula, $\widehat{\mathbf{R}}^{ur}_{\tau_0|\tau_k} \in \mathscr{R}^{n_u(k+1)\times n_r j}, \tau_{i-1} \le \tau_k \le \tau_l$.

Proof. For $\tau_k = \tau_{i-1}$, (17) is brought into (19). For $\tau_{i-1} < \tau_k$, for example, $\tau_k = \tau_{i-1+s}$, item $\mathbf{T}_{0|i-1}$ can be obtained by extending zero column to the right side of the matrix.

$$\mathbf{T}^s_{0|i-1} = \begin{bmatrix} \mathbf{D}_p & 0 & \cdots & 0 & 0 & \cdots & 0 \\ \mathbf{C}_p\mathbf{A}_p & \mathbf{D}_p & \cdots & 0 & 0 & \cdots & 0 \\ \vdots & \vdots & \ddots & \vdots & \vdots & \vdots & \vdots \\ \mathbf{C}_p\mathbf{A}_p^{i-2}\mathbf{B}_p & \mathbf{C}_p\mathbf{A}_p^{i-3}\mathbf{B}_p & \cdots & \mathbf{D}_p & 0 & \cdots & 0 \end{bmatrix}. \quad (20)$$

Similarly, $\widehat{\mathbf{R}}^{ur}_{\tau_0|\tau_{i-1}}$ can obtain $\widehat{\mathbf{R}}^{ur}_{\tau_0|\tau_{i-1+s}}$ by adding zero line vectors. For all $s \ge 0$, $\mathbf{T}^s_{0|i-1}$ is a fixed value, and then available

$$\mathbf{T}^s_{0|i-1}\widehat{\mathbf{R}}^{ur}_{\tau_0|\tau_{i-1+s}} = \mathbf{T}_{0|i-1}\widehat{\mathbf{R}}^{ur}_{\tau_0|\tau_{i-1}},$$
$$\mathbf{T}^s_{0|i-1}\widehat{\mathbf{R}}^{ur}_{\tau_0|\tau_{i-1+s}} \prod_{\tau_0|\tau_k}^{ur} = \mathbf{0}. \quad (21)$$

The required proof results can be obtained by replacing $\mathbf{T}^s_{0|i-1}\widehat{\mathbf{R}}^{ur}_{\tau_0|\tau_{i-1+s}}$ with $\mathbf{T}_{0|i-1}\widehat{\mathbf{R}}^{ur}_{\tau_0|\tau_{i-1+s}}$. $\tau_k > \tau_l$ is not allowed here to ensure that (17) satisfies the singularity.

3.3. System Dynamics Estimation. The direct result obtained in Theorem 1 is that the same projection $\prod^{ur}_{\tau_0|\tau_i}$ removes the fixed items including $\mathbf{T}_{0|i-1}$ and $\mathbf{T}_{1|i}$ from $\widehat{\mathbf{R}}^{yr}_{\tau_0|\tau_{i-1}}$ and $\widehat{\mathbf{R}}^{yr}_{\tau_0|\tau_i}$ by the right projection of the (11) and (15), and then the following equations can be obtained:

$$\widehat{\mathbf{R}}^{yr}_{\tau_0|\tau_{i-1}} \prod_{\tau_0|\tau_i}^{ur} = \boldsymbol{\Gamma}\mathbf{R}^{xr}_{\tau_0} \prod_{\tau_0|\tau_i}^{ur}, \quad (22)$$

$$\widehat{\mathbf{R}}^{yr}_{\tau_1|\tau_i} \prod_{\tau_0|\tau_i}^{ur} = \boldsymbol{\Gamma}\mathbf{A}_p\mathbf{R}^{xr}_{\tau_0} \prod_{\tau_0|\tau_i}^{ur}. \quad (23)$$

Therefore, from the (22) on the right side of the singular value decomposition, estimation is obtained by the extended observability matrix $\boldsymbol{\Gamma}$. The calculation form is

$$\widehat{\mathbf{R}}^{yr}_{\tau_0|\tau_{i-1}} \prod_{\tau_0|\tau_i}^{ur} = [\boldsymbol{U}_n \ \boldsymbol{U}_s] \begin{bmatrix} \sum_n & 0 \\ 0 & \sum_s \end{bmatrix} \begin{bmatrix} \boldsymbol{V}_n^T \\ \boldsymbol{V}_s^T \end{bmatrix}. \quad (24)$$

$$\widehat{\boldsymbol{\Gamma}} = \mathbf{U}_n \sum_n^{1/2}. \quad (25)$$

According to (22) and (23), the minimization problem of the least squares solution is as follows:

$$J = \arg\min_{\mathbf{A}_p} \left\| \mathbf{A}_p \mathbf{\Gamma}^{\dagger} \widehat{\mathbf{R}}^{yr}_{\tau_0|\tau_{i-1}} \prod^{ur}_{\tau_0|\tau_i} - \mathbf{\Gamma}^{\dagger} \widehat{\mathbf{R}}^{yr}_{\tau_1|\tau_i} \prod^{ur}_{\tau_0|\tau_i} \right\|.$$
$$\widehat{\mathbf{A}}_p = \sum_{n}^{-1/2} \mathbf{U}_n^T \widehat{\mathbf{R}}^{yr}_{\tau_1|\tau_i} \prod^{ur}_{\tau_0|\tau_i} \mathbf{V}_n \sum_{n}^{-1/2}. \tag{26}$$

The system matrix $\mathbf{C}_p$ can be calculated by using the n_y rows of the extended observation matrix.

$$\widehat{\mathbf{C}}_p = \widehat{\mathbf{\Gamma}}(1 : n_y, :). \tag{27}$$

Based on the estimation of $\mathbf{A}_p$ and $\mathbf{C}_p$, the entire problem becomes linear in the unknown $\mathbf{B}_p$ and $\mathbf{D}_p$. Then, the estimated form of the output $\mathbf{y}(k)$ is

$$\widehat{\mathbf{y}}(k) = \widehat{\mathbf{C}}_p \widehat{\mathbf{A}}_p \widehat{\mathbf{x}}(0) + \left(\mathbf{u}(k)^T \otimes \mathbf{I}_{n_y}\right) vec\left(\widehat{\mathbf{D}}_p\right) + \left(\sum_{t=0}^{k-1} \mathbf{u}(k)^T \otimes \widehat{\mathbf{C}}_p \widehat{\mathbf{A}}_p^{k-t-1}\right) vec\left(\widehat{\mathbf{B}}_p\right). \tag{28}$$

In the formula, the estimates of $\mathbf{B}_p$ and $\mathbf{D}_p$ can be obtained by solving the linear least squares method.

$$\min_{\mathbf{B}_p, \mathbf{D}_p} \frac{1}{N} \sum_{k=0}^{N-1} \left\| \mathbf{y}(k) - \boldsymbol{\varphi}^T(k) \boldsymbol{\theta} \right\|_2^2. \tag{29}$$

In the formula, where

$$\boldsymbol{\varphi}^T(k) = \left[\left(\mathbf{u}(k)^T \otimes \mathbf{I}_{n_y}\right)\left(\sum_{k=0}^{s-1} \mathbf{u}(k)^T \otimes \widehat{\mathbf{C}}_p \widehat{\mathbf{A}}_p^{k-s-1}\right)\right],$$
$$\boldsymbol{\theta} = \left[\mathrm{vec}\left(\widehat{\mathbf{D}}_p\right) \mathrm{vec}\left(\widehat{\mathbf{B}}_p\right)\right]^T. \tag{30}$$

3.4. LQ Computing Framework. According to (20), the expansion matrix, which constitutes a lower triangular matrix, that is, the matrix LQ decomposition, shows that the matrix expansion is used in the matrix of right to satisfy the characteristics. Therefore, select the LQ decomposition of matrix multiplication calculation. By constructing projection matrices and replacing (11) and (15) projections, $\widehat{\mathbf{R}}^{yr}_{\tau_0|\tau_{i-1}}$, $\prod^{ur}_{\tau_0|\tau_i}$, and $\widehat{\mathbf{R}}^{yr}_{\tau_1|\tau_i} \prod^{ur}_{\tau_0|\tau_i}$ can be computed more efficiently by following LQ decomposition:

$$\begin{bmatrix} \widehat{\mathbf{R}}^{ur}_{\tau_0|\tau_i} \\ \widehat{\mathbf{R}}^{yr}_{\tau_0|\tau_{i-1}} \end{bmatrix} = \begin{bmatrix} \mathbf{L}_{11} & 0 \\ \mathbf{L}_{21} & \mathbf{L}_{22} \end{bmatrix} \begin{bmatrix} \mathbf{Q}_1^T \\ \mathbf{Q}_2^T \end{bmatrix}, \tag{31}$$

$$\begin{bmatrix} \widehat{\mathbf{R}}^{ur}_{\tau_0|\tau_i} \\ \widehat{\mathbf{R}}^{yr}_{\tau_1|\tau_i} \end{bmatrix} = \begin{bmatrix} \tilde{\mathbf{L}}_{11} & 0 \\ \tilde{\mathbf{L}}_{21} & \tilde{\mathbf{L}}_{22} \end{bmatrix} \begin{bmatrix} \tilde{\mathbf{Q}}_1^T \\ \tilde{\mathbf{Q}}_2^T \end{bmatrix}. \tag{32}$$

Then the output $\widehat{\mathbf{R}}^{yr}_{\tau_0|\tau_{i-1}}$ can be calculated as

$$\widehat{\mathbf{R}}^{yr}_{\tau_0|\tau_{i-1}} \prod^{ur}_{\tau_0|\tau_i} = \mathbf{\Gamma} \mathbf{R}^{xr}_{\tau_0} + \mathbf{T}_{0|i-1} \mathbf{L}_{11} \mathbf{Q}_1^T = \mathbf{L}_{21} \mathbf{Q}_1^T + \mathbf{L}_{22} \mathbf{Q}_2^T. \tag{33}$$

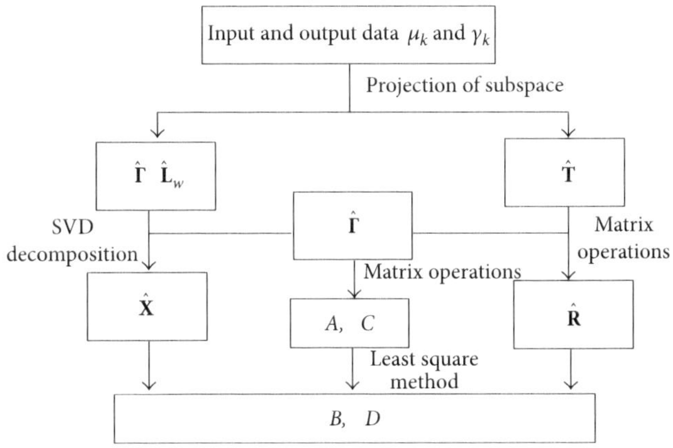

Figure 2: Closed-loop subspace identification process.

Formula (33) is multiplied by $\mathbf{Q}_2$ to get $\mathbf{L}_{22} = \mathbf{\Gamma} \mathbf{R}^{xr}_{\tau_0} \mathbf{Q}_2$ and $\mathbf{I}_m(\mathbf{L}_{22}) = \mathbf{I}_m(\mathbf{\Gamma})$. The algorithm then performs $\mathbf{L}_{22} \mathbf{Q}_2^T$ and $\tilde{\mathbf{L}}_{22} \tilde{\mathbf{Q}}_2^T$ at $\widehat{\mathbf{R}}^{yr}_{\tau_0|\tau_{i-1}} \prod^{ur}_{\tau_0|\tau_i}$ and $\widehat{\mathbf{R}}^{yr}_{\tau_1|\tau_{i-1}} \prod^{ur}_{\tau_0|\tau_i}$, respectively. In addition, the singular value decomposition is performed on $\mathbf{L}_{22}$ $\mathbf{Q}_2^T$ to achieve the estimation of the extended observation matrix $\mathbf{\Gamma}$.

3.5. Subspace Model Identification Process. The proposed closed-loop subspace identification algorithm for the concrete calculation process is shown in Figure 2. First, the closed-loop prediction of the parameters of the $\mathbf{\Gamma}$, $\widehat{\mathbf{L}}$, and $\widehat{\mathbf{G}}$ matrices is obtained based on the subspace projection method. Then, based on the matrix operation and singular value decomposition process, the values of $\widehat{\mathbf{X}}$, $\mathbf{A}$, $\mathbf{C}$, and $\widehat{\mathbf{R}}$ are obtained. Finally, the system parameter matrices $\mathbf{B}$ and $\mathbf{D}$ are solved by the least squares method.

In Figure 2, $\widehat{\mathbf{\Gamma}}$ can be obtained according to (25), and B can be calculated according to the (31-32). $\widehat{\mathbf{X}}$ is the state vector matrix. $\mathbf{A}$, $\mathbf{B}$, $\mathbf{C}$, and $\mathbf{D}$ are the state space matrices (for calculation process, see Section 2.3—system dynamic estimation part). $\widehat{\mathbf{T}}$ and $\widehat{\mathbf{R}}$ can be calculated according to (14) and (33), respectively.

4. Experimental Analysis

4.1. Model of UAV Servo Induction Motor. Several different induction UAV servo motor models have been described in the literature [6]. The model of the induction motor used in this experiment is

$$\frac{d\mathbf{X}}{dt} = \mathbf{AX} + \mathbf{DU},$$
$$\mathbf{Y} = \mathbf{CX} + \mathbf{BU}. \tag{34}$$

In the formula, $\mathbf{X} = [\bar{\mathbf{I}}, \bar{\boldsymbol{\phi}}]$ is the inductance matrix, $\mathbf{U} = \bar{\mathbf{V}}$ is the input matrix, and $\mathbf{Y} = \bar{\mathbf{I}}$ is the output matrix. And there are

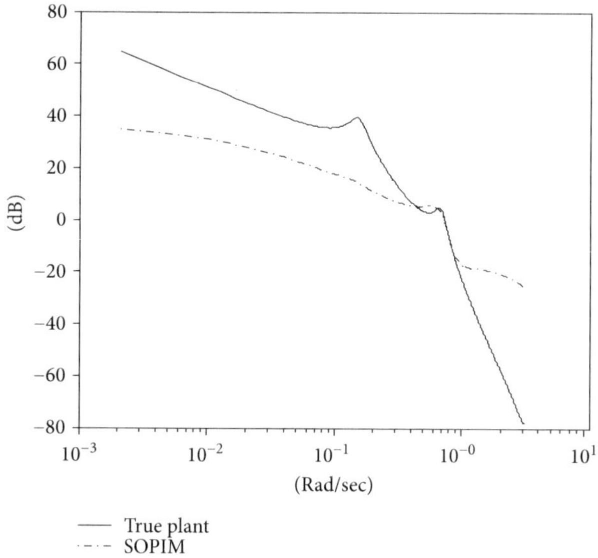

Figure 3: SOPIM Bode amplitude tracking curve.

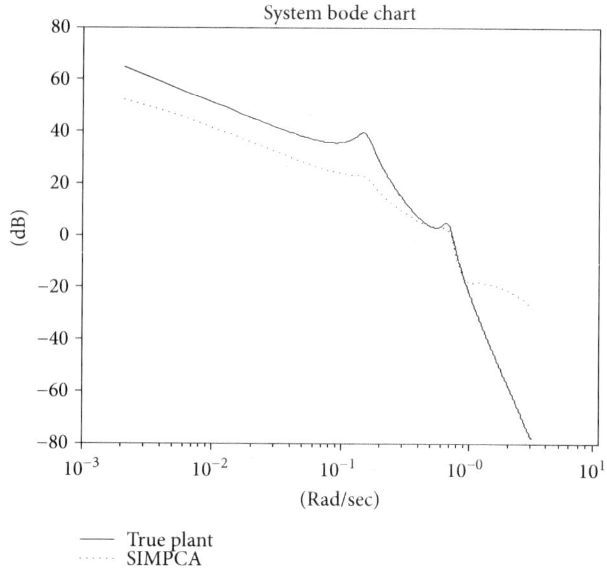

Figure 4: SIMPCA Bode amplitude tracking curve.

$$\mathbf{A} = \begin{bmatrix} -\left(\dfrac{1}{\sigma\tau_s} + \dfrac{1-\sigma}{\sigma\tau_r}\right) & \dfrac{L_m}{\sigma L_s L_r \tau_r} + j\dfrac{L_m}{L_s L_r} p\omega_r \\ \dfrac{L_m}{\tau_r} & -\dfrac{1}{\tau_r} + jp\omega_r \end{bmatrix}, \tag{35}$$

$$\mathbf{B} = \left(\frac{1}{\sigma\tau_s} 0\right)^T, \quad C = (1, 0), \quad D = 0.$$

In the formula, $\sigma = 1 - (L_m^2/L_s L_r)$, $\tau_s = L_s/R_s$, and $\tau_r = L_r/R_r$. L_m is excitation inductance, L_r is rotor inductance, L_s is stator inductance, R_s is stator resistance, and R_r is rotor resistance.

Here, the Bode amplitude diagram of the Gauss white noise of the induction motor under the Matlab environment is compared with the experimental performance. The hardware settings are the following: memory for the Kingston 8G ddr4-2400GHz processor, i7-6400HQ2.8GHz, and system simulation platform for the selection of win7 ultimate, matlab2014a. The input is generated by filtering a white noise with a fourth-order lowpass Butterworth filter that has a cutoff frequency of 0.8 times the Nyquist frequency and with the zero mean white noise signal disturbance. We select SOPIM and SIMPCA as comparison algorithms, which are shown in [23] and [24].

Experimental parameter settings are the following: $R_s = 10\,\Omega$, $R_r = 6.58\,\Omega$, $L_r = 0.31$ H, $L_s = 0.31$ H, and $\omega_r = 157.08$ rad/s.The sampling period used is equal to 10^{-4} s. According to the experimental parameters, the following induction motor model parameter matrices can be obtained:

$$A = \begin{pmatrix} -200.32 & 247.02 + j3656.16 \\ 5.73 & -21.23 + j314.16 \end{pmatrix}, \tag{36}$$

The number of conditions can be calculated as

$$\operatorname{cond}(A) = 322.47. \tag{37}$$

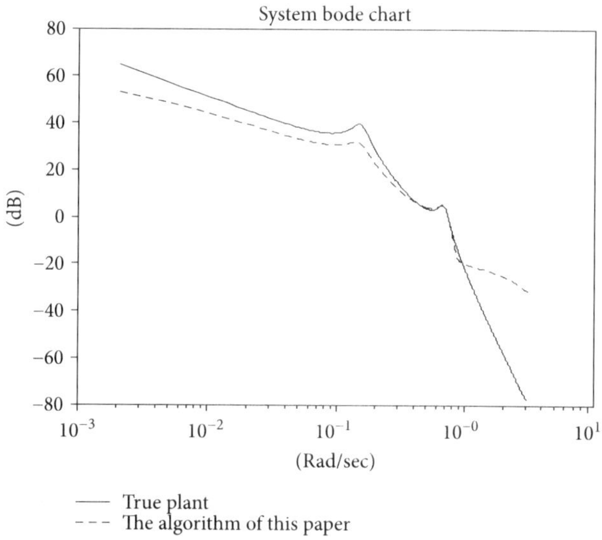

Figure 5: The algorithm of this paper Bode amplitude tracking curve.

It can be seen that the conditional number of the matrix is much larger than 1; therefore, the matrix is ill conditioned. In this case, the instant messaging model is chosen to illustrate the phenomenon of morbidity.

In Figures 3–5, three kinds of Bode amplitude tracking curves of the induction motor model are given. They are SOPIM estimation curve, SIMPCA estimation curve, and the estimation curve of our algorithm, respectively. In each figure, we perform 50 times Monte Carlo simulations of estimating the system to examine the accuracy of the algorithm in this paper.

As shown in Figures 3–5, the Bode amplitude tracking curve shows that the proposed algorithm outperforms the SOPIM estimation results and the SIMPCA estimation results in amplitude tracking. At the same time, SIMPCA

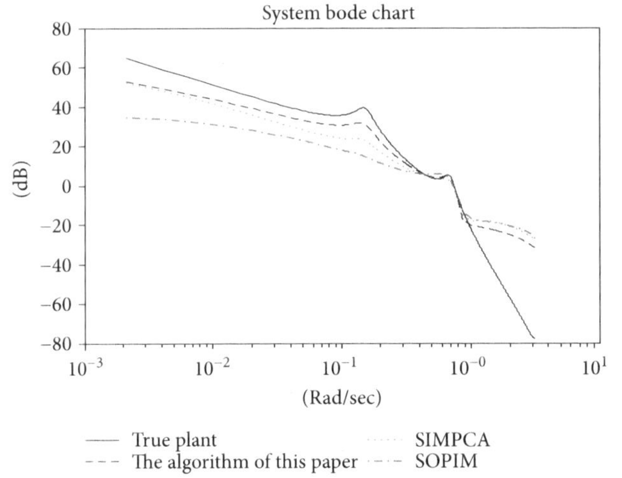

FIGURE 6: Comparison of Bode amplitude tracking curves among the three algorithms.

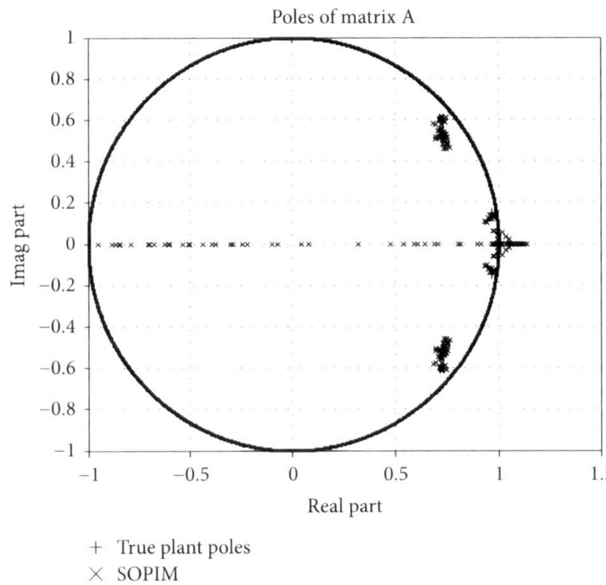

FIGURE 7: SOPIM simulation and comparison of root loci.

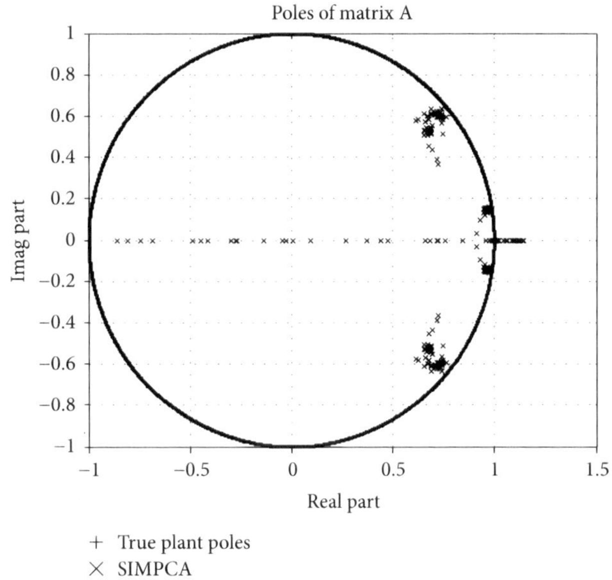

FIGURE 8: SIMPCA simulation and comparison of root loci.

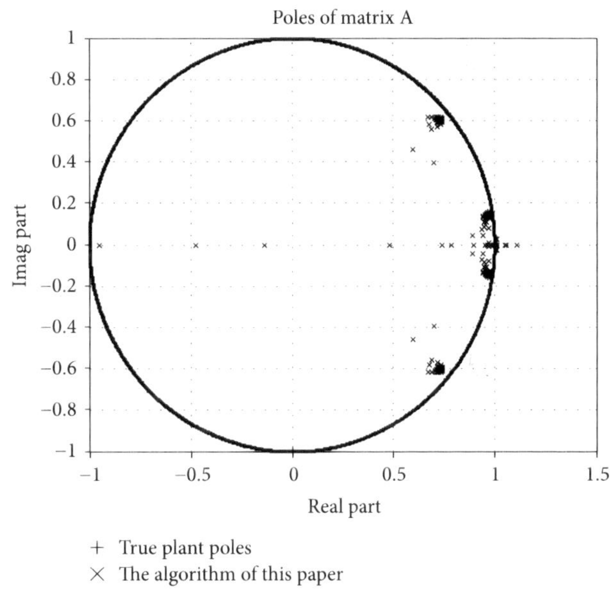

FIGURE 9: The algorithm of this paper simulation and comparison of root loci.

estimation results are better than SOPIM estimation results, mainly because Column weighting for SIMPCA is introduced, while SOPIM uses orthogonal projection only. This is the main reason for zero space projection method based on the estimation of correlation function to fill the block Hankel matrix being adopted in this paper. The Bode amplitude tracking curves among the three algorithms are depicted in Figure 6, which shows the effectiveness of our method.

To confirm that the Bode amplitude tracking results can be obtained, the root locus simulations of the SOPIM algorithm, the SIMPCA algorithm, and the proposed estimation algorithm are performed. The curves obtained are shown in Figures 7–9.

Figures 7–9 illustrate that the red "+" symbol is the real trajectory of the model, and the real trajectory is concentrated on the right side of the chart. In the comparative test of three algorithms, the SOPIM algorithm tracks the most points in the graph, which shows that the locus distribution is dispersed. There are relatively few points near the true trajectory of the "+" symbol, indicating the poor accuracy of trajectory tracking. The trajectory distribution of SIMPCA algorithm in Figure 8 is relatively concentrated. The distribution of our algorithm is the most concentrated, and the main results in the distribution of real track near the "+" sign, which indicates that this algorithm with respect to the root locus simulation has higher tracking accuracy compared with the SOPIM algorithm and SIMPCA algorithm, as shown in Figure 9.

4.2. Hardware Verification Experiment. The hardware test platform is under the condition of single-phase inverter unit

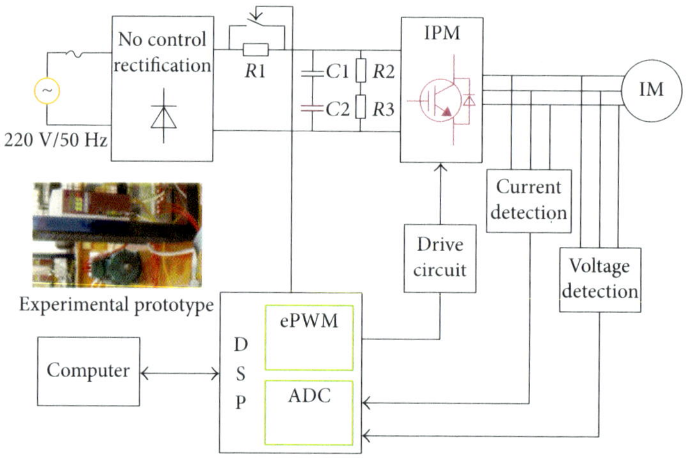

FIGURE 10: Hardware test connection diagram.

TABLE 1: Convergence of identification results.

Time (s)	Identification error Rotor resistance/Ω	LMAg/H
9	5.363	8.351
13	2.585	5.262
17	0.822	2.185
21	0.287	0.873

and controlled rectifier of model PM201CL1A061. Platform identification algorithm is DSP27334 main control chip. The production company is TI. The simulation time connection circuit diagram is shown in Figure 10. The following parameters are P_N = 250 MW motor rated power, rated power factor cos_ N = 0.9, rated voltage U_N = 15.75 kV, rated speed of n_N = 250 r/min D_a1 = 8500 mm, the outside diameter of the stator core, stator core diameter D_i1 = 7500 mm, slot number Z = 360, the stator slot width b_s = 24.7 mm, damping bar diameter d_B = 25 mm, damping number n_B = 7, and air cooling mode.

Because the model belongs to the serial working mode, the proposed algorithm can realize simultaneous identification of magnetizing inductance and rotor resistance with more relative storage. But with the development of science and technology of hard disk, the problem can be ignored. At the time $t = 8$ s, the motor speed changes from 750 r/min to 1200 r/min step by step. Moreover, the rotor resistance synchronous identification error and excitation inductance identification results at this time are shown in Table 1.

Table 1 shows the changes of motor speed with the step changing. The identification results for the algorithm of the motor rotor resistance identification error and the synchronous excitation inductance error change in a passage of time, because the speed of order at $t = 8$ s changes step by step. Here, we start recording the identification error from $t = 9$ s. According to the experimental results, the identification error of the algorithm decreases as time goes by, which shows that the algorithm has better convergence.

5. Concluding Remarks

This paper proposed a kind of zero space projection based on correlation function estimation of closed-loop subspace identification method. Through the correlation function estimation and zero space followed by projection, block Hankel matrix of identification framework was filled by the numerical calculation and LQ decomposition to achieve multiplication projection, get unknown device estimation of dynamic system model. The experimental results verify the effectiveness of the proposed method. The next step will focus on the closed-loop system combined with the subspace identification method for the control problem of UAV.

Conflicts of Interest

The authors declare that they have no conflicts of interest.

Acknowledgments

This study is supported by the Science and Technology Research Program of Chongqing Municipal Education Commission (Grant no. KJ1729403); Chongqing Science and Technology Communication and Popularization Project (no. cstc2017kp-ysczA0021); National Natural Science Foundation of China (no. 61403055); the Intelligent Robot Technology Research Center of Chongqing College of Electronic Engineering (no. XJPT201705); the Fundamental Research Funds for the Central Universities (Grant no. CDJXS12 17 11 01).

References

[1] S. Joe Qin and L. Ljung, "Closed-loop subspace identification with innovation estimation," *IFAC Proceedings*, vol. 36, no. 16, pp. 861–866, 2003.

[2] G. Mercère and M. Lovera, "Convergence analysis of instrumental variable recursive subspace identification algorithms," *Automatica*, vol. 43, no. 8, pp. 1377–1386, 2007.

[3] M. Verhaegen, "Application of a subspace model identification technique to identify LTI systems operating in closed-loop," *Automatica*, vol. 29, no. 4, pp. 1027–1040, 1993.

[4] J. Wang and S. J. Qin, "A new subspace identification approach based on principal component analysis," *Journal of Process Control*, vol. 12, no. 8, pp. 841–855, 2002.

[5] G. van der Veen, J. W. van Wingerden, M. Bergamasco, M. Lovera, and M. Verhaegen, "Closed-loop subspace identification methods: an overview," vol. 7, no. 10, pp. 1339–1358, 2013.

[6] I. Khan, D. Shan, and Q. Li, "Continuous modal parameter identification of a cable-stayed bridge based on Robustious decomposition and covariance-driven stochastic subspace identification," *Iranian Journal of Science and Technology, Transactions of Civil Engineering*, vol. 40, no. 1, pp. 11–22, 2016.

[7] K. Jalaleddini and R. E. Kearney, "Subspace identification of SISO Hammerstein systems: application to stretch reflex identification," *IEEE Transactions on Biomedical Engineering*, vol. 60, no. 10, pp. 2725–2734, 2013.

[8] A. Naitali and F. Giri, "Persistent excitation by deterministic signals for subspace parametric identification of MISO Hammerstein systems," *IEEE Transactions on Automatic Control*, vol. 61, no. 1, pp. 258–263, 2016.

[9] C. S. Rao and M. Chidambaram, "Experimental application of subspace model identification of an unstable system," *International Journal of Advances in Engineering Sciences and Applied Mathematics*, vol. 7, no. 1-2, pp. 70–76, 2015.

[10] T. Jiang, H. Yuan, H. Jia, N. Zhou, and F. Li, "Stochastic subspace identification-based approach for tracking inter-area oscillatory modes in bulk power system utilising synchrophasor measurements," *IET Generation, Transmission & Distribution*, vol. 9, no. 15, pp. 2409–2418, 2015.

[11] Z. Jianyuan, L. Xingfei, and T. Ziling, "The orthogonal decomposition of the recursive subspace identification method of closed loop based on," *Control and Decision*, vol. 30, no. 3, pp. 490–494, 2015.

[12] C. T. Chou and M. Verhaegen, "Subspace algorithms for the identification of multivariable dynamic errors-in-variables models," *Automatica*, vol. 33, no. 10, pp. 1857–1869, 1997.

[13] T. Katayama and H. Tanaka, "An approach to closed-loop subspace identification by orthogonal decomposition," *Automatica*, vol. 43, no. 9, pp. 1623–1630, 2007.

[14] L. Xinming, G. Xianwen, and L. Xinzhe, "Improved pressure identification method based on manifold subspace," *Chinese Journal of Instrumentation*, vol. 35, no. 5, pp. 1079–1085, 2014.

[15] A. G. C. Junior, J. A. Riul, and P. H. M. Montenegro, "Application of the subspace identification method using the N4SID technique for a robotic manipulator," *IEEE Latin America Transactions*, vol. 14, no. 4, pp. 1588–1593, 2016.

[16] Y. Xie, P. Liu, and G. P. Cai, "Modal parameter identification of flexible spacecraft using the covariance-driven stochastic subspace identification (SSI-COV) method," *Acta Mechanica Sinica*, vol. 32, no. 4, pp. 710–719, 2016.

[17] R. N. Cabrera, V. M. A. Martínez, and M. A. Medina, "Analysis of subspace identification methods based on the estimation of the system matrices," *IEEE Latin America Transactions*, vol. 13, no. 4, pp. 1068–1076, 2015.

[18] X. Wu, B. Huang, L. Wang, and J. Zhang, "GPU-based parallel design of the hyperspectral signal subspace identification by minimum error (HySime)," *IEEE Journal of Selected Topics in Applied Earth Observations and Remote Sensing*, vol. 9, no. 9, pp. 4400–4406, 2016.

[19] S. A. Nezam Sarmadi and V. Venkatasubramanian, "Electromechanical mode estimation using recursive adaptive stochastic subspace identification," *IEEE Transactions on Power Systems*, vol. 29, no. 1, pp. 349–358, 2014.

[20] D. Wang, F. Ding, and L. Ximei, "Least squares algorithm for an input nonlinear system with a dynamic subspace state space model," *Nonlinear Dynamics*, vol. 75, no. 1-2, pp. 49–61, 2014.

[21] J. M. Ni, C. Shen, and F. Liu, "Estimation of the electromechanical characteristics of power systems based on a revised stochastic subspace method and the stabilization diagram," *Science China Technological Sciences*, vol. 55, no. 6, pp. 1677–1687, 2012.

[22] T. Gustafsson, "Subspace identification using instrumental variable techniques," *Automatica*, vol. 37, no. 12, pp. 2005–2010, 2001.

[23] B. Huang, S. X. Ding, and S. J. Qin, "Closed-loop subspace identification: an orthogonal projection approach," *Journal of Process Control*, vol. 15, no. 1, pp. 53–66, 2005.

[24] J. Wang and S. J. Qin, "Closed-loop subspace identification using the parity space," *Automatica*, vol. 42, no. 2, pp. 315–320, 2006.

15

An Approach to Mathematical Modeling and Estimation of Probe-Drogue Docking Success Probability for UAV Autonomous Aerial Refueling

Xufeng Wang,[1,2,3] Jianmin Li,[2] Xingwei Kong,[1] Xinmin Dong,[1] and Bo Zhang[2]

[1]*School of Aeronautics and Astronautics Engineering, Air Force Engineering University, Xi'an 710038, China*
[2]*State Key Laboratory of Intelligent Technology and Systems, Tsinghua National Laboratory for Information Science and Technology, Department of Computer Science and Technology, Tsinghua University, Beijing 100084, China*
[3]*Unit 95835 of PLA, Urumqi 841700, China*

Correspondence should be addressed to Xufeng Wang; xf-wang14@mails.tsinghua.edu.cn and Xinmin Dong; dongxinmin@139.com

Academic Editor: Roger L. Davis

One of the keys to the success of aerial refueling for probe-drogue aerial refueling system (PDARS) is the successful docking between the probe and drogue. The study of probe-drogue docking success probability offers an important support to achieving successful docking. During the docking phase of PDARS, based on prior information and reasonable assumptions for the movements of the drogue under atmospheric disturbance, the probe-drogue docking success probability is converted to the probability of the drogue center located in a specific area. A model of the probe-drogue docking success probability is established with and without actuation error, respectively. The curves of the probe-drogue docking success probability with the standard deviation of the drogue central position, the maximum distance from the drogue center position to the equilibrium position, the actuation error, and the standard deviation of the actuation error are obtained through simulations. The study has referential value for the docking maneuver decision of aerial refueling for PDARS.

1. Introduction

Aerial refueling can greatly improve the ability of manned aircrafts and unmanned aerial vehicles (UAV) to execute remote and long endurance tasks [1–4]. Probe-drogue aerial refueling system (PDARS) has been widely concerned by the research institutions of various countries because of its flexibility and economy [5].

One critical stage of PDARS in the air refueling process is the docking of the probe and drogue [5]. Due to the shaking movement caused by the disturbance during the docking process of PDARS, the PDARS test and assessment are a complex task. The relative position and attitude between the probe and the drogue during the PDARS docking stage was studied in [6–10], and the relative position between the probe and the drogue was obtained accurately by the detection and recognition of the drogue. The hose-drogue movement of the PDARS during the aerial refueling process was analyzed in [11, 12]. In [13], the contacting dynamic behavior model of the probe-drogue during the PDARS docking stage was established, and the collision force and velocity response curve were obtained. The modeling and simulation of hose-drogue shaking phenomenon during the PDARS docking stage were studied in [14–16]. In [17, 18], the dynamic behavior of the hose-drogue whip phenomenon during the PDARS docking stage was modeled, and the influencing factors of the whip phenomenon were analyzed. A detailed study on the docking control during the PDARS aerial refueling process was carried out in [19, 20]. However, specific research on mathematical modeling and analysis of probe-drogue docking success probability for PDARS has not yet been reported. To the best of our knowledge, only some flight tests show that the docking success probability of the manned aircraft is 30%~40%, and the NASA UAV aerial refueling docking success probability is about 60% [5, 17, 21]. As mentioned above, the previous research mainly focused on drogue detection and recognition, drogue 3D locating, docking control, bow wave effect analysis, and hose whipping control, while it did not involve the probe-drogue docking success probability estimation, which will be of great guiding significance to the decision making of PDARS in the engineering application of probe-drogue aerial refueling [5].

One of the keys to the success of aerial refueling for PDARS is the successful docking between the probe and drogue. The study of probe-drogue docking success probability offers an important support to achieving successful docking. During the process of probe-drogue aerial refueling, the drogue moves faster than the UAV/probe. Therefore, probe-drogue aerial refueling can be regarded as the slow dynamic system (namely, the probe) tracking the fast dynamic target (namely, the drogue). In this case, point (namely, the probe) to point (namely, the drogue center) tracking is difficult to dock successfully, but point (namely, the probe) to area (namely, the range of movement of drogue center) tracking can realize successful docking to a certain probability.

Considering the facts that the receiver aircraft operates docking maneuver when the probe is in front of the drogue and the atmospheric disturbance affects the movement of the drogue, therefore, this paper focuses on the theoretical estimation of probe-drogue docking success probability before docking maneuver when the probe is in front of the equilibrium position of drogue center. Firstly, according to the prior information during the docking phase of PDARS, a reasonable assumption for the distribution of the drogue center position under atmospheric disturbance is provided. Then, by converting the probe-drogue docking success probability to the probability of the drogue center located in a specific area, the model of probe-drogue docking success probability is established. The simulation analysis shows the curves of the probe-drogue docking success probability with the standard deviation of the drogue central position, the maximum distance from the drogue central position to the equilibrium position, the actuation error, and the standard deviation of the actuation error, respectively, which can provide an important theoretical basis for the engineering application of PDARS docking maneuver decision during the aerial refueling.

2. System Design

We present an approach to mathematical modeling and simulation of probe-drogue docking success probability for UAV autonomous aerial refueling (AAR). Firstly, based on prior information and reasonable assumptions for the movements of the drogue under atmospheric disturbance, the probe-drogue docking success probability is converted to the probability of the drogue center located in a specific area. Then, considering the derivation of the control of UAV, a model of the probe-drogue docking success probability is established with and without actuation error, respectively. Finally, simulations on probe-drogue docking success probability are conducted, and the discussion about considering the relative attitude of probe-drogue for docking success probability is given. Figure 1 shows the flowchart of the estimation of probe-drogue docking success probability for UAV AAR.

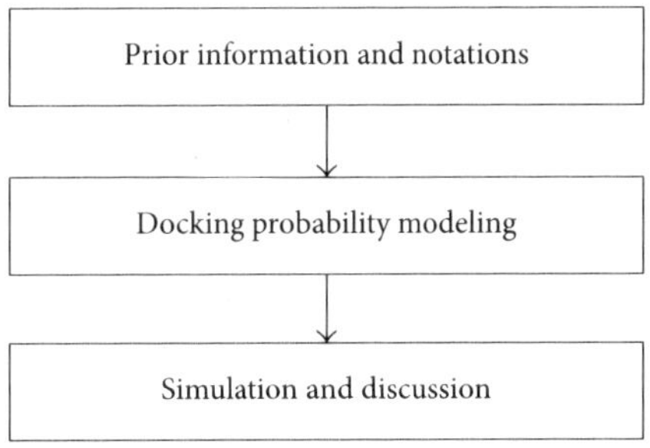

FIGURE 1: Estimation of probe-drogue docking success probability for UAV AAR.

3. Prior Information and Notations

For the simplification of the model of docking success probability, the prior information and notations of probe-drogue docking for PDARS are described as follows.

Assumption 1. During a potential successful probe-drogue docking, the drogue randomly meanders in the lateral and vertical displacement by as much as the drogue diameter.

Remark 2. The disturbance, such as atmospheric turbulence and tanker wave, can lead to drogue lateral and vertical displacement [6, 14, 22, 23]. A NASA flight test approach to an aerial refueling system indicates that the drogue randomly meanders in the lateral and vertical displacement by as much as the drogue diameter (approximately 61 cm), where the atmospheric turbulence is considered, during the docking phase of PDARS [22]. So, considering the reachability and safety of PDARS maneuvers, it is reasonable to assume that, during a potential successful probe-drogue docking, the lateral and vertical displacement of the drogue is less than the drogue diameter, as shown in Figure 2, the red dashed circle. Note that the other direction displacement of the drogue is also less than the drogue diameter, which is a reasonable approximation for estimating the probe-drogue docking success probability for UAV AAR. Thus, the area of disturbed movement of drogue (center) is a circle.

Assumption 3. The lateral and vertical motions of the disturbed drogue are independent.

Remark 4. Considering the shaking movement of the drogue under atmospheric disturbance, the lateral and vertical movements of the drogue have strong independence and weak coupling properties [16]. Therefore, during the docking phase of PDARS, it can be reasonably assumed that the lateral and vertical disturbed drogue motions are independent.

Assumption 5. For specific layout of probe located on the nose of aircrafts, it is reasonable to ignore the bow wave effect.

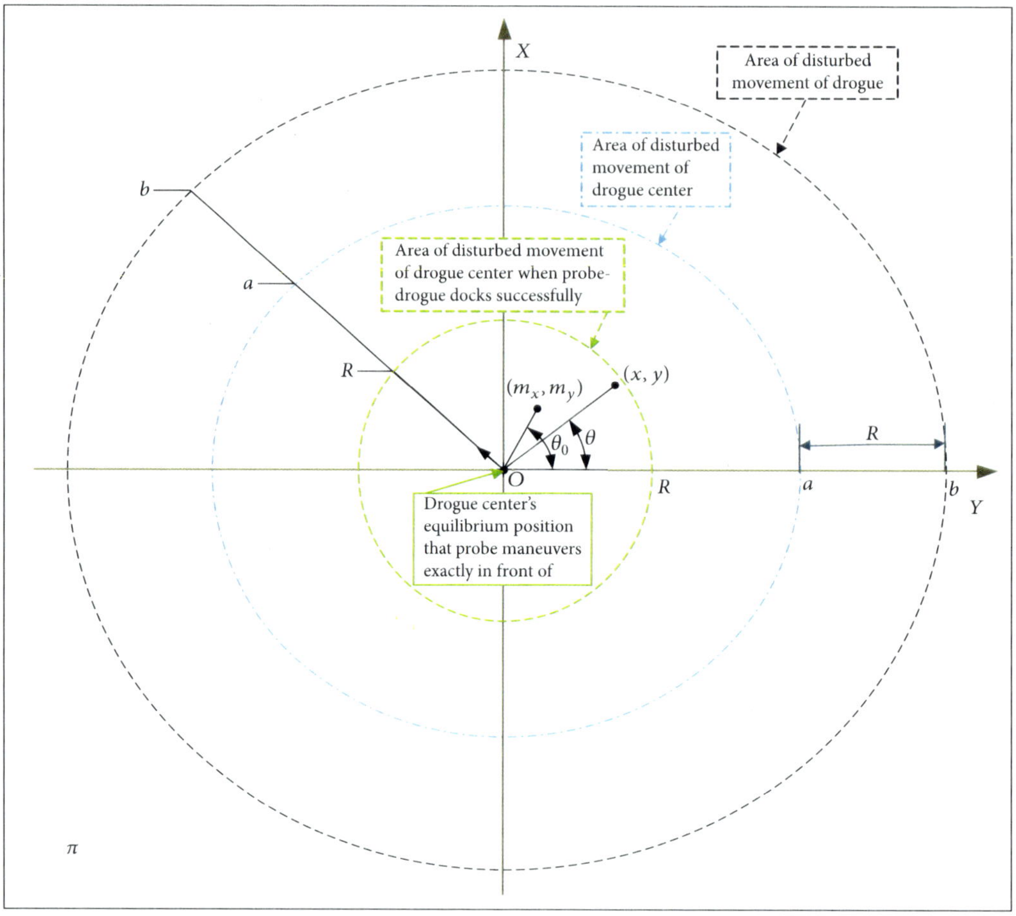

Figure 2: Disturbed movement of drogue and docking area without actuation error.

Remark 6. However, in addition to atmospheric turbulence and tanker wake, bow wave also has an effect on the drogue, leading to the drogue leaving the original position when the drogue is close to the receiver aircraft, and this perturbation is obvious for the layout of probe located on the side in manned aircrafts [24–26]. Fortunately, UAV with aerial refueling capability can use the layout of probe located on the nose of aircrafts, which has much less bow wave effect than that of probe located on the side [5]. Therefore, this paper mainly focuses on UAV with the layout of probe located on the nose, without considering bow wave effect. It should be noted that there are references about how to decrease bow wave effect by appropriately locating probe and drogue in practice, such as [27] that proposed a drogue dynamic model under bow wave effect in probe and drogue aerial refueling, which can describe the drogue dynamics during the docking stage and be applicable to a docking controller design to overcome the bow wave effect actively.

Assumption 7. The lateral and vertical components of the drogue center position coordinate are subject to two-dimensional normal distribution with isotropic.

Remark 8. The collected real probe-drogue aerial refueling videos, as shown in http://pan.baidu.com/s/1mi31gQ8, indicate that, during the docking phase of PDARS, frontal images of the drogue can be reasonably expected, and the probe can be approximately perpendicular to the drogue end plane. And in view of the layout of probe located on the nose, this paper mainly focuses on the motion of the drogue in the plane being perpendicular to the forward direction of the UAV. To further clarify the reasonability of the assumption of using a normal distribution to indicate the position of the drogue in the presence of disturbance, the provided real probe-drogue aerial refueling videos, as shown in http://pan.baidu.com/s/1mi31gQ8, can serve as a justification to indicate that assuming a normal distribution is appropriate.

4. Docking Probability Modeling

4.1. Model without Actuation Error Considered. According to the above discussion in Section 3, the disturbed movement of drogue and docking area without actuation error are depicted in Figure 2. Note that without actuation error means that the probe maneuvers exactly in front of the drogue equilibrium position, and actuation error $d = |OO'| = \sqrt{x_c^2 + y_c^2}$ is depicted in Figure 3. As shown in Figure 2, π is the drogue end plane, O is the equilibrium position of the drogue center,

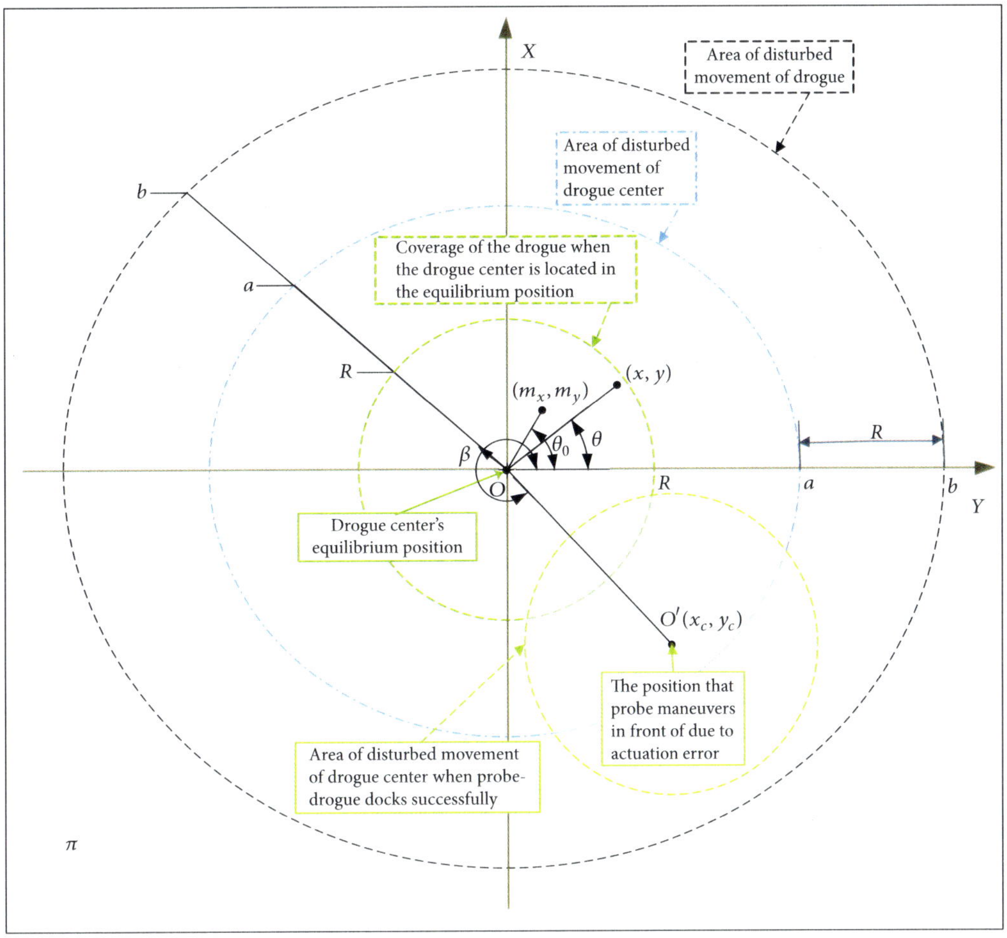

FIGURE 3: Disturbed movement of drogue and docking area with actuation error.

and the radius of the drogue is R. When the drogue center is located in the equilibrium position O, the coverage of the drogue is $x^2 + y^2 \leq R^2$. Because during the implementation of refueling docking process, the range of the disturbed movement of the drogue in X and Y direction of the plane π is generally not more than the diameter size of the drogue, the perturbed motion of the drogue center is in the range of $x^2 + y^2 \leq a^2, 0 \leq a \leq 2R$, and the perturbed motion of the drogue end is in the range of $x^2+y^2 \leq b^2, R \leq b = a+R \leq 3R$, accordingly.

Considering the engineering application of aerial refueling docking, the probe maneuvers gradually right in front of the equilibrium position O. Note that the probe maneuvers exactly in front of the equilibrium position O, according to corresponding control algorithm, without actuation error means. Therefore, when the drogue center located inside area $x^2 + y^2 < R^2$, the probe in front of (without actuation error) the equilibrium position O can ensure probe-drogue docking successfully. Namely, the probe-drogue docking success probability is equal to the probability of the drogue center locating in the region $x^2 + y^2 < R^2$.

The probability density function of the drogue center position (x, y) under disturbance is established as

$$f(x, y) = \frac{1}{2\pi\sigma_x\sigma_y} \exp\left\{-\frac{1}{2}\left[\frac{(x-m_x)^2}{\sigma_x^2} + \frac{(y-m_y)^2}{\sigma_y^2}\right]\right\}, \quad (1)$$

where m_x and m_y are the systematic deviation of drogue center in X and Y direction, respectively. σ_x^2 and σ_y^2 are the variance of the drogue center position in X and Y direction, respectively.

When the polar coordinates are used, there are

$$x = r\sin\theta, \quad y = r\cos\theta, \quad (2)$$

$$m_x = m\sin\theta_0, \quad m_y = m\cos\theta_0. \quad (3)$$

The Jacobian determinant of (2) is as follows:

$$\frac{\partial (x, y)}{\partial (r, \theta)} = \begin{vmatrix} \frac{\partial x}{\partial r} & \frac{\partial x}{\partial \theta} \\ \frac{\partial y}{\partial r} & \frac{\partial y}{\partial \theta} \end{vmatrix} = \begin{vmatrix} \sin\theta & r\cos\theta \\ \cos\theta & -r\sin\theta \end{vmatrix} = -r. \quad (4)$$

Therefore, we can obtain the polar coordinate form of $f(x, y)$ defined by

$$g(r, \theta) = f(x, y) \left| \frac{\partial (x, y)}{\partial (r, \theta)} \right|. \quad (5)$$

Based on (1)–(5), when $\sigma_x = \sigma_y = \sigma$, (1) can be transformed into the polar coordinate form

$$\begin{aligned} g(r, \theta) &= f(r\sin\theta, r\cos\theta) \left| \frac{\partial (x, y)}{\partial (r, \theta)} \right| = \frac{r}{2\pi\sigma^2} \\ &\cdot \exp\left\{ -\frac{1}{2} \left[\frac{(r\sin\theta - m\sin\theta_0)^2}{\sigma^2} \right.\right. \\ &\left.\left. + \frac{(r\cos\theta - m\cos\theta_0)^2}{\sigma^2} \right] \right\} = \frac{r}{2\pi\sigma^2} \exp\left\{ -\frac{1}{2\sigma^2} \left[r^2 \right.\right. \\ &\left.\left. + m^2 - 2rm\cos(\theta - \theta_0) \right] \right\}. \end{aligned} \quad (6)$$

To calculate the probe-drogue docking success probability, we then only need to consider the marginal distribution along r (the distance between the drogue center and the equilibrium position O) as follows:

$$\begin{aligned} p(r) &= \int_0^{2\pi} g(r, \theta) \, d\theta = \frac{r}{\sigma^2} \exp\left(-\frac{r^2 + m^2}{2\sigma^2} \right) \int_0^{2\pi} \frac{1}{2\pi} \\ &\cdot \exp\left[\frac{rm\cos(\theta - \theta_0)}{\sigma^2} \right] d\theta = \frac{r}{\sigma^2} \exp\left(-\frac{r^2 + m^2}{2\sigma^2} \right) \\ &\cdot I_0\left(\frac{rm}{\sigma^2} \right), \end{aligned} \quad (7)$$

where $I_0(\cdot)$ is the first kind zero-order modified Bessel function:

$$I_0(x) = \frac{1}{2\pi} \int_0^{2\pi} \exp\left[x\cos(\theta - \theta_0) \right] d\theta. \quad (8)$$

When the observation period is large enough, the systematic deviation of the center position of the drogue in the direction of X and Y can be approximately zero: namely, $m_x = 0$, $m_y = 0$, and $m = 0$. Then, we have $I_0(rm/\sigma^2) = I_0(0) = 1$, and therefore, (7) can be simply expressed as

$$p(r) = \frac{r}{\sigma^2} \exp\left(-\frac{r^2}{2\sigma^2} \right). \quad (9)$$

Then, based on the prior information, when the drogue center located inside area C_R, $x^2 + y^2 < R^2$, the probe in front of (without actuation error) the equilibrium position O can ensure probe-drogue docking successfully. Namely, the probe-drogue docking success probability is equal to the probability of the drogue center locating in the region C_R (with the equilibrium position O as the center and the drogue radius R as the radius):

$$\begin{aligned} P &= \iint_{(x,y)\in C_R} f(x, y) \, dx \, dy = \int_0^R p(r) \, dr \\ &= \int_0^R \frac{r}{\sigma^2} \exp\left(-\frac{r^2}{2\sigma^2} \right) dr = 1 - \exp\left(-\frac{R^2}{2\sigma^2} \right). \end{aligned} \quad (10)$$

Further, a more intuitive approximate representation between probe-drogue docking success probability and drogue central position is derived. Assuming that, in the observation period T (T is large enough), a total of n coordinates of drogue center position are observed, namely, (x_i, y_i) with $i = 1, 2, \ldots, n$ $(n > 2)$, $\bar{x} = (1/n)\sum_{i=1}^{n} x_i$, $\bar{y} = (1/n)\sum_{i=1}^{n} y_i$, the drogue equilibrium position $(\bar{x}, \bar{y}) \simeq (0, 0)$.

Then, we can obtain the variance of the drogue center position:

$$\begin{aligned} \sigma^2 &= \frac{1}{2} \left[\frac{1}{n-1} \sum_{i=1}^{n} (x_i - \bar{x})^2 + \frac{1}{n-1} \sum_{i=1}^{n} (y_i - \bar{y})^2 \right] \\ &\simeq \frac{1}{2(n-1)} \sum_{i=1}^{n} \left(x_i^2 + y_i^2 \right) = \frac{1}{2(n-1)} \sum_{i=1}^{n} r_i^2 \\ &\le \frac{n}{2(n-1)} \max_{i=1,2,\ldots,n} \left\{ r_i^2 \right\}. \end{aligned} \quad (11)$$

When $n \ge 2$, it follows that

$$\frac{1}{2} \le \frac{n}{2(n-1)} \le 1 \quad (12)$$

and $\lim_{n\to\infty}(n/2(n-1)) = 1/2$.

Let

$$\begin{aligned} \max_{i=1,2,\ldots,n} \left\{ r_i^2 \right\} &= r_{\max}^2, \\ \max_{i=1,2,\ldots,n} \left\{ r_i \right\} &= r_{\max}, \end{aligned} \quad (13)$$

where $r_{\max}$ represents the maximum distance between the drogue center position and the equilibrium position O in the observation period T.

Substituting expressions (12), (13) into expression (11) leads to

$$\sigma^2 \le \frac{1}{2} r_{\max}^2. \quad (14)$$

Equivalently,

$$r_{\max\_\inf} = \sqrt{2}\sigma, \quad (15)$$

where $r_{\max\_\inf}$ represents the lower-bound value of $r_{\max}$.

Substituting (15) into (10), we get that

$$P = 1 - \exp\left(-\frac{R^2}{r_{\max\_\inf}^2} \right). \quad (16)$$

Because $r_{\max}$ represents the maximum distance between the drogue center position and the equilibrium position O in the observation period T. The smaller $r_{\max}$ is, the smaller the disturbed motion of the drogue is, and thus the higher the probe-drogue docking success probability is. Therefore, statistically, the probe-drogue docking success probability satisfies the following relation:

$$\begin{aligned} P\left(r_{\max\_sup} = \xi\right) &= P\left(r_{\max} \leq \xi\right) \geq P\left(r_{\max} \geq \xi\right) \\ &= P\left(r_{\max\_inf} = \xi\right), \end{aligned} \quad (17)$$

where $r_{\max\_sup}$ represents the upper-bound value of $r_{\max}$, with ξ being the threshold.

The proof of expression (17) is as follows.

Proof. There exists $r_{\max\_sup} = \xi \Leftrightarrow r_{\max} \leq \xi$ and $r_{\max\_inf} = \xi \Leftrightarrow r_{\max} \geq \xi$, such that

$$\begin{aligned} P\left(r_{\max\_sup}\right) &= P\left(r_{\max} \leq \xi\right), \\ P\left(r_{\max} \geq \xi\right) &= P\left(r_{\max\_inf} = \xi\right). \end{aligned} \quad (18)$$

Moreover, because the docking success probability when $r_{\max} \leq \xi$ is bound to be no less than the docking success probability when $r_{\max} \geq \xi$, it can be deduced that

$$P\left(r_{\max} \leq \xi\right) \geq P\left(r_{\max} \geq \xi\right). \quad (19)$$

Thus, by expressions (18) and (19), expression (17) is established.

The proof is completed. □

Note that the probability estimation based on $r_{\max}$ is more intuitive and conservative, while the probability estimation based on σ is more comprehensive and accurate. Thus, the two estimation methods have their own advantages and the docking success probability should be estimated based on the two methods simultaneously.

4.2. Model with Actuation Error Considered. Under the condition with actuation error, because of the existence of the actuation error, the docking position of the probe-drogue deviates from $O(0,0)$, the equilibrium position of the movement of drogue center. The disturbed movements of drogue and docking area with actuation error are depicted in Figure 3. As shown in Figure 3, $O'(x_c, y_c)$ is the docking position, the actuation error $d = |OO'| = \sqrt{x_c^2 + y_c^2}$, β is the angle between d and y axis, $x_c = d\sin\beta$, and $y_c = d\cos\beta$. Thus, under the condition with actuation error, in order to make the probe-drogue docking successfully, the drogue center must be located within the region $(x - x_c)^2 + (y - y_c)^2 \leq R^2$. Therefore, under the condition with actuation error, the probe-drogue docking success probability is equal to the probability that the drogue center is located in the region $(x - x_c)^2 + (y - y_c)^2 \leq R^2$.

As mentioned above, under the condition with actuation error, the probe-drogue docking success probability is equal to the probability of the drogue center locating in the region C'_R: $(x - x_c)^2 + (y - y_c)^2 < R^2$ (with $O'(x_c, y_c)$ as the center and the drogue radius R as the radius):

$$\begin{aligned} P &= \iint_{(x,y)\in C'_R} f(x,y)\,dx\,dy \\ &= \iint_{(x-x_c)^2+(y-y_c)^2\leq R^2} \frac{1}{2\pi\sigma^2}\exp\left(-\frac{x^2+y^2}{2\sigma^2}\right)dx\,dy \\ &= \iint_{(x-d\sin\beta)^2+(y-d\cos\beta)^2\leq R^2} \frac{1}{2\pi\sigma^2}\exp\left(-\frac{x^2+y^2}{2\sigma^2}\right)dx\,dy \\ &= \iint_{x^2+(y-d)^2\leq R^2} \frac{1}{2\pi\sigma^2}\exp\left(-\frac{x^2+y^2}{2\sigma^2}\right)dx\,dy \\ &= \iint_{(x-d)^2+y^2\leq R^2} \frac{1}{2\pi\sigma^2}\exp\left(-\frac{x^2+y^2}{2\sigma^2}\right)dx\,dy \\ &= P_{\sigma,d}(\sigma, d), \end{aligned} \quad (20)$$

where σ^2 is the variance of the drogue center position in X and Y direction. And because R is a constant, (20) shows that the probe-drogue docking success probability $P_{\sigma,d}(\sigma, d)$ is a function of σ and d, under the condition with actuation error.

Further, from the view of statistics, the relationship between the probability of successful docking and the variance of the control deviation is derived.

Under the condition with actuation error, given that the variance of the control deviation d in X and Y direction is σ_c^2. Now we define $P_{\sigma,\sigma_c}(\sigma, \sigma_c)$ as the probe-drogue docking success probability, a function of σ and σ_c, under the condition with actuation error. Next, the relationship between $P_{\sigma,\sigma_c}(\sigma, \sigma_c)$ and $P_{\sigma,d}(\sigma, d)$ is derived, and the relationship of the probe-drogue docking success probability versus σ and σ_c is given. The derivation process of the relationship between $P_{\sigma,\sigma_c}(\sigma, \sigma_c)$ and $P_{\sigma,d}(\sigma, d)$ is as follows.

According to expression (14), we can obtain

$$\sigma_c \leq \frac{1}{\sqrt{2}} d_{\max}, \quad (21)$$

where $d_{\max}$ represents the maximum of the actuation error in the observation period T.

Due to the smaller bias of the control, the higher probability of probe-drogue docking success, we have

$$\begin{aligned} P_{\sigma,\sigma_c}(\sigma, \sigma_c) &= P_{\sigma,\sigma_c}\left(\sigma, \sigma_c \leq \frac{1}{\sqrt{2}} d_{\max}\right) \\ &\geq P_{\sigma,\sigma_c}\left(\sigma, \sigma_c = \frac{1}{\sqrt{2}} d_{\max}\right). \end{aligned} \quad (22)$$

It can be proved that

$$P_{\sigma,\sigma_c}\left(\sigma, \sigma_c = \frac{1}{\sqrt{2}} d_{\max}\right) \geq P_{\sigma,d}\left(\sigma, d = \sqrt{2}\sigma_c\right). \quad (23)$$

From the intuitive understanding, expression (23) was established because the docking success probability when $d \leq \sqrt{2}\sigma_c$ (namely, $P_{\sigma,\sigma_c}(\sigma, \sigma_c = (1/\sqrt{2})d_{\max})$) is bound to be no less than the docking success probability when $d = \sqrt{2}\sigma_c$ (namely, $P_{\sigma,d}(\sigma, d = \sqrt{2}\sigma_c)$).

The proof of expression (23) is as follows.

Proof. Let $\sigma_c = (1/\sqrt{2})d_{\max}$. Then there exists $d = d_1 \le d_{\max} = \sqrt{2}\sigma_c$ such that

$$P_{\sigma,\sigma_c}\left(\sigma, \sigma_c = \frac{1}{\sqrt{2}} d_{\max}\right) = P_{\sigma,d}\left(\sigma, d = d_1 \le d_{\max} = \sqrt{2}\sigma_c\right). \tag{24}$$

Moreover, it can be deduced that

$$P_{\sigma,d}\left(\sigma, d = d_1 \le d_{\max} = \sqrt{2}\sigma_c\right) \ge P_{\sigma,d}\left(\sigma, d = d_{\max} = \sqrt{2}\sigma_c\right). \tag{25}$$

Thus, by expressions (24) and (25), expression (23) is established.

The proof is completed. □

Then, from expressions (22) and (23), we get that

$$P_{\sigma,\sigma_c}(\sigma,\sigma_c) \ge P_{\sigma,d}\left(\sigma, d = \sqrt{2}\sigma_c\right). \tag{26}$$

Expression (26) is the relationship between $P_{\sigma,\sigma_c}(\sigma,\sigma_c)$ and $P_{\sigma,d}(\sigma,d)$. Namely,

$$P_{\sigma,\sigma_c}(\sigma,\sigma_c) \ge P_{\sigma,\sigma_c}(\sigma,\sigma_c)_{\inf}, \tag{27}$$

where $P_{\sigma,\sigma_c}(\sigma,\sigma_c)_{\inf}$ represents the lower-bound value of $P_{\sigma,\sigma_c}(\sigma,\sigma_c)$ and

$$\begin{aligned} P_{\sigma,\sigma_c}(\sigma,\sigma_c)_{\inf} &= P_{\sigma,d}\left(\sigma, d = \sqrt{2}\sigma_c\right) \\ &= \iint_{(x-\sqrt{2}\sigma_c \sin\beta)^2+(y-\sqrt{2}\sigma_c\cos\beta)^2 \le R^2} \frac{\exp\left(-\left(x^2+y^2\right)/2\sigma^2\right)dx\,dy}{2\pi\sigma^2} \\ &= \iint_{x^2+(y-\sqrt{2}\sigma_c)^2\le R^2} \frac{1}{2\pi\sigma^2}\exp\left(-\frac{x^2+y^2}{2\sigma^2}\right)dx\,dy \\ &= \iint_{(x-\sqrt{2}\sigma_c)^2+y^2\le R^2} \frac{1}{2\pi\sigma^2}\exp\left(-\frac{x^2+y^2}{2\sigma^2}\right)dx\,dy. \end{aligned} \tag{28}$$

5. Simulations and Analyses

According to [14, 22], we set the drogue diameter D to be 0.61 m, and, namely, the drogue radius R is 0.305 m.

5.1. Without Actuation Error Considered. Based on (10), the probe-drogue docking success probability P versus σ is shown in Figures 4 and 5.

Based on (16), the probe-drogue docking success probability P versus $r_{\max\_\inf}$ is depicted in Figures 6 and 7. Note that $P(r_{\max\_\sup} = \xi) = P(r_{\max} \le \xi) \ge P(r_{\max\_\inf} = \xi)$.

The variation ratio of the probe-drogue docking success probability P versus σ is depicted in Figure 8.

The variation ratio of the probe-drogue docking success probability P versus $r_{\max\_\inf}$ is depicted in Figure 9.

Through the analysis of Figures 4–9, we can get the following:

(1) During the probe-drogue docking process, the more stable the drogue central position, the larger the probe-drogue docking success probability.

(2) The probe-drogue docking success probability drops from 89.46% to 63.21% rapidly when $r_{\max\_\inf}$ increases from 20 cm to 30 cm, which provides a docking maneuver threshold to some extent. Because that aerial refueling is a high difficult task requiring stability, accuracy, and safety, to ensure the success of the aerial refueling, the value of $r_{\max\_\sup}$ can be used as a decision reference of probe-drogue docking maneuver. When $r_{\max\_\sup} = R$, the probe-drogue docking success probability $P(r_{\max\_\sup} = R) \ge 63.21\%$. When $r_{\max\_\sup} = (2/3)R$, the probe-drogue docking success probability $P(r_{\max\_\sup} = (2/3)R) \ge 89.46\%$. When $r_{\max\_\sup} = (1/2)R$, the probe-drogue docking success probability $P(r_{\max\_\sup} = (1/2)R) \ge 98.17\%$. Therefore, the docking maneuver should be made when $r_{\max\_\sup} < (2/3)R \approx 20$ cm for safety. Note that the decision threshold $r_{\max\_\sup} = \xi$ (namely, $r_{\max} \le \xi$) of the probe-drogue docking maneuver during the aerial refueling deserves further extensive flight tests and comprehensive assessments to determine a reasonable decision threshold.

(3) For the probe-drogue docking success probability/rate, a comparison between the theoretical analysis proposed in this paper and the flight test results of NASA AAR demonstration [21] is described as follows, with a detailed introduction of the NASA results and a visualized comparison in Figure 10. Note that the probe-drogue docking success probability is corresponding to the theoretical analysis proposed in this paper, and the probe-drogue docking success rate is corresponding to the flight test results of NASA AAR demonstration [21].

Figure 10 shows the visualized comparison between the flight test results of NASA AAR demonstration [21] and the theoretical analysis proposed in this paper for the probe-drogue docking success rate/probability. Figure 10(a) represents the left view of the drogue and R_C value, corresponding to a 90 percent probe-drogue docking success rate in the flight test of NASA AAR demonstration [21]. Figure 10(b) describes the main view of the drogue and $r_{\max\_\sup}$ value, corresponding to larger than 89.44 percent probe-drogue docking success probability in the theoretical analysis proposed in this paper.

As shown in Figure 10(a), according to the flight test results of NASA AAR demonstration [21], R_C is the capture radius that was defined as being 10.16 cm inside the outer ring of the drogue, which was suggested by the project pilot as a diameter that would result in a 90 percent probe-drogue docking success rate with minimal vertical and lateral velocity. Actually, R_C defines a tube coaxial (obtained by revolving the blue area 180° about the blue dash dot line) to the drogue, and when the probe is located within the capture radius R_C of the drogue, the probe-drogue docking success rate would reach 90 percent [21].

As shown in Figure 10(b), according to the theoretical analysis proposed in this paper, when $r_{\max\_\sup} = 20.34$ cm, which means the probe is located within the capture radius 20.34 cm of the drogue in practical engineering (namely, $r_{\max\_\sup} = 20.34$ cm is equal to $R_C = 20.34$ cm in practical

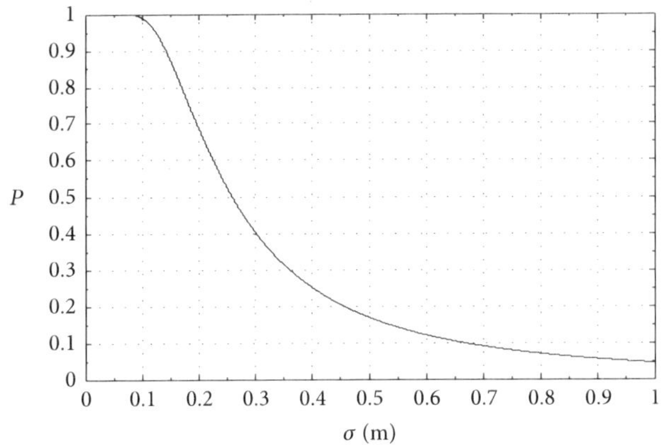

FIGURE 4: Probe-drogue docking success probability P versus σ.

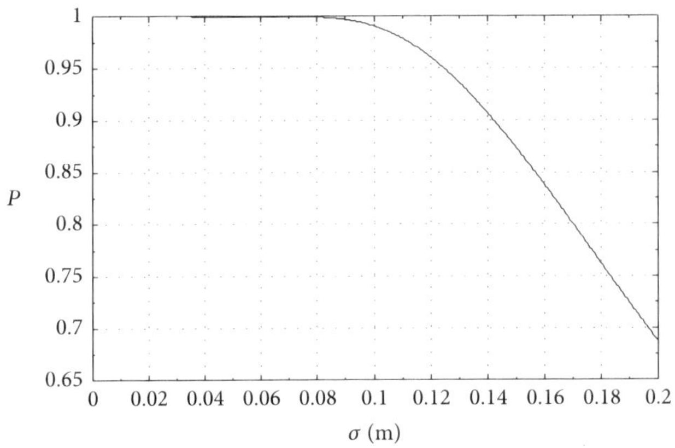

FIGURE 5: P versus σ (local magnification).

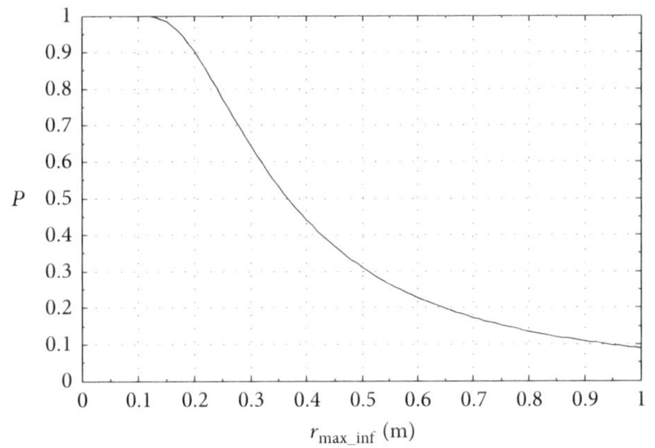

FIGURE 6: P versus $r_{\text{max\_inf}}$.

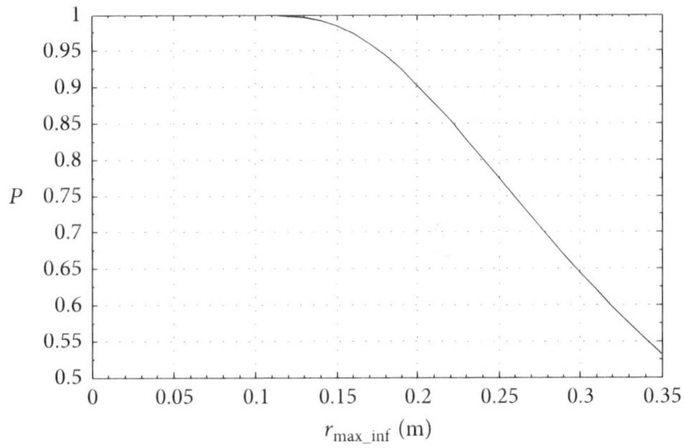

FIGURE 7: P versus $r_{\text{max\_inf}}$ (local magnification).

engineering), the probe-drogue docking success probability $P(r_{\text{max\_sup}} = 0.2034\,\text{m}) \geq 89.44\%$.

From the above discussion, it can be inferred that "R_C = 20.34 cm in Figure 10(a)" is equal to "$r_{\text{max\_sup}}$ = 20.34 cm in Figure 10(b)" in practical engineering, which result in 90 percent probe-drogue docking success rate in the flight test results of NASA AAR demonstration [21] and 89.44 percent probe-drogue docking success probability in the theoretical analysis proposed in this paper, respectively. Therefore, for the probe-drogue docking success probability/rate, the theoretical analysis proposed in this paper is consistent with the flight test results of NASA AAR demonstration [21].

5.2. With Actuation Error Considered. Under the condition with actuation error, based on (20), the probe-drogue docking success probability $P_{\sigma,d}(\sigma, d)$ versus σ and d is shown in Figure 11.

From Figure 11, the following is shown:

(1) Under the condition with actuation error, when $0 \leq d < R = 0.305\,\text{m}$, the probe-drogue docking success probability $P_{\sigma,d}(\sigma, d)$ decreases with the increase of σ and d. The larger the actuation error d is, the faster the probe-drogue docking success probability $P_{\sigma,d}(\sigma, d)$ decreases with the increase of σ.

(2) Under the condition with actuation error, $d = R = 0.305\,\text{m}$ is the separatrix. When $d > R = 0.305\,\text{m}$, the probe-drogue docking success probability $P_{\sigma,d}(\sigma, d)$ remains at a low level with $P_{\sigma,d}(\sigma, d)_{\max} < 0.5$.

(3) Because of the accuracy and safety requirements of the aerial refueling mission, the study of $P_{\sigma,d}(\sigma, d)$ with $0 \leq d < R = 0.305\,\text{m}$ will be more meaningful and helpful. Considering that when $0 \leq d < R = 0.305\,\text{m}$, the high value of $P_{\sigma,d}(\sigma, d)$ is basically in the range of $0 \leq \sigma \leq 0.2\,\text{m}$. In view of this, we choose the range of $0 \leq d < R = 0.305\,\text{m}$ with $0 \leq \sigma \leq 0.2\,\text{m}$ and further give the probe-drogue docking success probability $P_{\sigma,d}(\sigma, d)$ versus σ and d in local detail, as shown in Figure 12.

From Figure 12, it can be seen in more detail that, in $P_{\sigma,d}(\sigma, d)$ high value range ($0 \leq d < R = 0.305\,\text{m}$ with $0 \leq \sigma \leq 0.2\,\text{m}$), the existence of actuation error d can have a negative impact on the $P_{\sigma,d}(\sigma, d)$. When σ is certain (namely, under the same external environment), the greater the actuation error d, the larger the decline of $P_{\sigma,d}(\sigma, d)$. That is, the greater the actuation error d, the higher requirements of the external environment when making docking maneuver for UAV AAR. Consequently, it is necessary to take appropriate control strategy to reduce the

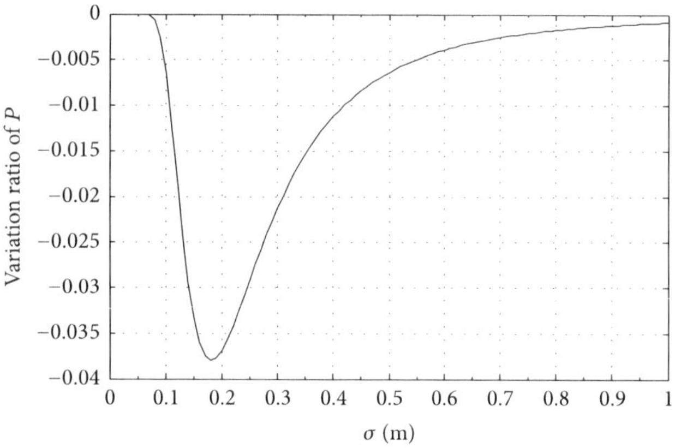

FIGURE 8: Variation ratio of P versus σ.

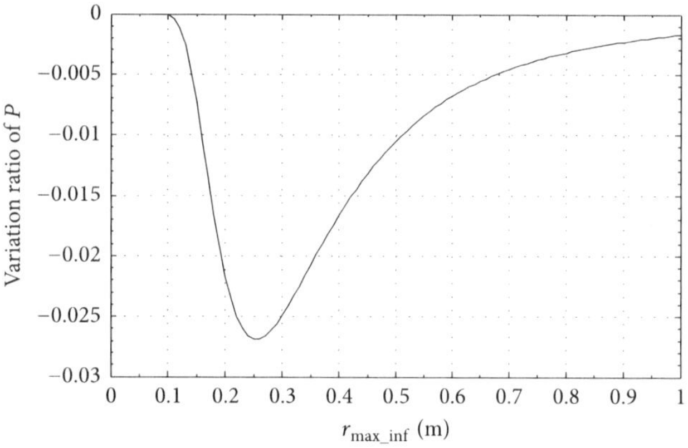

FIGURE 9: Variation ratio of P versus $r_{\text{max\_inf}}$.

actuation error d, and when actuation error d is small, we can ensure high probe-drogue docking success probability even when the external disturbance is relatively large, which increases the adaptability to the external environment of AAR.

Under the condition with actuation error, based on (28), the lower-bound value of probe-drogue docking success probability $P_{\sigma,\sigma_c}(\sigma,\sigma_c)_{\text{inf}}$ versus σ and σ_c is shown in Figure 13. Note that the probe-drogue docking success probability $P_{\sigma,\sigma_c}(\sigma,\sigma_c) \geq P_{\sigma,\sigma_c}(\sigma,\sigma_c)_{\text{inf}}$.

From Figure 13, the following is shown:

(1) Under the condition with actuation error, when $0 \leq \sigma_c < 0.2157$ m, the lower-bound value of probe-drogue docking success probability $P_{\sigma,\sigma_c}(\sigma,\sigma_c)_{\text{inf}}$ decreases with the increase of σ and σ_c. The larger σ_c is, the faster $P_{\sigma,\sigma_c}(\sigma,\sigma_c)_{\text{inf}}$ decreases with the increase of σ.

(2) Under the condition with actuation error, $\sigma_c = 0.2157$ m is the separatrix. When $\sigma_c > 0.2157$ m, $P_{\sigma,\sigma_c}(\sigma,\sigma_c)_{\text{inf}}$ remains at a low level with $P_{\sigma,\sigma_c}(\sigma,\sigma_c)_{\text{inf}} < 0.5$.

(3) Because of the accuracy and safety requirements of the aerial refueling mission, the study of $P_{\sigma,\sigma_c}(\sigma,\sigma_c)_{\text{inf}}$ with $0 \leq \sigma_c < 0.2157$ m will be more meaningful and helpful. Consider that when $0 \leq \sigma_c < 0.2157$ m, the high value of $P_{\sigma,\sigma_c}(\sigma,\sigma_c)_{\text{inf}}$ is basically in the range of $0 \leq \sigma \leq 0.2$ m. In view of this, we choose the range of $0 \leq \sigma_c < 0.2157$ m with $0 \leq \sigma \leq 0.2$ m and further give $P_{\sigma,\sigma_c}(\sigma,\sigma_c)_{\text{inf}}$ versus σ and σ_c in local detail, as shown in Figure 14. Note that the probe-drogue docking success probability $P_{\sigma,\sigma_c}(\sigma,\sigma_c) \geq P_{\sigma,\sigma_c}(\sigma,\sigma_c)_{\text{inf}}$.

From Figure 14, it can be seen in more detail that, in $P_{\sigma,\sigma_c}(\sigma,\sigma_c)_{\text{inf}}$ high value range ($0 \leq \sigma_c < 0.2157$ m with $0 \leq \sigma \leq 0.2$ m), the existence of σ_c can have a negative impact on $P_{\sigma,\sigma_c}(\sigma,\sigma_c)_{\text{inf}}$. When σ value is certain (namely, under the same external environment), the greater σ_c, the larger decline of $P_{\sigma,\sigma_c}(\sigma,\sigma_c)_{\text{inf}}$. That is, the greater σ_c, the higher requirements of the external environment when making docking maneuver for UAV AAR. Consequently, it is necessary to take appropriate control strategy to reduce σ_c, and when σ_c is small, we can ensure high probe-drogue docking success probability even when the external disturbance is relatively large, which increases the adaptability to the external environment of AAR.

6. Discussions

6.1. Considering the Relative Attitude of Probe-Drogue for Docking Success Probability. Note that indeed the movement of drogue center has an important effect on the success for docking, but there are many other factors that can produce effect on the success of docking. In order to make the proposed model of probe-drogue docking success probability more comprehensive and reasonable, we further consider the relative attitude between the drogue and the probe, which is an important factor for a successful docking. Namely, a more reasonable probe-drogue docking success probability *P_optimal* can be estimated, according to the relative attitude between the drogue and the probe, based on the initial probe-drogue docking success probability *P_initial* using the movement mathematical model of drogue center. Details are as follows.

Due to the rolling symmetry property of drogue, we only consider its pitch angle θ and yaw angle φ. Note that the theory, experiment, and precision of drogue attitude measurement have been described in our previous work [6] in detail. So, in this study, we assume that accurate parameters, such as the pitch angle θ and yaw angle φ of the drogue, can be obtained precisely.

During the process of aerial refueling, there is a setting that when the probe is right in front of the drogue and the probe is perpendicular to the end of the drogue, then the pitch angle θ and yaw angle φ of the drogue are all 0°. Then, theoretically, there exist $\theta \in [-\pi,\pi]$ and $\varphi \in [-\pi,\pi]$. We should note that, during the docking phase of aerial refueling, 30° deflection of the drogue can be regarded as large, and when the deflection is more than 90°, the probe and drogue cannot dock successfully.

When $\theta = 0^\circ$ and $\varphi = 0^\circ$, *P_optimal* is equal to *P_initial*; when the deflection of the drogue increases from $0^\circ$to 90°, *P_optimal* decreases from *P_initial* to 0; when the deflection

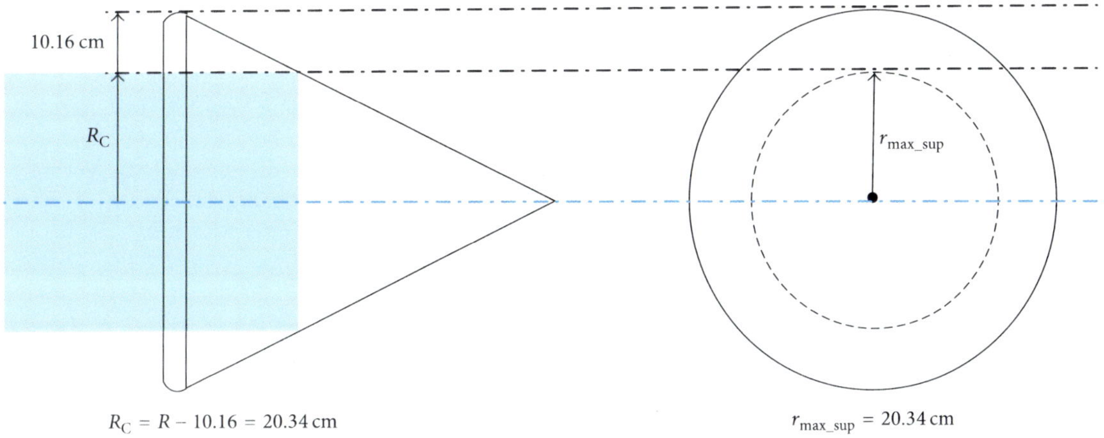

FIGURE 10: Visualized comparison between the flight test results of NASA AAR demonstration [21] and the theoretical analysis proposed in this paper for the probe-drogue docking success rate/probability. (a) Left view of the drogue and R_C value, corresponding to a 90 percent probe-drogue docking success rate in the flight test of NASA AAR demonstration [21]. (b) Main view of the drogue and r_{max_sup} value, corresponding to larger than 89.44 percent probe-drogue docking success probability in the theoretical analysis proposed in this paper.

of the drogue is more than 90°, *P_optimal* will be 0. Motivated by the above discussion, the pitch angle θ and yaw angle φ of the drogue can be regarded as the penalty factor c of *P_initial*. Thus, a concise and reasonable definition of penalty factor c is given as follows:

$$c(x) = \begin{cases} \dfrac{||x| - \pi/2|}{\pi/2}, & |x| \in \left[0, \dfrac{\pi}{2}\right], \\ 0, & |x| \in \left(\dfrac{\pi}{2}, \pi\right], \end{cases} \quad (29)$$

where $x \in [-\pi, \pi]$, x represents the deflection angle of the drogue, and $|x|$ represents the deflection amplitude of the drogue. When $|x|$ increases from $0 \rightarrow \pi/2$, penalty factor $c(x)$ decreases from $1 \rightarrow 0$; when $|x| \in (\pi/2, \pi]$, $c(x) = 0$.

Then, considering the relative attitude between the drogue and the probe, the *P_optimal* is defined as

$$P_{\text{optimal}} = c(\theta) \cdot c(\varphi) \cdot P_{\text{initial}} = \begin{cases} \dfrac{||\theta| - \pi/2|}{\pi/2} \cdot \dfrac{||\varphi| - \pi/2|}{\pi/2} \cdot P_{\text{initial}}, & |\theta| \in \left[0, \dfrac{\pi}{2}\right] \text{ and } |\varphi| \in \left[0, \dfrac{\pi}{2}\right], \\ 0, & \text{otherwise}, \end{cases} \quad (30)$$

where $\theta \in [-\pi, \pi]$ and $\varphi \in [-\pi, \pi]$. When $|\theta|$ and $|\varphi|$ increase from $0 \rightarrow \pi/2$ and $c(\theta)$ and $c(\varphi)$ decrease from $1 \rightarrow 0$, *P_optimal* decreases from *P_initial* $\rightarrow 0$; when $|\theta| \in (\pi/2, \pi]$ or $|\varphi| \in (\pi/2, \pi]$ and $c(\theta) = 0$ or $c(\varphi) = 0$, *P_optimal* $= 0$. We should note that, during the docking phase of aerial refueling, 30° deflection of the drogue can have much influence on the probe-drogue docking.

6.2. Considering the Accurate Measurements and UAV Control for Docking Success Probability. Note that a successful docking for probe-drogue aerial refueling is not only dependent on the movement of the drogue center, but also related to accurate measurements for the parameters between probe and the drogue and the control of UAV. Firstly, the theory, experiment, and measurement precision of drogue location and attitude have been described in our previous work [6] in detail. So, in this study, we assume that accurate parameters of the drogue can be obtained precisely. Secondly, the actuation error described in this paper can be regarded as a token of the control of UAV. Namely, without actuation error can be regarded as the high accuracy of the control of UAV, while with actuation error can be regarded as the deviation of the control of UAV.

6.3. Considering the Statistical Probability Model for Docking Success Probability. Note that it can be meaningful and helpful to model the random factors in PDARS and give its statistical probability model using statistical methods based on the experimental data.

Currently, due to a lack of flight test data of UAV AAR for some practical reasons, the statistical probability model for probe-drogue docking success probability cannot be built for the time being but deserves further research in future. Nevertheless, in order to make the proposed model of probe-drogue docking success probability more comprehensive and reasonable, the movements of the drogue (center) under

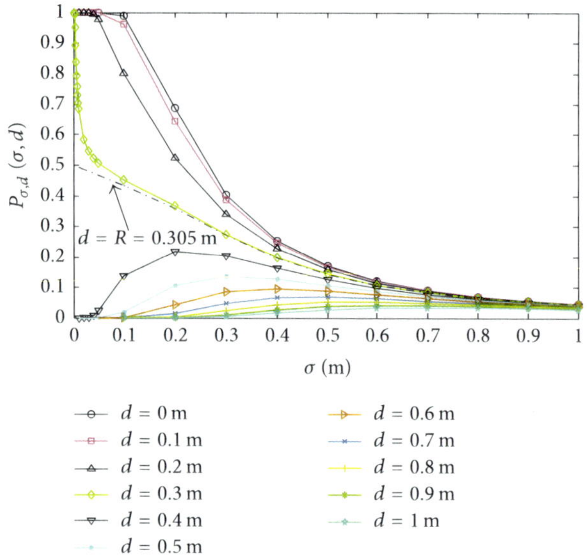

FIGURE 11: $P_{\sigma,d}(\sigma, d)$ versus σ and d.

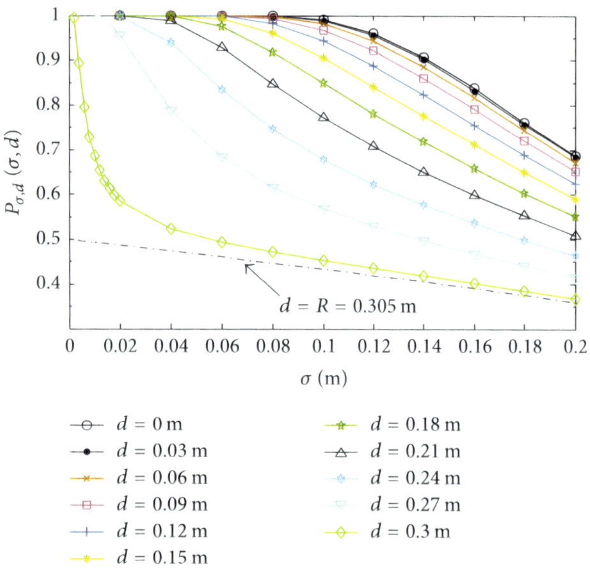

FIGURE 12: $P_{\sigma,d}(\sigma, d)$ versus σ and d (local detail).

disturbance, the actuation error (namely, the control of UAV), the relative attitude between the drogue and the probe, and the accurate measurements for the parameters between probe and the drogue are all taken into consideration. Note that the movements of the drogue (center), the actuation error (namely, the control of UAV), and the relative attitude between the drogue and the probe can be regarded as the intuitive reflection of the factors that affect probe-drogue docking success probability. Besides, a comparison between the theoretical analysis proposed in this paper and the flight test results of NASA AAR demonstration is reported. For the probe-drogue docking success probability, the flight test results of NASA AAR demonstration are consistent with the theoretical analysis proposed in this paper.

It is expected that, through an approach based on the mathematical model rather than human heuristic insight, UAV can obtain the probe-drogue docking success probability, which can be of some help for UAV AAR. Meanwhile, statistical probability model based on the flight test data of UAV AAR for probe-drogue docking success probability deserves further research and concern. And, whether artificial intelligence (such as deep learning) can be used to train the UAV to carry out aerial refueling also deserves concern and future study [28].

6.4. The Applicability of the Research. The research can provide a useful and valuable theoretical basis/reference, the probe-drogue docking success probability, for the practical engineering of UAV probe-drogue autonomous aerial refueling during the close docking stage. For example, during the close docking phase of UAV probe-drogue autonomous aerial refueling, (1) when the estimated probe-drogue docking success probability $P_{\text{estimated}} \geq \mathit{Threshold}$, the UAV can execute the docking maneuver under safe conditions; (2) when the estimated probe-drogue docking success probability $P_{\text{estimated}} < \mathit{Threshold}$, the docking maneuver can be regarded as unsafe, and the docking maneuver should be terminated. Without loss of generality, the *Threshold* can be set as 0.9. Note that the value of the *Threshold* should be determined by practical engineering.

7. Conclusions

(1) This paper numerically simulates the docking success probability during the docking phase of PDARS. Considering the facts that the receiver aircraft operates docking maneuver when the probe is in front of the drogue, and the atmospheric disturbance affects the movement of the drogue, this paper provides the theoretical estimation of probe-drogue docking success probability before docking maneuver when the probe is in front of the equilibrium position of drogue center.

(2) Reasonable assumptions for the distribution of the drogue center position under atmospheric disturbance are provided, according to the prior information during the docking phase of PDARS.

(3) A model of probe-drogue docking success probability is established, by converting the probe-drogue docking success probability to the probability of the drogue center located in a specific area.

(4) The simulation results can provide a theoretical basis for docking maneuver decision during the docking phase of PDARS. It is expected that, through an approach based on the mathematical model rather than human heuristic insight, UAV can obtain the probe-drogue docking success probability, which can be of some help for UAV AAR.

(5) The future work will include the following: (a) a more comprehensive model to estimate the docking success probability is currently under consideration;

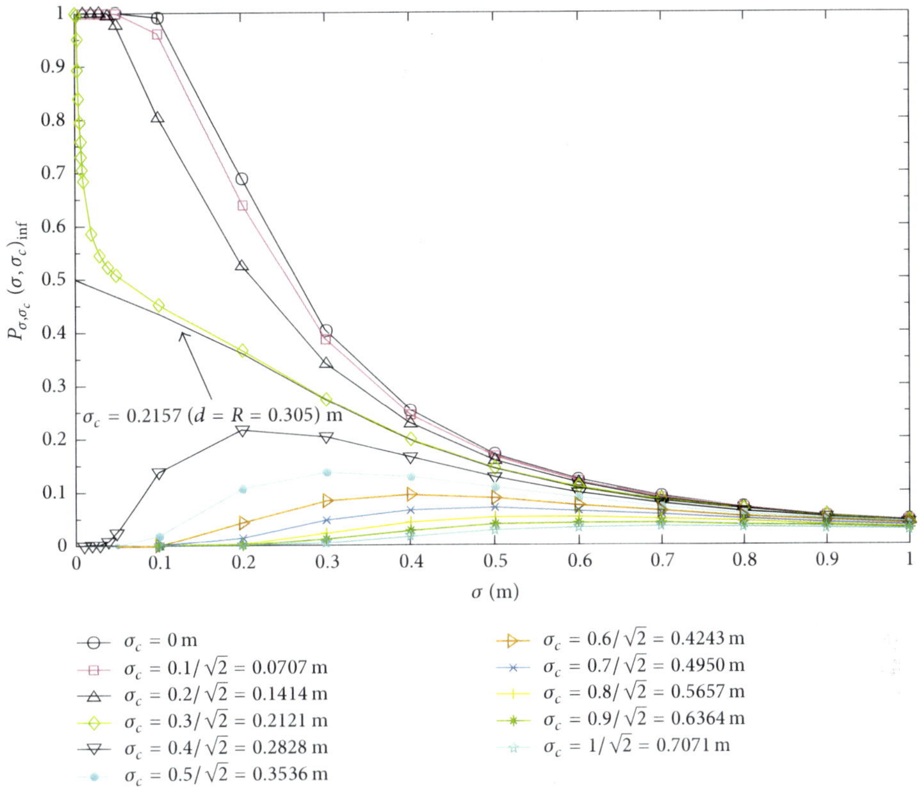

FIGURE 13: $P_{\sigma,\sigma_c}(\sigma, \sigma_c)_{\text{inf}}$ versus σ and σ_c.

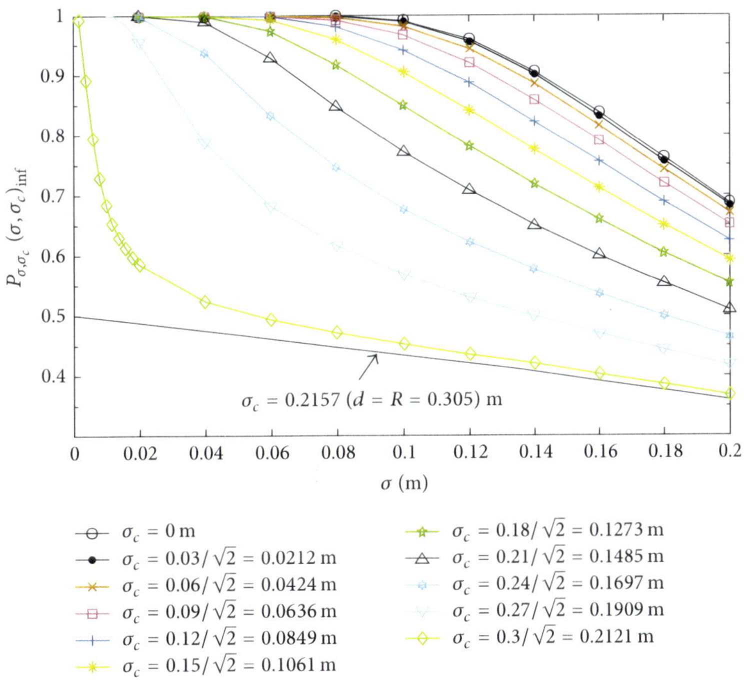

FIGURE 14: $P_{\sigma,\sigma_c}(\sigma, \sigma_c)_{\text{inf}}$ versus σ and σ_c (local detail).

(b) to further demonstrate the correctness of the proposed model, the simulated probability results will be compared with real data in future research; (c) the proposed approach needs formal verification and validation by the flight test of AAR.

Conflicts of Interest

The authors declare that there are no conflicts of interest regarding the publication of this paper.

Acknowledgments

This work was cosupported by the National Basic Research Program of China (nos. 2012CB316301 and 2013CB329403) and the National Natural Science Foundation of China (nos. 61473307, 61304120, 61603411, 61273023, and 61332007).

References

[1] J. P. Nalepka and J. L. Hinchman, "Automated aerial refueling: extending the effectiveness of UAVs," in *Proceedings of the AIAA Modeling and Simulation Technologies Conference and Exhibit*, pp. 2005–6005, San Francisco, Calif, USA.

[2] X. M. Dong, Y. J. Xu, and B. Chen, "Progress and challenges in automatic aerial refueling," *Journal of Air Force Engineering University: Natural Science Edition*, vol. 9, no. 6, pp. 1–5, 2008.

[3] C. D. Mu and B. R. Li, "Vision-based autonomous aerial refueling," *Journal of Tsinghua University (Science and Technology)*, vol. 52, no. 5, pp. 670–676, 2012.

[4] P. R. Thomas, U. Bhandari, S. Bullock, T. S. Richardson, and J. L. Du Bois, "Advances in air to air refuelling," *Progress in Aerospace Sciences*, vol. 71, pp. 14–35, 2014.

[5] Q. Quan, Z. B. Wei, J. Gao, R. F. Zhang, and K. Y. Cai, "A survey on modeling and control problems for probe and drogue autonomous aerial refueling at docking stage," *Acta Aeronautica et Astronautica Sinica*, vol. 35, no. 9, pp. 2390–2410, 2014.

[6] X. F. Wang, X. W. Kong, J. H. Zhi, Y. Chen, and X. M. Dong, "Real-time drogue recognition and 3D locating for UAV autonomous aerial refueling based on monocular machine vision," *Chinese Journal of Aeronautics*, vol. 28, no. 6, pp. 1667–1675, 2015.

[7] L. Pollini, G. Campa, F. Giulietti, and M. Innocenti, "Virtual simulation set-up for UAVs aerial refuelling," in *Proceedings of the AIAA Modeling and Simulation Technologies Conference and Exhibit*, pp. 2003–5682, Austin, Texas, USA, August 2003.

[8] R. Mati, L. Pollini, A. Lunghi, M. Innocenti, and G. Campa, "Vision-based autonomous probe and drogue aerial refueling," in *Proceedings of the 14th Mediterranean Conference on Control and Automation, (MED '06)*, Ancona, Italy, June 2006.

[9] H. W. Xie and H. L. Wang, "Binocular vision-based short-range navigation method for autonomous aerial refueling," *Journal of Beijing University of Aeronautics and Astronautics*, vol. 37, no. 2, pp. 206–209, 2011.

[10] C.-I. Chen and R. Stettner, "Drogue tracking using 3D flash lidar for autonomous aerial refueling," in *Proceedings of the SPIE 8037, Laser Radar Technology and Applications XVI, 80370Q*, Orlando, Florida, USA, June 2008.

[11] M. Q. Hu, X. Nie, and L. M. Wang, "Determination of hose static catenary shape in probe-drogue in-flight refueling system," *Journal of Air Force Engineering University: Natural Science Edition*, vol. 10, no. 5, pp. 22–26, 2009.

[12] M. Q. Hu, P. Liu, X. Nie, and R. X. Zhou, "Influence of air turbulence on the movement of hose-drogue," *Flight Dynamics*, vol. 28, no. 5, pp. 20–23, 2010.

[13] L. Y. Zhang, H. Zhang, Y. Yang, and L. Huang, "Dynamics modeling and simulation of docking process in aerial refueling," *Acta Aeronautica et Astronautica Sinica*, vol. 33, no. 7, pp. 1347–1354, 2012.

[14] K. Ro and J. W. Kamman, "Modeling and simulation of hose-paradrogue aerial refueling systems," *Journal of Guidance, Control, and Dynamics*, vol. 33, no. 1, pp. 53–63, 2010.

[15] H. T. Wang, X. M. Dong, H. F. Dou, and J. P. Xue, "Dynamic modeling and characteristics analysis of hose-paradrogue aerial refueling system," *Journal of Beijing University of Aeronautics and Astronautics*, vol. 40, no. 1, pp. 92–98, 2014.

[16] J. Zhang, S. Z. Yuan, and Q. Q. Gong, "Modeling and control of shaking motion of aerial refueling hose-drogue," *Journal of System Simulation*, vol. 28, no. 2, pp. 388–395, 2016.

[17] H. T. Wang, X. M. Dong, J. P. Xue, and J. L. Liu, "Dynamic modeling of a hose-drogue aerial refueling system and integral sliding mode backstepping control for the hose whipping phenomenon," *Chinese Journal of Aeronautics*, vol. 27, no. 4, pp. 930–946, 2014.

[18] H. T. Wang, X. M. Dong, J. Guo, J. L. Liu, and J. Wang, "Dynamics modeling and analysis of hose whipping phenomenon of aerial refueling hose-drogue assembly," *Acta Aeronautica et Astronautica Sinica*, vol. 36, no. 9, pp. 3116–3127, 2015.

[19] H. L. Wang, Y. Du, and W. D. Gai, "Precise docking control in unmanned aircraft vehicle automated aerial refueling," *Journal of Beijing University of Aeronautics and Astronautics*, vol. 37, no. 2, pp. 206–209, 2011.

[20] M. D. Tandale, R. Bowers, and J. Valasek, "Trajectory tracking controller for vision-based probe and drogue autonomous aerial refueling," *Journal of Guidance, Control, and Dynamics*, vol. 29, no. 4, pp. 846–857, 2006.

[21] R. P. Dibley, M. J. Allen, and N. Nabaa, "Autonomous airborne refueling demonstration phase I flight-test results," in *Proceedings of the AIAA Atmospheric Flight Mechanics Conference*, pp. 1131–1149, Reston, Virginia, USA, August 2007.

[22] J. L. Hansen, J. E. Murray, and N. V. Campos, "The NASA Dryden AAR project: a flight test approach to an aerial refueling system," in *Proceedings of the Collection of Technical Papers - AIAA Atmospheric Flight Mechanics Conference*, pp. 477–495, Reston, Virginia, USA, August 2004.

[23] M. J. Vachon, R. J. Ray, and C. Calianno, "Calculated drag of an aerial refueling assembly through airplane performance analysis," in *Proceedings of the 42nd AIAA Aerospace Sciences Meeting and Exhibit*, Reno, Nevada.

[24] A. Dogan, W. Blake, and C. Haag, "Bow wave effect in aerial refueling: computational analysis and modeling," *Journal of Aircraft*, vol. 50, no. 6, pp. 1856–1868, 2013.

[25] U. Bhandari, P. R. Thomas, S. Bullock, T. S. Richardson, and J. L. du Bois, "Bow wave effect in probe and drogue aerial refuelling," in *Proceedings of the AIAA Guidance, Navigation, and Control (GNC) Conference*, Boston, MA, USA, August 2013.

[26] X. H. Dai, Z. B. Wei, and Q. Quan, "Modeling and simulation of bow wave effect in probe and drogue aerial refueling," *Chinese Journal of Aeronautics*, vol. 29, no. 2, pp. 448–461, 2016.

16

Development of Design Methodology for a Small Solar-Powered Unmanned Aerial Vehicle

Parvathy Rajendran[1] **and Howard Smith**[2]

[1]*School of Aerospace Engineering, Universiti Sains Malaysia, Penang, Malaysia*
[2]*Aircraft Design Group, School of Aerospace, Transport and Manufacture Engineering, Cranfield University, Cranfield, UK*

Correspondence should be addressed to Parvathy Rajendran; aeparvathy@usm.my

Academic Editor: Mauro Pontani

Existing mathematical design models for small solar-powered electric unmanned aerial vehicles (UAVs) only focus on mass, performance, and aerodynamic analyses. Presently, UAV designs have low endurance. The current study aims to improve the shortcomings of existing UAV design models. Three new design aspects (i.e., electric propulsion, sensitivity, and trend analysis), three improved design properties (i.e., mass, aerodynamics, and mission profile), and a design feature (i.e., solar irradiance) are incorporated to enhance the existing small solar UAV design model. A design validation experiment established that the use of the proposed mathematical design model may at least improve power consumption-to-take-off mass ratio by 25% than that of previously designed UAVs. UAVs powered by solar (solar and battery) and nonsolar (battery-only) energy were also compared, showing that nonsolar UAVs can generally carry more payloads at a particular time and place than solar UAVs with sufficient endurance requirement. The investigation also identified that the payload results in the highest effect on the maximum take-off weight, followed by the battery, structure, and propulsion weight with the three new design aspects (i.e., electric propulsion, sensitivity, and trend analysis) for sizing consideration to optimize UAV designs.

1. Introduction

In recent years, awareness of the potential of unmanned aerial vehicles (UAVs) for multiple missions has increased the level of research into improving UAV design and technologies. Studies found that electric- or fossil fuel-powered UAVs have inadequate flying hours for various tasks. Increased focus has been placed recently on developing hybrid-powered electric UAVs, especially with the combination of solar energy and battery or fuel cells in a battery-powered system. The current study considers the pros and cons of various power systems [1–6] and examines the design and development of solar cells in battery-powered UAVs.

For more than a decade, small solar- and battery-powered electric UAVs were the subject of research and development [1, 5–19]. Seven small solar UAVs (i.e., So Long, Sky-Sailor, Sun-Sailor, Sun Surfer, AtlantikSolar AS-2, University of Minnesota's SUAV, and Cranfield University's Solar UAV) weighing less than 20 kg have been developed to date. As a prominent researcher in this field, Noth [14, 20] developed a mathematical design model for sizing solar-powered electric UAVs. However, this mathematical design model focuses only on the conceptual design stage in estimating the general sizing of UAV configurations.

Noth [14, 20] focused on the 3 sizing elements of mass, aerodynamics, and performance. Design characteristics, such as aircraft power consumption and cruise speed, must also be optimized to ensure that the required long endurance of the UAVs for 24 hours of real surveillance and mapping applications is achieved. The current study aims to improve the design properties, which requires further in-depth work, such as solar irradiance, mission profile, and aerodynamic characteristics. Research gaps in electric propulsion sizing, sensitivity studies, and trend analysis have also been identified.

A synthesized UAV mathematical design model suitable for sizing and configuration design of small solar-powered and battery-powered electric systems was developed to fulfill

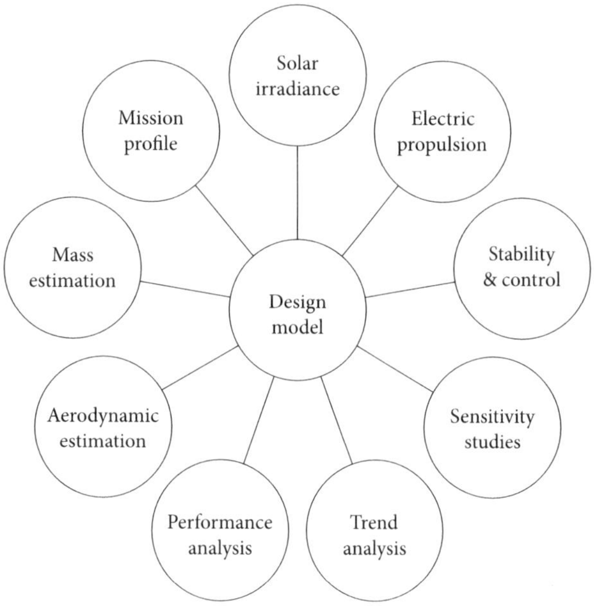

Figure 1: Solar-powered electric UAV design element.

the previously mentioned requirements. This comprehensive expansion work, based on various studies [21–23], considers nine design properties (i.e., mass sizing, aerodynamics, performance, stability and control, mission profile, solar irradiance, electric propulsion, sensitivity studies, and trend analysis) to improve the current design model.

The proposed mathematical model was initially quasi-validated using existing UAV data via rigorous computer modeling and analyses. The validity of the mathematical model was deemed satisfactory after achieving a lower error in the model's conceptual design than in the actual UAV design using the empirical database. An actual solar-powered electric UAV [1] was also designed, developed, and flight-tested to further quasi-validate this model. The comprehensive design model is developed in this study to further minimize assumptions and simplify previous designs.

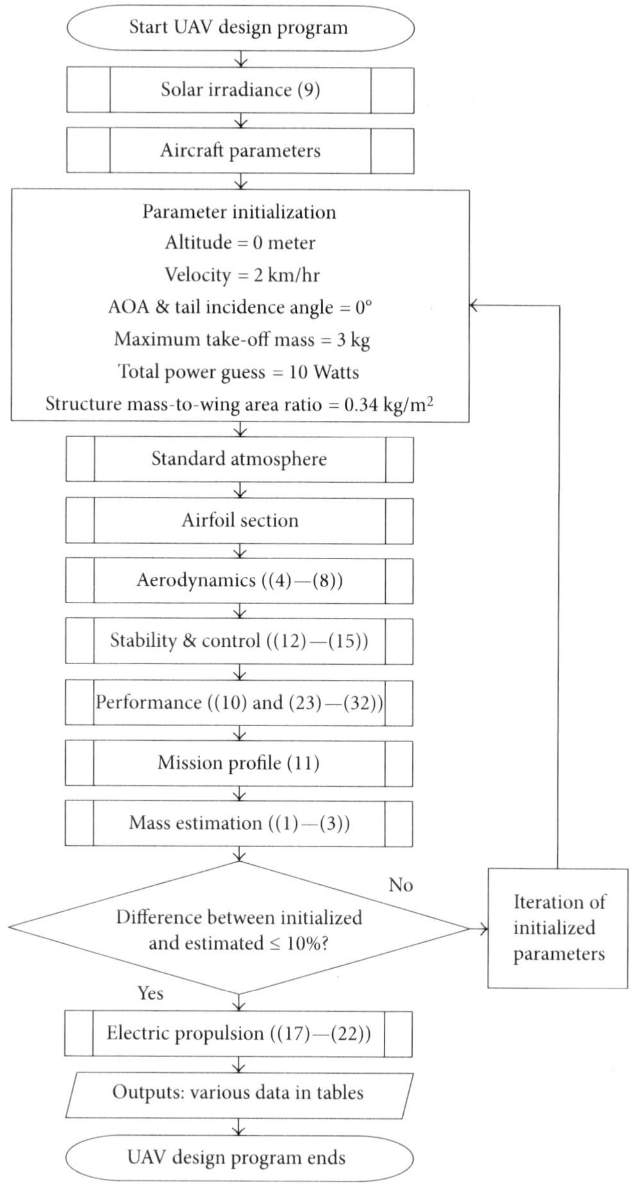

Figure 2: UAV design algorithm flowchart.

2. Mathematical Design Model

The developed solar-powered electric UAV mathematical design model and its algorithm flowchart are illustrated in Figures 1 and 2, respectively. The model contains the nine design components mentioned earlier. Three design components, namely, mass estimation, aerodynamic estimation, and performance analysis, were initially developed by Noth [14]. Performance analysis is the only design component that was maintained based on the specifications developed by Noth.

In the mass estimation, the study by Mueller et al. [22] was further adopted. The component mass was divided into the following basic elements, namely, structure, battery, solar, electric propulsion, control system, and payload, as shown in (1), respectively. The aircraft's total take-off weight (W_{TOmax}) may be expressed as a combination of the empty weight and payload weight, as shown in (2), respectively, because a pure electric UAV does not have a variable weight during flight.

$$W_{\text{TOmax}} = W_{\text{Struct}} + W_{\text{Batt}} + W_{\text{Solar}} + W_{\text{Electric}} + W_{\text{Ctrl}} + W_{\text{Pay\_Max}}, \quad (1)$$

$$W_{\text{TOmax}} = W_{\text{Empty}} + W_{\text{Pay\_Max}}. \quad (2)$$

The relevant coefficient in predicting the empty weight of an electric UAV that weighs less than 15 kg is given in (3). This equation is determined using regression analysis by collecting all possible measurements of 83 small electric UAVs. These 83 small UAVs, including solar-, battery-, fuel cell-, and hydrogen-powered electric UAVs, weigh less than 14 kg. The parameters gathered include weights, wing area (S), wing span (b), aspect ratio (AR), height, total length, root, and tip chord length of both the wing and tail surfaces.

$$W_{\text{Empty}} = 0.79 \times b^{18.9012} S^{-9.4755} \text{AR}^{-9.4558} W_{\text{TOmax}}^{0.99}. \quad (3)$$

Similarly, in aerodynamic estimation, specific lift and drag coefficient estimation was performed based on various wing and horizontal tail airfoil characteristics [21]. However, the fuselage and vertical tail characteristics have yet to be incorporated in this study. Additional power is incorporated into the mission power requirement, specifically by defining the maximum power consumption during maneuver (P_{Maneuver}) at total mission power, compensating for the drag increment that remains unaccounted due to the fuselage and vertical tail.

The Reynolds number (RN) is a major factor in aerodynamic analysis. The RN range for an aircraft depends on the mean aerodynamic chord length (c), airspeed (V), air density (ρ), and viscosity (μ). Equation (4) [21] defines the relation between these parameters.

$$\text{RN} = \frac{Vc\rho}{\mu}. \quad (4)$$

The aerodynamic submodel also predicts the lift coefficient by using a well-known theoretical estimation [21], as shown in (5). In level flight, the aircraft lift is equal to the take-off weight. Because the air density, airspeed, and wing area are known, the lift coefficient, C_L, may be estimated.

$$C_L = \frac{L}{1/2\rho V^2 S} = \frac{W_{\text{TOmax}}}{1/2\rho V^2 S}. \quad (5)$$

The drag coefficient C_D can be estimated using (6) [21], where the aircraft zero-lift-drag coefficient C_{Do} can be obtained from the airfoil polar data at each RN. A linear interpolation is performed on these airfoil data to obtain C_{Do} for a specific RN, angle of attack, and airspeed of an aircraft. Again, the vertical tail and fuselage drag component were not accounted since most high-endurance and lightweight UAVs have narrow fuselage. In addition, most of these UAVs are hand-launched, thus without a landing gear.

$$C_D = C_{\text{Do\_W}} + C_{\text{Di\_W}} + C_{\text{Do\_HT}} + C_{\text{Di\_HT}}. \quad (6)$$

Then, with the C_L calculated at each RN using (4) earlier, the lift and drag may also be determined using

$$L = \frac{1}{2}\rho V^2 S C_L, \quad (7)$$

$$D = \frac{1}{2}\rho V^2 S C_D. \quad (8)$$

Apart from the mass and aerodynamic estimation, the solar irradiance and mission profile were also incorporated. Integrations of these two design components have been detailed. Solar irradiance estimates include details such as specific time of the day, date of the year, and latitude and longitude coordinates [24] instead of a simplified average value of solar irradiance data for a particular day in a nearby major city. This estimation has enabled the prediction of the usability of a solar-powered aircraft in various parts of the world and its operation duration throughout the year.

The amount of solar irradiance (Ir_{max}) that strikes the Earth, as given in (9), is critical in designing a solar-powered UAV. The power available through the solar module facilitates the sufficiency of the solar energy harnessed for the power required. The equation is adopted from [23–27], which begins by determining the input parameters, including the day of the year, latitude, longitude, and altitude of the place of interest.

$$\text{Ir}_{\text{max}} = \frac{(1 + 0.033\cos(0.017203\text{DN}(\pi/180)))\sin(\text{SOLALT}(\pi/180))}{3.6(1000/4.8708)}. \quad (9)$$

No previous work has estimated in detail the UAV solar power usage of the mission profile (i.e., not only cruise but also the other mission profile phases, such as climb, loiter/maneuver, and descent). The effects of gust wind speed and direction were disregarded. The power required (P_{Required}) for a level flight determined by the definition of the wing area is shown in (10) [21]. This calculation has not taken into consideration the flight path angle, the efficiency of the electric propulsion system, and the additional power consumption of the aircraft from the control system and payload.

$$P_{\text{Required}} = \frac{C_D}{C_L}\sqrt{\frac{2\text{AR}(W_{\text{TOmax}})^3}{\rho C_L b^2}}. \quad (10)$$

The power consumption of each mission profile phase is determined based on the flight path angle, as proposed by Sherwin [28]. Detailed mission profile power sizing is crucial especially when designing an aircraft for high endurance. The percentage of time spent on a 24-hour mission during cruise may be higher in comparison with that on a one-hour mission. Regardless of mission endurance, the climb and descent phase is similar. Thus, an appropriate ratio portion of each mission profile phase has been considered which includes climb, cruise, maneuver, and descent, as shown in (11) in respective order, instead of assuming that the aircraft only flies in level flight. The time fraction defined is intended for a specific mission study. This mathematical design model is assumed to be applicable only for an unmanned aircraft without a landing gear to achieve maximum endurance or range by flying, primarily, in level flight.

$$P_{\text{Total-Mission}} = \frac{0.1}{60}P_{\text{Climb}} + \frac{59.5}{60}P_{\text{Cruise}} + \frac{0.3}{60}P_{\text{Maneuver}} + \frac{0.1}{60}P_{\text{Descent}}. \quad (11)$$

The stability and control analysis measures the tail plane and boom length to ensure that the UAV is statically stable. This design property ensures the viability of the overall aircraft design-sizing simulation results. Determining the stability and controllability of the designed UAV plays a significant role in developing a well-designed UAV. Only static stability and control have been considered in this research. No work has yet been done on aircraft dynamics, which is recommended for further studies.

In the proposed method, the parameters evaluated are the center of gravity (i.e., cg) range, trimmed elevator deflection angle, and trimmed angle of attack. The airfoil moment coefficient data are obtained by linear interpolation, as performed in the aerodynamic estimation, to determine these parameters. With these values and tail area (S_{HT}) and tail moment arm length (l_{HT}), the tail volume ratio (V_{HT}) is estimated using (12). Also, the tail zero-lift-pitching moment coefficient (C_{Mo_HT}) is then estimated using (13) [21], where $C_{L\alpha\_HT}$ is the lift curve slope of the tail, ε_o is the downwash angle at zero lift, i_W is the wing incidence angle, and i_{HT} is the tail incidence angle.

$$V_{HT} = \frac{l_{HT}}{c}\left(\frac{S_{HT}}{S}\right), \quad (12)$$

$$C_{Mo\_HT} = V_{HT} C_{L\alpha\_HT} (\varepsilon_o + i_W + i_{HT}). \quad (13)$$

The aircraft's trimmed angles of attack α_{trim} and elevator deflection angles $elev_{trim}$ are estimated. The aircraft's zero-lift-pitching moment coefficient (C_{Mo}) is first determined using (14) [21], where C_{Mo_W} is the wing zero-lift-pitching moment coefficient. The trimmed angle of attack and elevator deflection angle can be calculated using (15) and (16) [21], where $C_{Mcg\alpha}$ is the moment coefficient curve slope. Determining the aircraft's trimmed angle of attack at various elevator deflection angles gives a guideline to the required elevator deflection angle range that is within the allowable airplane angle of attack. Determining the elevator deflection range guarantees controllability by avoiding excessive pitch up (which can lead to stall) and pitch down.

$$C_{Mo} = C_{Mo\_W} + C_{Mo\_HT}, \quad (14)$$

$$\alpha_{trim} = \frac{C_{Mo}}{C_{Mcg\alpha}}, \quad (15)$$

$$elev_{trim} = \frac{C_{Mo} + C_{Mcg\alpha}\alpha_W}{V_{HT} C_{L\alpha\_HT}}. \quad (16)$$

The current practices of propulsion system selection for common electrically driven UAVs are imprecise. This propulsion sizing process gets more complex in a solar-augmented UAV. Yet, UAV designers commonly use the standard combination sets of the electric motor, gearbox, electronic speed controller, propeller, and battery recommended by model aircraft part manufacturers or suppliers for a specific UAV weight range. This process typically yields a margin of approximately 25% between the lower and upper weight for a given standard combination set.

Multiple combination sets are recommended, clearly indicating the wide range of flexibility and combination applicable to a specific unmanned aircraft. Thus, preevaluated sets of electric propulsion systems are obviously suboptimal options for aircraft designers. Electric propulsion analysis has been developed to optimize aircraft power and combination selection (i.e., battery, electronic speed controller, and electric motor and propeller sizes) of the propulsion system based on the desired aircraft performance.

In the current work, the electric propulsion system sizing is given by [29]. The electric motor efficiency (eff_{Motor}) can be estimated using (17) where $P_{Motor\text{-}Out}$ is the power out of the motor and $P_{Motor\text{-}In}$ is the power into the motor from the battery. Then, the power from the electric motor to the propeller ($P_{Prop\text{-}In}$) can be estimated using the motor voltage (V_{Motor}) and motor current (I_{Motor}) as shown in (18). Meanwhile, the required power from the propeller ($P_{Prop\text{-}Req}$) based on the simulated propeller diameter (Diam) and pitch (Pitch) size and propeller revolution per minute (RPM) can be evaluated as given in (19).

$$eff_{Motor} = \frac{P_{Motor\text{-}Out}}{P_{Motor\text{-}In}}, \quad (17)$$

$$P_{Prop\text{-}In} = eff_{Motor} V_{Motor} I_{Motor}, \quad (18)$$

$$P_{Prop\text{-}Req} = \frac{0.5\rho}{g}\left(\frac{RPM}{60}\right)^3 (0.0254 Diam)^4 (0.0254 Pitch). \quad (19)$$

Here, the error difference between the input and required propeller power can be determined, where within ±10% is preferred. This error difference limits and ensures that a suitable range of the propeller is obtained. In addition, the ideal performance occurs when the pitching velocity of the propeller ($V_{Prop\text{-}Pitch}$) given in (20) should be within 2.5 and three times the aircraft stall speed [29]. The thrust of the propeller ($Thrust_{Pr\ op}$) estimated using (21) must be more than 25% of the maximum take-off mass of the UAV to have reasonable climb and acceleration capabilities.

$$V_{Prop\text{-}Pitch} = \frac{RPM}{60}(0.0254 Pitch), \quad (20)$$

$$Thrust_{Prop} = \frac{P_{Prop\text{-}Req}}{g V_{Prop\text{-}Pitch}}. \quad (21)$$

Once a list of a simulated propeller's diameter and pitch size combination has been produced with the criteria listed above, the optimized propeller size is chosen based on the least propeller tip static velocity ($V_{Prop\text{-}Tip\text{-}Static}$). This velocity, estimated using (22), can provide the optimum propeller size combination for a better UAV performance in terms of high endurance and long range.

$$V_{Prop\text{-}Tip\text{-}Static} = \pi \frac{RPM}{60}(0.0254 Diam). \quad (22)$$

UAV performance analysis may be conducted to predict the effectiveness of the to-be-designed aircraft. The main performance criteria that are considered are velocity, stall velocity (V_{Stall}), required thrust ($T_{Required}$), available energy ($E_{Available}$), and required power ($P_{Required}$). The $V_{maxCell}$ is the maximum voltage of a battery cell, S_{Cell} is the number of battery cells in series, P_{Cell} is the number of battery cells in parallel, and Q_{Cell} is the capacity of each battery cell. These performance analyses are computed using (23), (24), (25), (26), and (27) in respective order.

$$V = \sqrt{\frac{W_{\text{TOmax}}}{(1/2)\rho C_L S}}, \tag{23}$$

$$V_{\text{Stall}} = \sqrt{\frac{W_{\text{TOmax}}}{(1/2)\rho C_{\text{Lmax\_W}} S}}, \tag{24}$$

$$T_{\text{Required}} = \frac{W_{\text{TOmax}}}{C_L/C_D}, \tag{25}$$

$$E_{\text{Available}} = V_{\text{max}_{\text{Cell}}} S_{\text{Cell}} P_{\text{Cell}} Q_{\text{Cell}}, \tag{26}$$

$$P_{\text{Required}} = T_{\text{Required}} V = \frac{C_D}{C_L}\sqrt{\frac{2W_{\text{TOmax}}{}^3}{\rho C_L S}}. \tag{27}$$

Stall velocity estimation is important prior to flight testing to prevent the aircraft from stalling. Ensuring that the available power is greater than the required power is crucial, not only for the aircraft to take-off and/or climb easily but for the anticipation of any gusts during flight. The excess available power (P_{Excess}) can be estimated using (28) where the initial power of a mission ($P_\text{I}$nitialise_Guess) is guessed. This initial power will be iterated to ensure that the error difference between the initial guess and actual mission power (11) is within ±10%.

$$P_{\text{Excess}} = P_{\text{Initialise\_Guess}} - P_{\text{Required}}. \tag{28}$$

The rate of climb and rate of optimal glide sink evaluation give a rough estimation of the required time to achieve the desired altitude. These parameters can be calculated using (29) and (30) in respective order [21].

$$\text{Rate of climb} = \frac{P_{\text{Excess}}}{W_{\text{TOmax}}}, \tag{29}$$

$$\text{Rate of optimal glide sink} = \frac{V}{L/D}. \tag{30}$$

Finally, the aircraft's endurance and range are determined. These two variables of aircraft electric propulsion capability for solar and battery flight are identified using an assumption of no wind and estimated using (31) and (32) [23–27]. Studies will be conducted using the performance criteria trade and optimization. This process will improve the aircraft design configuration and optimize its capabilities.

$$\text{Endurance}_{\text{S\&B}} = \frac{E_{\text{Available}}}{P_{\text{Required}}} + 0.7t_{\text{Daylight}}, \tag{31}$$

$$\text{Range}_{\text{S\&B}} = V \times \text{endurance}_{\text{S\&B}}. \tag{32}$$

In the mathematical design model, a new assessment scheme is developed to analyze the sensitivity of various parameters that exhibited the greatest and least effects on the UAV design. Evaluating how the entire design changes with parameter variations is important to ensure that the fidelity of specific models remains acceptable when change is introduced. Technologies were identified to considerably enhance the design or improve the system performance.

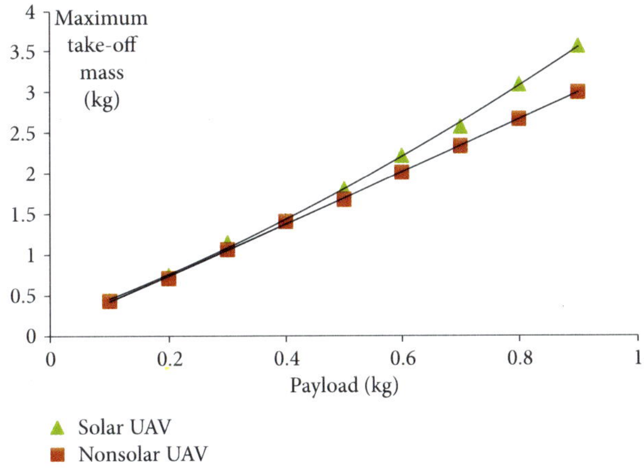

Figure 3: Payload versus maximum take-off mass of the solar and nonsolar UAVs.

Many components are evidently becoming smaller, more powerful, and more highly efficient than previous ones. A crucial approach is developed to assess the future trends in UAV designs. This approach evaluates the rate at which aircraft parts and component technologies are developing and projects their achievable or potential performance. The UAV mathematical design model can be used to identify the parts that have the most significant effect and that merit further research to yield the best prospects in UAV development.

3. Result and Discussion

A general analysis work was conducted on the mathematical design model that covers both solar (solar- and battery-powered) and nonsolar (battery-powered) UAV sizing. Although this design model can be used for design estimations for large UAVs, all calculations were explicitly focused on small hand-launchable UAVs weighing less than 4 kg. The simulation was also subject to UAVs that can perform a minimum of 9 hours of flight operation beginning at 12 noon on 22 June 2014 at Cranfield, United Kingdom, with an iteration sizing accuracy of 1 cm or 1 g.

Figure 3 shows the analysis results for both solar and nonsolar UAV configurations. The UAV take-off mass can be improved to sustain 9 hours of flight duration at a specified payload. The take-off mass of both solar and nonsolar UAVs nonlinearly increased with the payload. These nonlinear curves are mainly due to the dependence of the mass of the aircraft structure on the corresponding wing aspect ratio and span.

In this study, the feasible UAV design was restricted to the specified flight duration and payload. In this case, solar and nonsolar UAVs must have maximum take-off masses of 8.209 kg and 8.987 kg, respectively, and must be capable of supporting payloads weighing 1.258 kg and 3.04 kg, respectively, to achieve 9 hours of flight duration. The flight duration must be decreased to increase the payload or vice versa.

The relationship between the wing span and the mass of the structure at a constant payload of 0.35 kg and an aspect

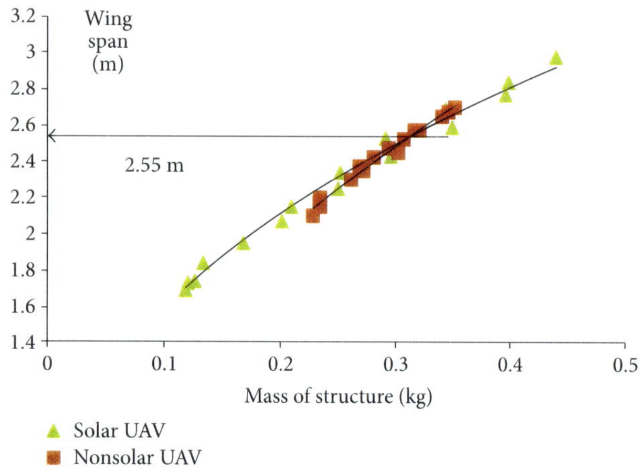

Figure 4: Mass of the structure versus wing span of the solar and nonsolar UAVs at 0.35 kg payload.

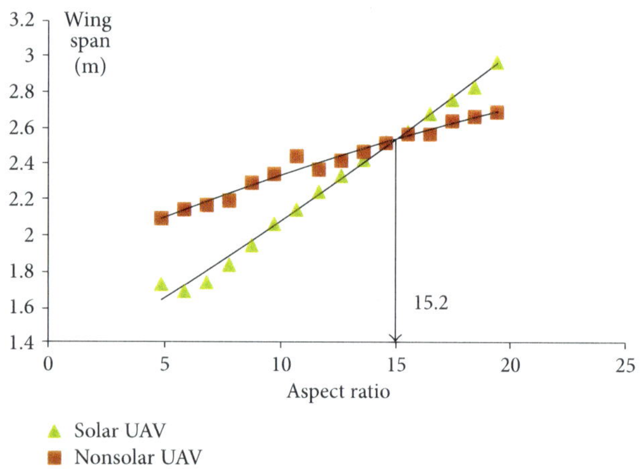

Figure 5: Aspect ratio versus wing span of the solar and nonsolar UAVs at 0.35 kg payload.

ratio of 12 is plotted in Figure 4. These data also identified that the wing span of 2.55 m is suitable for both types of flights at an aspect ratio of 12. This innovative design approach can be used to design and develop a single airframe structure suitable for both solar and nonsolar configurations.

The relationship between the wing span and aspect ratio at a constant payload of 0.35 kg is shown in Figure 5. The solar UAV configuration showed that the wing span increased more steeply than that of the nonsolar UAV configuration. This result is a clear sign of the structure rigidity issue. This issue is associated with the increase in wing span and decrease in wing chord to maintain the wing area. In this analysis criterion, the aspect ratio suitable for solar and nonsolar UAV configuration is 15.2. This value is fairly close to the aspect ratio of 16.4 that is suitable for the actual developed UAV [1], which was designed to carry a payload of 0.54 kg or 18% of the maximum take-off mass.

On a clear sunny day, approximately 55.5 W of the solar module power can be obtained per square meter wing area using an off-the-shelf 6% efficiency Powerfilm MP3-37 flexible solar cell specification at a constant aspect

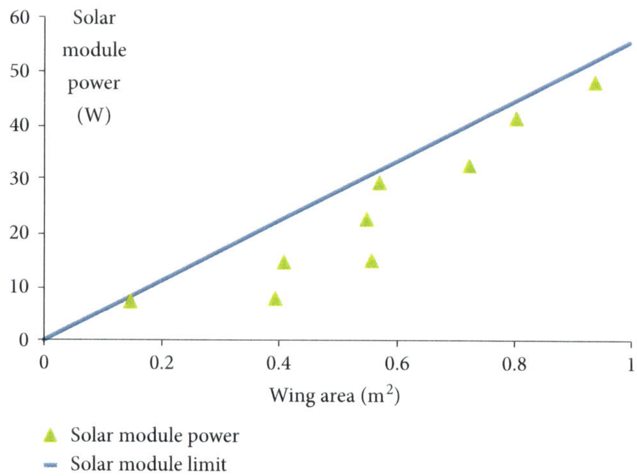

Figure 6: Wing area versus solar module power of the solar UAVs.

ratio of 12 (Figure 6). However, only some of the points were close to the maximum achievable solar power (blue line in Figure 6). The issue may be resolved by performing further simulations with different aspect ratios to optimize the solar module power achieved in solar UAV sizing.

The relationship between the mass of the payload and battery of both solar and nonsolar UAVs (Figure 7) indicates that the mass of the battery gradually increases with payload. This result can be attributed to the minimum size of the battery pack, which comes in multiples of 150 g ± 10%. This finding proves another advantage of applying the electric propulsion modeling as part of this design modeling, where sizing small UAVs provides an accurate weight prediction.

The curve also shows that at a fixed wing area, the battery mass of the solar UAV is almost a third larger than that of the nonsolar UAV at 0.9 kg payload. This clearly shows that at 0.9 kg payload, the solar UAV wing is fully covered by the solar module. Therefore, the mass of the battery has to be increased to compensate the limitation with the area available for solar module placement and its weight addition to the overall maximum take-off mass.

Similarly, Figure 8 shows the relationship between the payload and the mass of the propulsion for both solar and nonsolar UAVs. In this figure, the curve shows that the propulsion mass of the solar UAV is higher by almost a third than that of the nonsolar UAV (0.9 kg payload). This result clearly signifies the effect of propulsion mass which includes the propeller, electric motor, electronic speed controller, gear box, and propeller adapter at a specified payload.

The required power (total mission) for both solar and nonsolar UAVs as a function of maximum take-off mass is illustrated in Figure 9. Although the effect of the minimum battery pack mass can be observed for the low UAV mass, a smoother correlation is observed as the values increased. Again, this result can be attributed to the minimum size of the high-power-density lithium polymer cell, which comes in multiples of 50 g ± 10%. Since the system studied here operates at 3 battery cells in series, a minimum battery pack mass comes in multiples of 0.15 kg. Thus at an aircraft mass

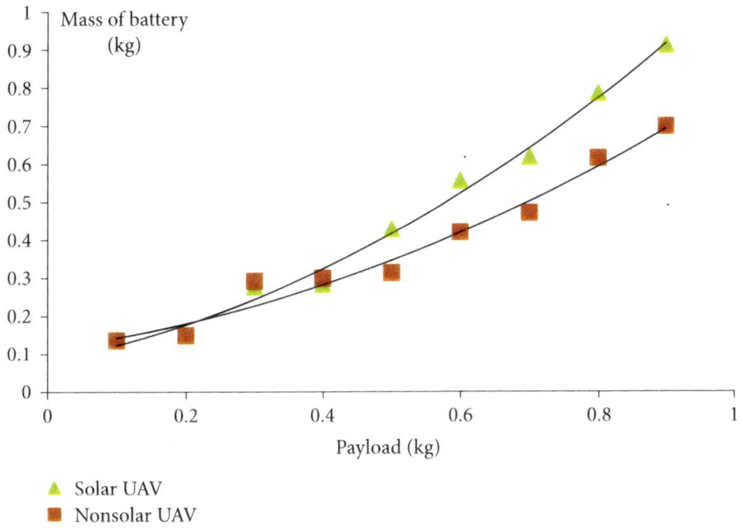

FIGURE 7: Mass of the payload versus battery of the solar and nonsolar UAVs.

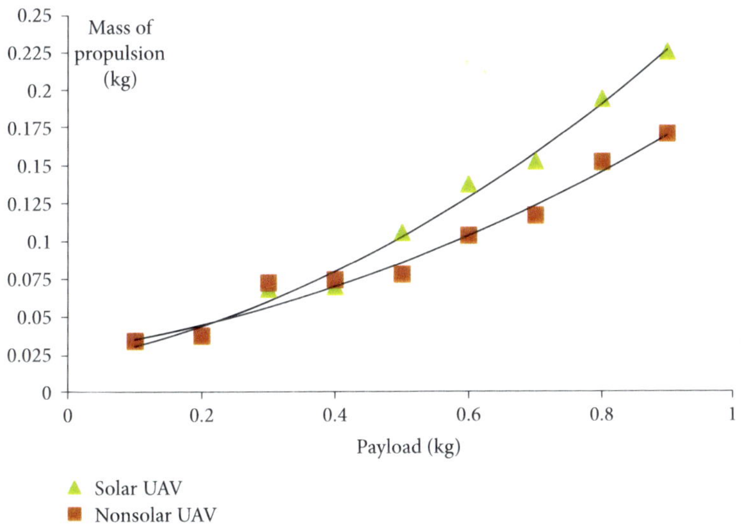

FIGURE 8: Mass of the payload versus propulsion of the solar and nonsolar UAVs.

of 1.5 kg, the power consumption rises slowly due to the increase in the control and propulsion system.

This UAV design model estimates that a nonsolar UAV can only carry a maximum of 3 kg for a flight of 9 hours or only 0.1 kg for a flight of almost 23 hours at an aspect ratio of 12. However, a solar UAV that operates in Cranfield, UK, during the summer can fly for 24 hours and 9 hours with 0.1 and 1.3 kg payloads, respectively.

Currently, nonsolar UAVs can carry more payload at any particular time and place than solar UAVs, at a given endurance requirement. However, the principal weakness of nonsolar UAVs is their poor endurance. This weakness can be addressed using a solar UAV, which has a high-endurance capability.

The choice of the electrically powered UAV type should be based on the technical requirements specified by the customer. A solar- and battery-powered electric UAV is preferred on missions that require flight durations of more than 24 hours regardless of the amount of payload it can carry. Otherwise, a battery-powered UAV may be opted for.

The needs of a solar-augmented battery-powered vehicle rather than of a battery-only flight vehicle must be evaluated carefully in terms of location and date of operation in the year. Determining the number of days a designed solar UAV engages in full operation is necessary for selecting the appropriate type of electrically powered flight vehicle. This simulation was conducted specifically for a UAV to be operated on a summer day in Cranfield. The developed solar irradiance submodel offers an efficient design optimization and customization process for various UAV operating locations.

The validation results of this study involve design modeling based on the three new and three improvised design properties that substantially improved the final UAV design

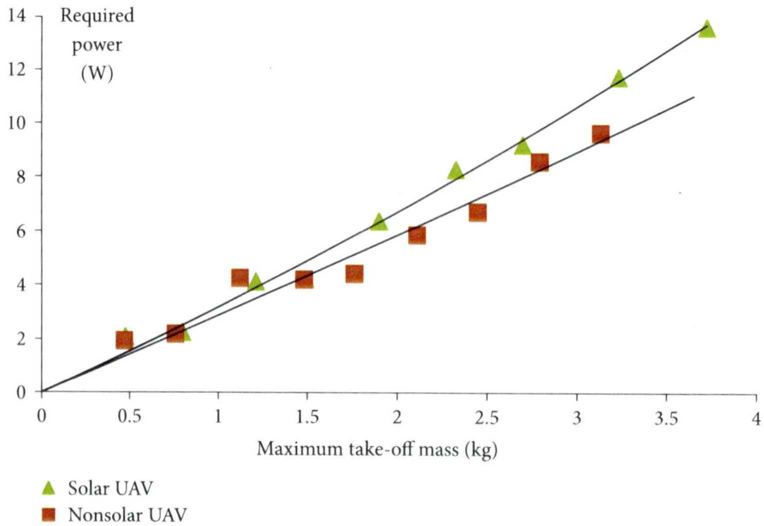

FIGURE 9: Maximum take-off mass versus required power of the solar and nonsolar UAVs.

TABLE 1: Comparison of current and previous small solar-powered UAV designs.

Research UAV	Aircraft	Mass (kg)	Wing span (m)	Endurance (hr)	Cruise power (W)	Power/take-off mass (W/kg)
AcPropulsion	So Long	12.8	4.75	48	95	7.42
Technion	Sun-Sailor	3.6	4.2	11	40	11.11
Swiss Federal Institute of Technology of Zurich	Sky-Sailor	2.4	3.2	27	15	6.25
Swiss Federal Institute of Technology of Zurich	Sun Surfer	0.12	1	—	1.77	14.75
Swiss Federal Institute of Technology of Zurich [10, 11]	AtlantikSolar AS-2 UAV	6.93	5.65	28	39.7	5.73
University of Minnesota [7, 8]	SUAV:Q hybrid	3.364	4	12	30	8.92
Cranfield University	Current work	3	3.7	23.0	15	5.00

characteristics and performance. Table 1 shows the comparison of all small solar-powered UAVs to the UAV designed using the current mathematical design model.

The UAV developed using the current design model had a 25% smaller ratio between power consumption and take-off mass than its closest competitor. Nevertheless, the AtlantikSolar AS-2 UAV design is a well-designed solar-powered UAV, yet the proposed design methodology may be used to further improve the design. This is because the power-to-mass ratio of the Cranfield Solar UAV is slightly better than that of AtlantikSolar AS-2 UAV. A detailed aircraft specification of the Cranfield Solar UAV is specified in Table 2.

The comparison further affirms the importance and need of the three new and three improvised design properties to optimize the UAV final design. The optimization of power and the propulsion system in UAVs based on the desired aircraft performance indicates that the new electric propulsion design properties are main contributors to the 25% reduction in the power consumption-to-take-off mass ratio. The overall enhancement of the developed mathematical design model ensures that the final UAV design is improved in all aspects of design optimization.

4. Conclusion

A comprehensive UAV mathematical design model with nine design properties was developed. This design model consists of three new aspects of design consideration and three improved design properties. This work substantially enhanced the final UAV design characteristics and performance for a high-endurance mission. Design validation showed a 25% smaller ratio between power consumption and take-off mass in the UAV designed using the current mathematical design model than in previous UAVs. Solar- (solar and battery) and nonsolar- (battery-only) powered UAVs were also compared. The nonsolar UAV can carry more payload at any particular time and place than a solar UAV, at a given endurance requirement. However, the endurance of a nonsolar UAV is its principal limitation. This feature can be addressed by using a solar UAV with a high-endurance performance. Payload has the highest effect on the maximum take-off mass, followed by the battery, structure, and propulsion mass. Ultimately, the type of an electrically powered vehicle to be selected will be in accordance with the technical requirements specified by the customer.

Table 2: Cranfield solar UAV specification.

Description	Specification
Wing airfoil	AG24, AG25, AG26
Tail airfoil	HT21
Length (m)	1.8
Wing span (m)	3.7
Wing aspect ratio	16.4
Wing area (dm^2)	83.6
PowerFilm MP3-37 solar module area (dm^2)	79.0
Peak electric motor operating power (W)	500
Structure mass (kg)	0.986
Hacker X-40 Pro electronic speed controller mass (kg)	0.050
Hacker A30-14L electric motor & mounting mass (kg)	0.160
$13' \times 8'$ folding propeller & propeller adapter mass (kg)	0.020
Eyecam 2.4 GHz micro wireless video camera mass (kg)	0.050
LC-18650-JP-2600 lithium-ion polymer battery pack mass (kg)	1.050
Futaba R617FS 7-channel 2.4 GHz receiver mass (kg)	0.010
Hitec HS-85MG Servos mass (kg)	0.044
Eagle Tree inboard system mass (kg)	0.200
Connector & wire mass (kg)	0.030
Total solar module mass (kg)	0.400
Take-off mass (kg)	3.0

Nomenclature

W_{TOmax}: Maximum take-off weight (N)
W_{Struct}: Structure weight (N)
W_{Empty}: Empty weight (N)
W_{Solar}: Solar power system weight (N)
W_{Batt}: Battery weight (N)
W_{Ctrl}: Control system weight (N)
$W_{\text{Pay\_Max}}$: Maximum payload weight (N)
eff_{Electric}: Electric propulsion system efficiency (%)
b: Wing span (m)
c: Mean aerodynamic chord length (m)
$C_{\text{Lmax\_W}}$: Wing maximum lift coefficient
$C_{\text{Mo\_HT}}$: Tail zero-lift-pitching moment coefficient
S_{Cell}: Number of battery cells in series in a battery pack
Q_{Cell}: Battery cell's capacity (A/hr)
P_{Cell}: Number of battery cells in parallel in a battery pack
V_{maxCell}: Battery cell's maximum voltage (V)
t_{Daylight}: Flight duration with daylight (hr)
DN: Day number
C_{Mo}: Aircraft zero-lift-moment coefficient
$C_{\text{Mo\_W}}$: Wing zero-lift-moment coefficient
V: Airspeed (m/s)
l_{HT}: Tail moment arm (m)
S_{HT}: Tail area (m^2)
P: Power (W)
i_{W}: Wing incidence angle (°)
i_{HT}: Tail incidence angle (°)
ρ: Air density (kg/m^3)
g: Gravitational acceleration at the chosen altitude (m/s^2)
μ: Viscosity (kg/ms)
S: Wing area (m^2)
$C_{\text{Do\_W}}$: Wing zero-lift-drag coefficient
$C_{\text{Di\_W}}$: Wing induced drag coefficient
AR: Aspect ratio
$P_{\text{Prop-In}}$: Power input to the propeller (W)
$P_{\text{Prop-Out}}$: Power output to the propeller (W)
$P_{\text{Motor-In}}$: Power input in the electric motor (W)
$P_{\text{Motor-Out}}$: Power output in the electric motor (W)
$P_{\text{Prop-Req}}$: Power required from the propeller (W)
$E_{\text{Available}}$: Energy available (W/hr)
SOLALT: Solar altitude
α_{trim}: Trimmed aircraft's angle of attack (°)
$\text{elev}_{\text{trim}}$: Trimmed elevator deflection (°)
V_{Motor}: Electric motor's operational voltage (V)
RPM: Electric motor's shaft rotation per minute (rpm)
I_{Motor}: Electric motor's current (A)
Diam: Propeller's diameter (in)
Pitch: Propeller's pitch (in)
$V_{\text{Prop-Tip-Static}}$: Propeller tip static velocity (m/s).

Conflicts of Interest

The authors declare no conflict of interest.

Authors' Contributions

Parvathy Rajendran and Howard Smith conceived and designed the experiments, Parvathy Rajendran performed the experiments, Parvathy Rajendran and Howard Smith analyzed the data, and Parvathy Rajendran and Howard Smith wrote the paper.

Acknowledgments

This publication was supported by Universiti Sains Malaysia Grant no. 304/PAERO/6315002.

References

[1] P. Rajendran and H. Smith, "The development of small solar powered electric unmanned aerial vehicle systems," in *4TH International Conference on Mechanical and Manufacturing Engineering 2013 (ICME2013)*, Applied Mechanics and Materials (ISI), Bangi, Selangor, 2013.

[2] H. M. G. Vidales, *Design, Construction and Test of the Propulsion System of a Solar UAV*, MSc, Aerospace Engineering, Technical University of Lisbon, Lisbon, Portugal, 2013.

[3] M. R. Bhatt, *Solar Power Unmanned Aerial Vehicle: High Altitude Long Endurance Applications (HALE-SPUAV)*, San Jose State University, San Jose State, 2012.

[4] S. Cowley, *Solar Powered UAV*, In NASA Tech Briefs 2013, New York, 2013.

[5] P. Rajendran and H. Smith, "Review of the battery & solar power system development of small solar powered electric unmanned aerial vehicles," *Advance Materials Research*, vol. 1125, pp. 641–647, 2014.

[6] P. Rajendran and H. Smith, "Review of the elementary aspect of solar-powered electric unmanned aerial vehicles," *Australian Journal of Basic and Applied Sciences*, vol. 8, pp. 252–259, 2014.

[7] R. D'Sa, D. Jenson, and N. Papanikolopoulos, "SUAV: Q-a hybrid approach to solar-powered flight," in *2016 IEEE International Conference on Robotics and Automation (ICRA)*, pp. 3288–3294, Stockholm, Sweden, May 2016.

[8] S. Morton and N. Papanikolopoulos, "Two meter solar UAV: Design approach and performance prediction for autonomous sensing applications," in *2016 IEEE/RSJ International Conference on Intelligent Robots and Systems (IROS)*, pp. 1695–1701, Daejeon, South Korea, October 2016.

[9] S. Morton, L. Scharber, and N. Papanikolopoulos, "Solar powered unmanned aerial vehicle for continuous flight: conceptual overview and optimization," in *2013 IEEE International Conference on Robotics and Automation*, pp. 766–771, Karlsruhe, Germany, May 2013.

[10] P. Oettershagen, A. Melzer, T. Mantel et al., "A solar-powered hand-launchable UAV for low-altitude multi-day continuous flight," in *2015 IEEE International Conference on Robotics and Automation (ICRA)*, pp. 3986–3993, Seattle, WA, USA, May 2015.

[11] P. Oettershagen, A. Melzer, T. Mantel et al., "Perpetual flight with a small solar-powered UAV: flight results, performance analysis and model validation," in *2016 IEEE Aerospace Conference*, Big Sky, MT, USA, March 2016.

[12] N. Diepeveen, "*The Sun-Surfer: design and construction of a solar powered MAV*," Autonomous Systems Lab, Swiss Federal Institute of Technology in Zurich, Switzerland, 2007.

[13] A. Noth, W. Engel, and R. Siegwart, "*Design of an ultra-lightweight autonomous solar airplane for continuous flight*," Proceedings of Field and Service Robotics, Port Douglas, Australia, 2005.

[14] A. Noth, "*Design of solar powered airplanes for continuous flight*," ETH, Zurich, Switzerland, 2008.

[15] A. Noth, "*Sky-Sailor solar powered airplane proved continuous flight*," 2008.

[16] P. Rajendran, "*Investigation in to the design space associated with electrically powered flight vehicles*," Cranfield University, Cranfield, UK, 2012.

[17] T. C. Tozer, D. Grace, J. Thompson, and P. Baynham, "UAVs and HAPs-potential convergence for military communications," in *IEE Colloquium on Military Satellite Communications (Ref. No. 2000/024)*, London, UK, June 2000.

[18] K. Valavanis, *Advances in Unmanned Aerial Vehicles: State of the Art and the Road to Autonomy*, vol. 33, Springer Science & Business Media, Berlin, Germany, 2008.

[19] O. K. Sahingoz, "Generation of bezier curve-based flyable trajectories for multi-UAV systems with parallel genetic algorithm," *Journal of Intelligent & Robotic Systems*, vol. 74, no. 1-2, pp. 499–511, 2014.

[20] A. Noth, "*Design of solar powered airplanes for continuous flight*," In Swiss Federal Institute of Technology of Zurich, Switzerland, 2006.

[21] J. D. Anderson, *Introduction to Flight*, McGraw Hill, New York, USA, 2005.

[22] T. Mueller, J. Kellogg, P. Ifju, and S. Shkkarayev, *Introduction to the Design of Fixed-Wing Micro Air Vehicles*, American Institute of Aeronautics and Astronautics Inc, Reston, Virginia, USA, 2006.

[23] T. Muneer, C. Gueymard, H. Kambezidis, and T. Muneer, *Solar Radiation and Daylight Models*, Elsevier Butterworth Heinemann: Oxford, Burlington, MA, USA, 2nd edition, 2004.

[24] T. Muneer and H. Kambezidis, *Solar Radiation and Daylight Models for the Energy Efficient Design of Buildings*, Architectural Pres, Oxford, 1997.

[25] P. Rajendran and H. Smith, "Implications of longitude and latitude on the size of solar-powered UAV," *Energy Conversion and Management*, vol. 98, pp. 107–114, 2015.

[26] P. Rajendran, H. Smith, and H. bin Masral, "Modeling and simulation of solar irradiance and daylight duration for a high-power-output solar module system," *Applied Mechanics and Materials*, vol. 629, pp. 475–480, 2014.

[27] P. Rajendran and H. Smith, "Modelling of solar irradiance and daylight duration for solar-powered UAV sizing," *Energy Exploration & Exploitation*, vol. 34, no. 2, pp. 235–243, 2016.

[28] K. Sherwin, *Man-Powered Flight*, In Modell& Allied Publications Ltd, Hertfordshire, UK, 1971.

[29] P. Rajendran, H. Smith, K. I. Yahaya, and N. M. Mazlan, "Electric propulsion system sizing for small solar-powered electric unmanned aerial vehicle," *International Journal of Applied Engineering Research*, vol. 11, pp. 9419–9423, 2016.

17

Antidisturbance Vibration Suppression of the Aerial Refueling Hose during the Coupling Process

Zikang Su[1,2,3] **and Honglun Wang**[1,3]

[1]*School of Automation Science and Electrical Engineering, Beihang University, Beijing 100191, China*
[2]*ShenYuan Honors College of Beihang University, Beijing 100191, China*
[3]*Science and Technology on Aircraft Control Laboratory, Beihang University, Beijing 100191, China*

Correspondence should be addressed to Honglun Wang; wang_hl_12@126.com

Academic Editor: Kenneth M. Sobel

In autonomous aerial refueling (AAR), the vibration of the flexible refueling hose caused by the receiver aircraft's excessive closure speed should be suppressed once it appears. This paper proposed an active control strategy based on the permanent magnet synchronous motor (PMSM) angular control for the timely and accurate vibration suppression of the flexible refueling hose. A nonsingular fast terminal sliding-mode (NFTSM) control scheme with adaptive extended state observer (AESO) is proposed for PMSM take-up system under multiple disturbances. The states and the "total disturbance" of the PMSM system are firstly reconstituted using the AESO under the uncertainties and measurement noise. Then, a faster sliding variable with tracking error exponential term is proposed together with a special designed reaching law to enhance the global convergence speed and precision of the controller. The proposed control scheme provides a more comprehensive solution to rapidly suppress the flexible refueling hose vibration in AAR. Compared to other methods, the scheme can suppress the flexible hose vibration more fleetly and accurately even when the system is exposed to multiple disturbances and measurement noise. Simulation results show that the proposed scheme is competitive in accuracy, global rapidity, and robustness.

1. Introduction

Unmanned aerial vehicles (UAVs) [1, 2] have been widely used in civilian or military fields in recent years, which include but are not limited to target tracking, combat radius increasing, flight range extending, and battle operational efficiency improving. As an effective method for increasing the endurance and range of aircraft by refueling them in flight [3], UAV autonomous aerial refueling (AAR) has attracted an increasing interest over the last decades from both theoretical and practical aspects covering aircraft control, sensor systems, and their integration [4–8]. Generally, there are two refueling methods [3]: flying boom method and probe-drogue refueling (PDR) [9–11]. And the PDR is getting to be a universal refueling equipment due to its low cost and wide operational aspects [11].

However, there have been a growing number of catastrophic failures involving the PDR [3, 12]. Among the failures, the flexible refueling hose vibration problem during autonomous aerial refueling (AAR) coupling [13], which is the result of excess closure speed caused by the receiver pilot, plays an extremely dangerous role to the AAR safety. The closure rate of the receiver aircraft must be constrained within a required range. The refueling coupling flight will fail to latch below the minimum speed, whereas a violent reverberation through the flexible hose (namely, the hose vibration, as shown in Figure 1) can lead to equipment severe damage at a higher closure rate. When the receiver aircraft probe couples the drogue at a higher closure rate, it will definitely cause excessive slack of the refueling flexible hose as the probe pushes the drogue forward. Then, the internal hose tension will rapidly decrease, and the

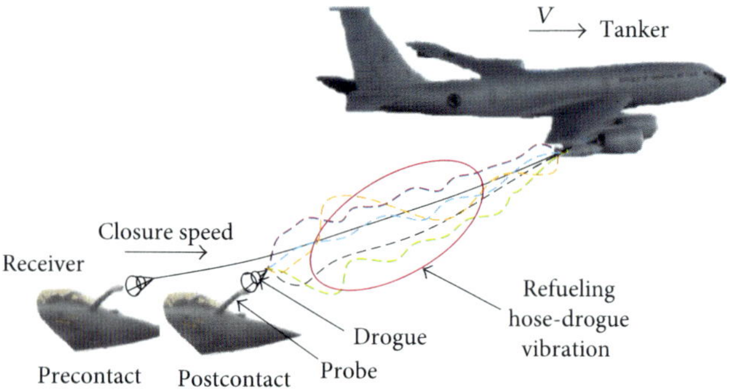

Figure 1: The refueling hose-drogue assembly vibration due to excessive closure speed.

hose will violently vibrate under the action of the unpredictable aerodynamic forces. Finally, tremendous high load torques on the hose and the probe rises, and equipment damage occurs. The refueling hose vibration greatly constraints the AAR's success rates and security. This drastic hose vibration requires some vibration suppression strategies to ensure the flexible refueling hose-drogue assembly stable even under unexpected excessive closure speed and multiple disturbances.

During coupling, the conventional pod is equipped with a tensator [13, 14] (spring-loaded take-up device) to retract any slack in the hose as the probe pushes the drogue forward. If the tensator malfunctions or being subjected to an excessive closure speed during coupling, slack in the hose can form. Then with the rapidly decreasing hose tension, the violent hose vibration occurs due to aerodynamic forces [15]. The vibration will generate high internal loads on the hose and exert high loads on the probe. While these extreme loads only exist for a fraction of a second, they result in critical damage to the refueling assembly and the receiver probe, which eventually lead to a potentially catastrophic accident. The refueling hose vibration has been a serious constraint to the success rates and security of AAR. Therefore, the vibration must be rapidly suppressed once it appears via specially designed hose take-up control system.

Although some experiments and model analysis are conducted to further research dynamic characteristics of the flexible refueling hose vibration [13–15], the conventional suppression methods do not show satisfactory performance in a more adverse condition. The Boeing Company confirmed that the reel take-up speed lagging behind the closure speed was responsible for the failure by numerical simulation [14]. That means, although every pod is equipped with a tensator to suppress the hose vibration at present, the tensator may not be the sufficient effective strategy. Alden and Vennero [16] invented a new refueling pod with a reel driven by a permanent magnet synchronous motor (PMSM), and it provides another chance for high-performance suppression methods for the vibration by integrating the PMSM and high-precision position sensors. Wang et al. [15] designed an active integral sliding-mode back-stepping control strategy for the PMSM to suppress the refueling hose vibration on the basis of the relative position between the tanker and the receiver. This strategy achieves satisfactory performance on the condition when the PMSM model is precise enough and the equipment's working environment is ideal enough. However, the internal and external disturbance of the PMSM control system and the measurement noise have not been taken into consideration.

Unfortunately, during the AAR, the PMSM installed in the pod is in an extremely hostile environment [3], multiple known or unknown strong disturbances can easily disable the PMSM control system designed based on the nominal model and ideal environment. The complex working environment may be caused by the drastic vibration of the tanker's wing, the uncertain load by the hose-drogue, and other unknown uncertainty and disturbance. Meanwhile, due to the rapidly drastic changing dynamic of the flexible hose vibration, an antidisturbance rapid response control system needs to be designed for the PMSM based take-up system to suppress the hose vibration.

Although there are many existing schemes for PMSM in practical [17, 18], we modestly find that few works focus on the PMSM tracking control and its extended applications if the following conditions are all considered at the same time: (1) only the position sensor is equipped and the speed sensor is unavailable; (2) the position sensor measurement signal is polluted by white noise; and (3) multiple internal and external disturbances concurrently exist in the PMSM drive system [19]. It should be specially pointed out that the dynamic of PMSM is an essential nonlinear subject to a wide range of disturbances/uncertainties in many high-performance applications. According to Yang et al. [19], the disturbances/uncertainties in PMSM systems can be classified as unmodeled dynamics, parametric uncertainties, and external disturbances. The unmodeled dynamics generally include the motor body structure-induced torque (such as cogging torque, flux harmonics, and other factors), dead-time effects, and measurement error effects. Parametric uncertainties contain the uncertainties of the mechanical parameters and electrical parameters. The load torque, friction torque, and other mechanical factors are usually classified as the external disturbance. In the AAR, the load torque caused by the hose-drogue assembly is actually the primary important part of the external disturbance. In order to simplify the description and to separate from the external disturbances, we regard the unmodeled dynamics and parametric uncertainties as the internal disturbance in this paper.

However, most of the existing literatures on the PMSM controller design generally focus on the handling of either condition (1) or (3) [18–22]. The condition (2) has never been synchronously considered together with (1) and (3) in the existing works. This paper focuses on the designing of the PMSM tracking control system which satisfies the rapidity and accuracy requirements of vibration suppression and capable of handling all of those scenarios together.

In recent decades, the sliding-mode control (SMC) [23, 24] has been successfully applied to many uncertain systems and areas [25, 26], due to its superiorities such as strong robustness, order reduction, easier implementation, and design simplification [27]. Among the SMC methods, the terminal sliding-mode (TSM) controller has drawn much attention. It has been further developed to the nonsingular terminal sliding mode (NTSM) [27] and has been applied to many plants [28, 29]. It has shown its fast finite time convergence, strong robustness, and preferable ability to handle the uncertainties and external disturbances. Besides, the adaptive extended state observer (AESO) [30], which possesses the excellent ability of state estimation and noise filtering, is proposed for a class of nonlinear systems with uncertain dynamics and sensor measurement noise. The gain of extended state observer (ESO) is automatically timely tuned to reduce the estimation errors of both states and "total disturbance" against the measurement noise. Most important of all, it provides another preferable way for the reconstruction of the PMSM system states and "total disturbance" on the condition when the sensor signals are polluted by noise.

With the NTSM controller and the AESO, an NTSM control scheme is proposed based on the "controller + observer" structure [24] for the PMSM tracking control with multiple disturbances (external and internal) and measurement noise. The designed control method provides another considerable choice for the problem of refueling hose vibration suppression which requires special consideration of the PMSM take-up control system's ability of the antidisturbance and rapid response. Based on those considerations, following works are investigated.

Firstly, to ensure the desired antidisturbance ability of the closed-loop system, the internal disturbances, external disturbances, and the items which are independent of the virtual control in the PMSM's speed subsystem are regarded as the "total disturbance," and then AESO is used to reconstitute the "total disturbance" and the system states in the presence of angular position measurement noise. This reconstituted "total disturbance" is further used for compensation during the controller design to improve the antidisturbance performance.

Secondly, to further improve the tracking rapidity and accuracy of the PMSM take-up system, a nonsingular fast terminal sliding-mode (NFTSM) controller with faster convergence performance is proposed for a class of uncertain nonlinear system with multiple disturbances. Some new strategies are introduced into both the sliding variable and reaching law. On one hand, a fast terminal sliding variable with tracking error exponential term is proposed to accelerate the convergence speed on the sliding surface. On the other hand, a terminal attractor with negative state exponential term is adopted for the reaching law to bring the sliding variable to the sliding surface and to eliminate the singular problem in the controller.

Thirdly, a new control scheme based on the NFTSM and AESO is designed and is well applied to the practical problem of the flexible aerial refueling hose suppression in AAR. The unexpected multiple disturbances and the measurement noise are all taken into consideration during controller design, and the angular speed sensor is needless for the feedback.

The paper is organized as follows. Section 2 describes the formulation of the PMSM tracking control problem. In Section 3, the nonsingular terminal sliding-mode control scheme with adaptive extended state observer is proposed with the relevant proofs. Section 4 applies the proposed control scheme to the PMSM tracking control for vibration suppression with a variable length link-connected dynamic model of the hose-drogue assembly, and then, comparative simulations are conducted. Finally, conclusions are given in Section 5.

2. Problem Formulation

2.1. Model of the PMSM System. The accurate mathematical model of the PMSM [15, 31], when there are no magnetic saturation, hysteresis, eddy current loss, or friction of the reducer, excluding sinusoidal magnetic field distribution, can be described in the d-q coordinate system as

$$\begin{aligned} \dot{\vartheta} &= \omega, \\ \dot{\omega} &= \frac{3P\psi_f}{2J} i_q - \frac{B}{J}\omega - \frac{T_L}{J}, \\ \dot{i}_q &= -\frac{R}{L_s} i_q - P\omega i_d - \frac{P\psi_f}{L_s}\omega + \frac{u_q}{L_s}, \\ \dot{i}_d &= -\frac{R}{L_s} i_d + P\omega i_q + \frac{u_d}{L_s}, \end{aligned} \tag{1}$$

where ϑ is the PMSM rotor angular position, ω is the PMSM rotor angular velocity, i_d, i_q, u_d, and u_q are d, q axis currents and voltages, respectively. R, L_s, P, ψ_f, J, B, and T_L are the stator resistance, stator inductance, the number of pole pairs, the stator flux linkage, the moment of inertia, the viscous friction coefficient of the PMSM, and the load torque, respectively.

When the internal disturbance, external disturbances, and the angular position measurement noise are considered, the PMSM's speed subsystem S_1 can be transformed into the form of

$$S_1 : \begin{cases} \dot{\vartheta} = \omega, \\ \dot{\omega} = \dfrac{3P\psi_f}{2J} i_q - \dfrac{B}{J}\omega - \dfrac{T_L}{J} + d_{in} + d_{ex} = F_1 + b_1 i_q, \\ y_1 = \theta + n_i, \end{cases} \tag{2}$$

where d_{in} denotes the internal disturbances, d_{ex} denotes the external disturbances except T_L/J, y_1 is the output, n_i is the measurement noise of the position sensor, $b_1 = 3P\psi_f/2J$, and $F_1 = -(B/J)\omega - (T_L/J) + d_{in} + d_{ex}$ is defined as the sum of the internal disturbance, external disturbances, and the items which are independent of the virtual control in the PMSM's speed subsystem and it is taken as the total disturbances of subsystem. The disturbance $d_1 = d_{in} + d_{ex}$ satisfies $d_1 < l_1$, where l_1 is a constant that is greater than zero.

Actually, the ideal model (1) always faces the challenge of many unavoidable disturbances when the PMSM is applied

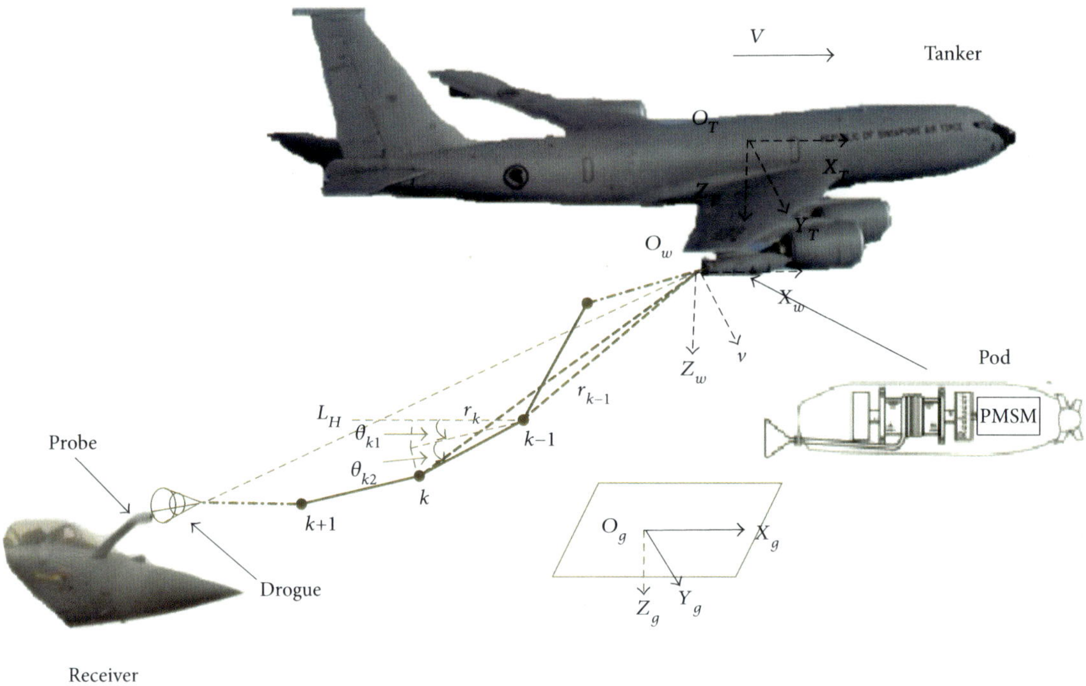

FIGURE 2: The configuration and variable length modeling of a hose-drogue aerial refueling system.

to vibration suppression of the flexible refueling hose in AAR. Generally, the PMSM-based refueling hose-drogue reel in/out servo system which is used to suppress the vibration is equipped at the long wingtip of the tanker, as shown in Figure 2. The PMSM system faces different disturbance existing concurrently under this tough work environment. On one hand, besides the unmodeled dynamic, the PMSM system works under the integrated effect of the strong wing vibration and strong electromagnetic interference. These may provoke the strong and fast-changing parametric uncertainties (belong to the internal disturbances according to above definition). On the other hand, the unpredictable dynamics of the hose-drogue assembly and the unpredictable vibrational states of the tanker wing will definitely lead to the unpredictable and fast-changing load torque and friction torque (belong to the external disturbances). Moreover, the demanding work environment will generally make the measurement noise of the sensors inevitable. According to (2), the total disturbances F_1 will directly affect the dynamics of the speed subsystem S_1, and the measurement noise n_i will definitely affect the closed-loop performance via the feedback channel. Ultimately, the total disturbances and the measurement noise will definitely impair the closed-loop performance. Thus, these disturbances and measurement noise should be considered during the controller design.

Generally, the two loop scheme [18, 21] (current loop and speed loop) is usually employed during the PMSM controller design due to different features and required qualities of these two control loops. Actually, the mechanic inertia of the speed loop is greater than that of the current loop. However, the electric inertia of the current loop is very small and it requires fast electromagnetic torque response, which makes its sampling frequency very high. Hence, the current control loop algorithm is usually desired to be as simple as the proportion integral (PI) controller or the proportion integral derivative (PID) controller [18]. Therefore, the controllers usually employ a structure of cascade control loops including a speed loop and two current loops. Here, two PI controllers are adopted in the two current loops to stabilize the current tracking errors of d-q axes, respectively. Usually, the reference i_d^* in the d current loop is set to be zero to maintain a constant flux operating condition. The reference current i_q^* is determined by S_1.

2.2. Command Transformation for the Vibration Suppression of Refueling Hose. In the AAR, it is assumed that the reel is driven by a PMSM to deploy or retrieve the hose through a reducer. A probe-drogue refueling system includes a reel mechanism, a refueling hose, and a drogue shown in Figure 2. A set of iterative equations of the hose's three-dimensional motion is derived subject to hose reeling in/out, tanker motion, gravity, and aerodynamic loads accounting for the effects of steady wind, atmospheric turbulence, and tanker vortex. The details of the model of the hose-drogue assembly can be found in [15].

Considering the mechanical inertia [15], the load torque T_L can be written as the transfer function of the first link hose tension T_1 of the hose-drogue assembly: $T_L = (r/i)(1/(s+1))$ T_1, where r is the radius of the reel and i is the reduction ratio of the reducer.

The most effective way to suppress the violate vibration of the refueling hose is to stabilize the hose tension by real-time active hose reeling in/out based on the relative position between the tanker and the receiver. The refueling hose

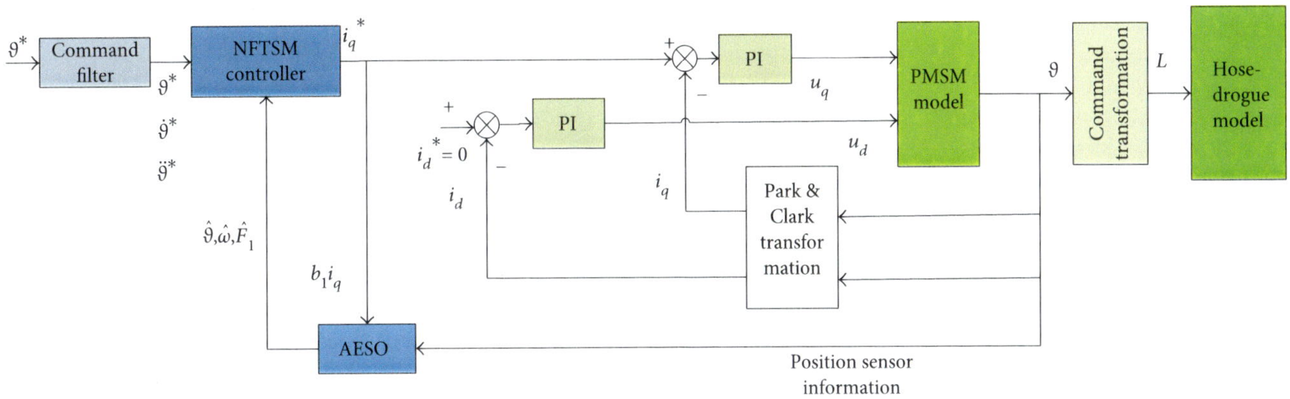

FIGURE 3: Block diagram of the PMSM control system for refueling hose vibration suppression in AAR.

length change L can be expressed as a function of the angular position ϑ of PMSM as follows:

$$L = \frac{\vartheta r}{i}, \tag{3}$$

where L is also a function of the relative distance between the probe and the pod during coupling, which is calculated as

$$L = \lambda\sqrt{\Delta x^2 + \Delta y^2 + \Delta z^2} - L_0, \tag{4}$$

where $\lambda = L_0/L_H$, as shown in Figure 2, reflects the degree of hose slack. L_0 is the initial length of the hose before coupling, and Δx, Δy, and Δz are three spatial components of the relative distance between probe and pod [16], which can be accurately measured by three position sensors, respectively, installed on the tanker, receiver, and drogue. Therefore, the length control of the hose (hose reeling in/out) can be transformed into the rotor angular position control of the PMSM through (1).

As the vibration suppression dynamic process which is decided by the excessive closure speed of the receiver aircraft during coupling is very short (usually, lasts only about 1 second to 2 seconds [13]), the control schemes which cannot work timely and rapidly will result in the shake of the hose-drogue becoming longer and more intense. During the hose vibration suppression process, one expects the fluctuations of the load and the hose to be as stable as possible, and this expectation can be achieved by designing a control scheme that can track the angular command rapidly and timely. Consequently, the task here is to design a controller for the speed loop that tracks the given angular signal (which is transformed from the relative distance between the drogue and the end of the pod after successful coupling) as soon as possible under the condition that various disturbances and measurement noise of the sensors exist.

3. The Nonsingular Fast Terminal Sliding-Mode (NFTSM) Control Scheme with Adaptive Extended State Observer (AESO)

As assumed in the introduction, the speed sensor is unavailable and measurement noise exists in the position sensor; hence, the states and disturbance should be reconstructed during the control system design. The recently proposed that AESO provides a befitting choice to reconstruct states and disturbance of the PMSM with measurement noise. With the reconstructed system by AESO, a new nonsingular fast terminal sliding-mode control method, which not only accelerates the error convergence but also reduces the overshoot, is proposed here. The structure of the proposed control scheme is depicted in Figure 3.

3.1. System Reconstitution via AESO. The adaptive extended state observer (AESO) is proposed for the nonlinear systems with the uncertainties both in the plant and in the sensor [30]. The gain of ESO is automatically timely tuned to reduce the estimation errors of both states and "total disturbance" against the measurement noise. The theory of AESO can be briefly described as follows [30].

Consider the following class of nonlinear uncertain continuous systems with measurement noise.

$$\begin{aligned} &\dot{X}(t) = AX(t) + BF(X(t), t) + BB_u U(ih),\\ &Y(ih) = X_1(ih) + n_i,\\ &\qquad t \in [ih, ih + h),\ i = 0, 1, \ldots, \end{aligned} \tag{5}$$

where

$$A = \begin{bmatrix} 0 & I & \cdots & 0 \\ \cdots & \cdots & \cdots & 0 \\ 0 & 0 & \cdots & I \\ 0 & 0 & \cdots & 0 \end{bmatrix}_{mn\times mn}, \quad B = \begin{bmatrix} 0 \\ 0 \\ \cdots \\ I \end{bmatrix}_{mn\times m}, \tag{6}$$

$X = [X_1{}^T\ X_2{}^T\ \cdots\ X_{n-1}{}^T\ X_n{}^T]^T$, $X_j{}^T \in R^m$, $j \in \underline{n} = \{1, 2, \ldots, n\}$ are continuous state vectors. $X_1(t)$ is the output to be controlled, $Y(ih) \in R^m$ is the measured output which contains the measurement noise vector $n_i \in R^m$, and $h > 0$ is the sampling period. B_u is a known nonsingular matrix and the control input $U(t)$ implemented by the zero-hold-device, that is, $U(t) \equiv U(ih)$ for $t \in [ih, ih + h)$. In addition, it is important to remark that $F(X(t), t) \in R^m$ in (5) includes both uncertain internal dynamics and external disturbances in the system (5), namely, the "total disturbance."

Remark 1. There are some assumptions for the system (5) that should be declared:

(A1) $\{n_i\}_0^\infty$ is a white random sequence and $E(n_i n_i{}^T) \le \bar{R}$, where $\bar{R}$ is a known matrix.

(A2) $E\begin{bmatrix} X(0) - \widehat{X}(0) \\ F(0) - \widehat{F}(0) \end{bmatrix}\begin{bmatrix} X(0) - \widehat{X}(0) \\ F(0) - \widehat{F}(0) \end{bmatrix}^T \le P_0$, where $\widehat{X}$ and $\widehat{F}$ are the estimation of X and F and P_0 is a known matrix.

(A3) $(F(X(t), t) - F(X(ih), ih))(F(X(t), t) - F(X(ih), ih))^T \le \bar{Q}$, where $t \in [ih, ih + 1)$ and $\bar{Q}$ is a known diagonal matrix.

A1-A2, which essentially assume the noise and the initial estimation errors to be bounded in the mean square, are reasonable due to the bounds of the sensor noise in practice. A3 implies that the varying of the uncertainty function $F(\cdot)$ in one sampling period is bounded.

The goal of the AESO is to get the precise estimations of the states $X_j{}^T$, $j = \{1, 2, \ldots, n\}$ and the total disturbance $F(X(t), t)$ despite the existence of various kinds of uncertainties and measurement noise, as the engineers usually desire. These precise estimations can be used to design the control input for the PMSM.

Note that the discrete error is usually decided by the sampling rate which is fixed by the hardware and the estimation error $\xi(ih) = [X(ih) - \widehat{X}(ih) F(ih) - \widehat{F}(ih)]^T$. Lemma 1 shows the formulas and the properties of the estimation error $\xi(ih)$.

Lemma 1 [30]. *Let A1–A3 hold, and the AESO for the general integral serial system is given as*

$$\begin{bmatrix} \widehat{X}(ih + h) \\ \widehat{F}(ih + h) \end{bmatrix} = \bar{A}\begin{bmatrix} \widehat{X}(ih) \\ \widehat{F}(ih) \end{bmatrix} + \widehat{L}_i(Y(ih) - \widehat{X}_1(ih)) + \begin{bmatrix} \bar{B}B_u U(ih) \\ 0 \end{bmatrix}, \tag{7}$$

$$\widehat{L}_i = \bar{A}P_i C\left(C^T P_i C + \frac{1}{1+\theta}\bar{R}\right)^{-1}, \tag{8}$$

$$P_{i+1} = (1+\theta)(\bar{A} - \bar{L}_i C^T)P_i(\bar{A} - \bar{L}_i C^T)^T + \bar{L}_i \bar{R} \bar{L}_i{}^T + \left(1 + \frac{1}{\theta}\right)Q. \tag{9}$$

Moreover, the estimation errors $\{\xi(ih)\}_{i=0}^\infty$ are uniformly bounded in the mean square and $E(\xi(ih)\xi(ih)^T) \le P_i$, $\forall i \ge 0$. There exists a positive definite matrix P^* so that $\lim_{i\to\infty} P_i = P^*$, where

$$\bar{A} = \exp\left(\begin{bmatrix} A & B \\ 0 & 0 \end{bmatrix}h\right) = \begin{bmatrix} I & \frac{h^1}{1!}I & \cdots & \frac{h^n}{n!}I \\ 0 & I & \cdots & \frac{h^{n-1}}{(n-1)!}I \\ \cdots & \cdots & \cdots & \cdots \\ 0 & 0 & \cdots & I \end{bmatrix},$$

$$\bar{B} = \int_0^h \exp\left(\begin{bmatrix} A & B \\ 0 & 0 \end{bmatrix}(h-\tau)\right)d\tau\begin{bmatrix} B \\ 0 \end{bmatrix} = \begin{bmatrix} \frac{h^n}{n!}I \\ \frac{h^{n-1}}{(n-1)!}I \\ \cdots \\ hI \\ 0 \end{bmatrix},$$

$$Q = m(n+1)\begin{bmatrix} h^{2n}\bar{Q} & 0 & \cdots & 0 \\ 0 & h^{2(n-1)}\bar{Q} & \cdots & 0 \\ \cdots & \cdots & \cdots & 0 \\ 0 & 0 & \cdots & \bar{Q} \end{bmatrix},$$

$$C = \begin{bmatrix} I \\ 0 \\ \cdots \\ 0 \end{bmatrix},$$

$$\theta = \sqrt{\frac{\sum_{j=1}^{(n+1)m} Q(j,j)}{\sum_{j=1} \mathbf{P}_0(j,j)}}. \tag{10}$$

Consequently, for the model (2) of S_1 with measurement noise used in AAR can be rewritten as

$$\begin{bmatrix} \dot{\vartheta}(t) \\ \dot{\omega}(t) \end{bmatrix} = \begin{bmatrix} 0 & 1 \\ 0 & 0 \end{bmatrix}\begin{bmatrix} \vartheta(t) \\ \omega(t) \end{bmatrix} + \begin{bmatrix} 0 \\ 1 \end{bmatrix}F_1(t) + \begin{bmatrix} 0 \\ 1 \end{bmatrix}b_1 i_q(ih),$$
$$Y(ih) = \vartheta(ih) + n_i,$$
$$t \in [ih, ih + h),\ i = 0, 1, \ldots. \tag{11}$$

The PMSM system can be reconstituted via the nominal AESO (12), according to (7), (8), and (9).

$$S_1: \bar{A}_1 = \begin{bmatrix} 1 & \frac{h^1}{1!} & \frac{h^2}{2!} \\ 0 & 1 & \frac{h^1}{1!} \\ 0 & 0 & 0 \end{bmatrix},$$

$$\bar{B}_1 = \begin{bmatrix} \frac{h^2}{2!} \\ \frac{h^1}{1!} \\ 0 \end{bmatrix}, \tag{12}$$

$$\bar{C}_1 = \begin{bmatrix} 1 \\ 0 \\ 0 \end{bmatrix},$$

$$Q_1 = 3\bar{Q}_1 \begin{bmatrix} h^4 & 0 & 0 \\ 0 & h^2 & 0 \\ 0 & 0 & 1 \end{bmatrix}.$$

With the AESO (12), the system states ϑ, ω and the total disturbance F_1 can be well estimated even with the output measurement noises n_i.

3.2. The Nonsingular Fast Terminal Sliding-Mode (NFTSM) Control. This section focuses on the rapid controller design of the PMSM take-up system based on a nonsingular terminal sliding model. Here, a new NFTSM surface and the sliding mode reaching law are given. Then, the NFTSM controller is designed for the speed loop and some of its properties are analyzed.

The conventional NTSM control is proposed to solve the controller singularity problem and improve the system tracking precision. For (2), the conventional NTSM is described by following nonlinear switching surface [27]:

$$s_0 = e_1 + \frac{1}{\beta} e_2^{p/q}, \tag{13}$$

where $e_1 = \vartheta - \vartheta^*$, $e_2 = \dot{e}_1 = \dot{\vartheta} - \dot{\vartheta}^* = \omega - \dot{\vartheta}^*$, $\beta > 0$ is the designed parameter, and p, q are the positive odd integers satisfying $1 < p/q < 2$.

Remark 2. When the sliding mode $s_0 = e_1 + (1/\beta)e_2^{p/q} = 0$ is reached, it will reach the equilibrium within the finite time

$$T_n = \int_0^{V_n(0)} \left(2^{(p+q)/2q} \beta V_n^{(p+q)/2q}\right)^{-p/q} dV_n, \tag{14}$$

where $V_n = 0.5e_1^2$ and the derivational process of (14) is given in the appendix.

For system (2) with NTSM (13), the controller can be designed as

$$u = -b_1^{-1}\left[F_1 + \beta \frac{q}{p} e_2^{2-p/q} + (l_1 + \eta_1)\operatorname{sgn}(s) - \ddot{\vartheta}^*\right], \tag{15}$$

where $\eta_1 > 0$ is the switching gain and ϑ^* is the angular position command signal to be tracked. And the NTSM surface function (15) will be reached in the finite time; furthermore, the state errors e_1 and e_2 will converge to zero.

However, for the NTSM controller (15), in order to achieve system convergence, the switching gain η_1 is generally selected for a bigger value [27]. It will obviously produce the chattering and influence the convergence precision during the steady state, and the chattering will further make the system unstable. From above analysis, it is very necessary to improve system convergence characteristics and eliminate system chattering based on the conventional NTSM controller.

The proposed NFTSM is described as (16), the state tracking error power item $(1/\alpha)e_1^{a/b}$ and the exponential term $(1/\eta)|e_1|\exp(|e_1|)$ to improve system convergence speed in the NFTSM.

$$s = e_1 + \frac{1}{\alpha} e_1^{a/b} + \frac{1}{\beta} e_2^{p/q} + \frac{1}{\eta}|e_1|\exp(|e_1|), \tag{16}$$

where a, b are the positive odd integers satisfying $a/b > p/q$.

Remark 3. For the sliding variable (16), when the states are far from the equilibrium point, the teams e_1 and $(1/\eta)|e_1| \exp(|e_1|)$ mainly affect the convergence efficiency, which can make the system trajectory converge quickly. When the states are near to the equilibrium point, the team $(1/\alpha)e_1^{a/b}$ largely affects the convergence efficiency, and the $(1/\eta)|e_1| \exp(|e_1|)$ can still make the system trajectory converge quickly. So the proposed NFTSM scheme can achieve the global convergence quickly, not just within the area adjacent to the equilibrium point.

Besides the sliding surface, the reaching law is also needed to bring the sliding variable to the sliding surface. Actually, the terminal attractor, which is widely used in many sliding-mode control methods, makes the system states approach the sliding surface in a finite time and improves the robustness to uncertainties and disturbances. An effective terminal attractor-based reaching law is proposed as

$$\dot{s} = -\phi s - \gamma s^{m/n}. \tag{17}$$

However, if we use (17) as the reaching law for the system (2) with the sliding surface (16), the item $e_2^{1-p/q}$ will arise in the control input. As $1 < p/q < 2$, thus $1 - p/q < 0$ and $e_2^{1-p/q}$ will case the singular problem. To eliminate $e_2^{1-p/q}$ from the control input, we design a modified reaching law with terminal attractor on the basis of (17) as

$$\dot{s} = \left(-\phi s - \gamma s^{m/n}\right) e_2^{p/q-1}, \tag{18}$$

where $\phi > 0, \gamma > 0$ and m, n are the positive odd integers satisfying $0 < m/n < 1$. And in Theorem 3, it can be found that the control input is absolutely nonsingular with the modified reaching law (18).

Remark 4. For the reaching law (18), $-\phi s$ plays the leading role to guarantee the faster-reaching speed primarily. When the system is near the sliding surface, $-\gamma s^{m/n}$ plays the leading role to reduce the reaching speed that makes the system states reach the sliding surface smoothly.

Based on (2), (16), and (18), then we design the NFTSM controller:

$$u = -b_1^{-1}\left[\beta\frac{q}{p}\left(\phi s + \gamma s^{m/n}\right) + \beta\frac{q}{p}e_2^{2-p/q} \cdot\left(1 + \frac{a}{\alpha b}e_1^{a/b-1} + \frac{1}{\eta}\exp(|e_1|)(1+|e_1|)\right) + \left(F_1 - \ddot{\vartheta}^*\right)\right]. \quad (19)$$

As p, q, a, b, m, and n are all positive odd numbers and $1 < p/q < 2$, $p/q < a/b$, and $0 < m/n < 1$, thus, $2 - p/q > 1$ and $a/b - 1 > 0$. Then, we know that the exponential terms of the errors e_1, e_2 are all positive, and no negative exponential terms appear in the controller (19). That is, no singular problem will not appear in the controller by the special designed reaching law.

The convergence characteristic of the closed-loop system is generally decided by the sliding surface and the reaching law. When the system is on the sliding phase ($s = 0$), the convergence characteristic is decided by the sliding surface; when the system is on the sliding phase, the convergence characteristic is decided by the reaching law.

Theorem 1 shows the convergence characteristic and the finite convergence time of the system (2) on the sliding surface $s = 0$. And Theorem 2 shows the convergence characteristic of the reaching phase.

Theorem 1. *For system (2) with the NFTSM (16), when the sliding mode $s = 0$ is reached, the system states will reach the equilibrium along the sliding surface $s = 0$ within the time T_{nf} and $T_{nf} < T_n$ is satisfied, where*

$$T_{nf} = \int_0^{V_n(0)}\left(\tau_1 V_n^{(p+q)/2q} + \tau_2 V_n^{(a/2b)+(p/2q)} + \tau_3 V_n^{p/2q}\left|V_n^{1/2}\right|\exp\left(\left|V_n^{1/2}\right|\right)\right)^{-q/p} dV_n. \quad (20)$$

Proof. When the system errors (e_1, e_2) reach the sliding surface $s = 0$ and with (16), we get that the system dynamics is determined by the following nonlinear differential equation:

$$s = e_1 + \frac{1}{\alpha}e_1^{a/b} + \frac{1}{\beta}e_2^{p/q} + \frac{1}{\eta}|e_1|\exp(|e_1|) = 0, \quad (21)$$

as $e_2 = \dot{e}_1 = \dot{\vartheta} - \dot{\vartheta}^* = \omega - \dot{\vartheta}^*$, that is,

$$e_2^{p/q} = \dot{e}_1^{p/q} = -\beta\left(e_1 + \frac{1}{\alpha}e_1^{a/b} + \frac{1}{\eta}|e_1|\exp(|e_1|)\right). \quad (22)$$

Define $V_n = 0.5e_1^2$; it can be got that $\dot{V}_n = e_1\dot{e}_1$ and then

$$\dot{V}_n^{p/q} = e_1^{p/q}\dot{e}_1^{p/q} = -\beta e_1^{p/q}\left(e_1 + \frac{1}{\alpha}e_1^{a/b} + \frac{1}{\eta}|e_1|\exp(|e_1|)\right) = -\beta\left(e_1^{p+q/q} + \frac{1}{\alpha}e_1^{(a/b)+(p/q)} + \frac{1}{\eta}e_1^{p/q}|e_1|\exp(|e_1|)\right). \quad (23)$$

As p, q, a, b are all positive odd numbers; hence, $V_n < 0$ is satisfied when $e_1 \neq 0$. Based on the Lyapunov's stability criterion, the dynamic $s = e_1 + (1/\alpha)e_1^{a/b} + (1/\beta)e_2^{p/q} + (1/\eta)|e_1| \exp(|e_1|) = 0$ is stable and the system states will converge to the equilibrium. And from (22), it can be known that the only equilibrium is $e_1 = 0$. To get the convergence time T_{nf}, we submit $e_1^2 = 2V_n$ into (23) and get

$$\dot{V}_n^{p/q} = -\left(\tau_1 V_n^{(p+q)/2q} + \tau_2 V_n^{(a/2b)+(p/2q)} + \tau_3 V_n^{p/2q}\left|V_n^{1/2}\right|\exp\left(\left|V_n^{1/2}\right|\right)\right), \quad (24)$$

where $\tau_1 = 2^{(p+q)/2q}\beta$, $\tau_1 = 2^{(a/2b)+(p/2q)}(\beta/\alpha)$, $\tau_1 = 2^{p/2q}(\beta/\eta)$.

As $\dot{V}_n = (dV_n)/(dt)$, the following can be obtained:

$$dt = \frac{dV_n}{\dot{V}_n} = \frac{dV_n}{-(-\dot{V}_n)} = -\left(\left(-\dot{V}_n\right)^{p/q}\right)^{-q/p} dV_n = -\left(\tau_1 V_n^{(p+q)/2q} + \tau_2 V_n^{(a/2b)+(p/2q)} + \tau_3 V_n^{p/2q}\left|V_n^{1/2}\right|\exp\left(\left|V_n^{1/2}\right|\right)\right)^{-q/p} dV_n. \quad (25)$$

Furthermore, as $V_n(T_{nf}) = 0$ and $V_n(0) = 0.5e_1^2(0) \geq 0$, we integrate (25) from both sides and get the time T_{nf}.

$$T_{nf} = \int_0^{V_n(0)}\left(\tau_1 V_n^{(p+q)/2q} + \tau_2 V_n^{(a/2b)+(p/2q)} + \tau_3 V_n^{p/2q}\left|V_n^{1/2}\right|\exp\left(\left|V_n^{1/2}\right|\right)\right)^{-q/p} dV_n,$$
$$T_{nf} = \int_0^{V_n(0)}\left(\tau_1 V_n^{(p+q)/2q} + \tau_2 V_n^{(a/2b)+(p/2q)} + \tau_3 V_n^{p/2q}\left|V_n^{1/2}\right|\exp\left(\left|V_n^{1/2}\right|\right)\right)^{-q/p} dV_n < T_n = \int_0^{V_n(0)}\left(\tau_1 V_n^{(p+q)/2q}\right)^{-q/p} dV_n. \quad (26)$$

Remark 5. From the proof above, it can also be known that the composition of the term $(1/\alpha)e_1^{a/b}$ and $(1/\eta)|e_1|\exp(|e_1|)$ is better than the choice of single $(1/\alpha)e_1^{a/b}$. Assume that the sliding variable is selected as $s_1 = e_1 + (1/\alpha)e_1^{a/b} + (1/\beta)e_2^{p/q}$, and then we can easily get that $T_{nf} < T_n = \int_0^{V_n(0)} (\tau_1 V_n^{(p+q)/2q} + \tau_2 V_n^{(a/2b)+(p/2q)})^{-q/p} dV_n$. Hence, (16) provides another good choice to obtain a faster convergence characteristic without increasing the gains in the sliding variable.

And Theorem 2 shows that the attractor in (18) will not affect the existence of the NFTSM (16) during the reaching phase. Moreover, the sliding mode $s = 0$ will be reached asymptotically.

Theorem 2. *For the system (2), if the closed-loop controller is designed as (19), then the NFTSM (16) exists globally and the system states from any initial positions will asymptotically converge to the sliding surface* $s = 0$.

Proof. Consider the following Lyapunov function:

$$V_0 = 0.5s^2. \tag{27}$$

Hence,

$$\dot{V}_0 = s\dot{s} = \left(-\phi s^2 - \gamma s^{m/n+1}\right) e_2^{p/q-1}. \tag{28}$$

Define $\mathbf{e} = (e_1, e_2)$, then we divide the $e_1 - e_2$ phase plane into two regions: $D_1 = \{\mathbf{e} | e_1 \neq 0, e_2 \neq 0\}$ and $D_2 = \{\mathbf{e} | e_1 \neq 0, e_2 = 0\}$. Indeed, it should be noted that D_2 is the axis e_1 in the $e_1 - e_2$ phase plane.

Suppose $l(e_2) = e_2^{p/q-1}$, then $l(e_2) > 0$ when $e_2 \neq 0$. Namely, when $\mathbf{e} = (e_1, e_2) \in D_1$, we get $l(e_2) > 0$, $s^{m/n+1} > 0$ and

$$\dot{V}_0 = s\dot{s} = \left(-\phi s^2 - \gamma s^{m/n+1}\right) l(e_2) \leq 0. \tag{29}$$

Thus, the sliding mode s will asymptotically converge to zero in the region D_1, according to Lyapunov stability criterion. That is, the NFTSM (16) exists in the region D_1.

Next, we will check whether the sliding mode s will still converge to zero in D_2. Indeed, substituting the control law (19) into the error dynamic equation of (2) yields

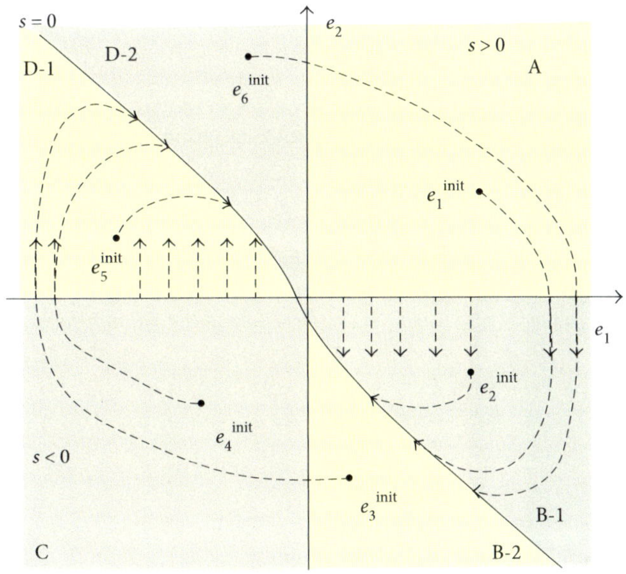

FIGURE 4: The relative coupling profile.

$$\begin{aligned} \dot{e}_1 &= e_2, \\ \dot{e}_2 &= F_1 - \ddot{\vartheta}^* + b_1 i_q = -\beta \frac{q}{p} \\ &\quad \cdot \left[\left(\phi s + \gamma s^{m/n}\right) + e_2^{2-p/q} \left(1 + \frac{a}{\alpha b} e_1^{a/b-1} \right.\right. \\ &\quad \left.\left. + \frac{1}{\eta} \exp(|e_1|)(1 + |e_1|) \right) \right]. \end{aligned} \tag{30}$$

From (30), we get the slope equation of the phase path.

$$\frac{de_2}{de_1} = \frac{\dot{e}_2}{\dot{e}_1} = \frac{-\beta(q/p)\left[\left(\phi s + \gamma s^{m/n}\right) + e_2^{2-p/q}\left(1 + (a/\alpha b)e_1^{a/b-1} + (1/\eta)\exp(|e_1|)(1 + |e_1|)\right)\right]}{e_2}. \tag{31}$$

As p, q, a, b, m, and n are all positive odd numbers and $1 < p/q < 2$, and $a/b > 1$, thus we get (32) when $e_1 > 0$ and $e_2 = 0$.

$$\begin{aligned} \frac{de_2}{de_1} &= -\infty, \\ \dot{e}_2 &= -\beta \frac{q}{p} \left[\left(\phi s + \gamma s^{m/n}\right) + e_2^{2-p/q} \right. \\ &\quad \left. \cdot \left(1 + \frac{a}{\alpha b} e_1^{a/b-1} + \frac{1}{\eta} \exp(|e_1|)(1 + |e_1|) \right) \right] < 0. \end{aligned} \tag{32}$$

Equation (32) means that the phase path is perpendicular to the positive axis e_1 when it crosses the positive axis e_1, and phase path direction is straight down from the positive axis e_1, as shown in Figure 4. Moreover, the motion speed of $e = (e_1, e_2)$ is not equal to zero. Similarly, the phase path is perpendicular to the negative axis e_1 when it crosses the negative axis e_1; the phase path direction is straight up, and the motion speed of $e = (e_1, e_2)$ is also not equal to zero.

The state trajectory in the reaching phase is shown in Figure 4. $e_i^{\text{init}} = (e_1, e_2)(e_1, e_2)_i^{\text{init}}, i = 1, \ldots, 6$ are the possible locations for the system initial states throughout the whole phase plane. The region D_1 is then separated into six regions (A, B-1, B-2, C, D-1, and D-2) to illustrate the state trajectory of any possible initial states during the reaching phase. Indeed, as concluded above, the system state's movement speed is not zero when it crosses the axis e_1 (namely, the system state is in D_2). Therefore, although the state trajectory may cross the axis e_1 (such as $\mathbf{e}_i^{\text{init}}$, $i = 1, 3, 4, 6$), the state $\mathbf{e}$ will not stay in D_2. In these cases, the states $\mathbf{e}_i^{\text{init}}$, $i = 1, 3, 4, 6$ will cross the axis e_1. Then, they will asymptotically converge to the sliding mode $s = 0$ like the initial states which do not cross axis e_1 (such as, $\mathbf{e}_i^{\text{init}}, i = 2, 5$).

3.3. The Proposed Control Scheme for PMSM Tracking Control. Based on the design and analysis of the AESO and NFTSM above, Theorem 3 gives the proposed controller for the PMSM angular position tracking control to suppress the hose vibration problem in AAR coupling.

Theorem 3. *Under Assumptions A1–A3, for the system (2) with the NFTSM (16), if the AESO is designed with (7), (8), and (9) and the closed-loop controller is designed as (33), then the system trajectory will fast converge to zero. Moreover, no occurring of singularity is ensured during the whole process.*

$$u = -b_1^{-1}\left[\beta\frac{q}{p}\left(\phi\widehat{s} + \gamma\widehat{s}^{m/n}\right) + \left(\widehat{F}_1 - \ddot{\vartheta}^*\right) + \beta\frac{q}{p}\widehat{e}_2^{\,2-p/q} \cdot \left(1 + \frac{a}{\alpha b}\widehat{e}_1^{\,a/b-1} + \frac{1}{\eta}\exp(|\widehat{e}_1|)(1+|\widehat{e}_1|)\right)\right], \tag{33}$$

where $\widehat{e}_1 = z_1 - \vartheta^*$ *and* $\widehat{e}_2 = z_2 - \dot{\vartheta}^*$, z_1 *and* z_2 *are the state estimations from AESO, and* $\widehat{F}_1$ *is the estimation of the "total disturbance" from AESO.* ϑ^* *and* $\dot{\vartheta}^*$ *are the reference from command filter [32], and* $\widehat{s} = \widehat{e}_1 + (1/\alpha)\widehat{e}_1^{\,a/b} + (1/\beta)\widehat{e}_2^{\,p/q} + (1/\eta)|\widehat{e}_1|\exp(|\widehat{e}_1|)$.

Proof. Consider the following Lyapunov function:

$$V = \frac{1}{2}s^2. \tag{34}$$

Therefore,

$$\dot{V} = s\dot{s} = s\left(e_2\left(1 + \frac{a}{\alpha b}\widehat{e}_1^{\,a/b-1} + \frac{1}{\eta}\exp(|\widehat{e}_1|)(1+|\widehat{e}_1|)\right) + \frac{p}{\beta q}e_2^{\,p/q-1}\dot{e}_2\right). \tag{35}$$

Substituting the control (33) and system (2) into (35) yields

$$\begin{aligned}\dot{V} &= s\Bigg(e_2\left(1 + \frac{a}{\alpha b}\widehat{e}_1^{\,a/b-1} + \frac{1}{\eta}\exp(|\widehat{e}_1|)(1+|\widehat{e}_1|)\right) + \frac{p}{\beta q}e_2^{\,p/q-1}\left(F_1 + b_1 u - \ddot{\vartheta}^*\right)\Bigg)\\
&= s\Bigg(e_2\left(1 + \frac{a}{\alpha b}\widehat{e}_1^{\,a/b-1} + \frac{1}{\eta}\exp(|\widehat{e}_1|)(1+|\widehat{e}_1|)\right) + \frac{p}{\beta q}e_2^{\,p/q-1}\Bigg(F_1 - \beta\frac{q}{p}\left(\phi\widehat{s}+\gamma\widehat{s}^{m/n}\right) - \beta\frac{q}{p}\widehat{e}_2^{\,2-p/q}\\
&\quad\cdot\left(1 + \frac{a}{\alpha b}\widehat{e}_1^{\,a/b-1} + \frac{1}{\eta}\exp(|\widehat{e}_1|)(1+|\widehat{e}_1|)\right) - \left(\widehat{F}_1 - \ddot{\vartheta}^*\right) - \ddot{\vartheta}^*\Bigg)\Bigg)\\
&= s\Bigg(e_2\left(1 + \frac{a}{\alpha b}\widehat{e}_1^{\,a/b-1} + \frac{1}{\eta}\exp(|\widehat{e}_1|)(1+|\widehat{e}_1|)\right) + \frac{p}{\beta q}e_2^{\,p/q-1}\Bigg(F_1 - \widehat{F}_1 - \beta\frac{q}{p}\left(\phi\widehat{s}+\gamma\widehat{s}^{m/n}\right) - \beta\frac{q}{p}\widehat{e}_2^{\,2-p/q}\\
&\quad\cdot\left(1 + \frac{a}{\alpha b}\widehat{e}_1^{\,a/b-1} + \frac{1}{\eta}\exp(|\widehat{e}_1|)(1+|\widehat{e}_1|)\right)\Bigg)\Bigg).\end{aligned} \tag{36}$$

Then,

$$\begin{aligned}\dot{V} &= s\Bigg(e_2\left(1 + \frac{a}{\alpha b}\widehat{e}_1^{\,a/b-1} + \frac{1}{\eta}\exp(|\widehat{e}_1|)(1+|\widehat{e}_1|)\right) + \frac{p}{\beta q}e_2^{\,p/q-1}\Bigg(\tilde{F}_1 - \beta\frac{q}{p}\left(\phi(s-\tilde{s}) + \gamma(s-\tilde{s})^{m/n}\right)\\
&\quad - \beta\frac{q}{p}\widehat{e}_2^{\,2-p/q}\left(1 + \frac{a}{\alpha b}\widehat{e}_1^{\,a/b-1} + \frac{1}{\eta}\exp(|\widehat{e}_1|)(1+|\widehat{e}_1|)\right)\Bigg)\Bigg)\\
&= l(e_2)s\Bigg(\left(e_2^{\,2-p/q} - \widehat{e}_2^{\,2-p/q}\right) + \frac{a}{\alpha b}\left(e_2^{\,2-p/q}e_1^{\,a/b-1} - \widehat{e}_2^{\,2-p/q}\widehat{e}_1^{\,a/b-1}\right)\\
&\quad + \left(e_2^{\,2-p/q}\exp(|e_1|)(1+|e_1|) - \widehat{e}_2^{\,2-p/q}\exp(|\widehat{e}_1|)(1+|\widehat{e}_1|)\right) + \frac{\beta q}{p}\tilde{F}_1 + \left(\phi\tilde{s}+\gamma\tilde{s}^{m/n}\right) - \left(\phi s + \gamma s^{m/n}\right)\Bigg),\end{aligned} \tag{37}$$

where $\tilde{F}_1 = F_1 - \widehat{F}_1$ and $\tilde{s} = s - \widehat{s}$. Note that $\tilde{e}_1 = e_1 - \widehat{e}_1$ and $\tilde{e}_2 = e_2 - \widehat{e}_2$, then according to the inequation theorem:

$$\begin{aligned}&e_2^{\,2-p/q} - \widehat{e}_2^{\,2-p/q} \le |\tilde{e}_2|^{2-p/q},\\
&e_2^{\,2-p/q}e_1^{\,a/b-1} - \widehat{e}_2^{\,2-p/q}\widehat{e}_1^{\,a/b-1}\\
&\quad = e_2^{\,2-p/q}e_1^{\,a/b-1} - \widehat{e}_2^{\,2-p/q}\left(e_1 - \tilde{e}_1\right)^{a/b-1}\\
&\quad \le e_1^{\,a/b-1}e_2^{\,2-p/q} - e_1^{\,a/b-1}\widehat{e}_2^{\,2-p/q} + \tilde{e}_1^{\,a/b-1}\widehat{e}_2^{\,2-p/q}\\
&\quad = e_1^{\,a/b-1}\left(e_2^{\,2-p/q} - \widehat{e}_2^{\,2-p/q}\right) + \tilde{e}_1^{\,a/b-1}\widehat{e}_2^{\,2-p/q}\\
&\quad \le e_1^{\,a/b-1}|\tilde{e}_2|^{2-p/q} + \widehat{e}_2^{\,2-p/q}\tilde{e}_1^{\,a/b-1},\\
&e_2^{\,2-p/q}\exp(|e_1|)(1+|e_1|) - \widehat{e}_2^{\,2-p/q}\exp(|\widehat{e}_1|)(1+|\widehat{e}_1|)\\
&\quad = \exp(|e_1|)(1+|e_1|)e_2^{\,2-p/q} - \exp(|\widehat{e}_1|)(1+|\widehat{e}_1|)\cdot\left(e_2 - \tilde{e}_2\right)^{2-p/q}\\
&\quad \le e_2^{\,2-p/q}\left(\exp(|e_1|)(1+|e_1|) - \exp(|\widehat{e}_1|)(1+|\widehat{e}_1|)\right) + \exp(|\widehat{e}_1|)(1+|\widehat{e}_1|)\tilde{e}_2^{\,2-p/q}\\
&\quad = e_2^{\,2-p/q}\left(\exp(|\widehat{e}_1+\tilde{e}_1|)(1+|\widehat{e}_1+\tilde{e}_1|) - \exp(|\widehat{e}_1|)(1+|\widehat{e}_1|)\right) + \exp(|\widehat{e}_1|)(1+|\widehat{e}_1|)\tilde{e}_2^{\,2-p/q}\\
&\quad = e_2^{\,2-p/q}\exp(|\widehat{e}_1|)\left((1+|e_1|)\exp(|\tilde{e}_1|) - (1+|\widehat{e}_1|)\right) + \exp(|\widehat{e}_1|)(1+|\widehat{e}_1|)\tilde{e}_2^{\,2-p/q}.\end{aligned} \tag{38}$$

Consequently,

$$\begin{aligned}\dot{V} \le l(e_2)s\Bigg(&|\tilde{e}_2|^{2-p/q} + \left(e_1^{\,a/b-1}|\tilde{e}_2|^{2-p/q} + \widehat{e}_2^{\,2-p/q}\tilde{e}_1^{\,a/b-1}\right)\\
&+ \left(e_2^{\,2-p/q}\exp(|\widehat{e}_1|)\left((1+|e_1|)\exp(|\tilde{e}_1|) - (1+|\widehat{e}_1|)\right) + \exp(|\widehat{e}_1|)(1+|\widehat{e}_1|)\tilde{e}_2^{\,2-p/q}\right)\\
&+ \left(\phi\tilde{s}+\gamma\tilde{s}^{m/n}\right) + \frac{p}{\beta q}\tilde{F}_1 - \left(\phi s + \gamma s^{m/n}\right)\Bigg).\end{aligned} \tag{39}$$

Indeed, the errors $\tilde{e}_1$, $\tilde{e}_2$, and $\tilde{F}_1$ are the estimation errors by the AESO. Moreover, from Lemma 1 in Section 3.1,

TABLE 1: Parameters of the hose-drogue assembly.

Parameters	Nomenclatures	Values
μ	Mass per unit length	4.11 kg/m
d_k	Diameter of hose	0.067 m
$c_{t,k}$	Skin friction coefficient	0.001
$c_{n,k}$	Hose drag coefficient	0.28
m_{drogue}	Mass of drogue	29.5 kg
d_{drogue}	Diameter of drogue	0.61 m
c_{drogue}	Drogue drag coefficient	0.831
N	Number of links	24
H_0	Flight altitude	7620 m
V_0	Airspeed	550 km/h

it can be certainly known that these estimation errors ($\tilde{e}_1$, $\tilde{e}_2$, and $\tilde{F}_1$) are uniformly bounded $|\tilde{e}_1| < \varepsilon_{e_1}$, $|\tilde{e}_2| < \varepsilon_{e_2}$, and $|\tilde{F}_1| < \varepsilon_{F_1}$, and the bounds ε_{e_1}, ε_{e_2}, and ε_{F_1} are deterministic if Assumption 1–3 are satisfied. Namely, although states and disturbance are estimated in the probabilistic framework by AESO, the bounds of the estimation errors are deterministic. Based on this conclusion, we continue the proof using the deterministic Lyapunov function.

Consequently, the estimation error $\tilde{F}_1$ can be small enough and uniformly bounded via properly turning the parameters of AESO according to Lemma 1. As $0.5 < p/q < 1$ and β is selected big enough generally, the $(p/(\beta q))\tilde{F}_1$ will be small enough.

Besides, $\tilde{e}_1$ and $\tilde{e}_2$ are the estimation errors of AESO, and they also are uniformly bounded according to Lemma 1. As $\tilde{s}$ is the bounded function of $\tilde{e}_1$ and $\tilde{e}_2$, it is also uniformly bounded. Assume Δ is the function of the residual error from AESO,

$$\begin{aligned}\Delta = {} & |\tilde{e}_2|^{2-p/q} + \left(e_1^{a/b-1}|\tilde{e}_2|^{2-p/q} + \hat{e}_2^{2-p/q}\tilde{e}_1^{a/b-1}\right) \\ & + \left(e_2^{2-p/q}\exp(|\hat{e}_1|)((1+|e_1|)\exp(|\tilde{e}_1|) - (1+|\hat{e}_1|))\right. \\ & \left. + \exp(|\hat{e}_1|)(1+|\hat{e}_1|)\tilde{e}_2^{2-p/q}\right) + \left(\phi\tilde{s} + \gamma\tilde{s}^{m/n}\right).\end{aligned} \tag{40}$$

Based on the analysis above, the residual error Δ can be bounded by properly turning the parameters of AESO. Therefore, (39) can be written as

$$\dot{V} \le l(e_2)\left(\left(\Delta + \frac{p}{\beta q}\tilde{F}_1\right)s - \left(\phi s^2 + \gamma s^{m/n+1}\right)\right). \tag{41}$$

Hence, $\dot{V} < 0$ can be satisfied if the ϕ and γ are chosen great enough. Then, $\vartheta \to \vartheta^*$ and $\omega \to \dot{\vartheta}^*$ can be achieved by the proposed control scheme for PMSM tracking control to suppress the hose vibration during AAR coupling. And from the controller (33), it can be further known that no occurring of singularity is ensured during the whole process.

TABLE 2: Parameters of the PMSM.

Parameters	Nomenclatures	Values
R	Stator resistance	1.65 Ω
L_s	Stator inductance	0.0092 H
P	Pole pairs	4
ψ_f	Stator flux linkage	0.175 Wb
J	Moment of inertia	0.001 kg·m^2
B	Viscous friction coefficient	$4.831e^{-5}$ N·m·s
i	Reduction ratio	10
r	Radius of reel	0.06 m

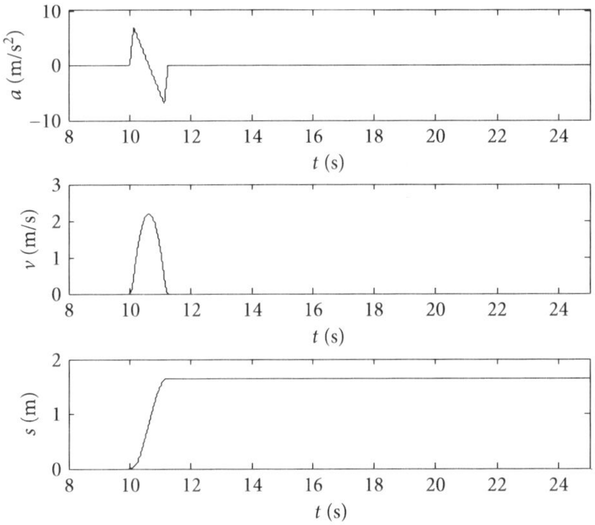

FIGURE 5: The relative coupling profile.

As described in Section 2, the current control loops are designed with proportion integral (PI) [18]. Here, two PI controllers are adopted in the two current loops to stabilize the current tracking errors of d-q axes, respectively.

$$\begin{aligned} u_q &= k_{qp}\left(i_q^* - i_q\right) + k_{qI}\int_0^t \left(i_q^* - i_q\right)dt, \\ u_d &= k_{dp}\left(i_d^* - i_d\right) + k_{dI}\int_0^t \left(i_d^* - i_d\right)dt. \end{aligned} \tag{42}$$

Generally, the reference i_d^* in the d current loop is set to be zero to maintain a constant flux operating condition, and the reference current i_q^* is determined by the subsystem S_1. Together with (33), the controller for PMSM in AAR is completed.

4. Application to the Vibration Suppression of the Flexible Refueling Hose

The vibration suppression problem of the refueling hose in AAR coupling is specially selected as an application example here for the proposed scheme due to its hostile working environment where all the assuming conditions presented in the

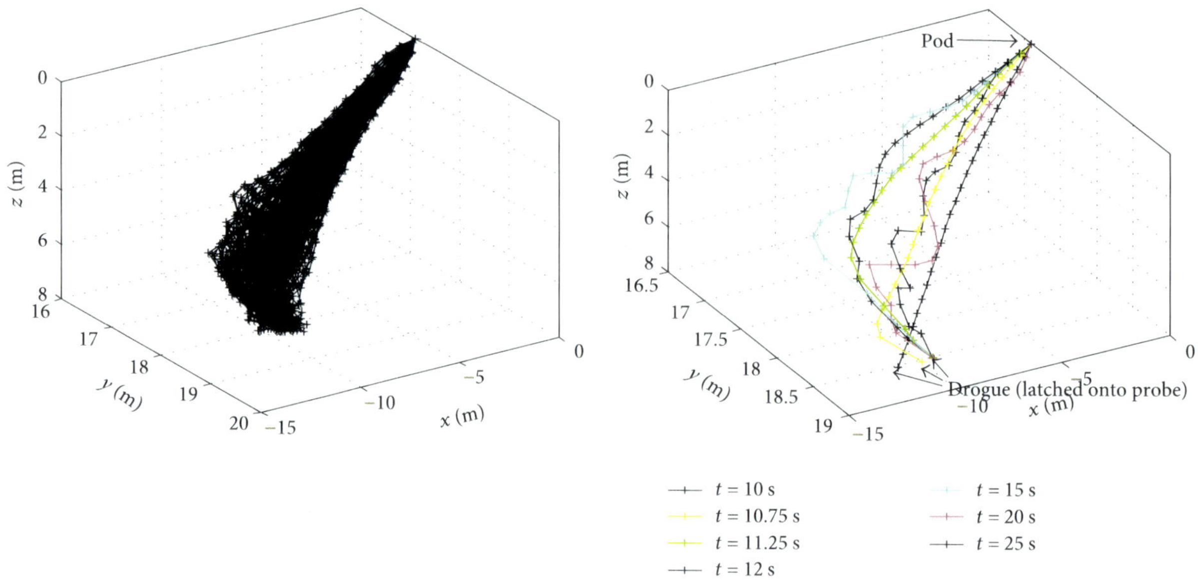

(a) Hose vibration history (10 s–25 s)—without vibration suppression

(b) Hose geometry history (selected time)—without vibration suppression

FIGURE 6: Flexible aerial refueling hose vibration during coupling without control scheme.

introduction section are satisfied. The features of the proposed method can be roundly verified in this application.

Compared with the traditional methods, in order to describe the complex dynamic characteristics of the host vibration more accurately, the hose-drogue assembly model with all links having variable length (by Wang et al.'s work [15] which is built based on the works of Ro et al. [14]) is adopted in this paper to model the hose-drogue. The parameters of the hose-drogue assembly and the flight condition are listed in Table 1. Table 2 lists the parameters of the PMSM used here.

Remark 6. For details of the model of the hose-drogue assembly (including the models for the skin friction, aerodynamic forces, and the kinematics and dynamics of the hose-drogue assembly under multiple flows in AAR), it can be found in [14] and [15]. Considering the primary motivation and purpose of this work, we will not repeat that in detail.

The initial hose length before the contact of the probe and drogue is set as 14.3 m here. And the longer distance the probe moves forward, the more violent the vibration goes. Therefore, in this paper, assume that the probe initially contacts the drogue at 10 s and then pushes the drogue forward along the relative high acceleration coupling profile [13], as illustrated in Figure 5.

The hose vibration without any suppression can be seen in Figure 6. Figure 6(a) shows the hose-drogue assembly vibration history from the time $t = 10$ s to $t = 25$ s. The hose vibration envelope during the simulation time interval can be clearly seen in Figure 6(a). It can be seen that the hose vibrates at a high frequency, and the vibration amplitudes involve a wide space range. The refueling hose will vibrate more and more drastically with the time going on. Figure 6(b) selects some specific time ($t = 10$, 10.75, 11.25, 12, 15, 20, and 25 s) to show the hose-drogue geometry history. It is obvious that the hose-drogue will shake execrably if no suppression scheme is introduced. It is a great threat to the safety of the AAR; the hose may even be broken off, and the receiver aircraft behind also may be damaged. The drastic shake of the hose indicates that there must be some control strategies to suppress the hose vibration for the safe AAR.

Figure 7 intuitively shows the shock forces on the hose-drogue assembly under vibration. As the drogue is pulled forward, the tension of the selected 24 joints (take the 1st, 13th, and 24th as examples) on the hose immediately decreases almost to zero owing to the instantaneous hose slackness. Then, the geometry of the hose changes, which will further increase the aerodynamic drags of the joints on the hose. As the aerodynamic drag choppily changes, the vibration goes more violent and travels toward the pod. As a result, the hose tension on the drogue and the force on the probe become more violent. The increasing forces will even break the PDR system and cause disastrous AAR accident. This drastic hose vibration caused by the excessive closure speed requires that there must be some vibration suppression strategies to ensure the flexible refueling hose-drogue assembly stable even under unexpected excessive closure speed and multiple disturbances.

The PMSM tracking control scheme above is applied to the vibration suppression of the refueling hose. To verify its effectiveness and superiority, the proposed NFTSM scheme with AESO and other two control schemes (linear ADRC [6, 33] with ESO and NTSM with AESO) are applied to the vibration suppression under strong disturbance and measurement noise. These methods are all designed based on the structure of cascade control loops (speed and current) shown in Section 2. The PI controller [21] is adopted for the current loop controller of these schemes. The

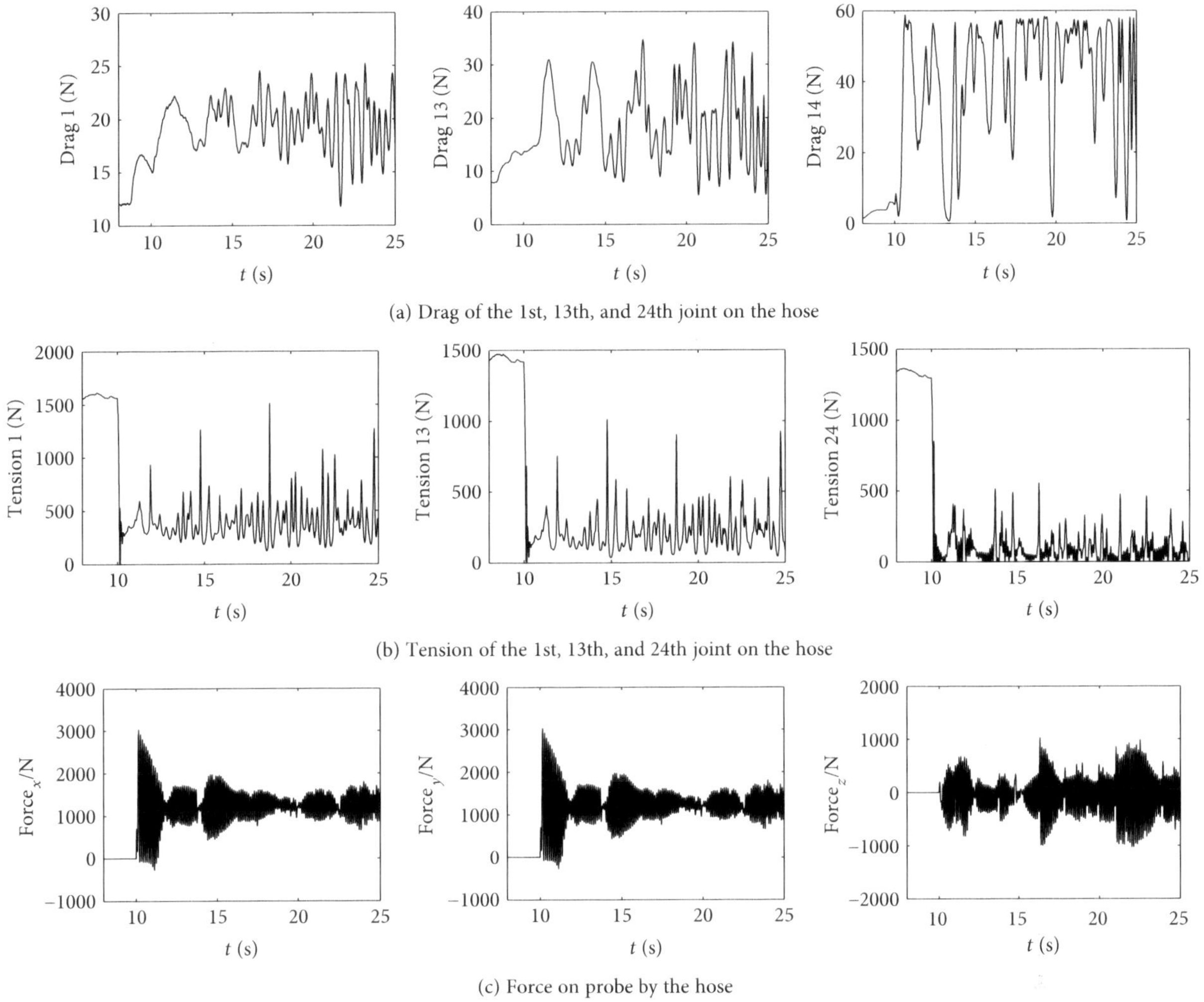

FIGURE 7: Forces on hose and probe during coupling without vibration suppression.

parameters of the PI may vary from 10^0 to 10^3 for different PMSMs [8, 18]. For comparison fairness, the PI parameters of these schemes k_p, k_I are kept same, as shown in Table 3.

For system (2), a strong disturbance with multiple frequency characteristics is set as $d_1 = 3e3(\sin(\pi t) - 0.5\cos(5\pi t) + 0.5\sin(10\pi t))$. Besides, the Gaussian random noise with its variance $E(n_{i1}^2) = (1)^2$ is assumed to be the polluted noise superposed in the output of the system (2). And it is also assumed that the position of the probe in the vertical direction is disturbed by a sine disturbance as $0.01\sin(0.2\pi t)$ after the coupling achieved due to the surrounding airflow disturbances, and the impact will finally appear in the PMSM system in the form of load disturbance and the fluctuation in the command signal. The other relevant parameters of these controllers are listed in Table 3. Contrastive simulations for these methods are conducted in MATLAB/SIMULINK 2011a and implemented on a personal computer with 3.2 GHz Core i5 CPU. The simulation sampling period is set as 0.001 s.

It assumes that the control laws all come to work immediately here when the contact of the drogue and probe successfully achieved.

TABLE 3: Parameters of the different control schemes.

Control schemes	Parameters	Values	Parameters	Values
LADRC with ESO	L_ϑ	200	K_ϑ	$[50, 100]^T$
	k_{qp}, k_{dp}	6	k_{qI}, k_{dI}	$[600, 600]^T$
NTSM with AESO	p	15	q	13
	β	100	l_1	5e3
	η_1	2e3	$k_{qp}, k_{qI}, k_{dp}, k_{dI}$	$[6, 600, 6, 600]^T$
Proposed NFTSM with AESO	$\bar{R}_1$	1^2	$\bar{Q}_1$	0.1^2
	p	15	q	13
	a	17	b	13
	m	11	n	15
	α	100	β	100
	ϕ	100	γ	100
	η	100	$k_{qp}, k_{qI}, k_{dp}, k_{dI}$	$[6, 600, 6, 600]^T$

L_ϑ is the bandwidth of the ESO in LADRC, and K_ϑ is the feedback controller gains of the PD in LADRC.

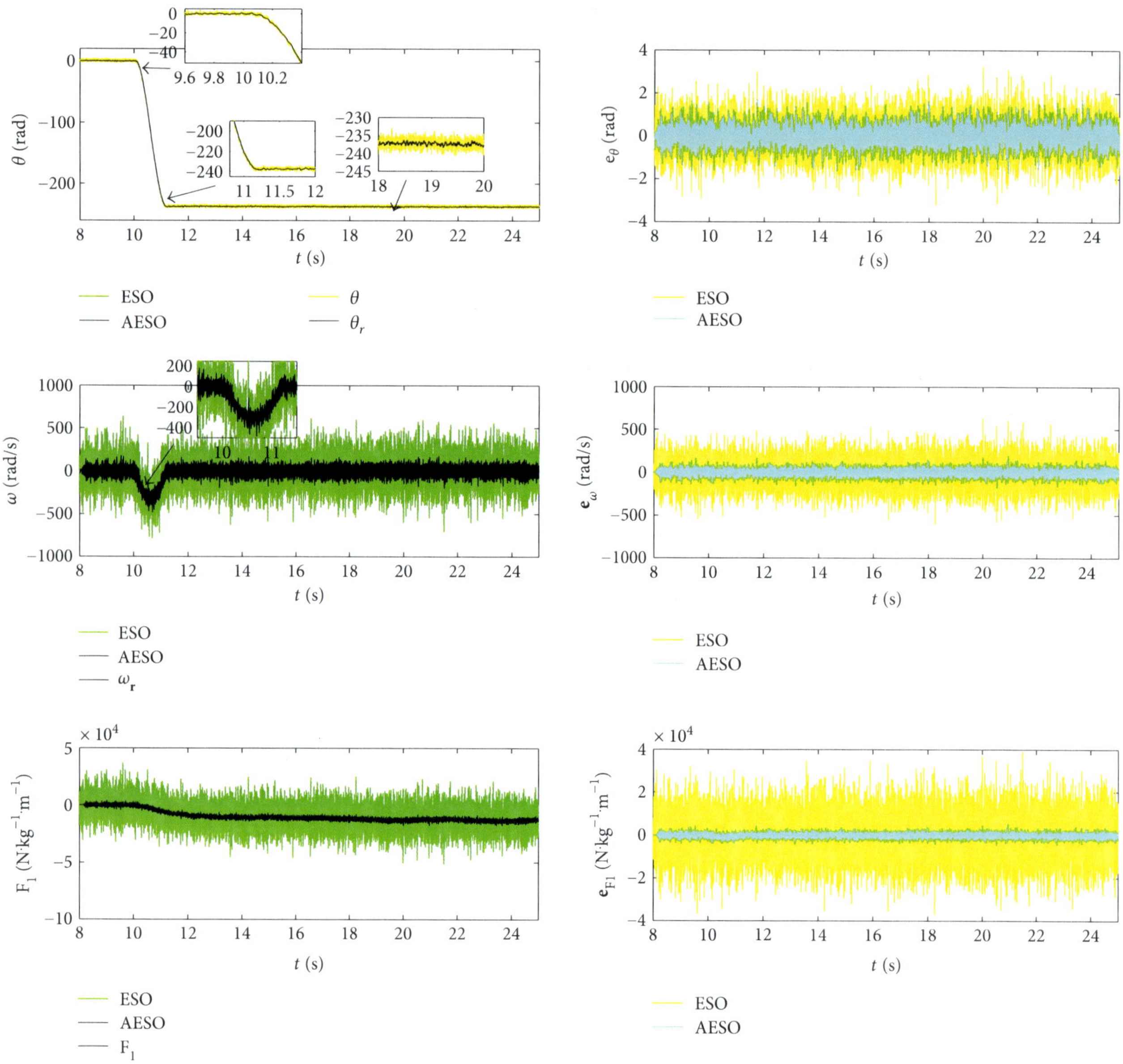

Figure 8: Estimation comparisons of ESO and AESO by the proposed NFTSM controller.

Firstly, to illustrate the superiority of AESO in reconstituting states and "total disturbance" of the system with measurement noise, the proposed NFTSM is used as the unified controller, and the comparison of ESO and AESO is conducted. Figure 7 shows the observation results of the two different observers.

In Figure 8, ϑ_r and ω_r are the real values of the angular position and the angular velocity, respectively. ϑ is the measured value from the angular position sensor. Based on the same controller, it is obvious that AESO performs better than ESO when the sensor measurement noise exists. The reconstituted states and total disturbance of AESO are closer to the real values of the systems, and the observation errors are much smaller for all the states and disturbance. Although the existence of the measurement noise whose variance is $(1\text{ rad})^2$, the angular tracking error e_θ is ensured within 2 ra d, which is much smaller than the estimation error by ESO. A similar result can be found in the estimation angular velocity. The estimation errors for the total disturbance are especially obvious. It can be seen that the amplitude of the estimated disturbance $\widehat{F}_1$ by ESO may be ten times of the real value F_1, while the $\widehat{F}_1$ by AESO is much closer to the real value F_1 and the estimation error for F_1 by AESO is 5 to 10 times smaller than the ESO's. Therefore, AESO possesses the ability to well reconstitute the system states and disturbance even with measurement noise and strong disturbance.

Then, the time history of the hose geometries by different control schemes is shown in Figure 8, and the side view in XOZ plane for each scheme is also attached to the figures. It is clear that the vibration in LADRC is still serious on the condition that the disturbances and measurement noise are taken into consideration. However, the NTSM with AESO can suppress the vibration effectively in the same condition,

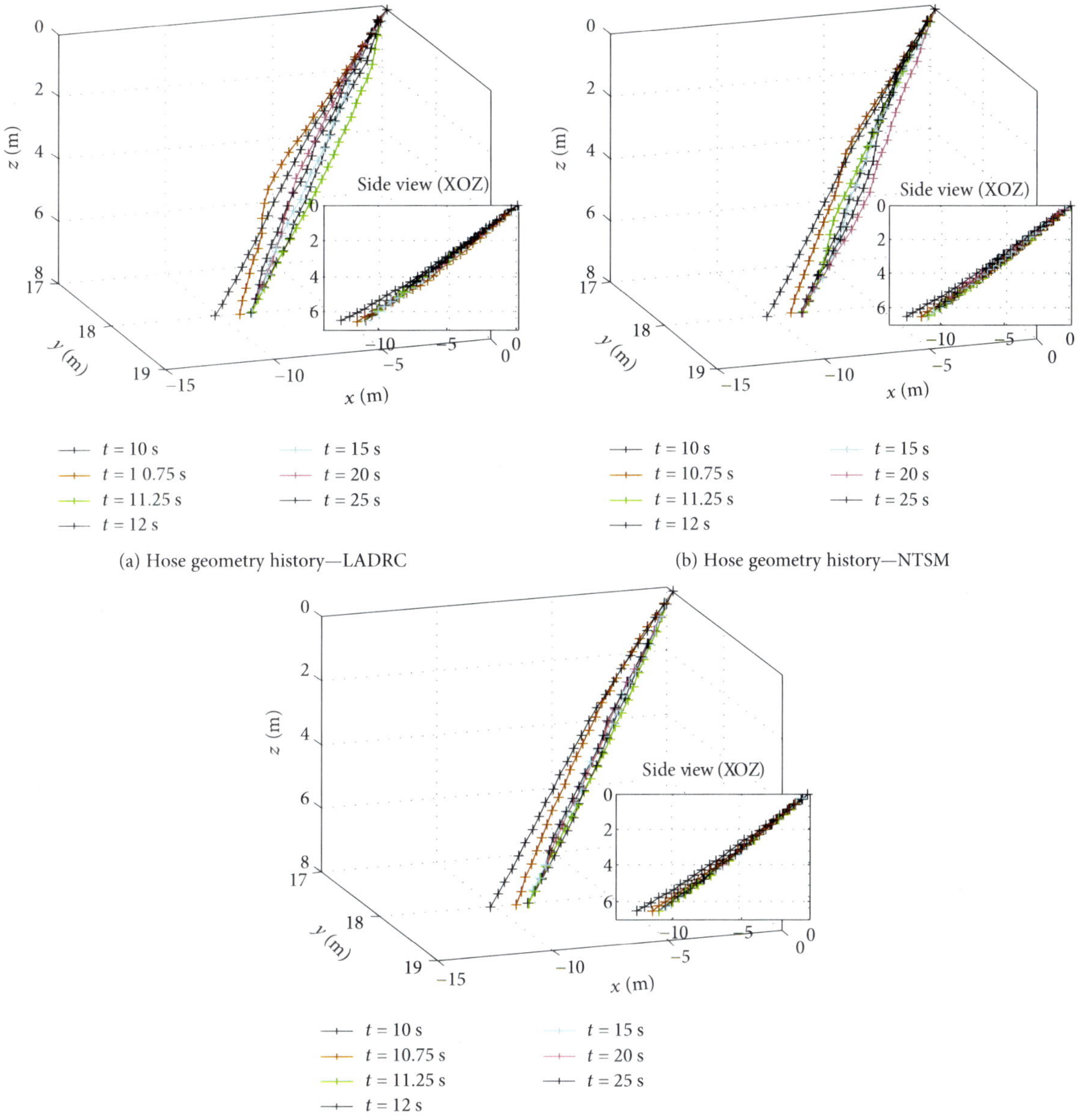

(a) Hose geometry history—LADRC

(b) Hose geometry history—NTSM

(c) Hose geometry history—NFTSM

FIGURE 9: The hose-drogue assembly's dynamic motions by different control scheme.

but its effectiveness is slightly poorer than the proposed NFTSM with AESO. From Figure 9, one can find that the hose geometries by the proposed NFTSM schemes look more stable and orderly. The proposed NFTSM with AESO cannot only track the command (transformed from the hose slack during the vibration) accurately in a shorter time. As the probe travels forward, the proposed controller can retract the hose rapidly and precisely and keep the catenary shape of the hose steady. The hose slack is maintained at the level just before coupling, and the hose vibration does not occur at all. With the compensation provided by the control laws, the catenary shape of the hose can always be kept steady. As a result, the vibration can be suppressed accurately and immediately. Figure 10 shows the drags, tensions on the 1st, 13th, and 24th joint on the drogue, and the forces on probe caused by the vibration hose-drogue assembly. Although the superiority of the proposed NFTSM scheme over the NTSM schemes has been verified via the hose geometry history in Figure 9, Figure 10 presents more visualized results to evidence that. From Figure 10, it can be seen that the proposed vibration scheme ensures the drag forces, tensions on the hose, and the forces on the probe all much smaller than those of the NTSM schemes. These smaller forces indicate that the hose-drogue assembly are more stable and organized under the proposed scheme. The vibration can be more effectively suppressed by the proposed NFTSM scheme.

From Figures 11, 12, 13, and 14, the states and relevant tracking errors of the PMSM subsystem are shown, respectively. The external load on the probe is shown in Figure 15. And the length histories of the hose reeling driven by the PMSM are shown in Figure 16.

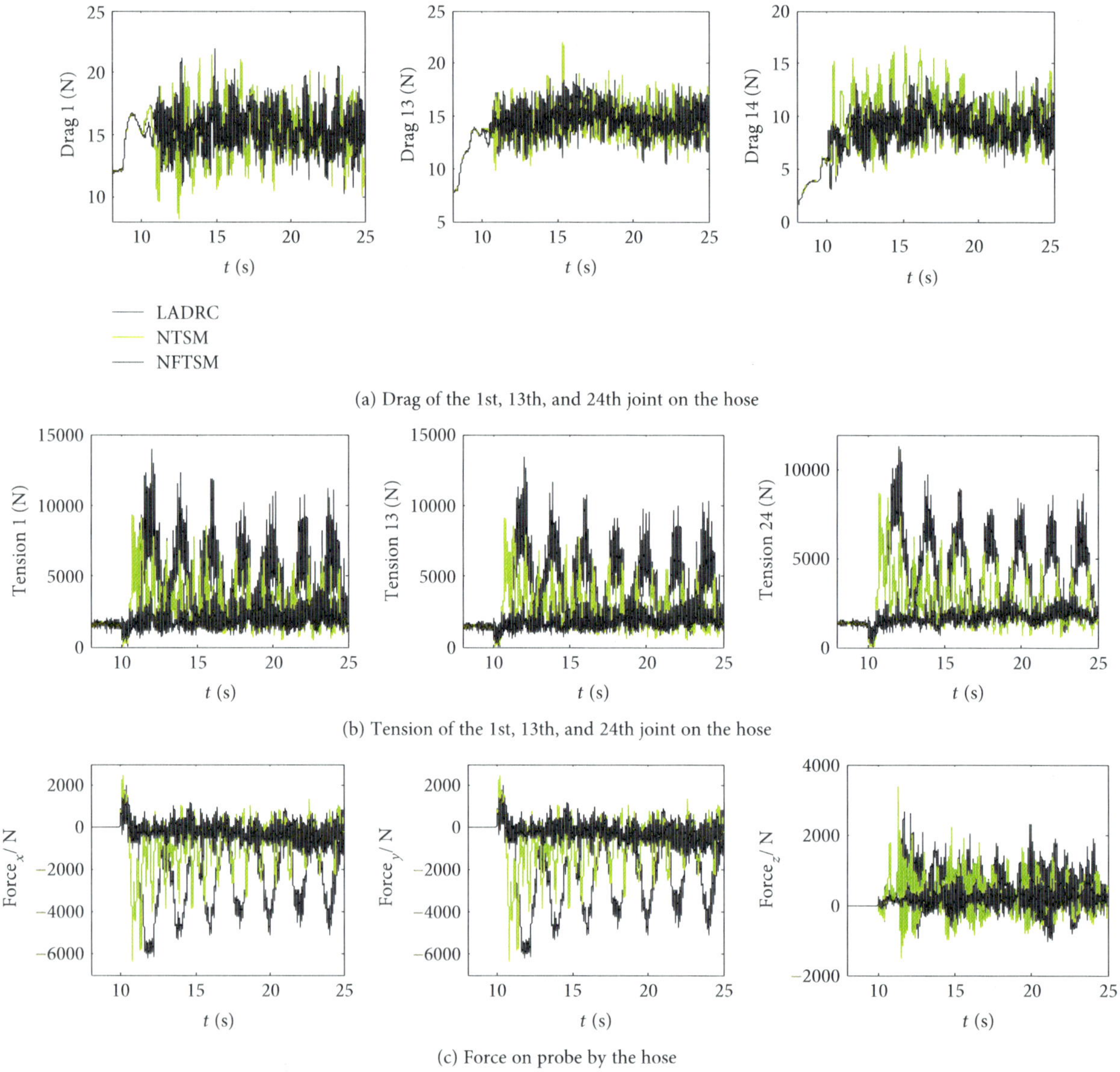

FIGURE 10: Forces on hose and probe by different vibration suppression schemes.

From Figures 11 and 12, the influence of the sensor measurement noise to the system is clearly presented. It can also be inferred that when the external disturbance is big enough, it cannot be expected to track the command accurately by a single controller. The strong total disturbance, including the internal and external disturbance, should be observed by an observer and then compensated by the control laws. By the composition from AESO, the LADRC can track the command signal with strong disturbance and measurement noise. However, the simple PD controller in the LADRC limits the tracking accuracy and rapidity. The NTSM with AESO shows better track performance and robustness than the LADRC. However, by obvious observation of the tracking curve in Figure 11 and the tracking error in Figure 12, a distinct overshoot still exists nearby $t = 10$ s. Namely, the mechanism of the NTSM still should be improved for better tracking performance especially the rapidity, because its global convergence is limited. By introducing the tracking

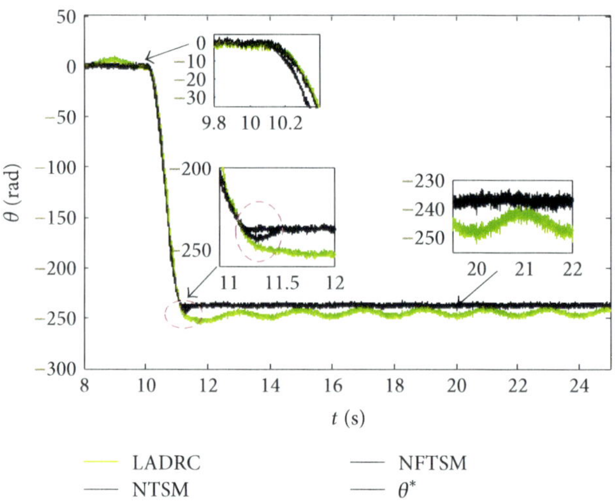

FIGURE 11: Angular position tracking results by different schemes.

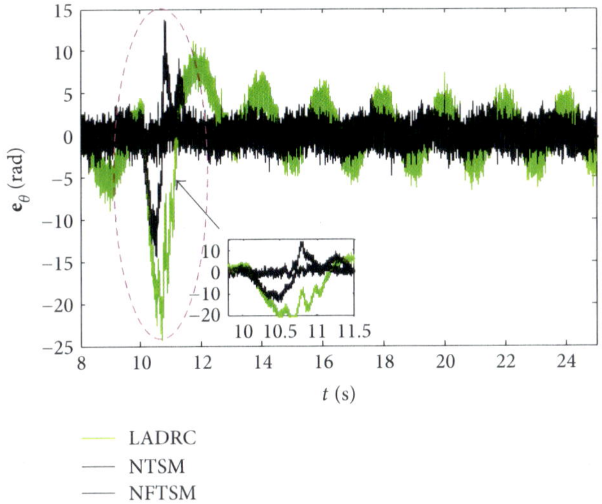

Figure 12: Angular position tracking errors by different schemes.

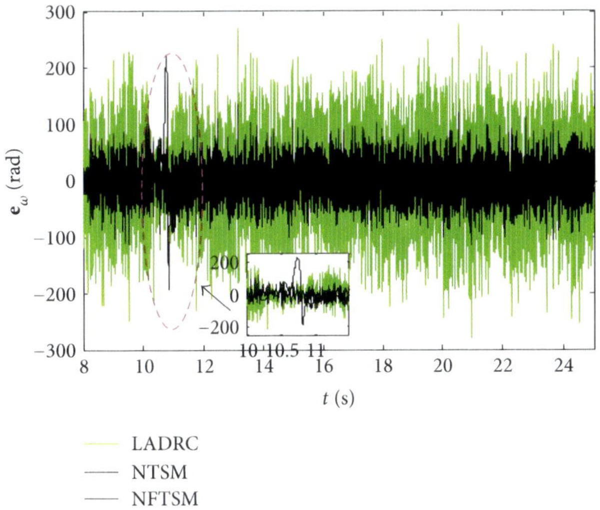

Figure 14: Angular velocity tracking errors by different schemes.

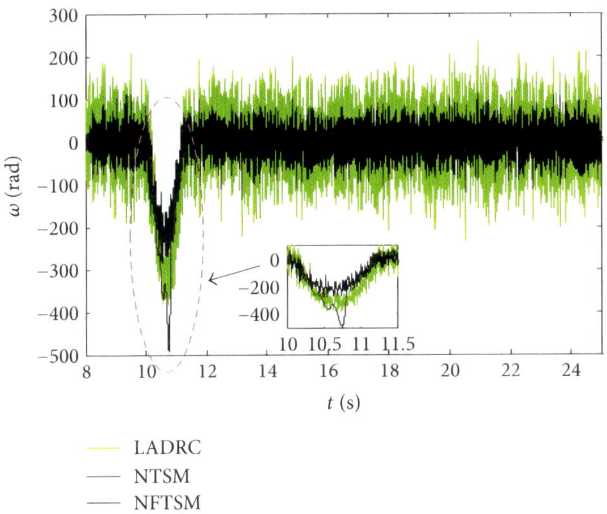

Figure 13: Angular velocity tracking results by different schemes.

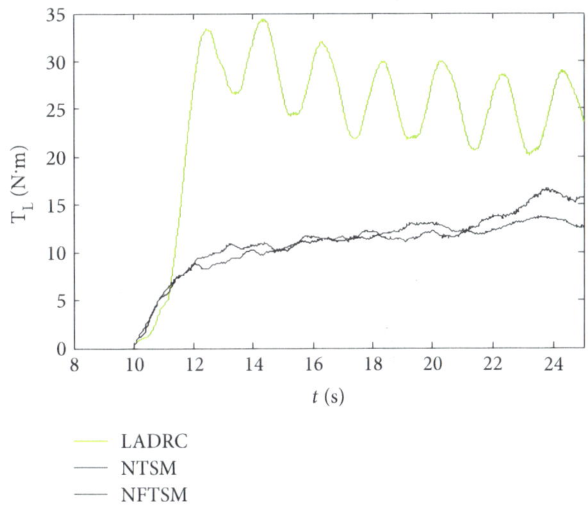

Figure 15: Load torques by different schemes.

error exponential term in sliding surface and specially designed reaching law, the proposed control scheme can accurately track the angular position command given by the hose in AAR with almost zero overshoot. Moreover, the angular tracking error is well guaranteed within 5 rad even under the existence of the angular sensor measurement noise with its variance 1^2 rad and the strong total disturbance, as shown in Figure 11. And the angular velocity tracking error can also be ensured much smaller than that of LADRC. From the tracking result in Figures 11–14, it can be found that the proposed NFTSM with AESO shows the best performance in rapidity and precision than the other two.

From Figure 15, the external load by the LADRC is greater than the other two, and the periodic fluctuation is caused by the poorer tracking performance of the PD controller in LADRC. The external load by NTSM with AESO is about 0.5 times of the LADRC's. However, with the relatively poor transient characteristics (slower tracking speed and greater overshoot), the NTSM does not track the command rapidly enough compared to the proposed NFTSM. This relatively poor performance during the coupling process will lead to greater fluctuation of the external load as time goes on, which will finally affect the stability of the hose-drogue. It can be seen that the external load by NFTSM is no more than 15 N·m in Figure 15 throughout the refueling contact process. And although the external load by NTSM is very close to NFTSM's, the curve of T_L is the smoothest, which indicates that the dynamic of the hose-drogue by this method is the most stable one. Indeed, this kind of tracking performance is expected, because we always expect the hose-drogue to keep stable even with the perturbed contact process of probe and drogue in AAR. It should especially be addressed that there is little difference between 21 and 25 seconds for the load torques by the NTSM and NFTSM methods. From Figures 9 and 10, we know that the hose vibration by the NTSM method is still more drastic than

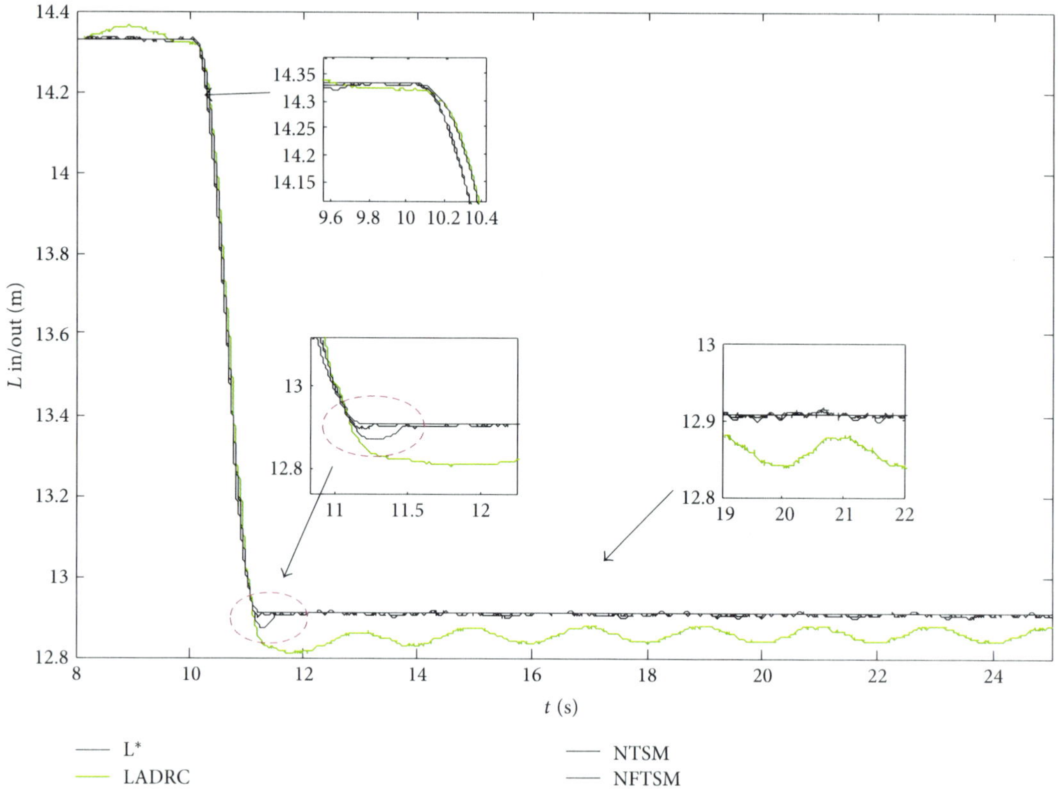

FIGURE 16: The hose length reeling in/out results by different schemes.

the NFTSM's. As time goes on, the vibration spreads along the flexible refueling hose under the impact of the unpredictable aerodynamic force behind the tanker aircraft. Then, according to the formulation $T_L = (r/i)(1/(s+1))T_1$ in Section 2.2, the load torque T_L of the NTSM methods will change greatly and may even become larger than the load torque by the NFTSM method with the hose vibration method goes on.

From Figure 16, the length histories of the hose reeling by LADRC wave intensely due to the strong disturbance. That means that the degree of hose slack induced by coupling is still high, and the slackness of the hose is still severe. The refueling hose vibration may still go violent under the multiple disturbance and measurement noise, and the tension at the end of the hose will change drastically and rapidly. Similar to the angular position in Figure 11, an obvious overshoot also exists in the length of the hose by NTSM, and this overshoot will last about 0.5 seconds, while it can accurately track the angular position command which is given by the hose in AAR with almost zero overshoot. The hose can be precisely reeled in/out by the PMSM with almost no error and lag. The tracking performance of the proposed NFTSM with AESO is the best. And after the coupling, as the relative distance between the receiver and the tanker does not change anymore, the hose length is also maintained to be a constant. And the aerodynamic drags and tensions of the hose gradually become stable as well, then the vibration does not occur anymore.

5. Conclusion

The permanent magnet synchronous motor- (PMSM-) based take-up system is adopted for the aerial refueling hose vibration suppression in autonomous aerial refueling (AAR). A nonsingular fast terminal sliding-mode (NFTSM) control scheme with adaptive extended state observer (AESO) is designed for PMSM take-up system considering external disturbances and measurement noise. A sliding variable with tracking error exponential term is proposed and a special reaching law is designed to enhance the global convergence speed as well as the precision of the controller. The AESO is used to reconstitute the states and "total disturbance" of PMSM with the uncertainties both in the plant and the sensors. Then, the proposed control scheme for PMSM is applied to the flexible refueling hose vibration suppression during AAR coupling. Simulations demonstrate that the proposed scheme is more competitive in global convergence speed and accuracy for vibration suppression with PMSM.

Appendix

The Derivational Process of (14)

When the system errors (e_1, e_2) reach the sliding surface $s = 0$ and with (13), we get

$$s = e_1 + \frac{1}{\beta} e_2^{p/q} = 0. \tag{A.1}$$

$e_2 = \dot{e}_1 = \dot{\vartheta} - \dot{\vartheta}^* = \omega - \dot{\vartheta}^*$; that is, $e_2{}^{p/q} = \dot{e}_1{}^{p/q} = -\beta e_1$. Define $V_n = 0.5e_1^2$; it can be got that $e_1^2 = 2V_n$ and $\dot{V}_n = e_1\dot{e}_1$ and then

$$\dot{V}_n{}^{p/q} = e_1^{p/q}\dot{e}_1^{p/q} = -\beta e_1^{p/q}e_1 = -\beta e_1^{(p+q)/q}. \quad \text{(A.2)}$$

As p and q are all positive odd numbers, hence, $\dot{V}_n < 0$ is satisfied when $e_1 \neq 0$. Based on the Lyapunov stability criterion, the dynamic $s = e_1 + (1/\beta)e_2{}^{p/q} = 0$ is stable, and the system states will converge to the equilibrium. And from $\dot{e}_1{}^{p/q} = -\beta e_1$ (A.2), it can be known that the only equilibrium is $e_1 = 0$. To get the convergence time T_n, we submit $e_1^2 = 2V_n$ into (A.2) and get

$$\dot{V}_n{}^{p/q} = -2^{(p+q)/2q}\beta V_n^{(p+q)/2q}. \quad \text{(A.3)}$$

As $\dot{V}_n = (dV_n)/(dt)$, the following is obtained:

$$\begin{aligned} dt &= \frac{dV_n}{\dot{V}_n} = \frac{dV_n}{-(-\dot{V}_n)} = -\left((-\dot{V}_n)^{p/q}\right)^{-q/p} dV_n \\ &= -\left(2^{(p+q)/2q}\beta V_n^{(p+q)/2q}\right)^{-q/p} dV_n. \end{aligned} \quad \text{(A.4)}$$

Furthermore, as $V_n(T_n) = 0$ and $V_n(0) = 0.5e_1{}^2(0) \geq 0$, we integrate (A.4) from both sides.

$$T_n = \int_0^{V_n(0)} 2^{(p+q)/2q}\beta V_n^{(p+q)/2q} dV_n. \quad \text{(A.5)}$$

Conflicts of Interest

The authors declare that they do not have any conflicts of interest to this work.

Acknowledgments

This work is supported by the National Natural Science Foundations of China under Grants 61673042 and 61473012, and Program for Changjiang Scholars and Innovative Research Team in University (Grant no. IRT13004). The authors especially like to thank Dr. Haitao Wang for his suggestions on the first version.

References

[1] G. Gramajo and P. Shankar, "An efficient energy constraint based UAV path planning for search and coverage," *International Journal of Aerospace Engineering*, vol. 2017, Article ID 8085623, 13 pages, 2017.

[2] Y. Zhu, X. Chen, and C. Li, "A moving frame trajectory tracking method of a flying-wing UAV using the differential geometry," *International Journal of Aerospace Engineering*, vol. 2016, Article ID 3406256, 9 pages, 2016.

[3] P. R. Thomas, U. Bhandari, S. Bullock, T. S. Richardson, and J. L. du Bois, "Advances in air to air refuelling," *Progress in Aerospace Sciences*, vol. 71, pp. 14–35, 2014.

[4] Z. Liu, J. Liu, and W. He, "Modeling and vibration control of a flexible aerial refueling hose with variable lengths and input constraint," *Automatica*, vol. 77, pp. 302–310, 2017.

[5] Z. Su, H. Wang, X. Shao, and Y. Huang, "A robust back-stepping based trajectory tracking controller for the tanker with strict posture constraints under unknown flow perturbations," *Aerospace Science and Technology*, vol. 56, pp. 34–45, 2016.

[6] Z. Su, H. Wang, P. Yao, Y. Huang, and Y. Qin, "Back-stepping based anti-disturbance flight controller with preview methodology for autonomous aerial refueling," *Aerospace Science and Technology*, vol. 61, pp. 95–108, 2017.

[7] Z. Su, H. Wang, N. Li, Y. Yu, and J. Wu, "Exact docking flight controller for autonomous aerial refueling with back-stepping based high order sliding mode," *Mechanical Systems and Signal Processing*, vol. 101, pp. 338–360, 2018.

[8] J. L. Hansen, J. E. Murray, and N. V. Campos, "The NASA Dryden AAR project: a flight test approach to an aerial refueling system," in *AIAA Atmospheric Flight Mechanics Conference and Exhibit*, pp. 2004–4939, Providence, RI, USA, August 2004, AIAA.

[9] H. Duan and Q. Zhang, "Visual measurement in simulation environment for vision-based UAV autonomous aerial refueling," *IEEE Transactions on Instrumentation and Measurement*, vol. 64, no. 9, pp. 2468–2480, 2015.

[10] K. Ro and J. W. Kamman, "Modeling and simulation of hose-paradrogue aerial refueling systems," *Journal of Guidance, Control, and Dynamics*, vol. 33, no. 1, pp. 53–63, 2010.

[11] Z. Liu, J. Liu, and W. He, "Vibration control of a flexible aerial refuelling hose with input saturation," *International Journal of Systems Science*, vol. 48, no. 5, pp. 971–983, 2016.

[12] Z. H. Zhu and S. A. Meguid, "Elastodynamic analysis of aerial refueling hose using curved beam element," *AIAA Journal*, vol. 44, no. 6, pp. 1317–1324, 2006.

[13] J. C. Vassberg, D. T. Yeh, A. J. Blair, and J. M. Evert, "Numerical simulations of KC-10 wing-mount aerial refueling hose-drogue dynamics with a reel take-up system," in *21st AIAA Applied Aerodynamics Conference*, pp. 2003–3508, Orlando, FL, USA, June 2003, AIAA.

[14] K. Ro, T. Kuk, and J. W. Kamman, "Dynamics and control of hose-drogue refueling systems during coupling," *Journal of Guidance, Control, and Dynamics*, vol. 34, no. 6, pp. 1694–1708, 2011.

[15] H. Wang, X. Dong, J. Xue, and J. Liu, "Dynamic modeling of a hose-drogue aerial refueling system and integral sliding mode backstepping control for the hose whipping phenomenon," *Chinese Journal of Aeronautics*, vol. 27, no. 4, pp. 930–946, 2014.

[16] R. E. Alden, G. G. Vennero, and Whittaker Controls Inc., assignee, "Aerial refueling system," US Patent 5141178, 1992.

[17] J. Linares-Flores, C. García-Rodríguez, H. Sira-Ramirez, and O. D. Ramirez-Cardenas, "Robust backstepping tracking controller for low-speed PMSM positioning system: design, analysis, and implementation," *IEEE Transactions on Industrial Informatics*, vol. 11, no. 5, pp. 1130–1141, 2015.

[18] S. Li, H. Liu, and S. Ding, "A speed control for a PMSM using finite-time feedback control and disturbance compensation," *Transactions of the Institute of Measurement and Control*, vol. 32, no. 2, pp. 170–187, 2010.

[19] J. Yang, W.-H. Chen, S. Li, L. Guo, and Y. Yan, "Disturbance/uncertainty estimation and attenuation techniques in PMSM drives—a survey," *IEEE Transactions on Industrial Electronics*, vol. 64, no. 4, pp. 3273–3285, 2017.

[20] D.-H. Ha, C.-S. Lim, and D.-S. Hyun, "Robust optimal nonlinear control with observer for position tracking of permanent magnet synchronous motors," *Journal of Power Electronics*, vol. 13, no. 6, pp. 975–984, 2013.

[21] S. Zheng, X. Tang, B. Song, S. Lu, and B. Ye, "Stable adaptive PI control for permanent magnet synchronous motor drive based on improved JITL technique," *ISA Transactions*, vol. 52, no. 4, pp. 539–549, 2013.

[22] Y. A.-R. I. Mohamed and E. F. el-Saadany, "A current control scheme with an adaptive internal model for torque ripple minimization and robust current regulation in PMSM drive systems," *IEEE Transactions on Energy Conversion*, vol. 23, no. 1, pp. 92–100, 2008.

[23] Y. Shtessel, C. Edwards, L. Fridman, and A. Levant, *Sliding Mode Control and Observation*, Springer, New York, 2014.

[24] H. Wang, W. Zhang, T. Yang, and Q. Qu, "Sliding mode control for diesel engine using extended state observer," *Studies in Informatics & Control*, vol. 24, no. 4, pp. 439–448, 2015.

[25] M. Bahrami, B. Ebrahimi, and G. R. Ansarifar, "Sliding mode observer and control design with adaptive parameter estimation for a supersonic flight vehicle," *International Journal of Aerospace Engineering*, vol. 2010, Article ID 474537, 9 pages, 2010.

[26] H. P. Wang, D. Zheng, and Y. Tian, "High pressure common rail injection system modeling and control," *ISA Transactions*, vol. 63, pp. 265–273, 2016.

[27] L. Yang and J. Yang, "Nonsingular fast terminal sliding-mode control for nonlinear dynamical systems," *International Journal of Robust and Nonlinear Control*, vol. 21, no. 16, pp. 1865–1879, 2011.

[28] M. Chen, Q.-X. Wu, and R.-X. Cui, "Terminal sliding mode tracking control for a class of SISO uncertain nonlinear systems," *ISA Transactions*, vol. 52, no. 2, pp. 198–206, 2013.

[29] B. Zhao and J. Zhou, "Smooth adaptive finite time guidance law with impact angle constraints," *International Journal of Aerospace Engineering*, vol. 2016, Article ID 5730168, 19 pages, 2016.

[30] W. Xue, W. Bai, S. Yang, K. Song, Y. Huang, and H. Xie, "ADRC with adaptive extended state observer and its application to air–fuel ratio control in gasoline engines," *IEEE Transactions on Industrial Electronics*, vol. 62, no. 9, pp. 5847–5857, 2015.

[31] X. Zhang, L. Sun, K. Zhao, and L. Sun, "Nonlinear speed control for PMSM system using sliding-mode control and disturbance compensation techniques," *IEEE Transactions on Power Electronics*, vol. 28, no. 3, pp. 1358–1365, 2013.

[32] J. A. Farrell, M. Polycarpou, M. Sharma, and W. Dong, "Command filtered backstepping," *IEEE Transactions on Automatic Control*, vol. 54, no. 6, pp. 1391–1395, 2009.

[33] X. Shao and H. Wang, "Active disturbance rejection based trajectory linearization control for hypersonic reentry vehicle with bounded uncertainties," *ISA Transactions*, vol. 54, pp. 27–38, 2015.

Permissions

All chapters in this book were first published in IJAE, by Hindawi Publishing Corporation; hereby published with permission under the Creative Commons Attribution License or equivalent. Every chapter published in this book has been scrutinized by our experts. Their significance has been extensively debated. The topics covered herein carry significant findings which will fuel the growth of the discipline. They may even be implemented as practical applications or may be referred to as a beginning point for another development.

The contributors of this book come from diverse backgrounds, making this book a truly international effort. This book will bring forth new frontiers with its revolutionizing research information and detailed analysis of the nascent developments around the world.

We would like to thank all the contributing authors for lending their expertise to make the book truly unique. They have played a crucial role in the development of this book. Without their invaluable contributions this book wouldn't have been possible. They have made vital efforts to compile up to date information on the varied aspects of this subject to make this book a valuable addition to the collection of many professionals and students.

This book was conceptualized with the vision of imparting up-to-date information and advanced data in this field. To ensure the same, a matchless editorial board was set up. Every individual on the board went through rigorous rounds of assessment to prove their worth. After which they invested a large part of their time researching and compiling the most relevant data for our readers.

The editorial board has been involved in producing this book since its inception. They have spent rigorous hours researching and exploring the diverse topics which have resulted in the successful publishing of this book. They have passed on their knowledge of decades through this book. To expedite this challenging task, the publisher supported the team at every step. A small team of assistant editors was also appointed to further simplify the editing procedure and attain best results for the readers.

Apart from the editorial board, the designing team has also invested a significant amount of their time in understanding the subject and creating the most relevant covers. They scrutinized every image to scout for the most suitable representation of the subject and create an appropriate cover for the book.

The publishing team has been an ardent support to the editorial, designing and production team. Their endless efforts to recruit the best for this project, has resulted in the accomplishment of this book. They are a veteran in the field of academics and their pool of knowledge is as vast as their experience in printing. Their expertise and guidance has proved useful at every step. Their uncompromising quality standards have made this book an exceptional effort. Their encouragement from time to time has been an inspiration for everyone.

The publisher and the editorial board hope that this book will prove to be a valuable piece of knowledge for researchers, students, practitioners and scholars across the globe.

List of Contributors

Bokyoung Oh
Agency for Defense Development, Daejeon, Republic of Korea

Junho Jeong, Jinyoung Suk and Seungkeun Kim
Department of Aerospace Engineering, Chungnam National University, Daejeon, Republic of Korea

Andrea Papaiz
WiPLi Lab, University of Udine, Via delle Scienze 206, 33100 Udine, Italy

Andrea M. Tonello
EcoSys Lab, Alpen-Adria-Universität Klagenfurt, Universitätsstraße 65-67, 9020 Klagenfurt, Austria

Yi Zhu, Xin Chen and Chuntao Li
College of Automation Engineering, Nanjing University of Aeronautics and Astronautics, Nanjing 210016, China

Olha Sushchenko
Aircraft Control Systems Department, Educational and Research Institute of Air Navigation, National Aviation University, Cosmonaut Komarova Avenue 1, Kyiv 03058, Ukraine

Andriy Goncharenko
Aircraft Airworthiness Department, Aircraft Faculty, Educational and Research Airspace Institute, National Aviation University, Cosmonaut Komarova Avenue 1, Kyiv 03058, Ukraine

Zhenchang Liu, Jie Guo, Mengting Li, Shengjing Tang and Xiao Wang
School of Aerospace Engineering, Beijing Institute of Technology, Beijing 100081, China

Yonghong Zhang and Tiantian Dong
B-DAT, C-MEIC, CICAEET, School of Information and Control, Nanjing University of Information Science & Technology, Nanjing, China

Yunping Liu
B-DAT, C-MEIC, CICAEET, School of Information and Control, Nanjing University of Information Science & Technology, Nanjing, China

Department of Mechanical and Industrial Engineering, Ryerson University, Toronto, ON, Canada

Yahui Qi, Shaolei Zhou, Yuhang Kang and Shi Yan
Department of Control Engineering, Naval Aeronautical and Astronautical University, Yantai 264000, China

Yu Gu, Jason N. Gross and Kyle Lassak
Department of Mechanical and Aerospace Engineering (MAE) at West Virginia University (WVU), Morgantown, WV 26506, USA

Matthew B. Rhudy
Division of Engineering, Pennsylvania State University, Reading, PA 19610, USA

Tegoeh Tjahjowidodo and Shian Lee
School of Mechanical and Aerospace Engineering, Nanyang Technological University, 50 Nanyang Avenue, North Spine (N3), Singapore 639798

N. Ramos-Pedroza and W. MacKunis
Department of Physical Sciences, Embry-Riddle Aeronautical University, Daytona Beach, FL 32114, USA

M. Reyhanoglu
Department of Engineering, University of North Carolina at Asheville, Asheville, NC 28804, USA

Jason N. Gross, Ryan M. Watson, Stéphane D'Urso and Yu Gu
Department of Mechanical and Aerospace Engineering, West Virginia University, Morgantown, WV 26506, USA

Kaijia Xue, Congqing Wang, Zhiyu Li and Hanxin Chen
College of Automation, Nanjing University of Aeronautics and Astronautics, Nanjing 210016, China

Donghun Park
Aerodynamics Research Team, Korea Aerospace Research Institute, Daejeon 34133, Republic of Korea

Yunggyo Lee, Taehwan Cho and Cheolwan Kim
Department of Aerospace Engineering, Pusan National University, Busan 46241, Republic of Korea

Minghong She
School of Automation, Chongqing University, Chongqing, China
Computer College, Chongqing College of Electronic Engineering, Chongqing, China

Pengju Zhao
Computer College, Chongqing College of Electronic Engineering, Chongqing, China

Jianmin Li and Bo Zhang
State Key Laboratory of Intelligent Technology and Systems, Tsinghua National Laboratory for Information Science and Technology, Department of Computer Science and Technology, Tsinghua University, Beijing 100084, China

Xingwei Kong and Xinmin Dong
School of Aeronautics and Astronautics Engineering, Air Force Engineering University, Xi'an 710038, China

Xufeng Wang
School of Aeronautics and Astronautics Engineering, Air Force Engineering University, Xi'an 710038, China
State Key Laboratory of Intelligent Technology and Systems, Tsinghua National Laboratory for Information Science and Technology, Department of Computer Science and Technology, Tsinghua University, Beijing 100084, China
Unit 95835 of PLA, Urumqi 841700, China

Parvathy Rajendran
School of Aerospace Engineering, Universiti Sains Malaysia, Penang, Malaysia

Howard Smith
Aircraft Design Group, School of Aerospace, Transport and Manufacture Engineering, Cranfield University, Cranfield, UK

Honglun Wang
School of Automation Science and Electrical Engineering, Beihang University, Beijing 100191, China
Science and Technology on Aircraft Control Laboratory, Beihang University, Beijing 100191, China

Zikang Su
School of Automation Science and Electrical Engineering, Beihang University, Beijing 100191, China
ShenYuan Honors College of Beihang University, Beijing 100191, China
Science and Technology on Aircraft Control Laboratory, Beihang University, Beijing 100191, China

Index